Lecture Notes in Mathematics 2068

For further volumes:
http://www.springer.com/series/304

Evgeny Spodarev
Editor

Stochastic Geometry, Spatial Statistics and Random Fields

Asymptotic Methods

Springer

Editor
Evgeny Spodarev
University of Ulm
Ulm, Germany

ISBN 978-3-642-33304-0 ISBN 978-3-642-33305-7 (eBook)
DOI 10.1007/978-3-642-33305-7
Springer Heidelberg New York Dordrecht London

Lecture Notes in Mathematics ISSN print edition: 0075-8434
ISSN electronic edition: 1617-9692

Library of Congress Control Number: 2012953997

Mathematics Subject Classification (2010): 60D05, 52A22, 60G55, 60G60, 60G57, 60F05, 60F15, 60J25, 62M30, 65C40

Printed on acid-free paper

Springer is part of Springer Science+Business Media (www.springer.com)

. . . Geometry is the knowledge of the eternally existent.

Plato, "Republic", 527.

Foreword

Geometric intuition is central to many areas of statistics and probabilistic arguments. This is particularly true in the areas covered by this book, namely stochastic geometry, random fields and random geometric graphs. Nevertheless, intuition must be followed with rigorous arguments if it is to become part of the general literature of probability and statistics. Many such arguments in this area deviate from traditional statistics, requiring special (and often beautiful) tools outside a working statistician's usual toolbox.

Professor Spodarev assembled a very impressive cast of instructors for a workshop in Söllerhaus in September 2009 in order to further the literature in this area and to introduce the topics to participating graduate students. I previously had the pleasure of hosting several of the contributors of this volume at workshop at BIRS in February 2009 on Random Fields and Stochastic Geometry and am certain the workshop in Söllerhaus was a tremendous success.

The success of the workshop is further evidenced from this volume of lecture notes. Professor Spodarev has managed to produce a volume that combines both introductory material and current research in these notes. This book will be a useful reference for myself in this area, as well as to all researchers with an interest in stochastic geometry.

Stanford, CA
March 2012

Jonathan Taylor

Preface

This volume is an attempt to write a textbook providing a modern introduction to stochastic geometry, spatial statistics, the theory of random fields and related topics. It has been made out of selected contributions to the Summer Academy on Stochastic Geometry, Spatial Statistics and Random Fields

http://www.uni-ulm.de/summeracademy09

which took place during 13–26 Sep 2009 at Söllerhaus, an Alpine conference centre of the University of Stuttgart and RWTH Aachen, in the village Hirschegg (Austria). It was organized by the Institute of Stochastics of Ulm University in cooperation with the Chair of Probability Theory of Lomonosov Moscow State University. In contrast with previous schools on this subject (Sandbjerg 2000, Martina Franca 2004, Sandbjerg 2007), this summer academy concentrated on the asymptotic theory of random sets, fields and geometric graphs. At the same time, it provided an introduction to more classical subjects of stochastic geometry and spatial statistics, giving (post)graduate students an opportunity to start their own research within a couple of weeks. The summer academy hosted 38 young participants from 13 countries (Australia, Austria, Denmark, Germany, France, Mongolia, Russia, Romania, Sweden, Switzerland, UK, USA and Vietnam). Twelve experts gave lectures on various domains of geometry, probability theory and mathematical statistics. Moreover, students and young researchers had the possibility to give their own short talks.

As it was pointed out above, this volume is focused on the asymptotic methods in the theory of random geometric objects (point patterns, sets, graphs, trees, tessellations and functions). It reflects advances in this domain made within the last two decades. This especially concerns the limit theorems for random tessellations, random polytopes, finite point processes and random fields.

The book is organized as follows. The first chapter provides an introduction to the theory of random closed sets (RACS). It starts with the foundations of geometric probability (Buffon needle problem, Bertrand's paradox) and continues with the classical theory of random sets by Matheron. Then it gives laws of large numbers and limit theorems for Minkowski sums and unions of independent identically distributed (i.i.d.) RACS. Chapter 2 provides basics of the classical integral geometry and its applications to stereology, a part of spatial stochastics which deals with the reconstruction of the higher-dimensional properties of geometric objects from lower-dimensional sections. In Chap. 3, principal classes of spatial point processes (Poisson-driven point processes, finite point processes) are introduced. Their simulation and statistical inference techniques (partially using the Markov Chain Monte Carlo (MCMC) methods) are discussed as well. Chapter 4 provides an account of the theory of marked point processes and the asymptotic statistics for them in growing domains. Ergodicity, mixing and m-dependence properties of marked point processes are studied in detail. Random tessellation models are the matter of Chap. 5. There, Poisson-driven tessellations as well as Cox processes on them and hierarchical networks constructed on their basis are considered. Scaling limits for some characteristics of these networks are found. Applications to telecommunication networks are also discussed. Distribution tail asymptotics and limit theorems for the characteristics of the (large) typical cell of Poisson hyperplane and Poisson–Voronoi tessellations are given in Chap. 6. The shape of large cells of hyperplane tessellations as well as limit theorems for some geometric functionals of convex hulls of a large number of i.i.d. points within a convex body and of random polyhedra are dealt with in Chap. 7. Weak laws of large numbers and central limit theorems for functionals of finite point patterns are discussed in Chap. 8. Additionally, their applications to various topics ranging from optimization to sequential packings of convex bodies are touched upon. Chapter 9 surveys the elementary theory of random functions with the focus on random fields. Basic classes of random field models as well as an account of the correlation theory, statistical inference and simulation techniques are provided. Special attention is paid to infinitely divisible random functions. Dependence concepts for random fields (such as mixing, association, (BL, θ)-dependence) as well as central limit theorems for weakly dependent random fields are the subject of Chap. 10. They are applied to establish the limiting behaviour of the volume of excursion sets of weakly dependent stationary random fields. Chapter 11 focuses on almost sure limit theorems for partial sums (or increments) of random fields on $\mathbb{N}^d$ such as laws of large numbers, laws of single or iterated logarithm and others. In the final chapter, the geometry of large rooted plane random trees with nearest neighbour interaction is studied. A law of large numbers, a large deviation principle for the branching type statistics and scaling limits of the tree are considered. Connections of these results with the solutions of some partial differential equations are discussed as well. Some of the chapters are written in a more formal and rigorous way than others which reflects the personal taste and style of the authors.

The topics of this volume are (almost) self-contained. Thus, we recommend Chaps. 1, 2 and 7 for the first acquaintance with the theory of random sets. A reader interested in the (asymptotic theory of) point processes might start reading with Chap. 3 and continue with Chaps. 4, 5 and 8 following the references to Chap. 1 if needed. Readers with an interest in tessellations and random polytopes might additionally read Chaps. 2, 6 and 7. To get a concise introduction to random fields and limit theorems for them, one could read Chaps. 9–11 occasionally following the references to earlier chapters. For random graphs and trees, Chaps. 8 and 12 are a good starting point.

All in all, the authors hope that the present volume will be helpful to graduates and PhD students in mathematics to get a first glance of the geometry of random objects and its asymptotical methods. Written in the spirit of a textbook (with a significant number of proofs and exercises for active reading), it might be also instrumental to lecturers in preparing their own lecture courses on this subject.

Söllerhaus at Hirschegg
September 2011

Evgeny Spodarev

Acknowledgements

I would like to thank the authors for their wonderful contributions that made this volume possible. Furthermore, I acknowledge the generous financial support of the International Office of the German Academic Exchange Service (DAAD) and of the Faculty of Mathematics and Economics of Ulm University (represented by its president Prof. K. Ebeling and the Dean Prof. W. Kratz). Finally, I am grateful to Florian Timmermann for his valuable help in producing most of the figures and the layout of this volume. Last but not least, my thanks go to Dr. Catriona Byrne, the editorial director of Springer who supported this book project from its very beginning.

Contents

Contributors

Adrian Baddeley CSIRO, Perth, Australia

Yuri Bakhtin Georgia Institute of Technology, Atlanta, GA 30332, USA

Alexander Bulinski Lomonosov Moscow State University, Moscow, Russia

Pierre Calka Université de Rouen, Rouen, France

Catherine Gloaguen Orange Labs, Issy-les-Moulineaux, France

Lothar Heinrich Augsburg University, Augsburg, Germany

Daniel Hug Karlsruhe Institute of Technology, Karlsruhe, Germany

Markus Kiderlen Aarhus University, Aarhus, Denmark

Ilya Molchanov University of Bern, Bern, Switzerland

Volker Schmidt Ulm University, Ulm, Germany

Evgeny Spodarev Ulm University, Ulm, Germany

Ulrich Stadtmüller Ulm University, Ulm, Germany

Florian Voss Boehringer-Ingelheim Pharma GmbH & Co., Ingelheim, Germany

Joseph Yukich Lehigh University, Bethlehem, PA 18015, USA

Acronyms

Sets

$\mathbb{C}$	Complex numbers
$\mathbb{N}$	Positive integer numbers
$\mathbb{Q}$	Rational numbers
$\mathbb{R}$	Real numbers
$\mathbb{Z}$	All integer numbers
$\mathcal{K}^d$	Class of all compact sets
$\mathcal{K}^d_{\text{conv}}$	Class of all convex bodies in $\mathbb{R}^d$
$\mathcal{E}^d_k$	Family of affine k-planes in $\mathbb{R}^d$
$\mathcal{B}_0(\mathbb{R}^d)$	Family of bounded Borel sets in $\mathbb{R}^d$
$\mathbb{S}^{d-1}$	Unit sphere in $\mathbb{R}^d$

Probability Theory

ξ, η	Random variables
X, Y	Random sets
$\mathbf{P}_\xi$	Probability measure of ξ
$\mathbf{E}\,\xi$	Expectation of ξ
$\mathbf{corr}(\xi, \eta)$	Correlation of ξ and η
$\mathbf{cov}(\xi, \eta)$	covariance of ξ and η
$\mathbf{var}\ \xi$	Variance of ξ
$\mathcal{N}$	Set of all locally finite simple point patterns
$\mathfrak{N}$	Set of all locally finite counting measures
$\mathbf{1}$	Indicator function
$\text{Ber}(p)$	Bernoulli distribution with parameter p
$\text{Binom}(n, p)$	Binomial distribution with parameters n and p
$\text{Exp}(\lambda)$	Exponential distribution with parameter λ
$N(\mu, \sigma^2)$	Normal distribution with parameters μ and σ^2

$\mathrm{Pois}(\lambda)$	Poisson distribution with parameter λ
$\mathrm{Unif}(a, b)$	Uniform distribution on $[a, b]$

Graph Theory

$\mathbb{G}$	Graph
$\mathbb{V}$	Set of vertices
$\mathbb{E}$	Set of edges

Other Notations

E	Energy
V	Potential
$\mathtt{T}$	Tessellation
diag	Diagonal matrix
dist	Distance function
conv	Convex hull
$\|A\|$	Cardinality of a set A
$\|a\|$	Absolute value of a number a
$\|x\|$, $\\|x\\|_2$	Euclidean norm of a vector x
diam	Diameter
SO	Rotation group
$\mathrm{Im}\, z$	Imaginary part of a complex number
$\mathrm{Re}\, z$	Real part of a complex number
sgn	Signum function
Lip	Lipschitz operator
span	Linear hull
ν_d	Lebesgue measure
$V_0, \ldots, V_d$	Intrinsic volumes
χ	Euler characteristic
K^r	r-neighbourhood of K
$\mathrm{Int}\, K$	Interior of K
$K\|L$	Orthogonal projection of K onto L
o	The origin of $\mathbb{R}^d$
$B_r(o)$	Ball of radius r centred around the origin
T_x	Shift operator

Chapter 1
Foundations of Stochastic Geometry and Theory of Random Sets

Ilya Molchanov

Abstract The first section of this chapter starts with the Buffon problem, which is one of the oldest in stochastic geometry, and then continues with the definition of measures on the space of lines. The second section defines random closed sets and related measurability issues, explains how to characterize distributions of random closed sets by means of capacity functionals and introduces the concept of a selection. Based on this concept, the third section starts with the definition of the expectation and proves its convexifying effect that is related to the Lyapunov theorem for ranges of vector-valued measures. Finally, the strong law of large numbers for Minkowski sums of random sets is proved and the corresponding limit theorem is formulated. The chapter is concluded by a discussion of the union-scheme for random closed sets and a characterization of the corresponding stable laws.

1.1 Geometric Probability and Origins of Stochastic Geometry

In this section we introduce basic notions of stochastic geometry: random points, lines, and random polytopes.

1.1.1 Random Points and the Buffon Problem

The first concepts of geometric probabilities usually appear at the very beginning of university probability courses. One of the very first problem of this type is the meeting problem: two persons come to meet at an agreed place, so that their arrival times

---

I. Molchanov (✉)
University of Bern, Bern, Switzerland
e-mail: ilya@stat.unibe.ch

E. Spodarev (ed.), *Stochastic Geometry, Spatial Statistics and Random Fields*,
Lecture Notes in Mathematics 2068, DOI 10.1007/978-3-642-33305-7_1,

are randomly chosen within some time interval $[0, T]$ and each person is supposed to wait for the other for some time t. The relevant probability space can be described as the square $[0, T]^2$, where the two arrivals are described as a uniformly distributed point (x_1, x_2). The two persons meet exactly in case $|x_1 - x_2| \leq t$, which singles out the "favourable" part A of the square. The probability that $(x_1, x_2) \in A$ is then given by the ratio of the areas of A and the whole square, so the answer is $1 - (T-t)^2/T^2$.

Note that the whole question become pointless if T is not given, indeed it is not possible to sample a point according to the Lebesgue measure on the plane, since the whole measure is infinite. This situation quite often appears when defining measure on families of geometric objects—the whole measure is infinite and one has to restrict it to a subset of finite measure. What is particularly important is that the measure defined on the whole space satisfies some invariance conditions, e.g. like the Lebesgue measure which is motion invariant.

This simple example already shows the first principles: a point is chosen from some set W, an event corresponds to the fact that this point x belongs to some subset A. Finally, the probability of this event is calculated as the ratio of the measures of A and W. If all points are equally likely and W is a subset of $\mathbb{R}^d$, then one takes the ratio of the Lebesgue measures of A and W. If the point is equally likely chosen on the sphere, the ratio of the surface measures appears.

In general, a random point ξ in $W \subset \mathbb{R}^d$ can be defined by taking W with its Borel σ-algebra as the probability space, where the probability measure is the Lebesgue measure ν_d normalized by its value $\nu_d(W)$. This model defines a random point whose positions are equally likely, in other words, ξ is uniformly distributed on W. Such a construction clearly fails if W has infinite Lebesgue measure.

Several random points can be used to construct more complicated geometric objects. For instance two points determine a segment, which brings us to the famous problem formulated by Buffon in 1777 and published under the title "Essai d'arithmétique morale". The *Buffon problem* was originally formulated as finding the probability that a rod thrown on a parquet floor crosses one of the parallel lines formed by the parquet.

A possible approach to its solution starts with a parametrization of a segment using its (say lower) end-point ξ and the angle θ it makes with the (say) x-axis. The parquet is a grid of lines formed by orthogonal to x-axis parallel lines with distance D between them. Throwing such a segment (of length L) at random means that all its positions are equally likely, i.e. θ is uniformly distributed on $[0, \pi)$ and the first coordinate ξ_1 of $\xi = (\xi_1, \xi_2)$ is uniformly distributed on $[0, D]$. The fact whether or not the segment intersects one of the lines does not depend on the value of the second coordinate ξ_2, and so it can be ignored. Then we need to identify the subset of the rectangle $[0, D) \times [0, \pi)$ (or its closure) such that the segments parametrized by its points hit the line, find the Lebesgue measure of this set and divide it by πD, which results at

$$\mathbf{P}(\text{crossing}) = \frac{2}{\pi} \int_0^{\pi/2} \frac{L \sin \alpha}{D} \, d\alpha = \frac{2L}{\pi D}$$

if $L \leq D$, the case of a short needle, see Fig. 1.1.

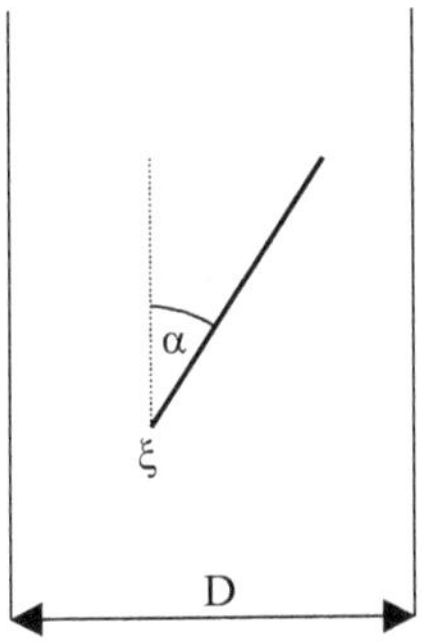

Fig. 1.1 Position of a random segment with respect to the line grid

Exercise 1.1. Find the crossing probability for the case of a long needle, i.e. for $L > D$.

An alternative very nice solution of the Buffon problem was suggested by Barbier in 1860, who argued as follows. Assume that our needle is a (piecewise linear) curve being a union of a finite number of line segments. Let p_k, $k \geq 0$, be the probability of exactly k crossings between the needle and the grid. Let $E(x)$ be the expected number of crossings by a segment of length x. Then $E(x + y) = E(x) + E(y)$ for all $x, y > 0$. Since $E(\cdot)$ is a monotone function, $E(x) = cx$. By approximating a circle with polygons, we see that the expected number of crossings by the circle is proportional to its boundary length. Since the circle of diameter D produces exactly two crossings, we obtain $2 = c\pi D$, and thus $c = 2/\pi D$. Assume that the needle (being again a line segment) is short, i.e. $L \leq D$. Then $p_2 = p_3 = \cdots = 0$, so that the expected number of crossings equals the probability of a crossing. Thus the segment of length $L \leq D$ crossed the grid with probability $cL = 2L/\pi D$.

The Buffon problem deals with a random object taking value from the family of segments such that all sampled objects are equally likely. Such definitions however require a careful consideration, which is shown by the following *Bertrand's paradox*. Consider the (say unit) disk on the plane with the aim to define a random chord in this disk. First, it is possible to define a random chord by joining two points uniformly distributed on the circle (Fig. 1.2a). Second, it is possible to choose a uniform direction on $[0, \pi)$ and then draw an orthogonal chord, which intersects the chosen direction at a point uniformly distributed on $[-1, 1]$ (Fig. 1.2b). Finally, it is possible to take a uniform direction on $[0, 2\pi)$ and draw an orthogonal chord to it with distance to the origin uniformly distributed on $[0, 1]$ (Fig. 1.2c). It is easy to calculate the probability that the chord length exceeds $\sqrt{3}$. It equals $1/3$, $1/2$ and $1/4$ respectively in each of the three definitions formulated above.

1.1.2 Random Lines

For simplicity consider random lines on the plane, noticing that similar arguments and constructions apply to define random affine subspaces of Euclidean spaces

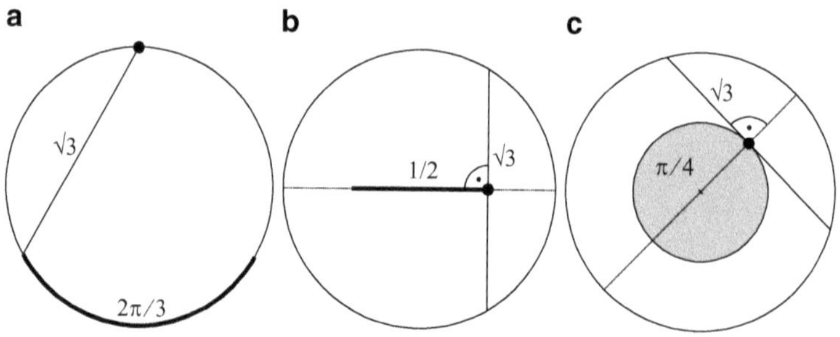

Fig. 1.2 Three possibilities for the choice of a random chord in the unit disk

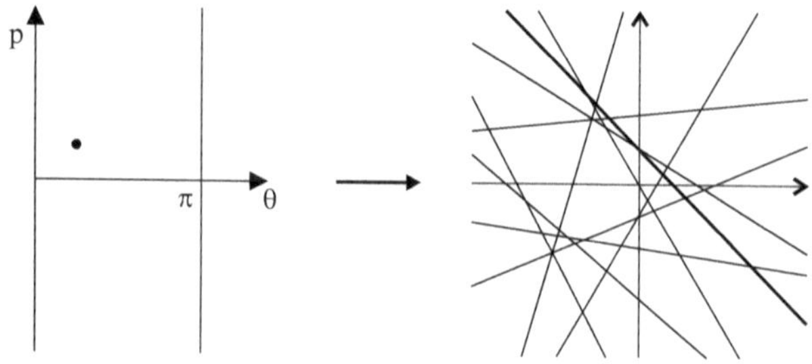

Fig. 1.3 Parametrization of the line on the plane

with general dimensions. A common idea that is used to define random geometric objects is to parametrize them by one or more real numbers. If these real numbers become random, one obtains a random geometric object. A line on the plane can be parametrized by its signed distance p to the origin and the direction of the normal vector $\theta \in [0, \pi)$.

A set of lines then is a subset of the parameter space $[0, \pi) \times \mathbb{R}$, see Fig. 1.3. Measurable sets of lines are exactly those which correspond to Borel sets in the parameter space. Each measure on the strip $[0, \pi) \times \mathbb{R}$ defines a measure on the family of lines. The most important case appears when the measure on the family of lines is defined as the image of the Lebesgue measure on $[0, \pi) \times \mathbb{R}$ or proportional to the Lebesgue measure. The resulting measure has density written as $\lambda\, dp\, d\theta$ for a constant $\lambda > 0$.

Let us show that the obtained measure on the family of lines is motion-invariant. The line parametrized as (θ, p) has equation $x \cos\theta + y \sin\theta - p = 0$. A rigid motion transforms the coordinates as

$$\begin{cases} x' = x\cos\alpha - y\sin\alpha + a \\ y' = x\sin\alpha + y\cos\alpha + b\,, \end{cases}$$

so that the transformed line has equation

$$x\cos(\theta - \alpha) + y\sin(\theta - \alpha) - (p - a\cos\theta - b\sin\theta) = 0\,.$$

Thus, the rigid motion corresponds to the following transformation on the parameter space:

$$\begin{cases} \theta \mapsto (\theta - \alpha)\,, \\ p \mapsto (p - a\cos\theta - b\sin\theta)\,, \end{cases}$$

and it remains to notice that the Lebesgue measure on $[0, \pi) \times \mathbb{R}$ is invariant with respect to this transformation.

Often it is useful to consider the measure on the space of lines with density $dp\,\mathcal{R}(d\theta)$, where $\mathcal{R}$ is a certain measure on $[0, \pi)$ which controls the *directional distribution* for lines. The obtained measure is then translation (but not necessarily rotation) invariant.

The total measure of all lines is clearly infinite. In order to obtain some interesting results we need to calculate some integrals with respect to the measure on the family of lines. Denote this family as $\mathcal{E}_1^2$, noticing that in general $\mathcal{E}_k^d$ denotes the family of affine k-planes in $\mathbb{R}^d$. A number of interesting quantities can be written as integrals

$$\int_{\mathcal{E}_1^2} f(E)\,\mu(dE)$$

with respect to the introduced measure $\mu(dE) = dp\,d\theta$ on $\mathcal{E}_1^2$. For instance $f(E)$ can be the indicator of the event that a line E hits a given set or can represent the length of the intersection of a line with a given set, etc.

Let K be a convex set on the plane. Consider the functional

$$\varphi(K) = \int_{\mathcal{E}_1^2} \chi(K \cap E)\,\mu(dE)\,,$$

where $\chi(K \cap E)$ is one if $K \cap E \neq \emptyset$ and zero otherwise. The value of φ determines the measure of all lines that intersect the set K. More general functionals of this type are considered in Sect. 2.1.1. The functional φ is monotone in K, is motion invariant and also continuous and additive on convex sets, meaning that

$$\varphi(K \cup M) + \varphi(K \cap M) = \varphi(K) + \varphi(M)$$

for all convex compact sets K and M such that $K \cup M$ is also convex. Note that the Lebesgue measure satisfies these properties in a much stronger form, namely for all measurable sets. Relaxing the condition to the family of convex sets enriches the family of such functionals φ. The Hadwiger theorem from convex geometry (see Theorem 2.1) establishes that all such functionals can be written as non-negative linear combinations of so-called intrinsic volumes. In the planar case, the intrinsic volumes are proportional to the area $A(K)$, perimeter $U(K)$ and the *Euler–Poincaré characteristic* $\chi(K)$. The latter is one if convex K is non-empty and is zero if K is empty. Thus

$$\varphi(K) = c_1 A(K) + c_2 U(K) + c_3 \chi(K)\,.$$

Noticing that φ is homogeneous of degree one, while the area is homogeneous of degree two and $\chi(\cdot)$ is homogeneous of degree zero (cf. Sect. 2.1.1 for the definition of homogeneity), we immediately obtain that $c_1 = c_3 = 0$, i.e. $\varphi(K)$ is proportional to the perimeter of K. If we take the unit ball instead of K then

$$\varphi(K) = \int_0^{2\pi} \int_0^1 dp\, d\theta = 2\pi\,,$$

while $U(K) = 2\pi$. Thus $c_2 = 1$ and

$$\int_{\mathcal{E}_1^2} \chi(K \cap E)\mu(dE) = U(K)\,. \tag{1.1}$$

In other words, the measure of all lines that hit K equals the perimeter of K. If $K \subset K_0$ then one arrives at the following result

$$\mathbf{P}(\text{line hits } K \mid \text{line hits } K_0) = \frac{U(K)}{U(K_0)}\,.$$

Exercise 1.2. Show that

$$\int_{\mathcal{E}_1^2} L(K \cap E)\, \mu(dE) = \pi A(K)\,, \tag{1.2}$$

where $L(E)$ is the length of intersection of K and line E, see also [451] for further reading on integral and convex geometry in view of applications in geometric probability.

Exercise 1.3. Show that the probability that two lines intersecting K cross at a point from K is given by $2\pi A(K)U(K)^{-2}$. Hint: calculate

$$\frac{\int_{\mathcal{E}_1^2} \int_{\mathcal{E}_1^2} \chi(K \cap E_1 \cap E_2)\mu(dE_1)\mu(dE_2)}{\left(\int_{\mathcal{E}_1^2} \chi(K \cap E)\mu(dE)\right)^2}$$

using also the result of the previous exercise. The isoperimetric inequality $U(K)^2 \geq 4\pi A(K)$ with the equality achieved for discs K implies that the above probability takes the largest value $1/2$ for K being a disc.

1.1.3 *Sets Constructed from Random Points*

A finite set $\{\xi_1, \dots, \xi_n\}$ of random points (not necessarily uniformly distributed) in a subset of $\mathbb{R}^d$ or in the whole space can be used to construct a random polytope.

We will see in the following lecture that such polytopes can be described as random sets.

Over the last 50 years substantial attention has been devoted to the studies of geometric characteristics for the polytope P_n defined as the *convex hull* of $\{\xi_1, \dots, \xi_n\}$. In such studies the points are often assumed to have a uniform distribution in a convex set K. Then P_n, $n \geq 1$, is a non-decreasing sequence of polytopes that converges to K as $n \to \infty$. Much more interesting (and complicated) questions concern limit theorems for various geometric functionals defined on P_n, e.g. for the Lebesgue measure, surface area or the number of vertices, see Chap. 7.

Another typical example of a set constructed from random points is a random closed ball $B_\xi(\eta)$ of radius ξ centred at $\eta \in \mathbb{R}^d$. Numerous examples appear from solutions of inequalities, e.g. of the type $\{x \in \mathbb{R}^d : g_1(x) \leq \xi_1, \dots, g_n(x) \leq \xi_n\}$. Further examples of random sets (called tessellations) can be produced from collection of random lines, see Chaps. 5 and 6.

1.2 Distributions of Random Sets

The purpose of this section is to introduce basic ideas from the theory of random sets with emphasis on their distributions and measurability issues.

1.2.1 Definition of a Random Closed Set

The concept of a random set was mentioned for the first time together with the mathematical foundations of probability theory. Kolmogorov wrote in 1933 in his book on foundations of probability theory:

> Let G be a measurable region of the plane whose shape depends on chance; in other words, let us assign to every elementary event ξ of a field of probability a definite measurable plane region G. We shall denote by J the area of the region G and by $\mathbf{P}(x, y)$ the probability that the point (x, y) belongs to the region G. Then
>
> $$\mathbf{E}(J) = \iint \mathbf{P}(x, y)\, dx\, dy\,.$$

The formal definition of random sets appeared much later, namely in 1974 when Kendall published a paper [292] on foundations of random sets and in 1975 when Matheron [345] laid out the modern approach to this theory. The current state of the art of the random sets theory is described in [363], where all results mentioned below can be found together with references to the original papers and various generalizations. Applications and statistical issues for random closed sets are presented in [366, 489, 494]. It is common to work with closed random sets, which can be naturally defined as maps from the probability space $(\Omega, \mathcal{A}, \mathbf{P})$ to the family $\mathcal{F}$ of closed sets in $\mathbb{R}^d$ (or more general topological space E). One often speaks about set-valued maps from Ω to $\mathbb{R}^d$.

These maps should also satisfy certain measurability conditions, which should be carefully chosen. Indeed, too strict measurability conditions unnecessarily restrict the possible examples of random sets. On the other hand, too weak measurability conditions do not ensure that important functionals of a random set become random variables.

Definition 1.1. A map $X : \Omega \mapsto \mathcal{F}$ is said to be a *random closed set* if $\{\omega : X(\omega) \cap K \neq \emptyset\}$ is a measurable event for each K from the family $\mathcal{K}^d$ of compact subsets of $\mathbb{R}^d$.

In other words, a random closed set is a random element with values in the space $\mathcal{F}$ equipped with the σ-algebra $\sigma_{\mathcal{F}}$ generated by the families of sets

$$\{F \in \mathcal{F} : \ F \cap K \neq \emptyset\}$$

with $K \in \mathcal{K}^d$. A random compact set is a random closed set which almost surely takes compact values. Similarly one defines a random convex set.

From Definition 1.1 it is easily seen that a random singleton $X = \{\xi\}$ is a random closed set. The support points of a point process (see Sect. 4.1.1) also constitute a random closed set. If ξ_t is an almost surely continuous stochastic process indexed by $t \in \mathbb{R}^d$, then its excursion set $A_u(\xi, \mathbb{R}^d) = \{t : \xi_t \geq u\}$ is a random closed set. Indeed,

$$\{A_u(\xi, \mathbb{R}^d) \cap K \neq \emptyset\} = \{\sup_{t \in K} \xi_t \geq u\}$$

is a measurable event. It is possible to prove directly from the definition that a random disk with random centre and random radius is a random closed set. More economical ways to check the measurability conditions will be explained later on.

If X is a random closed set, then $\mathbf{1}(x \in X)$ is a random variable. Moreover, $\mathbf{1}(x \in X)$, $x \in \mathbb{R}^d$, is a random field on $\mathbb{R}^d$ (see Sect. 9.1), which takes values zero or one. On the contrary, each two-valued random field gives rise to a (possibly non-closed) random set. However, random fields method cannot always help to handle random sets. For instance, let $X = \{\xi\}$ be a random singleton with non-atomically distributed ξ. Then the corresponding indicator random field is non-separable and is not distinguishable from the random field that always vanishes. Indeed, the finite-dimensional distributions of the indicator field miss the mere existence of a random singleton with non-atomic distribution.

It is easy to check that the norm

$$\|X\| = \sup\{\|x\| : \ x \in X\},$$

depicted in Fig. 1.4a is a (possibly infinite) random variable. For this, one should note that the event $\{\|X\| < r\}$ corresponds to the fact that X misses the compact set $\{x : \ r \leq \|x\| \leq n\}$ for all sufficiently large natural numbers n. Another important

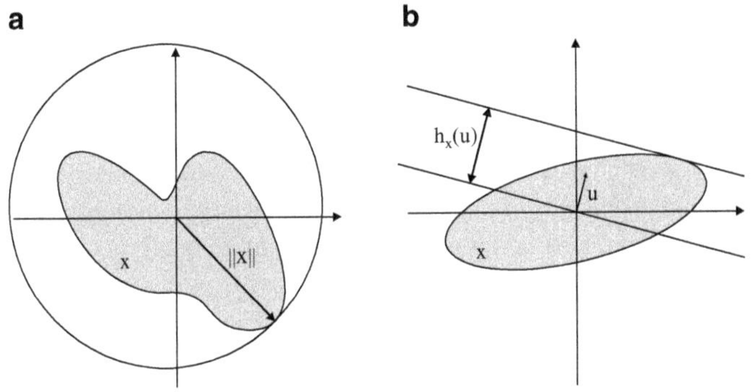

Fig. 1.4 The norm $\|X\|$ and the support function $h_X(u)$ of X

random variable associated with X is its *support function*

$$h_X(u) = \sup_{x \in X} \langle x, u \rangle ,$$

i.e. the supremum of the scalar products of $x \in X$ and the argument $u \in \mathbb{R}^d$, see Fig. 1.4b.

The distance

$$\rho(a, X) = \inf\{\rho(a, x) : x \in X\}$$

from $a \in \mathbb{R}^d$ to the nearest point $x \in X$ is a random variable, since $\rho(a, X) > t$ iff the closed ball of radius t centred at a does not hit X.

Furthermore, if μ is a locally finite measure, then $\mu(X)$ is a random variable. Indeed, Fubini's theorem applies to the integral of $\mathbf{1}(x \in X)$ with respect to $\mu(dx)$ and leads to

$$\mathbf{E}(\mu(X)) = \mathbf{E} \int \mathbf{1}(x \in X)\, \mu(dx) = \int \mathbf{E}(\mathbf{1}(x \in X))\, \mu(dx) = \int \mathbf{P}(x \in X)\, \mu(dx) .$$

The fact that the expected value of $\mu(X)$ for a locally finite μ equals the integral of the probability $\mathbf{P}(x \in X)$ is known under the name of the Robbins theorem formulated by Kolmogorov in 1933 and then independently by Robbins in 1944–1945. It should be noted that this fact does not hold for a general measure μ. For instance, if X is a singleton with an absolutely continuous distribution and μ is the counting measure, then $\mathbf{E}(\mu(X)) = 1$, while $\mathbf{P}(x \in X)$ vanishes identically. Note that limit theorems for the measure $\mu(X \cap W)$ with growing W can be deduced from limit theorems for random fields, see Sect. 10.3.

Exercise 1.4. Find an expression for $\mathbf{E}(\mu(X)^n)$ for a locally finite measure μ.

Exercise 1.5 (more difficult). Show that the Hausdorff dimension of a random closed set is a random variable.

1.2.2 Capacity Functional and the Choquet Theorem

Before discussing the construction of distributions for random closed sets recall that the distribution of a random variable ξ is defined by its cumulative distribution function $F(x) = \mathbf{P}(\xi \leq x)$ which is characterized by normalization conditions ($F(-\infty) = 0$ and $F(+\infty) = 1$), the right-continuity and monotonicity properties. These three properties are still required (in a slightly modified form) for distributions of random vectors and they also appear as characteristic properties for distributions of random closed sets.

Definition 1.2. The *capacity functional* of a random closed set X is defined as

$$T(K) = \mathbf{P}(X \cap K \neq \emptyset), \quad K \in \mathcal{K}^d .$$

The capacity functional is therefore defined on a family of compact sets. Since events $\{X \cap K \neq \emptyset\}$, $K \in \mathcal{K}^d$, generate the σ-algebra on the family of closed sets, it is easy to see that T uniquely determines the distribution of X. For instance, T is a restriction of a Borel probability measure on $\mathcal{K}^d$ iff X is a singleton. A closely related functional is the *avoidance functional* $Q(K) = 1 - T(K)$ that gives the probability that X misses compact set K. The capacity functional can be properly extended to a functional on all (even non-measurable!) subsets of $\mathbb{R}^d$, see [363, p. 9].

Exercise 1.6. Find a random closed set whose capacity functional is the restriction on $\mathcal{K}^d$ of a sub-probability measure, i.e. T is a measure with total mass less than or equal to one.

Another uniqueness issue is related to the concept of a point process from Sect. 4.1.1. Namely, a simple point process φ can be viewed as the random closed set X of its support points. Since

$$\mathbf{P}(X \cap K = \emptyset) = \mathbf{P}(\varphi(K) = 0),$$

the distribution of a simple point process is identically determined by its avoidance probabilities (i.e. probabilities that a given compact set contains no point of the process). Indeed, by the Choquet theorem these avoidance probabilities determine the distribution of the random closed set X and so φ as well. For instance, a random closed set with the capacity functional

$$T(K) = 1 - \mathrm{e}^{-\Lambda(K)}, \quad K \in \mathcal{K}^d,$$

with Λ being a locally finite measure on $\mathbb{R}^d$ is exactly the Poisson process with intensity measure Λ, i.e. it is a point process which fulfills the three conditions in Definition 3.4 of Sect. 3.1.2.

Exercise 1.7. Find out which random closed set on $\mathbb{R}^d$ has the capacity functional

$$T(K) = 1 - \mathrm{e}^{-\nu_d(K)^\alpha},$$

where ν_d is the Lebesgue measure and $\alpha \in (0, 1)$ (easy for $\alpha = 1$).

It remains to identify the properties of a functional $T(K)$ defined on a family of compact sets that guarantee the existence of a random closed set X having T as its capacity functional. This is done in the following Choquet theorem formulated in the current form by Matheron and proved in a slightly different formulation by Kendall.

Theorem 1.1 (Choquet–Kendall–Matheron). *A functional $T : \mathcal{K}^d \mapsto [0,1]$ defined on the family of compact subsets of a locally compact second countable Hausdorff space E is the capacity functional of a random closed set X in E iff $T(\emptyset) = 0$ and*

1. *T is* upper semicontinuous, *i.e. $T(K_n) \downarrow T(K)$ whenever $K_n \downarrow K$ as $n \to \infty$ with $K, K_n, n \geq 1$, being compact sets, i.e. $K_{n+1} \subset K_n$, $K = \cap_{n=1}^{\infty} K_n$.*
2. *T is* completely alternating, *i.e. the following successive differences*

$$\begin{aligned} \Delta_{K_1} T(K) &= T(K) - T(K \cup K_1)\,, \\ \Delta_{K_n} \cdots \Delta_{K_1} T(K) &= \Delta_{K_{n-1}} \cdots \Delta_{K_1} T(K) \\ &\quad - \Delta_{K_{n-1}} \cdots \Delta_{K_1} T(K \cup K_n)\,, \quad n \geq 2\,. \end{aligned}$$

are all non-positive for all compact sets $K, K_1, \ldots, K_n$.

The Euclidean space $\mathbb{R}^d$ is a locally compact second countable Hausdorff space, so that the Choquet theorem applies there.

There are three standard proofs of this theorem, see [363]. One derives it from the first principles of extension of measures from algebras to σ-algebras. For this, one notices that the events of the form $\{X \cap V = \emptyset, X \cap W_1 \neq \emptyset, \ldots, X \cap W_k \neq \emptyset\}$ form an algebra, where $V, W_1, \ldots, W_k$ are obtained by taking finite unions of open and compact sets and $k \geq 0$. The probabilities of these events are given by

$$\begin{aligned} \Delta_{W_1} T(V) &= \mathbf{P}(X \cap V \neq \emptyset) - \mathbf{P}(X \cap (V \cup W_1) \neq \emptyset) \\ &= -\mathbf{P}(X \cap W_1 \neq \emptyset,\ X \cap V = \emptyset)\,. \end{aligned}$$

and further by induction

$$-\Delta_{W_k} \cdots \Delta_{W_1} T(V) = \mathbf{P}(X \cap V = \emptyset,\ X \cap W_i \neq \emptyset,\ i = 1, \ldots, k)\,,$$

so that non-positivity of the successive differences corresponds to the non-negativity of the probabilities.

An alternative proof relies on powerful techniques of harmonic analysis on semigroups [62] by noticing that the complete alternative condition is related to the positive definiteness property on the semigroup of compact sets with the union operation. In a sense, the avoidance functional $Q(K)$ on the family of compact sets with union operation plays the same role as the function $\mathbf{P}(\xi > t)$ for random variables considered on real numbers with maximum operation and the characteristic function considered as a function on the real line with conventional addition. Finally, the lattice theoretic proof applies also in the case of non-Hausdorff space E, see [384].

Exercise 1.8. Identify a random closed set whose capacity functional is given by

$$T(K) = \sup_{x \in K} f(x)$$

for an upper semicontinuous function f with values in $[0, 1]$.

Since T determines uniquely the distribution of X, properties of X can be expressed as properties of T. For instance, X is stationary (i.e. $X + a$ coincides in distribution with X for all translations a) iff the capacity functional of X is translation invariant.

Exercise 1.9. Prove that a random closed set is convex iff its capacity functional satisfies

$$T(K_1 \cup K_2) + T(K_1 \cap K_2) = T(K_1) + T(K_2)$$

for all convex compact sets K_1 and K_2 such that $K_1 \cup K_2$ is also convex.

1.2.3 Selections and Measurability Issues

A random point ξ is said to be a selection of random set X if $\xi \in X$ almost surely. In order to emphasize the fact that ξ is measurable itself, one often calls it a measurable selection. A possibly empty random set clearly does not have a selection. Otherwise, the fundamental selection theorem establishes the existence of a selection of a random closed set under rather weak conditions. It is formulated below for random closed sets in $\mathbb{R}^d$.

Theorem 1.2 (Fundamental Selection Theorem [113]). *If $X : \Omega \mapsto \mathcal{F}$ is an almost surely non-empty random closed set in $\mathbb{R}^d$, then X has a (measurable) selection.*

Since the family of selections depends on the underlying σ-algebra, two identically distributed random closed sets might have different families of selections. However, it is known that the weak closures of the families of selections coincide if the random closed sets are identically distributed.

The following result by Himmelberg establishes equivalences of several measurability concepts. It is formulated in a somewhat restrictive form for Polish (complete separable metric) spaces. For any $X : \Omega \mapsto \mathcal{F}$, define $X^{-}(B) = \{\omega \in \Omega : X(\omega) \cap B \neq \emptyset\}$, where B is a subset of E.

Theorem 1.3 (Fundamental Measurability Theorem [246]). *Let E be a Polish space endowed with metric ρ and let $X : \Omega \mapsto \mathcal{F}$ be a function defined on a complete probability space $(\Omega, \mathcal{A}, \mathbf{P})$ with values being non-empty closed subsets of E. Then the following statements are equivalent.*

1. *$X^-(B) \in \mathcal{A}$ for every Borel set $B \subset E$.*
2. *$X^-(F) \in \mathcal{A}$ for every $F \in \mathcal{F}$.*
3. *$X^-(G) \in \mathcal{A}$ for every open set $G \subset E$ (in this case X is said to be* Effros measurable*).*
4. *The distance function $\rho(y, X) = \inf\{\rho(y, x) : x \in X\}$ is a random variable for each $y \in E$.*
5. *There exists a sequence $\{\xi_n, n \geq 1\}$ of measurable selections of X such that*

$$X = \text{cl}\{\xi_n,\ n \geq 1\}.$$

6. *The graph of X*

$$\text{graph}(X) = \{(\omega, x) \in \Omega \times E : x \in X(\omega)\}$$

is measurable in the product σ-algebra of $\mathcal{A}$ and the Borel σ-algebra on E.

Note that compact sets do not appear in the Fundamental Measurability Theorem. If $E = \mathbb{R}^d$ (or more generally if E is locally compact), then all above measurability conditions are equivalent to $X^-(K) \in \mathcal{A}$ for all compact sets K.

Exercise 1.10. Let X be *regular closed*, i.e. suppose X almost surely coincides with the closure of its interior. Show that all measurability properties of X are equivalent to $\{x \in X\} \in \mathcal{A}$ for all $x \in E$.

In particular, Statement 5 of Theorem 1.3 means that X can be obtained as the closure of a countable family of random singletons, known as the Castaign representation of X. This is a useful way to extend concepts defined for points to their analogues for random sets.

The most common distance on the family $\mathcal{K}^d$ of compact sets is the Hausdorff distance defined as

$$\rho_H(K, L) = \inf\{r > 0 : K \subset L^r,\ L \subset K^r\},$$

where K^r denotes the closed r-neighbourhood of K, i.e. the set of all points within distance r from K. This definition can be extended to not necessarily compact K and L, so that it no longer remains a metric but takes finite values if both K and L are bounded.

The fundamental measurability theorem helps to establish measurability of set-theoretic operations with random sets.

Theorem 1.4 (Measurability of set-theoretic operations [363, Theorem 1.2.25]). *If X is a random closed set in a Polish space E, then the following multifunctions are random closed sets:*

1. *the closed convex hull of X;*
2. *αX if α is a random variable;*
3. *the closed complement to X, the closure of the interior of X, and ∂X, the boundary of X.*

If X and Y are two random closed sets, then

1. *$X \cup Y$ and $X \cap Y$ are random closed sets;*
2. *the closure of $X \oplus Y = \{x + y : x \in X, y \in Y\}$ is a random closed set (if E is a Banach space);*
3. *if both X and Y are bounded, then the* Hausdorff distance *$\rho_H(X, Y)$ is a random variable.*

If $\{X_n, n \geq 1\}$ is a sequence of random closed sets, then

5. *$\mathrm{cl}(\cup_{n \geq 1} X_n)$ and $\cap_{n \geq 1} X_n$ are random closed sets;*
6. *$\limsup_{n\to\infty} X_n$ and $\liminf_{n\to\infty} X_n$ are random closed sets.*

1.3 Limit Theorems for Random Sets

This section deals with limit theorems for Minkowski sums and unions of random sets.

1.3.1 Expectation of a Random Set

The space of closed sets is not linear, which causes substantial difficulties in defining the expectation for a random set. One way described below is to represent a random set using a family of its selections.

Let X be a random closed set in $\mathbb{R}^d$. If X possesses at least one integrable selection then X is called integrable. For instance, if X is almost surely non-empty compact and its norm $\|X\|$ is integrable (in this case X is said to be integrably bounded) then all selections of X are integrable and so X is integrable too. Later on we usually assume that X is integrably bounded.

Definition 1.3. The *(selection or Aumann) expectation* $\mathbf{E}X$ of an integrable random closed set X is closure of the family of all expectations for its integrable selections.

If X is an integrably bounded subset of $\mathbb{R}^d$, then the expectations of all its selections form a closed set and there is no need to take an additional closure. The so defined expectation depends on the probability space where X is defined. For instance, the deterministic set $X = \{0, 1\}$ defined on the trivial probability space $\{\emptyset, \Omega\}$ has expectation $\mathbf{E}X = \{0, 1\}$, since it has only two trivial (deterministic) selections. However, if X is defined on a non-atomic probability space, then its selections are $\xi = \mathbf{1}(A)$ for all events $A \subset \Omega$, so that $\mathbf{E}\xi = \mathbf{P}(A)$ and the range of possible values for $\mathbf{E}\xi$ constitutes the whole interval $[0, 1]$.

The following result shows that the expectation of a random compact set defined on a non-atomic probability space is a convex set whose support function equals the expected support function of X, i.e. the expectation convexifies random compact sets.

Theorem 1.5. *If an integrably bounded X is defined on a non-atomic probability space, then $\mathbf{E}X$ is a convex set and*

$$h_{\mathbf{E}X}(u) = \mathbf{E}h_X(u), \quad u \in \mathbb{R}^d. \tag{1.3}$$

Proof. The convexity of the Aumann expectation can be derived from the Lyapunov theorem which says that the range of any non-atomic vector-valued measure is a convex set (to see this in the one-dimensional case note that real-valued measures do not have gaps in the ranges of their values). Let ξ_1 and ξ_2 be two integrable selections of X. Define the vector-valued measure

$$\lambda(A) = \mathbf{E}(\xi_1 \mathbf{1}(A), \xi_2 \mathbf{1}(A^c))$$

for all measurable events A. The closure of its range is convex, $\lambda(\emptyset) = (0,0)$ and $\lambda(\Omega) = (\mathbf{E}\xi_1, \mathbf{E}\xi_2)$. Let $\alpha \in (0,1)$. Thus, there exists an event A such that

$$\|\alpha \mathbf{E}\xi_i - \mathbf{E}(\xi_i \mathbf{1}(A))\| < \varepsilon/2, \quad i = 1,2.$$

Define the selection

$$\eta = \xi_1 \mathbf{1}(A) + \xi_2 \mathbf{1}(A^c).$$

Then

$$\|\alpha \mathbf{E}\xi_1 + (1-\alpha)\mathbf{E}\xi_2 - \mathbf{E}\eta\| < \varepsilon$$

for arbitrary $\varepsilon > 0$, whence $\mathbf{E}X$ is convex.

Now establish the relationship to support functions. Let $x \in \mathbf{E}X$. Then there exists a sequence ξ_n of selections such that $\mathbf{E}\xi_n \to x$ as $n \to \infty$. Furthermore

$$h_{\{x\}}(u) = \lim_{n\to\infty} \langle \mathbf{E}\xi_n, u\rangle = \lim_{n\to\infty} \mathbf{E}\langle \xi_n, u\rangle \le \mathbf{E}h_X(u),$$

and hence $h_X(u) = \sup_{x \in X} h_{\{x\}}(u) \le \mathbf{E}h_X(u)$. Finally, for each unit vector u and $\varepsilon > 0$ define a half-space as

$$Y_\varepsilon = \{x : \langle x, u\rangle \ge h_X(u) - \varepsilon\}.$$

Then $Y_\varepsilon \cap X$ is a non-empty random closed set, which has a selection ξ_ε, such that

$$h_{\{\xi_\varepsilon\}}(u) \ge h_X(u) - \varepsilon.$$

Taking the expectation confirms that $h_{\mathbf{E}X}(u) \ge \mathbf{E}h_X(u)$. □

The convexifying effect of the selection expectation limits its applications in such areas like image analysis, where it is sometimes essential to come up with averaging scheme for images, see [363, Sect. 2.2] for a collection of further definitions of expectations. However, it appears very naturally in the law of large numbers for random closed sets as described in the following section.

Example 1.1. Let $X = B_\xi(\eta)$ be the closed ball of radius $\xi > 0$ centred at $\eta \in \mathbb{R}^d$, where both ξ and η are integrable. Then its expectation is the ball of radius $\mathbf{E}\xi$ centred at $\mathbf{E}\eta$.

Exercise 1.11. Show that $\mathbf{E}X = \{a\}$ is a singleton iff X is a random singleton itself, i.e. $X = \{\xi\}$.

Exercise 1.12. Assume that X is isotropic, i.e. X coincides in distribution with its arbitrary rotation around the origin. Identify $\mathbf{E}X$.

1.3.2 *Law of Large Numbers and the Limit Theorem for Minkowski Sums*

Recall that the *Minkowski sum* of two compact sets K and L is defined as

$$K \oplus L = \{x + y : x \in K, y \in L\}.$$

In particular K^r is the Minkowski sum of K and the closed ball of radius r, centred at the origin. The same definition applies if one of the summands is compact and the other is closed. However if the both summands are closed (and not necessarily compact), then the sum is not always closed and one typically inserts the closure in the definition.

Support functions linearize the Minkowski sum, i.e.

$$\begin{aligned} h_{\alpha K}(u) &= \alpha h_K(u), \\ h_{K \oplus L}(u) &= h_K(u) + h_L(u), \quad u \in \mathbb{R}^d \end{aligned}$$

for convex compact sets K and L. The homogeneity property of support functions makes it possible to define them only on the unit sphere $\mathbb{S}^{d-1}$ in $\mathbb{R}^d$. Then the uniform metric for support functions on the sphere turns into the Hausdorff distance between compact sets. Namely

$$\rho_H(K, L) = \sup_{u \in \mathbb{S}^{d-1}} |h_K(u) - h_L(u)|$$

and also

$$\|K\| = \rho_H(K, \{o\}) = \sup_{u \in \mathbb{S}^{d-1}} |h_K(u)|.$$

Consider a sequence of i.i.d. random compact sets $X_1, X_2, \ldots$ all distributed as a random compact set X. It should be noted that the mere existence of such sequence implies that the probability space is non-atomic.

Theorem 1.6 (Law of large numbers for random sets [17]). *If $X, X_1, X_2, \ldots$ are i.i.d. integrably bounded random compact sets and $S_n = X_1 \oplus \cdots \oplus X_n$, $n \geq 1$, are their Minkowski sums, then*

$$\rho_H(n^{-1}S_n, \mathbf{E}X) \xrightarrow{a.s.} 0 \quad \textit{as } n \to \infty .$$

Proof. Let us prove the result assuming that X is almost surely convex. Then

$$h(n^{-1}S_n, u) = \frac{1}{n}\sum_{i=1}^{n} h(X_i, u) \xrightarrow{\text{a.s.}} \mathbf{E}h(X, u) = h(\mathbf{E}X, u)$$

by a strong law of large numbers in a Banach space specialized for the space of continuous functions on the unit ball with the uniform metric, see [372]. The uniform metric on this space corresponds to the Hausdorff metric on convex compact sets, whence the strong law of large numbers holds.

In order to get rid of the convexity assumption we rely on the following result known as Shapley–Folkman–Starr Theorem. If $K_1, \ldots, K_n$ are compact subsets of $\mathbb{R}^d$ and $n \geq 1$, then

$$\rho_H(K_1 \oplus \cdots \oplus K_n, \operatorname{conv}(K_1 \oplus \cdots \oplus K_n)) \leq \sqrt{d} \max_{1 \leq i \leq n} \|K_i\| .$$

Note that the number of summands does not appear in the factor on the right-hand side. For instance, if $K_1 = \cdots = K_n = K$, then one obtains that the distance between the sum of n copies of K and the sum of n copies of the convex hull of K is at most $\sqrt{d}\|K\|$.

A not necessarily convex X can be replaced by its convex hull $\operatorname{conv}(X)$, so that it remains to show that

$$n^{-1}\rho_H(X_1 \oplus \cdots \oplus X_n, \operatorname{conv}(X_1 \oplus \cdots \oplus X_n)) \leq \frac{\sqrt{d}}{n} \max_{1 \leq i \leq n} \|X_i\| \xrightarrow{\text{a.s.}} 0 \quad \text{as } n \to \infty .$$

The latter follows from the integrable boundedness of X. Indeed, then we have $n\mathbf{P}(\|X\| > n) \to 0$ as $n \to \infty$, and

$$\mathbf{P}(\max_{i=1,\ldots,n} \|X_i\| \geq nx) = 1 - (1 - \mathbf{P}(\|X\| \geq nx))^n \to 0 .$$

The proof is complete. □

Numerous generalizations of the above strong law of large numbers deal with random subsets of Banach spaces and possibly unbounded random closed sets in the Euclidean space, see [363].

The formulation of the central limit theorem is complicated by the fact that random sets with expectation o are necessarily singletons. Furthermore, it is not possible to define Minkowski subtraction as the opposite operation to the addition.

For instance, it is not possible to find a set that being added to a ball produces a triangle. Therefore, its not possible in general to normalize successive sums of random compact sets.

Note that the classical limit theorem may be (a bit weaker) formulated as the convergence of the normalized distance between the empirical mean and the expectation to the absolute value of a normally distributed random variable.

In order to formulate a limit theorem for random closed sets we need to define a centred Gaussian random field $\zeta(u)$ on the unit sphere $\mathbb{S}^{d-1}$ in $\mathbb{R}^d$ which shares the covariance structure with the random closed set X, i.e.

$$\mathbf{E}(\zeta(u)\zeta(v)) = \mathbf{cov}(h_X(u), h_X(v)), \quad u, v \in \mathbb{S}^{d-1}.$$

Since the support function of a compact set is Lipschitz, it is possible to show that the random field ζ has a continuous modification.

Theorem 1.7 (Central Limit Theorem [508]). *Let $X_1, X_2, \ldots$ be i.i.d. copies of a random closed set X in $\mathbb{R}^d$ such that $\mathbf{E}\|X\|^2 < \infty$. Then $\sqrt{n}\rho_H(n^{-1}S_n, \mathbf{E}X)$ converges in distribution as $n \to \infty$ to $\sup\{|\zeta(u)| : u \in \mathbb{S}^{d-1}\}$.*

Proof. For convex random sets the result follows from the central limit theorem for continuous random functions on the unit sphere, see [15, Corollary 7.17]. The general non-convex case is proved by an application of the Shapley–Folkman–Starr Theorem. □

It is not clear how to interpret geometrically the limit $\zeta(u)$, $u \in \mathbb{S}^{d-1}$, in the central limit theorem for random sets. It is possible to define Gaussian random sets as those whose support function is a Gaussian process on $\mathbb{S}^{d-1}$. All such sets have however degenerate distributions. Namely, X is a Gaussian random set iff $X = \xi + M$, where ξ is a Gaussian random vector in $\mathbb{R}^d$ and M is a deterministic convex compact set. This is seen by noticing that the so-called *Steiner point*

$$s(X) = \frac{1}{\kappa_d}\int_{\mathbb{S}^{d-1}} h(X,u)\,u\,du$$

is a Gaussian random vector that a.s. belongs to X, where κ_d is the volume of the d-dimensional unit ball. Thus, $M = X - \xi$ with $\xi = s(X)$ has Gaussian non-negative support function, which is then necessarily degenerate, so that M is deterministic.

1.3.3 Unions of Random Sets

While the arithmetic summation scheme for random variables gives rise to the Gaussian distribution in the limit, the maximum of random variables gives rise to extreme value distributions. Along the same limit, the Minkowski summation scheme for random sets being singletons reduces to the classical limit theorem for sums of random vectors, while taking unions of random sets generalizes the maximum (or minimum) scheme for random variables. Notice that if $X_i = (-\infty, \xi_i]$, $i = 1, 2 \ldots$, then

$$X_1 \cup \cdots \cup X_n = (-\infty, \max(\xi_1, \ldots, \xi_n)] .$$

Let $X, X_1, X_2, \ldots$ be a sequence of i.i.d. random closed sets in $\mathbb{R}^d$ and let $a_n > 0$, $n \geq 1$, be a sequence of non-negative normalizing constants. The weak convergence of the random set

$$Z_n = a_n(X_1 \cup \cdots \cup X_n)$$

to a random closed set Z is defined by specializing the general concept of weak convergence of probability measures for the space $\mathcal{F}$ of closed sets. In particular, a necessary and sufficient condition for this is the convergence of capacity functionals on sets K such that Z touches the boundary of K with probability zero, i.e.

$$\mathbf{P}(Z \cap K \neq \emptyset,\ Z \cap \mathrm{Int}(K) = \emptyset) = 0,$$

where $\mathrm{Int}(K)$ denotes the interior of K. The capacity functional of the set Z_n is easy to find as

$$T_{Z_n}(K) = 1 - (1 - T_X(a_n^{-1} K))^n .$$

Various convergence results for unions of random sets can be found in [363, Chap. 4].

Here we shall discuss properties of random sets that can appear in the limit. In more general triangular array schemes of building the n-fold union the limits are union-infinitely-divisible, while in the above described scheme the limit Z is necessarily union-stable, see [363, Sect. 2.3].

Definition 1.4. A random closed set is said to be *union-infinitely divisible* if Z coincides in distribution with the union of i.i.d. random closed sets $Z_{11}, \ldots, Z_{nn}$ for each $n \geq 2$.
A random closed set Z is said to be *union-stable* if Z coincides in distribution with $a_n^{-1}(Z_1 \cup \cdots \cup Z_n)$ for each $n \geq 2$ with normalizing constants $a_n > 0$, where $Z_1, \ldots, Z_n$ are i.i.d. copies of Z.

Exercise 1.13. Assume that Z coincides in distribution with the union of its n i.i.d. copies for some $n \geq 2$. Show that such Z is necessarily deterministic.

For the following it is essential to single out the deterministic part of a random set. A point x is said to be a *fixed point* of X if $x \in X$ with probability one. The set of fixed points is denoted by F_X. For instance, if $X = (-\infty, \xi]$ with exponentially distributed ξ, then $F_X = (-\infty, 0]$, while F_X is empty if ξ is normally distributed.

Theorem 1.8 ([363, Theorem 4.1.6]). *A random closed set X is union-infinitely-divisible iff there exists a completely alternating upper semicontinuous functional $\psi : \mathcal{K}^d \to [0, \infty]$ such that*

$$T(K) = 1 - e^{-\psi(K)}, \quad K \in \mathcal{K}^d,$$

and $\psi(K) < \infty$ *whenever* $K \cap F_X = \emptyset$.

Example 1.2. Let X be the Poisson point process with intensity measure Λ. Then $T(K) = 1 - e^{-\Lambda(K)}$ and X is union-infinitely-divisible. Indeed, X equals in distribution the union of n i.i.d. Poisson processes, each with intensity measure $n^{-1}\Lambda$.

The functional ψ shares nearly the same properties as T apart from the fact that the values of ψ are no longer required to lie in $[0, 1]$. The functional ψ defines a locally finite measure μ on $\mathcal{F}$, such that $\mu(\{F \in \mathcal{F} : F \cap K \neq \emptyset\}) = \psi(K)$. The measure μ defines a Poisson process on $\mathcal{F}$ such that the "points" of this process are actually closed sets. Then X is the union of the sets from the support of this Poisson process on $\mathcal{F}$ with intensity measure μ.

Theorem 1.9 ([363, Theorem 4.1.12]). *A random closed set is union-stable iff its capacity functional admits representation* $T(K) = 1 - e^{-\psi(K)}$ *with* ψ *being homogeneous, i.e.*

$$\psi(sK) = s^{\alpha}\psi(K), \quad K \in \mathcal{K}^d, \ K \cap F_X = \emptyset,$$

for some $\alpha \neq 0$ *and all* $s > 0$, *and also* $F_X = sF_X$ *for all* $s > 0$.

The proof of the above theorem relies on solving some functional equations for capacity functionals of random sets, quite similar to the corresponding characterization of max-stable random variables. The major complication stems from the fact that for any random variable ξ the equivalence of the distributions of ξ and $c\xi$ immediately implies that $c = 1$. This is however not the case for random sets, e.g. the set of zeros for the standard Brownian motion $X = \{t \geq 0 : W(t) = 0\}$, coincides in distribution with cX for each $c > 0$. Another example of such is a randomly rotated cone in $\mathbb{R}^d$. The key step in the proof of Theorem 1.9 aims to show that the union-stability property rules out all such self-similar random sets.

Exercise 1.14. Let $X = (-\infty, \xi]$ in $\mathbb{R}$. Prove that all distributions of ξ that correspond to union-stable X are exactly extreme value distributions of Fréchet and Weibull type.

If X is a Poisson process, then its union-stability property implies that the intensity measure of the process is homogeneous, i.e. $\Lambda(sA) = s^{\alpha}\Lambda(A)$ for all $A \in \mathcal{B}(\mathbb{R}^d)$ and $s > 0$.

Chapter 2
Introduction into Integral Geometry and Stereology

Markus Kiderlen

Abstract This chapter is a self-contained introduction into integral geometry and its applications in stereology. The most important integral geometric tools for stereological applications are kinematic formulae and results of Blaschke–Petkantschin type. Therefore, Crofton's formula and the principal kinematic formula for polyconvex sets are stated and shown using Hadwiger's characterization of the intrinsic volumes. Then, the linear Blaschke–Petkantschin formula is proved together with certain variants for flats containing a given direction (vertical flats) or contained in an isotropic subspace. The proofs are exclusively based on invariance arguments and an axiomatic description of the intrinsic volumes.

These tools are then applied in model-based stereology leading to unbiased estimators of specific intrinsic volumes of stationary random sets from observations in a compact window or a lower dimensional flat. Also, Miles-formulae for stationary and isotropic Boolean models with convex particles are derived. In design-based stereology, Crofton's formula leads to unbiased estimators of intrinsic volumes from isotropic uniform random flats. To estimate the Euler characteristic, which cannot be estimated using Crofton's formula, the disector design is presented. Finally we discuss design-unbiased estimation of intrinsic volumes from vertical and from isotropic sections.

2.1 Integral Geometric Foundations of Stereology

In the early 1960s stereology was a collection of mathematical methods to extract spatial information of a material of interest from sections. Modern stereology may be considered as "sampling inference for geometrical objects" ([31, Chap. 5]

M. Kiderlen (✉)
Aarhus University, Aarhus, Denmark
e-mail: kiderlen@imf.au.dk

E. Spodarev (ed.), *Stochastic Geometry, Spatial Statistics and Random Fields*,
Lecture Notes in Mathematics 2068, DOI 10.1007/978-3-642-33305-7_2,

and [142, pp. 56–57]) thus emphasizing the two main columns stereology rests upon: sampling theory and geometry. In this section, we shall discuss two of the most important geometric concepts for stereology: kinematic integral formulae and results of Blaschke–Petkantschin type. The proofs are exclusively based on invariance arguments and an axiomatic description of the intrinsic volumes. Section 2.2 is then devoted to stereology and describes in detail, how geometric identities lead to unbiased estimation procedures. The influence from sampling theory will also be mentioned in that later section.

2.1.1 *Intrinsic Volumes and Kinematic Integral Formula*

We start which a deliberately vague question of how to sample a set $K \subset \mathbb{R}^d$. In order to avoid possibly costly measurements on the whole of K, we sample K with a "randomly moved" sampling window $M \subset \mathbb{R}^d$ and consider only the part of K that is inside the moved window. To fix ideas, we assume that K and M are elements of the family $\mathcal{K}^d_{\text{conv}}$ of convex bodies (compact convex subsets) in $\mathbb{R}^d$, that the (orientation preserving) motion is the composition of a translation with a random vector $\xi \in \mathbb{R}^d$ and a random rotation $\rho \in \mathsf{SO}$ (special orthogonal group). Assume further that $f : \mathcal{K}^d_{\text{conv}} \to \mathbb{R}$ is a functional which gives, for each observation, the measured value (think of the volume). What is the expected value of $f(K \cap \rho(M + \xi))$?

To make this question meaningful, we have to specify the distributions of ρ and ξ. One natural condition would be that the distribution of $K \cap \rho(M + \xi)$ is independent of the location and orientation of M. In particular, this implies that ρ should be *right invariant*: $\rho \circ R$ and ρ have the same distribution for any deterministic $R \in \mathsf{SO}$. The space SO, identified with the family of all orthonormal matrices in $\mathbb{R}^{d \times d}$ with determinant 1 and endowed with the induced topology, becomes a compact topological group. The theory of invariant measures [451, Chap. 13] implies that there is a unique right invariant probability measure on SO, which we denote by ν. This measure, also called *normalized Haar measure*, has even stronger invariance properties: it is inversion invariant in the sense that ρ and ρ^{-1} have the same distribution. Together with the right invariance, this implies that ν is also left invariant in the obvious sense. The measure ν is therefore the natural measure on SO, its role being comparable to the one of the Lebesgue measure ν_d on $\mathbb{R}^d$. We shall therefore just write $dR = \nu(dR)$ when integrating with respect to this measure. The matrix corresponding to the random rotation ρ can be constructed explicitly by applying the Gram–Schmidt orthonormalization algorithm to a d-tuple $(\eta_1, \ldots, \eta_d)$ formed by random i.i.d. uniform vectors in the unit sphere $\mathbb{S}^{d-1}$. Note that the vectors $\eta_1, \ldots, \eta_d$ are almost surely linearly independent.

Similar considerations for the random translation vector ξ lead to the contradictory requirement that the distribution of ξ should be a multiple of the Lebesgue measure ν_d on $\mathbb{R}^d$. This was already pointed out in Chap. 1; see the first paragraph on page 2. In contrast to SO, the group $\mathbb{R}^d$ is only locally compact but not compact.

We therefore have to modify our original question. In view of applications we assume that the moved window hits a fixed reference set $A \in \mathcal{K}^d_{\mathrm{conv}}$, which contains K. Using the invariant measures defined above we then have

$$\mathbf{E}(f(K \cap \rho(M + \xi))) = \frac{\int_{\mathsf{SO}} \int_{\mathbb{R}^d} f(K \cap R(M + x))\, dx\, dR}{\int_{\mathsf{SO}} \int_{\mathbb{R}^d} \mathbf{1}(A \cap R(M + x) \neq \emptyset)\, dx\, dR}, \tag{2.1}$$

where we assumed $f(\emptyset) = 0$. Numerator and denominator of this expression are of the same form and we first consider the special case of the denominator with $M = B_r(o)$, $r > 0$, and $A = K \in \mathcal{K}^d_{\mathrm{conv}}$:

$$\int_{\mathsf{SO}} \int_{\mathbb{R}^d} \mathbf{1}(K \cap R(B_r(o) + x) \neq \emptyset)\, dx\, dR = \nu_d(K \oplus B_r(o)).$$

By a fundamental result in convex geometry [451, Sect. 14.2], this volume is a polynomial of degree at most d in $r > 0$, usually written as

$$\nu_d(K \oplus B_r(o)) = \sum_{j=0}^{d} r^{d-j} \kappa_{d-j} V_j(K), \tag{2.2}$$

where κ_j denotes the volume of the j-dimensional unit ball. This result is the *Steiner formula*. It defines important functionals, the *intrinsic volumes* $V_0, \dots, V_d$. They include the *volume* $V_d(K) = \nu_d(K)$, the *surface area* $2V_{d-1}(K)$ of the boundary of K (when $\operatorname{int} K \neq \emptyset$) and the trivial functional $V_0(K) = \mathbf{1}(K \neq \emptyset)$, also called *Euler characteristic* $\chi(K)$. The Steiner formula implies

$$\int_{\mathsf{SO}} \int_{\mathbb{R}^d} V_0(K \cap R(B_r(o) + x))\, dx\, dR = \sum_{j=0}^{d} r^{d-j} \kappa_{d-j} V_j(K).$$

Already in this special case with M being a ball, the intrinsic volumes play an essential role to express kinematic integrals explicitly. We shall soon see that this even holds true when V_0 is replaced by a function $f : \mathcal{K}^d_{\mathrm{conv}} \to \mathbb{R}$ satisfying some natural properties. To do so, we first clarify the basic properties of V_j.

It is easily seen from the Steiner formula that $V_j : \mathcal{K}^d_{\mathrm{conv}} \to \mathbb{R}$ is invariant under rigid motions and is homogeneous of degree j. Here, we call a function $f : \mathcal{K}^d_{\mathrm{conv}} \to \mathbb{R}$

1. *Invariant under rigid motions* if $f(R(K + x)) = f(K)$ for all $K \in \mathcal{K}^d_{\mathrm{conv}}$, $R \in \mathsf{SO}$, and $x \in \mathbb{R}^d$.
2. *Homogeneous of degree j* if $f(\alpha K) = \alpha^j f(K)$ for all $K \in \mathcal{K}^d_{\mathrm{conv}}$, $\alpha \geq 0$.

Using convexity properties, V_j can be shown to be additive and monotone (with respect to set inclusion). Here $f : \mathcal{K}^d_{\mathrm{conv}} \to \mathbb{R}$ is

3. *Additive* if $f(\emptyset) = 0$ and

$$f(K \cup M) = f(K) + f(M) - f(K \cap M)$$

for $K, M \in \mathcal{K}^d_{\mathrm{conv}}$ with $K \cup M \in \mathcal{K}^d_{\mathrm{conv}}$ (implying $K \cap M \neq \emptyset$).
4. *Monotone* if $f(K) \leq f(M)$ for all $K, M \in \mathcal{K}^d_{\mathrm{conv}}$, $K \subset M$.

Exercise 2.1. Use the Steiner formula to show that

$$V_j(B_r(o)) = \binom{d}{j} \frac{\kappa_d}{\kappa_{d-j}} r^j$$

holds for $j = 0, \ldots, d$.

Exercise 2.2. Show that if $K \in \mathcal{K}^d_{\mathrm{conv}}$ is contained in a k-dimensional subspace L_0, then $V_j(K) = 0$ for $j > k$ and $V_k(K)$ is the k-dimensional volume of K.

Hint. Due to rotation invariance we may assume that L_0 is spanned by the first k vectors of the standard basis of $\mathbb{R}^d$. Then Fubini's theorem implies

$$\nu_k(K)\kappa_{d-k} r^{d-k} \leq \nu_d(K \oplus B_r(o)) \leq \nu_k\left(K \oplus (B_\varepsilon(o) \cap L_0)\right) \kappa_{d-k} r^{d-k}$$

for all $0 \leq r \leq \varepsilon$.

Exercise 2.3. Show that the invariance under rigid motions and the homogeneity property of the intrinsic volumes are immediate consequences of the Steiner formula.

Already a selection of the above defined properties is sufficient to characterize intrinsic volumes axiomatically. This is the content of Hadwiger's famous characterization theorem; see [220], where a corresponding result is also shown with a continuity assumption replacing monotonicity. A simplified proof (for the characterization based on continuity) can be found in [301], see also [24] or [302].

Theorem 2.1 (Hadwiger). *Suppose $f : \mathcal{K}^d_{\mathrm{conv}} \to \mathbb{R}$ is additive, motion invariant and monotone. Then there exist $c_0, \ldots, c_d \geq 0$ with*

$$f = \sum_{j=0}^{d} c_j V_j.$$

This shows that the intrinsic volumes are essentially the only functionals that share some natural properties with the volume. We shall use this result without proof. It implies in particular, that under the named assumptions on f, we only have to consider the numerator of (2.1) for $f = V_j$. In view of Hadwiger's characterization the following result gives a complete answer to our original question for a large class of measurement functions f.

Theorem 2.2 (Principal kinematic formula). *Let $j \in \{0, \dots, d\}$ and $K, M \in \mathcal{K}^d_{\text{conv}}$. Then*

$$\int_{\text{SO}} \int_{\mathbb{R}^d} V_j(K \cap R(M+x))\, dx\, dR = \sum_{k=j}^{d} c_{j,d}^{k,d-k+j} V_k(K) V_{d-k+j}(M),$$

where the constants are given (for $m = 2$) by

$$c_{s_1,\dots,s_m}^{r_1,\dots,r_m} = \prod_{i=1}^{m} \frac{r_i! \kappa_{r_i}}{s_i! \kappa_{s_i}}, \quad m \in \mathbb{N}. \tag{2.3}$$

In certain cases the formula remains valid even when the rotation integral is omitted. This is trivially true for $M = B_r(o)$, but also for $j = d$ and $j = d-1$.

Proof. We denote the left hand side of the principal kinematic formula by $f(K, M)$. The functional $f(K, M)$ is symmetric in K and M due to the invariance properties of ν_d, ν and V_j. The homogeneity of V_j and a substitution yield

$$f(\alpha K, \alpha M) = \alpha^{d+j} f(K, M), \quad \alpha > 0.$$

As $f(K, \cdot)$ is additive, motion invariant and monotone, Hadwiger's characterization theorem implies the existence of constants $c_0(K), \dots, c_d(K) \geq 0$ (depending on K) with $f(K, \cdot) = \sum_{k=0}^{d} c_k(K) V_k$. Hence, for $\alpha > 0$,

$$\begin{aligned}
\sum_{k=0}^{d} c_k(K) V_k(M) \alpha^k &= \sum_{k=0}^{d} c_k(K) V_k(\alpha M) = f(K, \alpha M) \\
&= \alpha^{d+j} f\left(\frac{1}{\alpha} K, M\right) = \alpha^{d+j} f\left(M, \frac{1}{\alpha} K\right) \\
&= \sum_{k=0}^{d} c_k(M) V_k(K) \alpha^{d-k+j}.
\end{aligned}$$

Comparison of the coefficients of these polynomials yields $c_k(M) = 0$ for $k < j$ and that $c_k(M)$ is proportional to $V_{d-k+j}(M)$ for $k \geq j$. This gives the principal kinematic formula with unknown constants. The constants are then determined by appropriate choices for K and M, for which the integrals can be calculated explicitly; see also the comment after Theorem 2.4. □

This solves our original question for $f = V_j$. Formula (2.1) now gives

$$\mathbf{E}(V_j(K \cap \rho(M+\xi))) = \frac{\sum_{k=j}^{d} c_{j,d}^{k,d-k+j} V_k(K) V_{d-k+j}(M)}{\sum_{k=0}^{d} c_{0,d}^{k,d-k} V_k(A) V_{d-k}(M)}$$

if (ρ, ξ) has its natural distribution on $\{(R, x) : R(M+x) \cap A \neq \emptyset\}$.

Exercise 2.4. Show the following general basic integral formula using Fubini's theorem. If μ is a σ-finite measure on $\mathcal{B}(\mathbb{R}^d)$ then

$$\int_{\mathbb{R}^d} \mu(A \cap (B + x))dx = \mu(A)\nu_d(B)$$

for all $A, B \in \mathcal{B}(\mathbb{R}^d)$.

Conclude that the special case $j = d$ of Theorem 2.2 holds even when the rotation integral is omitted.

As invariant integration like in Theorem 2.2 does always lead to functionals in the linear span of $V_0, \ldots, V_d$, an iterated version for $k + 1$ convex bodies can be shown by induction.

Theorem 2.3 (Iterated principal kinematic formula). *Let $j \in \{0, \ldots, d\}, k \geq 1$, and $K_0, \ldots, K_k \in \mathcal{K}^d_{\text{conv}}$. Then*

$$\int_{\mathsf{SO}} \int_{\mathbb{R}^d} \cdots \int_{\mathsf{SO}} \int_{\mathbb{R}^d} V_j(K_0 \cap R_1(K_1 + x_1) \cap \ldots \cap R_k(K_k + x_k))$$
$$\times\, dx_1\, dR_1 \cdots dx_k\, dR_k$$
$$= \sum_{\substack{m_0,\ldots,m_k=j \\ m_0+\ldots+m_k=kd+j}}^{d} c^{m_0,\ldots,m_k}_{j,d,\ldots,d} V_{m_0}(K_0) \cdots V_{m_k}(K_k)$$

with constants given by (2.3).

For $j = d$ and $j = d - 1$ the rotation integrals on the left hand side can be omitted.

Theorem 2.2 has a counterpart where M is replaced by an affine subspace. For $k \in \{0, \ldots, d\}$ let $\mathcal{L}^d_k$ be the *Grassmannian* of all k-dimensional linear subspaces of $\mathbb{R}^d$. The image measure of ν on SO under $R \mapsto RL_0$, $L_0 \in \mathcal{L}^d_k$ fixed, is a rotation invariant probability measure on $\mathcal{L}^d_k$, and integration with respect to it is denoted by dL. It is the only rotation invariant distribution on $\mathcal{L}^d_k$. Similarly, let $\mathcal{E}^d_k$ be the space of all (affine) k-dimensional flats in $\mathbb{R}^d$. The elements $E \in \mathcal{E}^d_k$ are called *k-flats* and can be parametrized in the form $E = R(L_0 + x)$ with $R \in \mathsf{SO}$, $x \in L_0^\perp$, and a fixed space $L_0 \in \mathcal{L}^d_k$. The function $(R, x) \mapsto R(L_0 + x)$ on $\mathsf{SO} \times L_0^\perp$ maps the measure $\nu \otimes \nu_{d-k}$ to a motion invariant measure on $\mathcal{E}^d_k$. Integration with this measure is denoted by dE. Up to a factor, this is the only motion invariant measure on $\mathcal{E}^d_k$.

We shall later need families of subspaces containing or contained in a given space $L \in \mathcal{L}^d_k, 0 \leq k \leq d$:

$$\mathcal{L}^L_r = \begin{cases} \{M \in \mathcal{L}^d_r : M \subset L\}, & \text{if } 0 \leq r \leq k, \\ \{M \in \mathcal{L}^d_r : M \supset L\}, & \text{if } k < r \leq d. \end{cases}$$

Again, there is a uniquely determined invariant probability measure on $\mathcal{L}^L_r$. We shall write dM when integrating with respect to it; the domain of integration will always be clear from the context. For $r \leq k$ existence and uniqueness of this probability measure follow from identifying L with $\mathbb{R}^k$. For $r > k$, this measure is obtained as image of $\int_{\mathcal{L}_{r-k}^{L\perp}\cap(\cdot)} dM$ under $M \mapsto M \oplus L$. An invariance argument also shows that

$$\int_{\mathcal{L}_k^d} \int_{\mathcal{L}_r^L} f(L,M)\, dM\, dL = \int_{\mathcal{L}_r^d} \int_{\mathcal{L}_k^M} f(L,M)\, dL\, dM \tag{2.4}$$

holds for any measurable $f : \{(L,M) \in \mathcal{L}_k^d \times \mathcal{L}_r^d : M \in \mathcal{L}_r^L\} \to [0,\infty)$. The corresponding family of incident flats will only be needed for $0 \leq r \leq k \leq d$ and is defined as the space

$$\mathcal{E}_r^E = \{F \in \mathcal{E}_r^d : F \subset E\}$$

of all r-flats contained in a fixed k-flat E. Integration with respect to the invariant measure on this space will again be denoted by dF and can be derived by identifying E with $\mathbb{R}^k$ as in the linear case.

We can now state the announced counterpart of Theorem 2.2 with M replaced by a k-flat.

Theorem 2.4 (Crofton formula). *For $0 \leq j \leq k < d$ and $K \in \mathcal{K}_{\text{conv}}^d$ we have*

$$\int_{\mathcal{E}_k^d} V_j(K \cap E)\, dE = c_{j,d}^{k,d-k+j}\, V_{d-k+j}(K)$$

with $c_{j,d}^{k,d-k+j}$ given by (2.3).

This follows (apart from the value of the constant) directly from Hadwiger's characterization theorem, as the left hand side is additive, motion invariant, monotone and homogeneous of degree $d-k+j$. The constant is derived by setting $K = B_1(o)$. Particular formulae for the planar case ($d = 2$) are given in (1.1) and (1.2). That the same constants also appear in the principal kinematic formula is not coincidental, but a consequence of a deeper connection between the principal kinematic formula and Crofton integrals: Hadwiger [220] showed a general kinematic formula, where the intrinsic volume V_j in the principal kinematic formula is replaced by an additive continuous functional f on $\mathcal{K}_{\text{conv}}^d$ and the right hand side involves Crofton-type integrals with f as functional. In particular, this shows that the constants in the principal kinematic formula are the same as in corresponding Crofton formulae, facilitating their calculation.

Exercise 2.5 (more difficult). Show that the Crofton formula can directly be derived from the principal kinematic formula by letting M be a ball $B_r(o) \cap L_0$ in a q-dimensional space L_0, dividing both sides of the principal kinematic formula by $V_q(M)$ and taking the limit $r \to \infty$.

The results for V_j can be extended to *polyconvex sets*, i.e. sets in

$$\mathcal{R} = \{K \subset \mathbb{R}^d : \exists\, m \in \mathbb{N},\ K_1,\ldots,K_m \in \mathcal{K}_{\text{conv}}^d \text{ with } K = \bigcup_{i=1}^m K_i\}.$$

In fact, additivity suggests how to define $V_j(K \cup M)$ for two convex bodies K, M, which not necessarily satisfy $K \cup M \in \mathcal{K}^d_{\text{conv}}$. Induction then allows extension of V_j on $\mathcal{R}$. That such an extension is well-defined (it does not depend on the representation of $K \in \mathcal{R}$ as a union of convex bodies) follows from a result of Groemer [205, p. 408]. We denote the extension of V_j on $\mathcal{R}$ again by V_j. Using induction on m, additivity implies the *inclusion–exclusion principle*

$$V_j(K_1 \cup \ldots \cup K_m) = \sum_{r=1}^{m} (-1)^{r+1} \sum_{1 \le i_1 < \ldots < i_r \le m} V_j(K_{i_1} \cap \ldots \cap K_{i_r})$$

for all $m \in \mathbb{N}$ and $K_1, \ldots, K_m \in \mathcal{R}$. This principle in particular implies that Theorems 2.2–2.4 remain valid with the convexity assumption replaced by the assumption that all occurring sets are polyconvex.

There are numerous generalizations of the principal kinematic formula and the Crofton formula. Local versions exist, where the intrinsic volumes are replaced by support measures (generalized curvature measures). When the averaging with respect to rotations is omitted, one obtains translative integral formulae; see [450]. For instance, the principal kinematic formula in its translative form still allows on the right hand side for a sum of $d - j + 1$ summands distinguishable by their homogeneity properties, but these summands depend on the relative position of K and M. Iterated versions of the principal translative formula exist, but in contrast to Theorem 2.3 new functionals appear when the number of convex bodies is increased; see [511], where a translative formula of Crofton-type and for half-spaces is derived as well. Integral geometric formulae for convex cylinders can be seen as joint generalizations of the principal kinematic formula and Crofton's formula. Details can be found in [451].

We discussed integral geometric formulae for polyconvex sets. However, they are valid for considerably larger set classes. Already Federer [171] showed that the principal kinematic formula holds for sets of positive reach. Zähle [527] and Rother and Zähle [426] extended kinematic integral formulae to even larger set classes containing the class of so-called U_{PR}-sets. A set is an element of U_{PR} if it can be written as locally finite union of sets of positive reach such that any finite nonempty intersection of them has again positive reach. The mentioned results even hold locally, that is, for curvature measures.

Crofton's formula allows to derive mean values like in (2.1), where the moved convex body is replaced by a k-flat. A random k-flat E intersecting a given reference set $A \in \mathcal{B}(\mathbb{R}^d)$ with the natural distribution

$$\mathbf{P}(E \in \cdot) = \frac{\int_{(\cdot)} \mathbf{1}(E' \cap A \neq \emptyset)\, dE'}{\int_{\mathcal{E}^d_k} \mathbf{1}(E' \cap A \neq \emptyset)\, dE'} \qquad (2.5)$$

is called an *IUR (isotropic uniform random) k-flat hitting A*. For $K \in \mathcal{R}$ with $K \subset A$, Crofton's formula for an IUR k-flat hitting $A \in \mathcal{R}$ gives the mean value

$$\mathbf{E}(V_j(K \cap E)) = \frac{c_{j,d}^{k,d-k+j} V_{d-k+j}(K)}{c_{0,d}^{k,d-k} V_{d-k}(A)} \tag{2.6}$$

for $0 \leq j \leq k < d$. Hence, up to a known multiplicative constant depending on A, the random variable $V_j(K \cap E)$ is an unbiased estimator of $V_{d-k+j}(K)$. The relations (2.6) are sometimes called *fundamental stereological formulae*. We shall discuss them and related stereological results in more detail in Sect. 2.2.

It is not difficult to construct an IUR k-flat E hitting a compact set A. For $k = 0$ the flat E is a point, uniformly distributed in A. For $k > 0$ choose an $r > 0$ with $A \subset B_r(o)$ and a linear space $L_0 \in \mathcal{L}_k^d$. If $\rho \in \mathsf{SO}$ is a random rotation with distribution ν and η is independent of ρ with the uniform distribution on $B_r(o) \cap L_0^\perp$, then $E = \rho(L_0 + \eta)$ is an IUR k-flat hitting $B_r(o)$. Conditioning on the event $E \cap A \neq \emptyset$ yields an IUR k-flat hitting A. The construction of an IUR k-flat can be simplified when $k = 1$ (IUR line) or $k = d - 1$ (IUR hyperplane). To obtain an IUR line in $B_r(o)$ one can choose a uniform vector $\eta \in \mathbb{S}^{d-1}$ and, given η, a uniform point $\xi \in B_r(o) \cap \eta^\perp$. The line E parallel to η passing through ξ then is an IUR line hitting $B_r(o)$. In a similar way, an IUR hyperplane can be constructed by representing it by one of its normals and its closest point to o. For $d = 2$ and $d = 3$, which are the most important cases in applications, the construction of the random rotation ρ can thus be avoided.

It should be noted that an IUR k-flat hitting $B_1(o)$ cannot be obtained by choosing $k + 1$ i.i.d. uniform points in $B_1(o)$ and considering their affine hull H. Although H has almost surely dimension k, its distribution is not coinciding with the natural distribution of E in (2.5). The k-flat H is called *point weighted* k-flat, and its (non-constant) density with respect to $\int_{(\cdot)} \mathbf{1}(E \cap B_1(o) \neq \emptyset)\, dE$ can be calculated explicitly using the affine Blaschke–Petkantschin formula. As formulae of Blaschke–Petkantschin type play an important role in stereology, we discuss them in detail in the next section.

2.1.2 *Blaschke–Petkantschin Formulae*

Suppose we have to integrate a function of q-tuples $(x_1, \ldots, x_q)$ of points in $\mathbb{R}^d$ with respect to the product measure $(\nu_d)^q$. In several applications computations can be simplified by first integrating over all q-tuples of points in a q-dimensional linear subspace L (with respect to ν_q^q) and subsequently integrating over all linear subspaces with respect to the Haar measure $\int_{\mathcal{L}_q^d \cap (\cdot)} dL$. The case $q = 1$, $d = 2$ corresponds to the well-known integration in the plane using polar coordinates. The Jacobian appearing in the general transformation formula turns out to be a power of

$$\nabla_q(x_1, \ldots, x_q) = \nu_q([0, x_1] \oplus \ldots \oplus [0, x_q]),$$

where $[0, x_1] \oplus \ldots \oplus [0, x_q]$ is the parallelepiped spanned by the vectors $x_1, \ldots, x_q$. To simplify notation, we shall just write dx for integration with respect to Lebesgue measure in $\mathbb{R}^k$, as the appropriate dimension k can be read off from the domain of integration under the integral sign.

Theorem 2.5 (Linear Blaschke–Petkantschin formula). *Let $q \in \{1, \ldots, d\}$ and $f : (\mathbb{R}^d)^q \to [0, \infty)$ be measurable. Then*

$$\int_{(\mathbb{R}^d)^q} f(x)\, dx = b_{dq} \int_{\mathcal{L}_q^d} \int_{L^q} f(x)\, \nabla_q^{d-q}(x)\, dx\, dL,$$

with

$$b_{dq} = \frac{\omega_{d-q+1} \cdots \omega_d}{\omega_1 \cdots \omega_q},$$

where $\omega_j = j\kappa_j$ denotes the surface area of the unit ball in $\mathbb{R}^j$.

For the proof, which is by induction on q, we use a generalization of the polar coordinate formula.

Lemma 2.1. *Let $r \in \{0, \ldots, d-1\}$, $L_0 \in \mathcal{L}_r^d$ be fixed and $f : \mathbb{R}^d \to [0, \infty)$ be measurable. Then*

$$\int_{\mathbb{R}^d} f(x)\, dx = \frac{\omega_{d-r}}{2} \int_{\mathcal{L}_{r+1}^{L_0}} \int_M f(x) d(x, L_0)^{d-r-1}\, dx\, dM,$$

where $d(x, L_0)$ is the distance between x and L_0.

Proof. Let $L_0(u) = \{L_0 + \alpha u : \alpha \geq 0\}$ be the positive hull of L_0 and u. Then Fubini's theorem and spherical coordinates (in $L_o^\perp$) yield

$$\begin{aligned}
\int_{\mathbb{R}^d} f(z)\, dz &= \int_{L_0} \int_{L_0^\perp} f(x+y)\, dy\, dx \\
&= \int_{L_0} \int_0^\infty \int_{\mathbb{S}^{d-1} \cap L_0^\perp} f(x + \alpha u) \alpha^{d-r-1}\, du\, d\alpha\, dx \\
&= \int_{\mathbb{S}^{d-1} \cap L_0^\perp} \int_{L_0(u)} f(x) d(x, L_0)^{d-r-1}\, dx\, du \\
&= \frac{\omega_{d-r}}{2} \int_{\mathcal{L}_{r+1}^{L_0}} \int_M f(x) d(x, L_0)^{d-r-1}\, dx\, dM.
\end{aligned}$$

This concludes the proof of Lemma 2.1. □

We now prove Theorem 2.5. Amazingly, this can be achieved by a relatively simple induction on q and a suitable use of spherical coordinates in subspaces of $\mathbb{R}^d$.

Proof (of Theorem 2.5). For $q = 1$ the assertion reduces to Lemma 2.1 with $r = 0$. We assume now that the assertion is true for some $q \in \mathbb{N}$ and all dimensions d, and use the fact that

$$\nabla_{q+1}(x_1, \dots, x_{q+1}) = \nabla_q(x_1, \dots, x_q) d(x_{q+1}, L), \tag{2.7}$$

if $x_1, \dots, x_q \in L$, $L \in \mathcal{L}_q^d$. Fubini's theorem, the induction hypothesis and Lemma 2.1 with $r = q$ give

$$\begin{aligned}
I &:= \int_{(\mathbb{R}^d)^{q+1}} f(z)\, dz = \int_{\mathbb{R}^d} \int_{(\mathbb{R}^d)^q} f(x, y)\, dx\, dy \\
&= b_{dq} \int_{\mathbb{R}^d} \int_{\mathcal{L}_q^d} \int_{L^q} f(x, y) \nabla_q^{d-q}(x)\, dx\, dL\, dy \\
&= b_{dq} \int_{\mathcal{L}_q^d} \int_{L^q} \int_{\mathbb{R}^d} f(x, y)\, dy \nabla_q^{d-q}(x)\, dx\, dL \\
&= b_{dq} \frac{\omega_{d-q}}{2} \int_{\mathcal{L}_q^d} \int_{L^q} \int_{\mathcal{L}_{q+1}^L} \int_M f(x, y) d(y, L)^{d-q-1}\, dy\, dM \nabla_q^{d-q}(x)\, dx\, dL \\
&= b_{dq} \frac{\omega_{d-q}}{2} \int_{\mathcal{L}_{q+1}^d} \int_M \int_{\mathcal{L}_q^M} \int_{L^q} f(x, y) d(y, L)^{d-q-1} \nabla_q^{d-q}(x)\, dx\, dL\, dy\, dM,
\end{aligned}$$

where the integrals over q and $(q + 1)$-dimensional subspaces may be interchanged due to (2.4). From (2.7) and an application of the induction hypothesis for a q-fold integral over the $(q + 1)$-dimensional space M with function $f(\cdot, y)\nabla_{q+1}(\cdot, y)^{d-q-1}$, we get

$$\begin{aligned}
I &= b_{dq} \frac{\omega_{d-q}}{2} \int_{\mathcal{L}_{q+1}^d} \int_M \int_{\mathcal{L}_q^M} \int_{L^q} f(x, y) \nabla_{q+1}(x, y)^{d-q-1} \nabla_q(x)\, dx\, dL\, dy\, dM \\
&= \frac{b_{dq}\omega_{d-q}}{2 b_{(q+1)q}} \int_{\mathcal{L}_{q+1}^d} \int_M \int_{M^q} f(x, y) \nabla_{q+1}(x, y)^{d-q-1}\, dx\, dy\, dM \\
&= b_{d(q+1)} \int_{\mathcal{L}_{q+1}^d} \int_{M^{q+1}} f(z) \nabla_{q+1}(z)^{d-q-1}\, dz\, dM.
\end{aligned}$$

This concludes the proof. □

There are many formulae of Blaschke–Petkantschin type in the literature. Following [451, Sect. 7.2] we can describe their common feature: Instead of integrating q-tuples of geometric objects (usually points or flats) directly, a "pivot" is associated to this tuple (usually span or intersection) and integration of the q-tuple is first restricted to one pivot, followed by an integration over all possible pivots. For integrations the natural measures are used, and a Jacobian comes in. In Theorem 2.5, the pivot is the linear space (almost everywhere) spanned by the q points $x_1, \ldots, x_q$. As an affine subspace of dimension q is spanned by $q+1$ affine independent points, a similar formula for affine q-flats is to be expected.

Theorem 2.6 (Affine Blaschke–Petkantschin formula). *Let $q \in \{1, \ldots, d\}$ and assume that $f : (\mathbb{R}^d)^{q+1} \to [0, \infty)$ is measurable. Then*

$$\int_{(\mathbb{R}^d)^{q+1}} f(x)dx = b_{dq}(q!)^{d-q} \int_{\mathcal{E}_q^d} \int_{E^{q+1}} f(x)\Delta_q^{d-q}(x)dxdE,$$

where b_{dq} is the constant defined in Theorem 2.5, and

$$\Delta_q(x_0, \ldots, x_q) = (q!)^{-1}\nabla_q(x_1 - x_0, \ldots, x_q - x_0)$$

is the q-dimensional volume of $\mathrm{conv}\{x_0, \ldots, x_q\}$.

The affine Blaschke–Petkantschin formula can be directly derived from Theorem 2.5; see [451, Theorem 7.2.7].

Exercise 2.6. Show Crofton's chord formula in the plane and in three-dimensional space using the affine Blaschke–Petkantschin formula: For $d = 2, 3$ we have

$$\int_{\mathcal{E}_1^d} V_1^{d+1}(K \cap E)dE = \frac{3}{\pi} V_d^2(K)$$

for any convex body $K \subset \mathbb{R}^d$.

Hint. Show that if s is a line segment of length ℓ in $\mathbb{R}^d$, then

$$\int_{s^2} \Delta_1^i(x)d\nu_2(x) = \frac{\ell^{2+i}}{3i}$$

for $i = 1, 2$.

Exercise 2.7. Let H be a point weighted line in the plane. (Recall that H is the affine hull of two i.i.d. uniform points in $B_1(o)$.) Show that the distribution of H has a non-constant density with respect to the distribution (2.5). Conclude that H cannot be an IUR line hitting $B_1(o)$.

We give an example of another Blaschke–Petkantschin formula, where there is only one initial geometric element, namely an affine k-flat. The pivot is a linear space of dimension $r > k$ containing it.

Theorem 2.7. *Let $1 \le k < r \le d-1$ and let $f : \mathcal{E}_k^d \to [0, \infty)$ be measurable. Then*

$$\int_{\mathcal{E}_k^d} f(E)\,dE = \frac{\omega_{d-k}}{\omega_{r-k}} \int_{\mathcal{L}_r^d} \int_{\mathcal{E}_k^L} f(E)d(o, E)^{d-r}\,dE\,dL.$$

Proof. If $L \in \mathcal{L}_k^d$ is fixed, the restriction of the measure

$$\int_{\mathcal{L}_r^L} \int_{\mathbb{S}^{d-1} \cap L^\perp \cap M} \mathbf{1}(u \in (\cdot))\, du\, dM$$

on $\mathbb{S}^{d-1} \cap L^\perp$ is invariant with respect to all rotations of $L^\perp$ (leaving L fixed), and must thus be a multiple of $\int_{\mathbb{S}^{d-1} \cap L^\perp} \mathbf{1}(u \in (\cdot))\, du$. The factor is $\omega_{r-k}/\omega_{d-k}$.

Hence, integrating

$$\int_0^\infty f(\alpha(\cdot) + L)\alpha^{d-k-1}\, d\alpha$$

with respect to this measure, and using spherical coordinates in $L^\perp$ gives

$$\frac{\omega_{r-k}}{\omega_{d-k}} \int_{L^\perp} f(x + L)\, dx = \int_{\mathcal{L}_r^L} \int_{\mathbb{S}^{d-1} \cap L^\perp \cap M} \int_0^\infty f(\alpha u + L)\alpha^{d-k-1}\, d\alpha\, du\, dM.$$

A back-transformation of spherical coordinates appearing on the right in the $(r-k)$-dimensional space $L^\perp \cap M$ yields

$$\frac{\omega_{r-k}}{\omega_{d-k}} \int_{L^\perp} f(x + L)\, dx = \int_{\mathcal{L}_r^L} \int_{L^\perp \cap M} f(x + L)\, \|x\|^{d-r}\, dx\, dM.$$

Integration with respect to $L \in \mathcal{L}_k^d$ leads to

$$\begin{aligned}
\int_{\mathcal{E}_k^d} f(E)\, dE &= \frac{\omega_{d-k}}{\omega_{r-k}} \int_{\mathcal{L}_k^d} \int_{\mathcal{L}_r^L} \int_{L^\perp \cap M} f(x + L)\, \|x\|^{d-r}\, dx\, dM\, dL \\
&= \frac{\omega_{d-k}}{\omega_{r-k}} \int_{\mathcal{L}_r^d} \int_{\mathcal{L}_k^M} \int_{L^\perp \cap M} f(x + L) d(o, x + L)^{d-r}\, dx\, dL\, dM \\
&= \frac{\omega_{d-k}}{\omega_{r-k}} \int_{\mathcal{L}_r^d} \int_{\mathcal{E}_k^M} f(E) d(o, E)^{d-r}\, dE\, dM,
\end{aligned}$$

where (2.4) was used. This completes the proof. □

We also notice an example of a Blaschke–Petkantschin formula, where the pivot is spanned by an initial geometric element and a fixed subspace. We only consider initial geometric elements and fixed subspaces of dimension one here, although versions for higher dimensional planes (and q-fold integrals, $q > 1$) exist. The Jacobian appearing in the following relation is a power of the *generalized determinant* $[L, L_0]$ of two subspaces L and L_0. In the special case we consider here, L and L_0 are lines and $[L, L_0] = \sin(\angle(L, L_0))$ depends only on the angle $\angle(L, L_0)$ between them.

Lemma 2.2. *Let $L_0 \in \mathcal{L}_1^d$ be a fixed line. Then*

$$\int_{\mathcal{L}_1^d} f(L)\, dL = \frac{\omega_2 \omega_{d-1}}{\omega_1 \omega_d} \int_{\mathcal{L}_2^{L_0}} \int_{\mathcal{L}_1^M} f(L)[L, L_0]^{d-2}\, dL\, dM$$

holds for any measurable $f : \mathcal{L}_1^d \to [0, \infty)$.

Proof. An invariance argument implies

$$\int_{\mathbb{R}^d} f(\operatorname{span}\{x\}) \mathbf{1}(\|x\| \le 1)\, dx = \kappa_d \int_{\mathcal{L}_1^d} f(L)\, dL,$$

where $\operatorname{span}\{x\}$ is the line containing x and o. Using this and Lemma 2.1 twice, first with $r = 1$ in $\mathbb{R}^d$ and then with $r = 0$ in M, we get

$$\begin{aligned}
\int_{\mathcal{L}_1^d} f(L)\, dL &= \kappa_d^{-1} \int_{\mathbb{R}^d} f(\operatorname{span}\{x\}) \mathbf{1}(\|x\| \le 1)\, dx \\
&= \frac{\omega_{d-1}}{2\kappa_d} \int_{\mathcal{L}_2^{L_0}} \int_M f(\operatorname{span}\{x\}) \mathbf{1}(\|x\| \le 1) d(x, L_0)^{d-2}\, dx\, dM \\
&= \frac{\omega_2 \omega_{d-1}}{4\kappa_d} \int_{\mathcal{L}_2^{L_0}} \int_{\mathcal{L}_1^M} f(L) \int_L \mathbf{1}(\|x\| \le 1) d(x, L_0)^{d-2} \|x\|\, dx\, dL\, dM.
\end{aligned}$$

The innermost integral is

$$[L, L_0]^{d-2} \int_{L \cap B_1(o)} \|x\|^{d-1}\, dx = \frac{2}{d} [L, L_0]^{d-2},$$

and the claim follows. □

An affine version of Lemma 2.2 is obtained by replacing $f(L)$ by $\int_{L^\perp} f(x+L)\, dx$, where now f is a nonnegative measurable function on $\mathcal{E}_1^d$. Lemma 2.2 and Fubini's theorem then imply

$$\int_{\mathcal{E}_1^d} f(E)\, dE = \frac{\omega_2 \omega_{d-1}}{\omega_1 \omega_d} \int_{\mathcal{L}_2^{L_0}} \int_{M^\perp} \int_{\mathcal{E}_1^{M+x}} f(E)[E, L_0]^{d-2}\, dE\, dx\, dM, \quad (2.8)$$

where $[E, L_0] := [L, L_0]$, if $L \in \mathcal{L}_1^d$ is parallel to $E \in \mathcal{E}_1^d$.

The idea to base proofs of Blaschke–Petkantschin formulae on invariance arguments is due to Miles [360]. We followed mainly the presentation of his and Petkantschin's [404] results in [451, Sect. 7.2]. Santaló's monograph [433] is a general reference for Blaschke–Petkantschin formulae. His proofs use differential forms.

2.2 Stereology

The purpose of this section is to give an introduction into stereology with a special emphasis on the usefulness of integral geometric tools. *Stereology* (gr.: "stereos" meaning solid) is a sub-area of stochastic geometry and spatial statistics dealing with the estimation of geometric characteristics (like volume, area, perimeter or particle number) of a structure from samples. Typically samples are sections with or projections onto flats, intersections with full-dimensional test sets or combinations of those.

2.2.1 *Motivation*

Unlike tomography, stereology does not aim for a full-dimensional reconstruction of the geometry of the structure, but rather tries to assess certain key properties. This is what makes stereology extremely efficient and explains its widespread use in many applied sciences. As estimation is based on samples of the structure, one has to assure that these samples are in a certain sense representative for the structure as a whole—at least concerning the geometric characteristics of interest. Stereologists therefore assume that the structure is "statistically homogeneous", a property that only was vaguely defined in the early literature. The former East German stochastics school of J. Mecke, D. Stoyan and collaborators (see [489] and the references therein) made this concept rigorous by considering the structure $Z \subset \mathbb{R}^d$ as a random closed set which is stationary (i.e. the distribution of $Z + x$ is independent of $x \in \mathbb{R}^d$). Some authors prefer to call such random sets *homogeneous* rather than stationary. Often it was also assumed that Z is isotropic (the distribution of RZ is independent of $R \in \mathsf{SO}$). As (weak) model assumptions on Z are needed, this approach is called the *model-based approach*. Besides the monograph of Schneider and Weil [451] on stochastic geometry and integral geometry, the classical book of Stoyan et al. [489] is recommended as reference for the model-based approach. The stationarity assumption in the classical model-based approach is appropriate in many applications in geology, metallurgy and materials science. It is, however, often hard to check in other disciplines and certainly inappropriate in anatomy and soil science. In these cases the *design-based approach* has to be used, where the structure of interest is considered deterministic, and the selection of the sample is done in a controlled randomized way.

The Australian statisticians R.E. Miles and P.J. Davy [141, 142, 361] made this rigorous by pointing out the strong analogy between stereology and sample surveys. Sample surveys (think of opinion polls) infer properties of the whole population (for example the total number of citizens voting for the democratic party) from a randomized sample of the population. In a simplified stereological situation, where a feature of interest K is contained in a reference set A, the space A corresponds to the total population, the intersection with a set L corresponds to a sample, and K corresponds to the subpopulation of interest to us (Fig. 2.1).

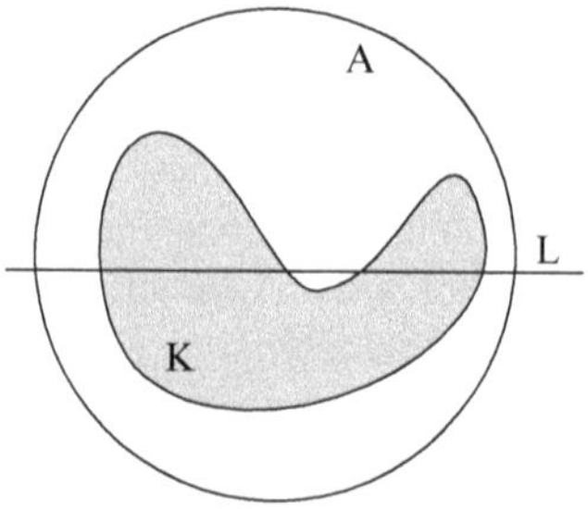

Fig. 2.1 Analogy between survey sampling and stereology: the feature of interest K is contained in a reference set A. The intersection with the line L corresponds to taking a sample

This analogy is more than a formal one and allows among other things to transfer variance reducing methods like systematic random sampling, unequal probability sampling and stratification to stereology. Concerning design-based stereology, Baddeley and Jensen's monograph [31] includes also recent developments. We return to design-based stereology in a later section, and start with model-based methods.

2.2.2 Model-Based Stereology

In model-based stereology we assume that the structure of interest $Z \subset \mathbb{R}^d$ is a stationary random closed set (see Definition 1.1). We want to use integral geometric formulae from Sect. 2.1.1, which we only have shown for polyconvex sets. The assumption that Z is stationary and polyconvex is not suitable, as a stationary set $Z \neq \emptyset$ is known to be almost surely unbounded [451, Theorem 2.4.4]. Instead we assume that Z is almost surely locally polyconvex, i.e. $Z \cap K$ is polyconvex for all $K \in \mathcal{K}^d_{\text{conv}}$, almost surely. We denote by $N(Z \cap K)$ the minimal number of convex bodies that is needed to represent $Z \cap K$ as their union and assume the integrability condition

$$\mathbf{E}\left(2^{N(Z\cap[0,1]^d)}\right) < \infty. \tag{2.9}$$

Following [451, p. 397] we call a random set $Z \subset \mathbb{R}^d$ a *standard random set* if

(a) Z is stationary
(b) Z is a.s. locally polyconvex
(c) Z satisfies (2.9)

The class of standard random sets forms the most basic family of random sets which is flexible enough to model real-world structures reasonably. To define *mean intrinsic volumes per unit volume*, one might consider $\mathbf{E}(V_j(Z \cap W))/\nu_d(W)$ for an observation window $W \in \mathcal{K}^d_{\text{conv}}$ with $\nu_d(W) > 0$. But this definition is inapt, as can already be seen in the special case $j = d - 1$ corresponding to surface area estimation: in addition to the surface area of ∂Z in W also the surface area of ∂W in Z is contributing, leading to an overestimation of the mean surface area per unit volume. In order to eliminate such boundary effects one defines

$$\overline{V}_j(Z) = \lim_{r\to\infty} \frac{\mathbf{E}(V_j(Z \cap rW))}{\nu_d(rW)},$$

where $W \in \mathcal{K}^d_{\mathrm{conv}}$, $\nu_d(W) > 0$ as before. If Z is a standard random set then $\overline{V}_j(Z)$ exists and is independent of W. It is called the *j-th specific intrinsic volume of Z*. It is shown in [512] that corresponding specific φ-values of Z exist whenever φ is an additive, translation invariant functional on $\mathcal{R}$ satisfying a certain boundedness condition. Specific intrinsic volumes can also be defined as Lebesgue densities of average curvature measures of Z; see for example [451, Corollary 9.4.1] and the references therein. As intrinsic volumes are also called quermass integrals, one finds the notion *quermass densities* for the specific intrinsic volumes in the earlier literature.

Exercise 2.8. Let Z be a standard random set. Use Fubini's theorem and the stationarity of Z to show that

$$\overline{V}_d(Z) = \mathbf{P}(x \in Z) = \mathbf{P}(o \in Z)$$

for all $x \in \mathbb{R}^d$.

If Z is stationary and isotropic, the principal kinematic formula holds for $\overline{V}_j(Z)$.

Theorem 2.8. *Let Z be an isotropic standard random set, and $j \in \{0, \ldots, d\}$. Then*

$$\mathbf{E}(V_j(Z \cap W)) = \sum_{k=j}^{d} c_{j,d}^{k,d-k+j} \overline{V}_k(Z) V_{d-k+j}(W), \quad W \in \mathcal{K}^d_{\mathrm{conv}}.$$

If W is a ball, $j = d$ or $j = d-1$, the isotropy assumption can be dropped.

Proof. Let Z be defined on the abstract probability space $(\Omega, \mathcal{A}, \mathbf{P})$. Fix $W \in \mathcal{K}^d_{\mathrm{conv}}$ and $r > 0$. It can be shown that $f : \mathbb{R}^d \times \Omega \to \mathbb{R}$ which maps (x, ω) to $V_j((Z(\omega) \cap W) \cap (B_r(o) + x))$ is measurable, and that the integrability condition (2.9) implies integrability of f with respect to $\nu_d \otimes \mathbf{P}$. This will allow us to use Fubini's theorem later in the proof. The motion invariance of V_j and stationarity and isotropy of Z imply

$$\begin{aligned}\mathbf{E}(V_j((Z \cap W) \cap (B_r(o) + x))) &= \mathbf{E}(V_j((RZ + x) \cap W \cap (RB_r(o) + x))) \\ &= \mathbf{E}(V_j((Z \cap B_r(o)) \cap R^{-1}(W - x)))\end{aligned}$$

for $x \in \mathbb{R}^d$, $R \in \mathsf{SO}$. Fubini's theorem and the invariance properties of ν_d and ν imply

$$\begin{aligned}&\mathbf{E}\int_{\mathsf{SO}}\int_{\mathbb{R}^d} V_j((Z \cap W) \cap R(B_r(o) + x))\, dx\, dR \\ &= \mathbf{E}\int_{\mathsf{SO}}\int_{\mathbb{R}^d} V_j((Z \cap B_r(o)) \cap R(W + x))\, dx\, dR,\end{aligned}$$

so the sets W and $B_r(o)$ can be interchanged. The principal kinematic formula, applied on both sides, yields

$$\sum_{k=j}^{d} c_{j,d}^{k,d-k+j} \mathbf{E}\, V_k(Z \cap W) V_{d-k+j}(B_r(o))$$

$$= \sum_{k=j}^{d} c_{j,d}^{k,d-k+j} \mathbf{E}\, V_k(Z \cap B_r(o)) V_{d-k+j}(W).$$

Now we divide both sides by $\nu_d(B_r(o))$ and let r tend to infinity. As

$$\frac{V_{d-k+j}(B_r(o))}{\nu_d(B_r(o))} = r^{j-k} \frac{V_{d-k+j}(B_1(o))}{\kappa_d},$$

the claim follows. In the cases where the principal kinematic formula holds even without averaging over all rotations, isotropy is not needed in the above proof. □

Theorem 2.8 shows that

$$\mathbf{E} \begin{pmatrix} V_0(Z \cap W) \\ \vdots \\ V_d(Z \cap W) \end{pmatrix} = A \cdot \begin{pmatrix} \overline{V}_0(Z) \\ \vdots \\ \overline{V}_d(Z) \end{pmatrix}$$

with a triangular matrix $A \in \mathbb{R}^{(d+1)\times(d+1)}$, which is regular if $\nu_d(W) > 0$. Hence $A^{-1}\begin{pmatrix} V_0(Z\cap W) \\ \vdots \\ V_d(Z\cap W) \end{pmatrix}$ is an unbiased estimator of $\begin{pmatrix} \overline{V}_0(Z) \\ \vdots \\ \overline{V}_d(Z) \end{pmatrix}$ and can be determined from observations of Z in the full-dimensional window W alone.

Exercise 2.9. An isotropic standard random set $Z \subset \mathbb{R}^2$ is observed in a square W of side length one: $Z \cap W$ is connected, has area $1/2$ and perimeter 5. The complementary set $\mathbb{R}^2 \setminus (Z \cap W)$ is also connected. Find unbiased estimators for the three specific intrinsic volumes.

Hint. Use that the Euler–Poincaré characteristic $V_0(Z \cap W)$ is the number of connected components minus the number of "holes". ("Holes" are the bounded connected components of the complement.)

If E is a k-flat, and Z is a standard random set in $\mathbb{R}^d$, then $Z \cap E$ is a standard random set in E (in particular, stationarity refers to invariance of $\mathbf{P}_{Z\cap E}$ under all translations in E). If Z is isotropic, then $Z \cap E$ is isotropic in E.

Theorem 2.9 (Crofton's formula for random sets). *If Z is an isotropic standard random set and $E \in \mathcal{E}_k^d$ with $0 \le j \le k < d$, then*

$$\overline{V}_j(Z \cap E) = c_{j,d}^{k,d-k+j} \overline{V}_{d-k+j}(Z).$$

Theorem 2.9 follows readily from Theorem 2.8. Due to stationarity one may assume $o \in E$. Then set $W = B_r(o) \cap E$ in the principal kinematic formula for random sets, divide by $\nu_k(W)$ and let r tend to infinity.

Exercise 2.10. Let Z be a standard random set in $\mathbb{R}^3$. Depending on the choice of the index j and the section dimension k, Theorem 2.9 yields three formulae to estimate the specific volume $\overline{V}_3(Z)$ of Z, two formulae to estimate the specific surface area $2\overline{V}_2(Z)$, and one relation to estimate the specific integrated mean curvature $\overline{V}_1(Z)$. Determine the constants in all these cases. Notice also that the specific Euler–Poincaré characteristic cannot be estimated using Crofton's formula.

The concept of standard random sets is not suited for simulation purposes, as it cannot be described by a finite number of parameters. To obtain more accessible random sets, germ-grain models are employed. If $\varphi = \{x_1, x_2, \ldots\}$ is a stationary point process in $\mathbb{R}^d$ and $K_0, K_1, \ldots$ are i.i.d. nonempty compact random sets, independent of φ, the random set

$$Z = \bigcup_{i=1}^{\infty} (x_i + K_i)$$

is called a stationary *germ-grain model*. The points of φ are considered as germs to which the grains K_i are attached. The set K_0 is called the *typical grain* and its distribution will be denoted by $\mathbf{Q}$. If K_0 is almost surely convex, Z is called a *germ-grain model with convex grains*. We shall always assume convexity. To assure that Z is a.s. closed, a condition on $\mathbf{Q}$ is required. We assume throughout

$$\overline{V}_j(K_0) = \mathbf{E}(V_j(K_0)) < \infty \quad \text{for all } j = 0, \ldots, d.$$

This condition is equivalent to saying that the mean number of grains $x_i + K_i$ that hit any bounded window is finite.

We shall consider stationary germ-grain models for which the underlying point process is a Poisson point process, $\varphi = \Pi_\lambda$ (cf. Sect. 3.1.2). They are called *stationary Boolean models*.

It can be shown that any stationary Boolean model with convex grains is a standard random set. The iterated principal kinematic formula implies a wonderful result for the specific intrinsic volumes of Boolean models.

Theorem 2.10. *Let Z be a Boolean model in $\mathbb{R}^d$ with convex typical grain K_0, based on a stationary Poisson point process Π_λ with intensity λ. Then*

$$\overline{V}_d(Z) = 1 - \mathrm{e}^{-\lambda \overline{V}_d(K_0)} \tag{2.10}$$

and

$$\overline{V}_{d-1}(Z) = \lambda \overline{V}_{d-1}(K_0) \mathrm{e}^{-\lambda \overline{V}_d(K_0)}.$$

If $j \in \{0, \ldots, d-2\}$ and K_0 is isotropic we have

$$\overline{V}_j(Z) = \lambda e^{-\lambda \overline{V}_d(K_0)} \left[\overline{V}_j(K_0) - c_j^d \sum_{s=2}^{d-j} \frac{(-1)^s}{s!} \lambda^{s-1} \sum_{\substack{m_1,\dots,m_s=j+1 \\ m_1+\dots+m_s=(s-1)d+j}}^{d-1} \prod_{i=1}^{s} c_d^{m_i} \overline{V}_{m_i}(K_0) \right].$$

The constants appearing in the previous theorem are again given by (2.3). Note that they are slightly different from the incorrect constants in [451, Theorem 9.1.4].

Proof (Sketch). To avoid technicalities we assume that K_0 is almost surely contained in a ball $B_\delta(o)$ for some fixed $\delta > 0$. Then $Z \cap W = \bigcup_{i=1}^{\infty}[(\xi_i + K_i) \cap W]$ only depends on the Poisson process $\Pi_\lambda = \{\xi_1, \xi_2, \dots\}$ in the bounded window $W^\delta = W \oplus B_\delta(o)$. The number of points of $\Pi_\lambda \cap W^\delta$ is Poisson distributed with parameter $\lambda V_d(W^\delta)$, and, given this number is n, the n points of $\Pi_\lambda \cap W^\delta$ are i.i.d. uniform in W^δ. If these points are denoted by $\xi_1, \dots, \xi_n$, the inclusion–exclusion principle gives

$$\begin{aligned}
&\mathbf{E}\left(V_j(Z \cap W) \big| |\Pi_\lambda \cap W^\delta| = n\right) \\
&= \mathbf{E}\left(V_j\left(\bigcup_{i=1}^{n}[(\xi_i + K_i) \cap W]\right) \Big| |\Pi_\lambda \cap W^\delta| = n\right) \\
&= \sum_{m=1}^{n}(-1)^{m+1} \sum_{1 \le i_1 < \dots < i_m \le n} \frac{\Phi_{i_1,\dots,i_m}(j)}{V_d(W^\delta)^m}
\end{aligned}$$

with

$$\begin{aligned}
\Phi_{i_1,\dots,i_m}(j) &= \mathbf{E}_{K_{i_1},\dots,K_{i_m}} \int_{W^\delta} \cdots \int_{W^\delta} V_j(W \cap (K_{i_1} + x_{i_1}) \cap \dots \cap (K_{i_m} + x_{i_m})) \\
&\quad dx_{i_1} \cdots dx_{i_m} \\
&= \mathbf{E}_{K_1,\dots,K_m} \int_{\mathbb{R}^d} \cdots \int_{\mathbb{R}^d} V_j(W \cap (K_1 + x_1) \cap \dots \cap (K_m + x_m)) \\
&\quad dx_1 \cdots dx_m.
\end{aligned}$$

Here we used that $K_1, K_2, \dots$ are i.i.d., and contained in $B_\delta(o)$. Hence

$$\begin{aligned}
\mathbf{E}(V_j(Z \cap W)) &= \sum_{n=1}^{\infty} \frac{(\lambda V_d(W^\delta))^n}{n!} e^{-\lambda V_d(W^\delta)} \sum_{m=1}^{n}(-1)^{m+1}\binom{n}{m} \frac{\Phi_{1,\dots,m}(j)}{V_d(W^\delta)^m} \\
&= e^{-\lambda V_d(W^\delta)} \sum_{m=1}^{\infty} \frac{(-1)^{m+1}}{m!} \Phi_{1,\dots,m}(j) V_d(W^\delta)^{-m} \sum_{n=m}^{\infty} \frac{(\lambda V_d(W^\delta))^n}{(n-m)!} \\
&= \sum_{m=1}^{\infty} \frac{(-1)^{m+1}}{m!} \lambda^m \Phi_{1,\dots,m}(j).
\end{aligned}$$

For $j = d$ and $j = d-1$, the iterated principal kinematic formula without the average over all rotations can be applied to simplify $\Phi_{1,\dots,m}(j)$. For the volume, we have

$$\Phi_{1,\dots,m}(d) = \mathbf{E}_{K_1,\dots,K_m} V_d(W)V_d(K_1)\cdots V_d(K_m) = V_d(W)(\overline{V}_d(K_0))^m,$$

and for half the surface area we get

$$\Phi_{1,\dots,m}(d-1) = V_{d-1}(W)(\overline{V}_d(K_0))^m + mV_d(W)\overline{V}_{d-1}(K_0)(\overline{V}_d(K_0))^{m-1}.$$

Thus

$$\mathbf{E}(V_d(Z\cap W)) = \sum_{m=1}^{\infty} \frac{(-1)^{m+1}}{m!}(\lambda \overline{V}_d(K_0))^m V_d(W) = V_d(W)\left(1 - \mathrm{e}^{-\lambda \overline{V}_d(K_0)}\right)$$

and

$$\mathbf{E}(V_{d-1}(Z\cap W)) = V_{d-1}(W)\left(1 - \mathrm{e}^{-\lambda \overline{V}_d(K_0)}\right) + V_d(W)\lambda\overline{V}_{d-1}(K_0)\mathrm{e}^{-\lambda \overline{V}_d(K_0)}.$$

Replacing W by rW, dividing by $\nu_d(rW)$ and letting r tend to infinity yields the claim for $j = d$ and $j = d-1$. For $j < d-1$, isotropy of K_0 implies that

$$\Phi_{i_1,\dots,i_m}(j) = \mathbf{E}_{K_1,\dots,K_m} \int_{\mathsf{SO}}\int_{\mathbb{R}^d}\cdots\int_{\mathsf{SO}}\int_{\mathbb{R}^d} V_j(W\cap R_1(K_1+x_1)\cap\ldots\cap R_m(K_m+x_m))\,dx_1\,dR_1\cdots dx_m dR_m.$$

The claim then follows in a similar way as before by applying the iterated principal kinematic formula and sorting the resulting expressions according to their homogeneity. In the final result s is the number of terms with homogeneity smaller than d. This concludes the sketch of the proof. □

Specialized to two dimensions, the formulae in Theorem 2.10 read

$$\overline{V}_2(Z) = 1 - \mathrm{e}^{-\lambda\overline{V}_2(K_0)} \qquad \text{(specific area)}$$

$$2\overline{V}_1(Z) = 2\lambda\overline{V}_1(K_0)\cdot \mathrm{e}^{-\lambda\overline{V}_2(K_0)} \qquad \text{(specific perimeter)}$$

$$\overline{V}_0(Z) = \mathrm{e}^{-\lambda\overline{V}_2(K_0)}\left(\lambda - \frac{1}{\pi}(\lambda\overline{V}_1(K_0))^2\right) \qquad \text{(specific Euler characteristics)}$$

The last relation requires isotropy. If all the quantities on the left side are known, these relations can be used to determine the mean intrinsic volumes of K_0 and the intensity λ of Π_λ. Hence, measurement (estimation) of the specific intrinsic volumes allows to estimate $\overline{V}_2(K_0)$, $\overline{V}_1(K_0)$ and λ, which determine all the parameters of Z if $\mathbf{Q}$ is a suitable distribution with at most two real parameters.

Based on translative integral formulae, Theorem 2.8 is generalized to curvature measures of standard random sets that are not necessarily isotropic in [509]. In [510] Theorem 2.10 is generalized to stationary Boolean models that are not necessarily isotropic. It is shown that at least for small dimensions ($d \le 4$), the underlying intensity is still determined by the Boolean model, but an estimation procedure would require more than just the measurement of the specific intrinsic volumes.

Exercise 2.11. One summary statistic that is often used to analyze stationary random sets Z is the *spherical contact distribution function*

$$H_Z(r) = \mathbf{P}(\rho(o, Z) \le r | o \notin Z),$$

$r \ge 0$. Recall that the distance $\rho(o, Z) = \inf\{|z-o| : z \in Z\}$ of Z from the origin is measurable due to Theorem 1.3 in Chap. 1. Hence H_Z is the cumulative distribution function of the random variable $\rho(o, Z)$ conditioned on the event $o \notin Z$. Show the following:

(a) $H_Z(r) = 1 - \dfrac{\mathbf{P}(o \notin Z \oplus B_r(o))}{\mathbf{P}(o \notin Z)}$.

(b) If Z is a stationary Boolean model with typical convex grain K_0 and underlying intensity γ then

$$H_Z(r) = 1 - \exp\left(-\gamma \sum_{j=0}^{d-1} \kappa_{d-j} \mathbf{E} V_j(K_0) r^{d-j}\right).$$

Hint. Start with (a) and use Exercise 2.8 and formula (2.10) for the two Boolean models Z and $Z \oplus B_r(o)$.

2.2.3 *Design-Based Stereology*

We now turn to design-based stereology, where the structure of interest is assumed to be a deterministic set, and the sampling is randomized in a suitable way. We have already derived the set of fundamental stereological formulae (2.6) from Crofton's formula, where the set $K \in \mathcal{R}$ was sampled by IUR k-flats. Recall that if K is contained in the compact reference set A, and E is an IUR k-flat hitting A, then

$$\left[\frac{c_{0,d}^{k,d-k}}{c_{j,d}^{k,d-k+j}} V_{d-k}(A)\right] V_j(K \cap E) \tag{2.11}$$

is an unbiased estimator for $V_{d-k+j}(K)$ for $0 \le j \le k < d$. This shows that $V_m(K)$ can unbiasedly be estimated from k-dimensional sections if $m \ge d-k$. For $m < d-k$, unbiased estimation of $V_m(K)$ from IUR k-flat sections is impossible: If K is

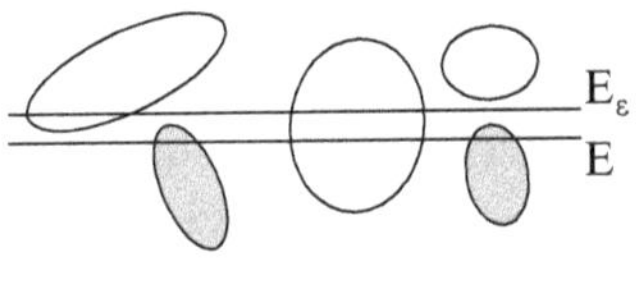

Fig. 2.2 The disector technique: only the two shaded particles are counted

a set of dimension $m < d - k$ (meaning that its affine hull is a flat of dimension m) then $V_m(K) > 0$, but $K \cap E = \emptyset$ almost surely. In particular, the Euler-characteristic $V_0(K)$ cannot be estimated from IUR sections. Therefore, the *disector technique* has been suggested in [484]. The basic idea is to work with hyperplanes and to replace the sectioning flat by a pair of parallel $(d-1)$-flats (E, E_ε) of distance $\varepsilon > 0$ apart. The flats must be randomized, but averaging with respect to rotations is not required, so it is enough to choose E as an *FUR (fixed orientation uniform) k-flat hitting A* with $k = d - 1$. An *FUR* k-flat E hitting A is obtained by uniformly translating a fixed subspace $L_0 \in \mathcal{L}_k^d$ with a translation vector in $x \in L_0^\perp$ such that $E = L_0 + x$ hits A. In other words, E has distribution

$$\mathbf{P}(E \in \cdot) = c(A)^{-1} \int_{A|L_0^\perp} \mathbf{1}(L_0 + x \in \cdot)\, dx,$$

where $c(A) = \nu_{d-k}(A|L_0^\perp)$ is the $(d-k)$-dimensional content of the orthogonal projection $A|L_0^\perp$ of A on $L_0^\perp$. To describe the disector let E be an FUR $(d-1)$-flat hitting A, parallel to some deterministic $L_0 \in \mathcal{L}_{d-1}^d$, and let $E_\varepsilon = E + \varepsilon u$, where $u \in L_0^\perp$ is a unit vector (Fig. 2.2).

To fix ideas let K be a union of m disjoint convex particles $K_1, \ldots, K_m$. Let N_{E,E_ε} be the number of particles that hit E, but not E_ε. Then $V_0(K) = m$ is the number of particles and can be estimated unbiasedly by

$$\widehat{V}_0 = \frac{c(A)}{\varepsilon} N_{E,E_\varepsilon}$$

if, almost surely, none of the particles is located between E and E_ε, that is, if the projected height of K_i on a line orthogonal to E is at least ε for all $i = 1, \ldots, m$. If the approximate size of the particles is known, this can be achieved choosing ε small enough. The unbiasedness follows from

$$\varepsilon \mathbf{E}(\widehat{V}_0) = \sum_{i=1}^{m} \int_{-\infty}^{\infty} \mathbf{1}((L_0 + tu) \cap K_i \neq \emptyset) \mathbf{1}((L_0 + (t+\varepsilon)u) \cap K_i = \emptyset)\, dt = m\varepsilon,$$

as the integrand is one exactly on an interval of length ε. In applications N_{E,E_ε} is often approximated by a comparison of $K \cap E$ and $K \cap E_\varepsilon$ using a priori information on the particles. However, strictly speaking, this estimator requires more information than just these intersections. To decide whether two profiles in E and E_ε originate from the same particle, the part of K between E and E_ε

must be known. In a typical biological application, this is achieved using confocal microscopy. By continuously moving the focal plain from E_ε to E, one obtains N_{E,E_ε} by counting all particles that come into focus during this process. The method can be extended to sets K in more general set classes, but then, tangent points between the planes with normal u have to be counted according to whether they are convex, concave or of saddle type.

Exercise 2.12. This example combines the model-based approach with design-based methods. Let X be a random convex body in the plane.

Assume that X is almost surely contained in $B_1(o)$, and that E is an IUR line hitting $B_1(o)$ that is stochastically independent of X.

(a) Find an unbiased estimator of $\mathbf{E} V_2(X)$ depending only on $X \cap E$.
(b) Use Exercise 2.6 to determine an unbiased estimator of $\mathbf{E} V_2^2(X)$ depending only on $X \cap E$.

We return to the fundamental stereological formulae and discuss possible improvements. For illustration we restrict ourselves to perimeter estimation of $K \in \mathcal{R}$ from linear sections ($k = 1$) in the plane ($d = 2$). By (2.11) with $j = 0$ the random number

$$\hat{V}_1 = 2V_1(A)V_0(K \cap E) \tag{2.12}$$

is an unbiased estimator of the perimeter $2V_1(K)$ of $K \subset A \in \mathcal{R}$, if $E \in \mathcal{L}_1^2$ is IUR hitting A. To reduce the variance, one could repeat the measurements with n i.i.d. random lines E and consider the arithmetic mean of the corresponding estimates (2.12). However, the variance reduction is generally only of order $1/n$, as the estimates are uncorrelated. It may happen that some of the sampling lines are close to one another, and the corresponding section counts are therefore very similar and contain redundant information. It would be desirable to work with section counts that are negatively correlated. In classical survey sampling one uses *systematic random sampling* in such situations: sampling from a linearly ordered population of units can generally be improved by choosing every m-th unit in both directions from a randomly selected starting unit, $m > 1$. This way, units that are close to one another (and tend to be similar) are not in the same sample. This concept, transferred to the random translation of $E \subset \mathbb{R}^2$ leads to sampling with a IUR grid of lines of distance $h > 0$ apart:

$$G = \{\eta^\perp + (\xi + mh)\eta : m \in \mathbb{Z}\},$$

where η is uniform in $\mathbb{S}^1$, and ξ is independent of η and uniform in $[0, h]$. It is not difficult to show that

$$\mathbf{E}(V_0(K \cap G)) = \frac{1}{h}\int_{\mathcal{E}_1^2} V_0(K \cap E)\, dE = \frac{2}{\pi h} V_1(K), \qquad K \in \mathcal{R},$$

where we used Crofton's formula. Hence $\pi h V_0(K \cap G)$ is an unbiased estimator for the perimeter of K. This estimator is called *Steinhaus estimator* and does not involve any reference set A. Similar variance reduction procedures are possible in the case of sampling with k-flats in $\mathbb{R}^d$.

The assumption of IUR section planes is sometimes too strong: it is either impracticable or not desired to use fully randomized sections. For instance, when analyzing sections of the skin in biology it is natural to use planar sections parallel to a fixed axis, the normal of the skin surface. This way, different layers of tissue in the section can be distinguished more easily. The common axis is usually thought to be the vertical direction, and the samples are therefore called *vertical sections*. We restrict considerations to planar vertical sections in three-dimensional space to avoid technicalities.

Let $L_0 \in \mathcal{L}_1^3$ be the vertical axis and A a bounded Borel set in $\mathbb{R}^3$. A random 2-flat H in $\mathbb{R}^3$ is called a *VUR (vertical uniform random) 2-flat hitting* A if it has the natural distribution on

$$\{E \in \mathcal{E}_2^3 : E \cap A \neq \emptyset,\ E \text{ is parallel to } L_0\}.$$

Explicitly, $\mathbf{P}(H \in \cdot)$ coincides up to a normalizing constant with

$$\int_{\mathcal{L}_2^{L_0}} \int_{A|L^\perp} \mathbf{1}(L + x \in \cdot)\, dx\, dL.$$

For $A \in \mathcal{K}_{\text{conv}}^d$ the normalizing constant is $\pi/(2V_1(A|L_0^\perp))$. This can be seen as follows: the convexity of A implies that

$$\mathbf{1}_{A|L^\perp}(x) = V_0(A \cap (x + L)) = V_0((A|L_0^\perp) \cap (x + L))$$

for all $L \in \mathcal{L}_2^{L_0}$, $x \in L^\perp$. The definition of the invariant distribution on $\mathcal{L}_2^{L_0}$, and Crofton's formula (applied in $L_0^\perp$) yield

$$\begin{aligned}\int_{\mathcal{L}_2^{L_0}} \int_{A|L^\perp} dx\, dL &= \int_{\mathcal{L}_1^{L_0^\perp}} \int_{(L_0+L)^\perp} V_0((A|L_0^\perp) \cap (x + L))\, dx\, dL \\ &= \int_{\mathcal{E}_1^{L_0^\perp}} V_0((A|L_0^\perp) \cap E) dE = \frac{2}{\pi} V_1(A|L_0^\perp).\end{aligned}$$

As vertical flats all contain the vertical axis, they are surely not IUR, so Crofton's formula cannot be applied directly. The key idea is to choose a random line E in H in such a way that E is IUR in $\mathbb{R}^3$ and apply Crofton's formula to E. Given H, this random line E will have a density with respect to the natural measure on $\mathcal{E}_1^H$, and this density can be determined using Blaschke–Petkantschin formulae.

Let $K \in \mathcal{R}$ be contained in the reference set $A \in \mathcal{K}_{\text{conv}}^d$, and fix a vertical axis $L_0 \in \mathcal{L}_1^3$. From (2.8) with $d = 3$, $f(E) = V_0(K \cap E)\mathbf{1}(E \cap A \neq \emptyset)$, and Crofton's formula

$$\int_{\mathcal{L}_2^{L_0}} \int_{A|L^\perp} \int_{\mathcal{E}_1^{L+x}} V_0(K \cap E)[E, L_0]\, dE\, dx\, dL$$
$$= \frac{2}{\pi} \int_{\mathcal{E}_1^3} V_0(K \cap E)\, dE = \frac{1}{\pi} V_2(K).$$

Hence, if H is a VUR 2-flat hitting A with vertical axis L_0,

$$W(K, H) = \int_{\mathcal{E}_1^H} V_0(K \cap E)[E, L_0]\, dE \tag{2.13}$$

is an unbiased estimator for $1/(2V_1(A|L_0^\perp))V_2(K)$. Instead of using a single VUR 2-flat hitting A, one often applies a randomized stack of serial vertical sections

$$V = \{v^\perp + (\zeta + mt)v : m \in \mathbb{Z}\},$$

where $t > 0$ is the distance between neighboring flats, v is uniformly distributed in the circle $S^2 \cap L_0^\perp$, and ζ is independent of v and uniform in $[0, t]$. Then

$$(2\pi t)W(K, V) = (2\pi t) \sum_{m \in \mathbb{Z}} W(K, v^\perp + (\zeta + mt)v) \tag{2.14}$$

is an unbiased estimator of the surface area $2V_2(K)$ of K.

There are several possibilities to measure or estimate the quantity $W(K, H)$ in (2.13), which only depends on K through the section $K \cap H$. If the boundary of $K \cap H$ has a piecewise differentiable parametrization, $W(K, H)$ can be written as a curve integral along this boundary; see [31, p. 181]. Alternatively, a modified Steinhaus estimator in the plane H can be used. Construct a random grid of lines in H of distance $h > 0$ apart:

$$G_H = \{(\eta^\perp \cap H) + (\xi + mh)\eta : m \in \mathbb{Z}\}.$$

Here η has uniform distribution on the unit circle in the linear space parallel to H, and ξ is independent of η and uniform in $[0, h]$. The value of $V_0(K \cap G_H)$ is obtained by counting the number of line segments in the intersection of G_H with the profile $K \cap H$. This count has to be sine-weighted in accordance with (2.13):

$$hV_0(K \cap G_H)[\eta^\perp \cap H, L_0] \tag{2.15}$$

is an unbiased estimator for $W(K, H)$. Using this estimator in each of the planes of V in (2.14) therefore leads to an unbiased estimator of the surface area of K.

The subsequent weighting in (2.15) with the sine function can be avoided by using a non-uniform orientation distribution for G_H. More precisely, if η is chosen with density $(\pi/2)[\eta^\perp \cap H, L_0]$ with respect to the uniform distribution then no numerical weighting factor is required and $(2h/\pi)V_0(K \cap G_H)$ is an unbiased

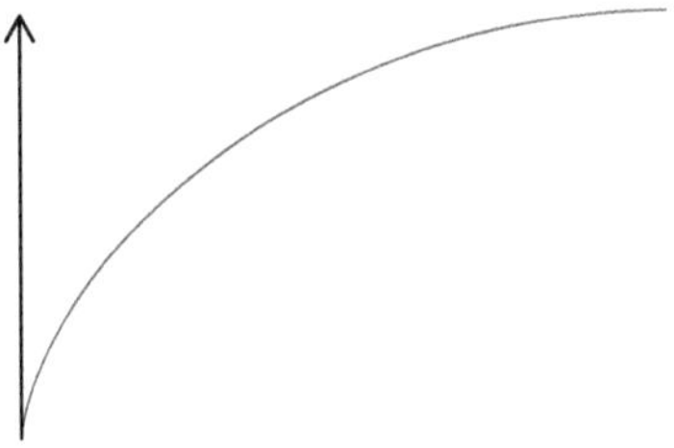

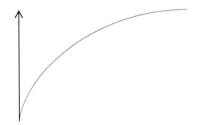

Fig. 2.3 Cycloid curve γ together with a "vertical" arrow of length 2

estimator for $W(K,H)$. In applications, in order to obtain estimators of (2.13), one usually counts the number of intersections of E with the boundary of K. In fact, under the assumption that K doesn't contain any lower dimensional parts (K is the topological closure of its interior), we have

$$|\partial K \cap E| = 2V_0(K \cap E)$$

almost everywhere. The randomization of the orientation of the test system can be omitted altogether, if lines are replaced by appropriate curves whose orientation distribution (this is the distribution of the tangent in a uniformly chosen point on the curve) has a sine-weighted density with respect to the uniform distribution. The *cycloid*, a curve traced by a point on the rim of a rolling wheel, is such a curve, if it is appropriately oriented with respect to the vertical axis L_0. This is illustrated in Fig. 2.3, where the cycloid γ has parametric equation $x = t - \sin t, y = 1 - \cos t$, $0 \le t \le \pi$, and curve length 4. If the direction of the arrow in this figure is aligned with the direction of L_0, then this curve has an orientation distribution with uniform density $(\pi/2)[\cdot, L_0]$. To estimate $W(K,H)$ for a given H using the cycloid curve $\gamma \subset H$ in Fig. 2.3, a compact reference set A in H can be chosen that contains the set $\{x \in H : (x+\gamma) \cap K \cap H \neq \emptyset\}$ of all translation vectors such that the translation of γ meets $K \cap H$. If ξ is uniform random in A, it can be shown that

$$\frac{V_2(A)}{4\pi}|\partial K \cap (\xi + \gamma)|$$

is an unbiased estimator for $W(K,H)$. This can then, again, be substituted into (2.14) to obtain an unbiased estimator of the surface area of K.

In applications, one prefers to work with a stationary systematic grid of cycloid curves; see [30], where also the practical sampling procedures are explained. Vertical section designs in general dimensions are developed in [28].

The last stereological concept that we shall discuss here is the so-called *local design*. It is again motivated by applications: When sampling a biological cell it is convenient to consider only sections of the cell with planes through a given reference point, which usually is the cell nucleus or the nucleolus. For a mathematical description we assume that the reference point is the origin. The branch of stereology dealing with inference on $K \in \mathcal{R}$ from sections $K \cap L$, $L \in \mathcal{L}_r^d$, $1 \le r \le d-1$, is called *local stereology*. Like in the case of vertical sections, Crofton's formula cannot be applied directly, but only after a sub-sampling in L with a suitably weighted affine plane. Theorem 2.7 and Crofton's formula imply for $0 \le j \le k < r \le d-1$

$$\int_{\mathcal{L}_r^d} \int_{\mathcal{E}_k^L} V_j(K \cap E) d(E,o)^{d-r} \, dE \, dL$$

$$= \frac{\omega_{r-k}}{\omega_{d-k}} \int_{\mathcal{E}_k^d} V_j(K \cap E) \, dE = \frac{\omega_{r-k}}{\omega_{d-k}} c_{j,d}^{k,d-k+j} V_{d-k+j}(K).$$

Stereologically this can be interpreted as follows: Let $K \in \mathcal{R}$ be contained in some reference set A. In order to focus on the essentials, we assume that $A = B_s(o)$ is a ball with radius $s > 0$. Let $L \in \mathcal{L}_r^d$ be an isotropic random plane. Given L, let $E \in \mathcal{E}_k^L, k < r$, be a random flat in L with density proportional to $\mathbf{1}(E \cap B_s(o) \neq \emptyset) \, d(E,o)^{d-r}$ with respect to the invariant measure on $\mathcal{E}_k^L$. Then $cV_j(K \cap E)$ is an unbiased estimator for $V_{d-k+j}(K)$, where the constant is given by

$$c = \binom{r}{k} \frac{\omega_{d-k}}{\omega_{r-k}} c_{0,r,d-k+j}^{j,r-k,d} \frac{\kappa_r}{\kappa_k} s^{r-k}.$$

Note that $(K \cap L) \cap E = K \cap E$, so the estimator depends on K only through $K \cap L$. The intrinsic volume $V_m(K)$ can be estimated from r-dimensional isotropic sections with the above formula only if $m > d - r$. That there cannot exist any unbiased estimation procedure for $m \le d - r$ is clear: for an m-dimensional ball K contained in a m-dimensional linear subspace, we have $K \cap E = \{o\}$ almost surely, so the radius of K is almost surely invisible in the sections.

The monograph [140] is an excellent introduction to local stereology, focusing on formulae for k-dimensional Hausdorff measures instead of intrinsic volumes. Such relations are based on generalized Blaschke–Petkantschin formulae for Hausdorff measures. A local stereological formula for the intrinsic volumes, as presented here, is a relatively recent development taken from [19, 208] based on ideas in [137].

Exercise 2.13. This exercise parallels Exercise 2.12 in a local stereological setting. Assume that the random convex body $X \subset \mathbb{R}^3$ is almost surely contained in $B_1(o)$.

(a) Using Crofton's chord formula (Exercise 2.6) show that

$$\int_{\mathcal{L}_2^3} \int_{\mathcal{E}_1^L} V_1^4(X \cap E) d(E,o) \, dE \, dL = \frac{3}{\pi^2} V_3^2(X).$$

(b) Use this to derive an unbiased estimator of $\mathbf{E} V_3^2(X)$.

Chapter 3
Spatial Point Patterns: Models and Statistics

Adrian Baddeley

Abstract This chapter gives a brief introduction to spatial point processes, with a view to applications. The three sections focus on the construction of point process models, the simulation of point processes, and statistical inference. For further background, we recommend [Daley et al., Probability and its applications (New York). Springer, New York, 2003/2008; Diggle, Statistical analysis of spatial point patterns, 2nd edn. Hodder Arnold, London, 2003; Illian et al., Statistical analysis and modelling of spatial point patterns. Wiley, Chichester, 2008; Møller et al., Statistical inference and simulation for spatial point processes. Chapman & Hall, Boca Raton, 2004].

Introduction

Spatial point patterns—data which take the form of a pattern of points in space—are encountered in many fields of research. Currently there is particular interest in point pattern analysis in radioastronomy (Fig. 3.1), epidemiology (Fig. 3.2a) and prospective geology (Fig. 3.2b).

Under suitable conditions, a point pattern dataset can be modelled and analysed as a realization of a spatial point process. The main goals of point process analysis are to

1. Formulate "realistic" stochastic models for spatial point patterns
2. Analyse, predict or simulate the behaviour of the model
3. Fit models to data

These three goals will be treated in three successive sections.

A. Baddeley (✉)
CSIRO, Perth, Australia
e-mail: Adrian.Baddeley@csiro.au

E. Spodarev (ed.), *Stochastic Geometry, Spatial Statistics and Random Fields*, Lecture Notes in Mathematics 2068, DOI 10.1007/978-3-642-33305-7_3,

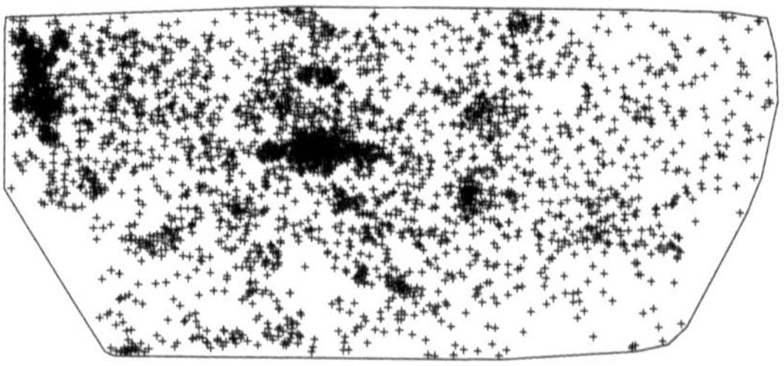

Fig. 3.1 Sky positions of 4,215 galaxies observed in a radioastronomical survey [161]

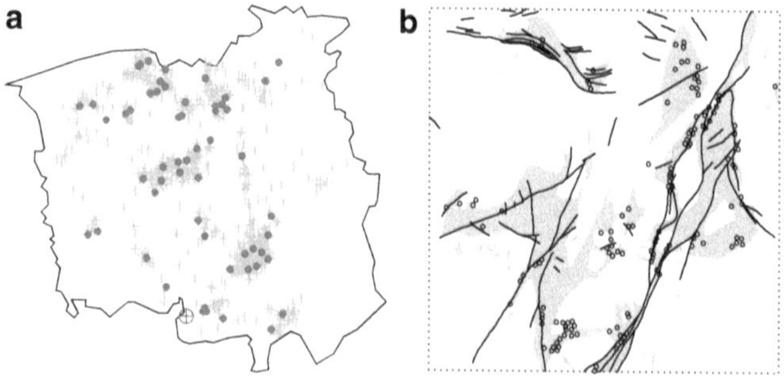

Fig. 3.2 Examples of point pattern data. (**a**) Locations of cases of cancer of the lung (*plus*) and larynx (*filled circle*), and a pollution source (*oplus*), in a region of England [153]. (**b**) Gold deposits (*circle*), geological faults (*lines*) and rock type (*grey shading*) in a region of Western Australia [507]

3.1 Models

In this section we cover some basic notions of point processes (Sect. 3.1.1), introduce the Poisson process (Sect. 3.1.2), discuss models constructed from Poisson processes (Sect. 3.1.4), and introduce finite Gibbs point processes (Sect. 3.1.5).

3.1.1 Point Processes

In one dimensional time, a point process represents the successive instants of time at which events occur, such as the clicks of a Geiger counter or the arrivals of customers at a bank. A point process in time can be characterized and analysed using several different quantities. One can use the *arrival times* $T_1 < T_2 < \dots$ at which the events occur (Fig. 3.3a), or the *waiting times* $S_i = T_i - T_{i-1}$ between successive arrivals (Fig. 3.3b). Alternatively one can use the *counting process* $N_t = \sum_i \mathbf{1}(T_i \leq t)$ illustrated in Fig. 3.4, or the *interval counts* $N(a, b] = N_b - N_a$.

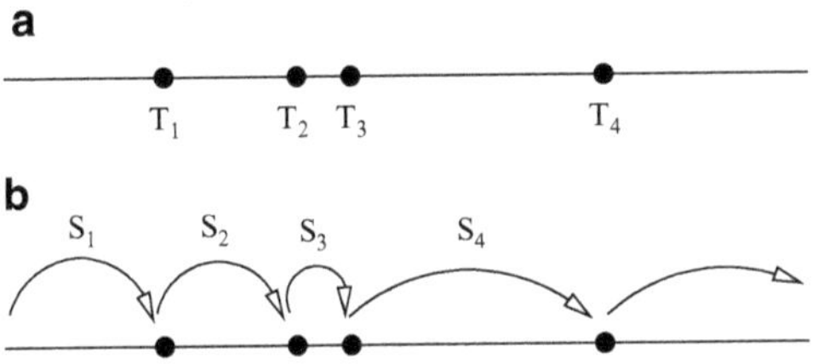

Fig. 3.3 Arrival times (**a**) and waiting times (**b**) for a point process in one dimensional time

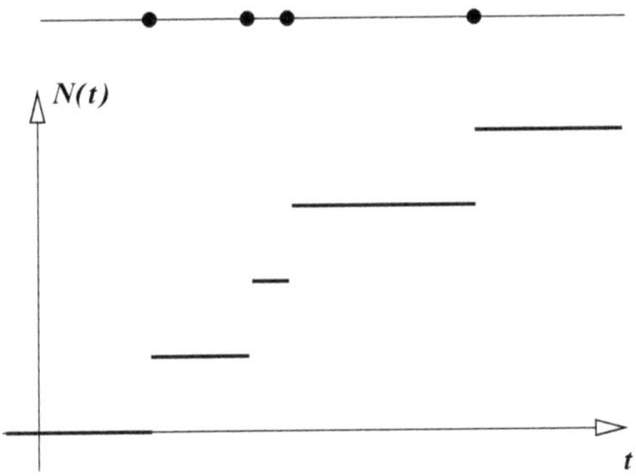

Fig. 3.4 The counting process N_t for a point process in one-dimensional time

For example, the homogeneous Poisson process with intensity parameter λ in $[0, \infty)$ has

1. Poisson counts: $N_t \sim \text{Pois}(\lambda t)$ and $N(a, b] \sim \text{Pois}(\lambda(b - a))$.
2. Independent increments: if the intervals $(a_1, b_1], \ldots, (a_m, b_m]$ are disjoint then $N(a_1, b_1], \ldots, N(a_m, b_m]$ are independent.
3. Independent exponential waiting times: $S_1, S_2, \ldots$ are i.i.d. $\text{Exp}(\lambda)$ variables.

In higher dimensional Euclidean space $\mathbb{R}^d$ for $d > 1$, some of these quantities are less useful than others. We typically define a point process using the counts

$$N(B) = \text{number of points falling in } B$$

for bounded sets $B \subset \mathbb{R}^d$.

Definition 3.1. Let S be a complete, separable metric space. Let $\mathcal{N}$ be the set of all nonnegative integer valued measures μ on S such that $\mu(K) < \infty$ for every compact $K \subseteq \mathsf{S}$. Define $\mathfrak{N}$ to be the smallest σ-field on $\mathcal{N}$ containing $\{\mu : \mu(K) = n\}$ for every compact $K \subseteq \mathsf{S}$ and every integer $n \geq 0$. A *point process* Ψ on S is a random element of $(\mathcal{N}, \mathfrak{N})$.

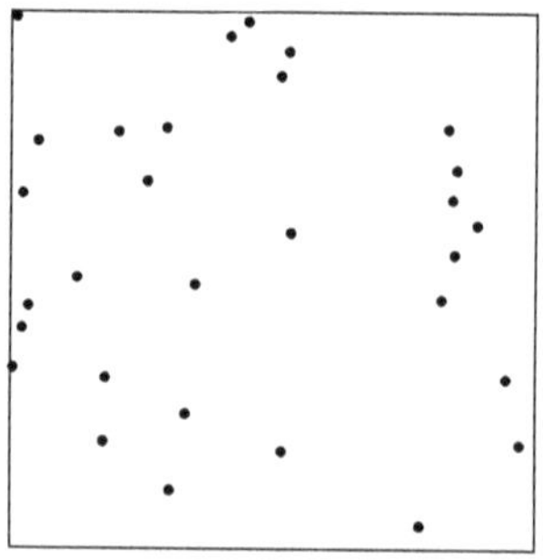

Fig. 3.5 Realization of a binomial process

Thus $\Psi(K)$ is a random variable for every compact $K \subseteq \mathsf{S}$.

Example 3.1. Let $X_1, \ldots, X_n$ be independent, identically distributed (i.i.d.) random points in $\mathbb{R}^d$, uniformly distributed in a bounded set $W \subset \mathbb{R}^d$. For any Borel set $B \subset \mathbb{R}^d$, let

$$\Psi(B) = \sum_{i=1}^{n} \mathbf{1}(X_i \in B)$$

be the number of random points that fall in B. For each realisation of $(X_1, \ldots, X_n)$ it is clear that $B \mapsto \Psi(B)$ is a nonnegative integer valued measure, and that $\Psi(B) \leq n < \infty$ for all B, so that Ψ is an element of $\mathcal{N}$. Furthermore, for each compact set K, $\mathbf{1}(X_i \in K)$ is a random variable for each i, so that $\Psi(K)$ is a random variable. Thus Ψ defines a point process on $\mathbb{R}^d$ (Fig. 3.5).

Exercise 3.1. The point process in the previous Example is often called the "binomial process". Why?

Definition 3.2. A point process Ψ is *simple* if

$$\mathbf{P}(\Psi(\{s\}) \leq 1 \text{ for all } s \in \mathsf{S}) = 1.$$

Exercise 3.2. Prove that the binomial process (Exercise 3.1) is simple. (*Hint:* prove $\mathbf{P}(X_i = X_j) = 0$ for $i \neq j$.)

A simple point process can be regarded as a locally-finite random set. Hence there are many connections between point process theory and stochastic geometry. One of the interesting connections is that the distribution of a point process is completely determined by its *vacancy probabilities* $V(K) = \mathbf{P}(\Psi(K) = 0)$, i.e. the probability that there are no random points in K, for all compact sets K.

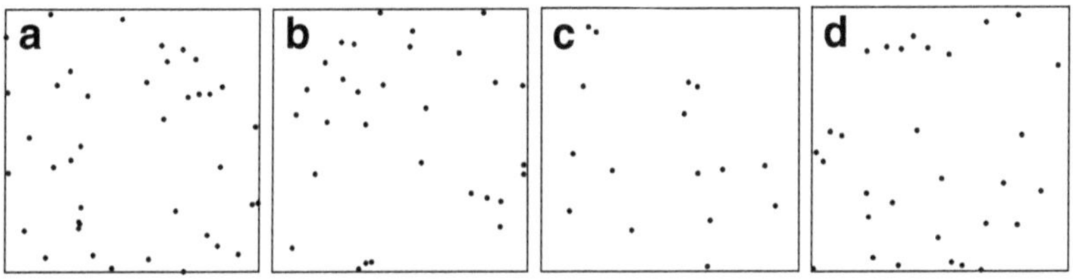

Fig. 3.6 Four realizations of the Poisson point process

3.1.2 Poisson Processes

Homogeneous Poisson Process

Definition 3.3. The *homogeneous Poisson point process* Π_λ on $\mathbb{R}^d$ with intensity $\lambda > 0$ (Fig. 3.6) is characterized by the properties

1. $\Pi_\lambda(B)$ has a Poisson distribution, for all bounded Borel $B \subset \mathbb{R}^d$;
2. $\mathbf{E}\Pi_\lambda(B) = \lambda \nu_d(B)$, for all bounded Borel $B \subset \mathbb{R}^d$;
3. $\Pi_\lambda(B_1), \ldots, \Pi_\lambda(B_m)$ are independent when $B_1, \ldots, B_m$ are disjoint bounded Borel sets.

A pivotal property is the following distributional relation.

Lemma 3.1. *If* $\eta \sim \text{Pois}(\mu)$ *and* $(\xi \mid \eta = n) \sim \text{Binom}(n, p)$ *then* $\xi \sim \text{Pois}(p\mu)$, $\eta - \xi \sim \text{Pois}((1-p)\mu)$ *and* ξ *and* $\eta - \xi$ *are independent.*

Exercise 3.3. Prove Lemma 3.1.

In one dimension, the homogeneous Poisson process has *conditionally uniform arrivals*: given $N_t = n$, the arrival times in $[0, t]$

$$T_1 < T_2 < \ldots < T_n < t$$

are (the order statistics of) n i.i.d. uniform random variables in $[0, t]$. Similarly in higher dimensions we have the following property.

Lemma 3.2 (Conditional property of Poisson process). *For the homogeneous Poisson process on* $\mathbb{R}^d$ *with intensity* $\lambda > 0$, *given the event* $\{\Pi_\lambda(B) = n\}$ *where* $B \subset \mathbb{R}^d$, *the restriction of* Π_λ *to* B *has the same distribution as a binomial process (n i.i.d. random points uniformly distributed in B).*

In one dimension, the arrival time of the first event in a Poisson process is exponentially distributed. Similarly in higher dimensions (Fig. 3.7):

Lemma 3.3 (Exponential waiting times of Poisson process). *Let* Π_λ *be a homogeneous Poisson process in* $\mathbb{R}^2$ *with intensity* λ. *Let* R *be the distance from the origin* o *to the nearest point of* Π_λ. *Then* πR^2 *has an exponential distribution with parameter* λ.

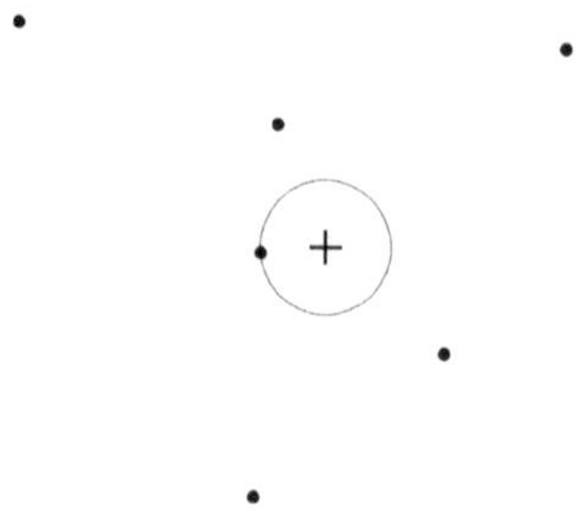

Fig. 3.7 Distance from the origin (*plus*) to the nearest point of a Poisson process (*filled circle*)

To prove this, observe that $R > r$ iff there are no points of Π_λ in $B_r(o)$. Thus R has distribution function

$$F(r) = \mathbf{P}(R \leq r) = 1 - \mathbf{P}(R > r) = 1 - \mathbf{P}(N(B_r(o)) = 0)$$
$$= 1 - \exp\{-\lambda \nu_d(B_r(o))\} = 1 - \exp\{-\lambda \pi r^2\}.$$

This implies that the *area* of the disc $B_R(o)$ is exponentially distributed.

Similar properties hold for other "waiting sets" [358, 370].

General Poisson Process

There is a more general version of the Poisson process, which has a spatially-varying density of points.

Definition 3.4. Suppose Λ is a measure on $(\mathbb{R}^d, \mathcal{B}(\mathbb{R}^d))$ such that $\Lambda(K) < \infty$ for all compact $K \subset \mathbb{R}^d$ and $\Lambda(\{x\}) = 0$ for all $x \in \mathbb{R}^d$. A *Poisson point process* Π on $\mathbb{R}^d$ with intensity measure Λ (Fig. 3.8) is characterized by the properties

1. $\Pi(B)$ has a Poisson distribution, for all bounded Borel $B \subset \mathbb{R}^d$;
2. $\mathbf{E}\Pi(B) = \Lambda(B)$, for all bounded Borel $B \subset \mathbb{R}^d$;
3. $\Pi(B_1), \ldots, \Pi(B_m)$ are independent when $B_1, \ldots, B_m$ are disjoint bounded Borel sets.

This definition embraces the homogeneous Poisson process of intensity $\lambda > 0$ when we take $\Lambda(\cdot) = \lambda \nu_d(\cdot)$.

Exercise 3.4. Show that the vacancy probabilities $\mathbf{P}(\Xi(K) = 0)$ of an inhomogeneous Poisson point process in $\mathbb{R}^d$, if known for all compact sets $K \subset \mathbb{R}^d$, completely determine the intensity measure Λ.

Transformation of a Poisson Process

Suppose Π is a Poisson process in $\mathbb{R}^d$ with intensity measure Λ. Let $T : \mathbb{R}^d \to \mathbb{R}^k$ be a continuous mapping. Consider the point process $T\Pi$ obtained applying T to each point of Π, sketched in Fig. 3.9a.

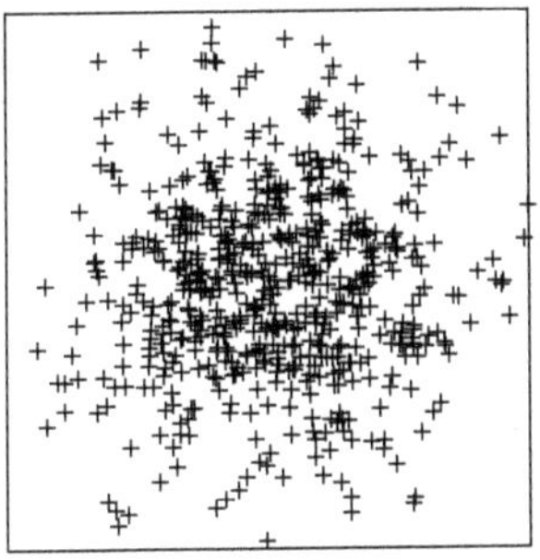

Fig. 3.8 Realization of an inhomogeneous Poisson process with intensity proportional to a Gaussian density on the plane

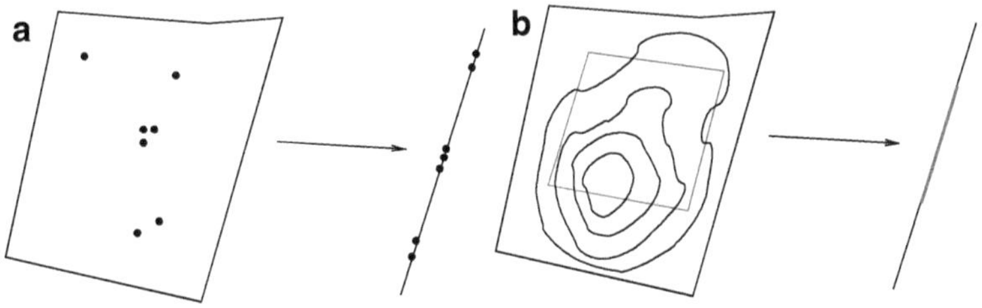

Fig. 3.9 (**a**) Transformation of a Poisson point process by a mapping. (**b**) Image of a measure

For any measure μ on $\mathbb{R}^d$ we can define a measure $T\mu$ on $\mathbb{R}^k$ by

$$(T\mu)(B) = \mu(T^{-1}(B))$$

for all Borel $B \subseteq \mathbb{R}^k$. See Fig. 3.9b.

Theorem 3.1. *Suppose Π is a Poisson process in $\mathbb{R}^d$ with intensity measure Λ. Let $T : \mathbb{R}^d \to \mathbb{R}^k$ be a continuous mapping such that $(T\Lambda)(K) < \infty$ for all compact sets $K \subset \mathbb{R}^k$, and $(T\Lambda)(\{x\}) = 0$ for all $x \in \mathbb{R}^k$. Then the image of Π under T is a Poisson process on $\mathbb{R}^k$ with intensity measure $T\Lambda$.*

Example 3.2 (Waiting times). Let Π_λ be a homogeneous Poisson process in $\mathbb{R}^2$ with intensity λ. Let $T(x) = \|\mathbf{x}\|^2$. We have $\Lambda(T^{-1}([0,s])) = \Lambda(B_{\sqrt{s}}(o)) = \lambda\pi s < \infty$ for all $0 < s < \infty$. So $T\Pi_\lambda$ is a Poisson process on $[0,\infty)$ with intensity measure $\lambda\pi$. Let R_k = distance from o to the k-th nearest point of Π_λ. Then $R_1^2, R_2^2 - R_1^2, R_3^2 - R_2^2, \ldots$ are i.i.d. exponential random variables with rate $\lambda\pi$. See Fig. 3.10.

Exercise 3.5. In Example 3.2, find the distribution of R_k^2 for each k and use it to verify that $\Pi_\lambda(B_r(o))$ has a Poisson distribution.

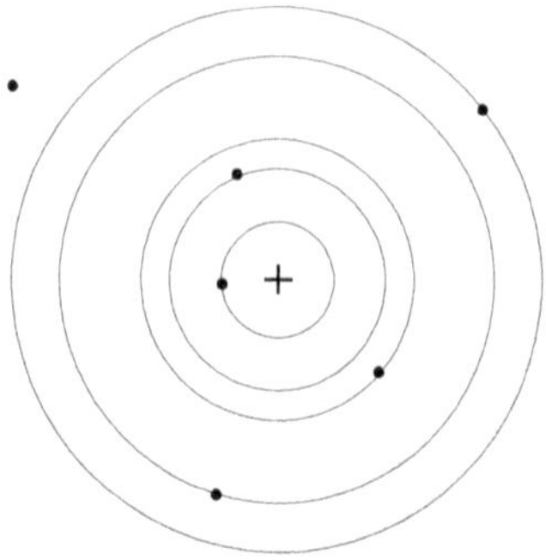

Fig. 3.10 The consecutive nearest neighbours of the origin in a Poisson point process. The areas of the rings are independent and exponentially distributed

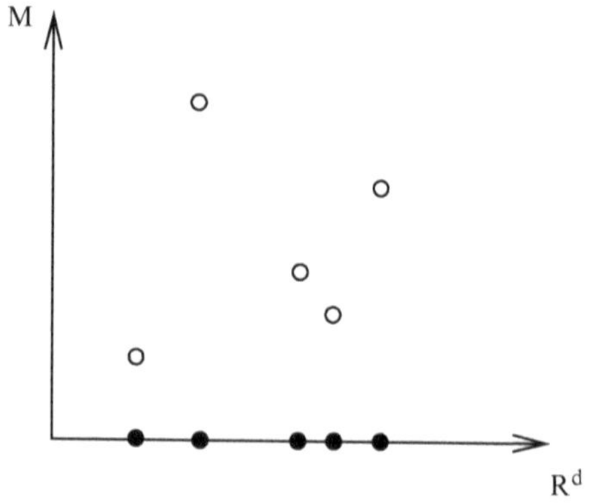

Fig. 3.11 Projection of a marked Poisson point process to an unmarked Poisson point process

Projection

Consider the projection $T(x_1, x_2) = x_1$ from $\mathbb{R}^2$ to $\mathbb{R}$. Let Π_λ be a homogeneous Poisson point process in $\mathbb{R}^2$ with intensity λ. The projection of Π_λ is not a point process (in our sense) because the number of points projected onto a compact interval $[a, b]$ is infinite:

$$(T(\Pi_\lambda))([a,b]) = \Pi_\lambda(T^{-1}([a,b])) = \Pi_\lambda([a,b] \times \mathbb{R}) = \infty \quad \text{a.s.}$$

Independent Marking

Consider a homogeneous Poisson process Π_λ in $\mathbb{R}^d \times [0, a]$ with intensity λ. This can be viewed as a marked point process of points x_i in $\mathbb{R}^d$ with marks m_i in $[0, a]$, see Sect. 4.1 for more details on marked point processes. The projection of Π_λ onto $\mathbb{R}^d$ is a bona fide Poisson process with intensity λa (Fig. 3.11).

Thinning

Let Π_λ be a homogeneous Poisson point process in $\mathbb{R}^d$ with intensity λ. Suppose we randomly delete or retain each point of Π_λ, with retention probability p for each point, independently of other points. The process $\Pi_{p\lambda}$ of retained points is a Poisson process with intensity measure $p\lambda$.

Exercise 3.6. Prove this, using Lemma 3.1.

Conditioning

The conditional probability distribution of a point process Ψ, given that an event A occurs, is generally different from the original distribution of Ψ. This is another way to construct new point process models.

However, for a Poisson process Ψ, the independence properties often imply that the conditional distribution of Ψ is another Poisson process. For example, a Poisson process Ψ with intensity function λ, conditioned on the event that there are no points of Ψ in $B \subset \mathbb{R}^d$, is a Poisson process with intensity $\lambda(u)\mathbf{1}(u \notin B)$.

A related concept is the *Palm distribution* P^x of Ψ at a location $x \in \mathbb{R}^d$. Roughly speaking, P^x is the conditional distribution of Ψ, given that there is a point of Ψ at the location x. For a Poisson process, *Slivnyak's Theorem* states that the Palm distribution P^x of Ψ is equal to the distribution of $\Psi \cup \{x\}$, that is, the same Poisson process with the point x added. See [265, 281].

3.1.3 *Intensity*

For a point process in one-dimensional time, the average rate at which points occur, i.e. the expected number of points per unit time, is called the "intensity" of the process. For example, the intensity of the point process of clicks of a Geiger counter is a measure of radioactivity. This concept of intensity can be defined for general point processes.

Definition 3.5. Let Ψ be a point process in a complete separable metric space S. Suppose that the expected number of points in any compact set $K \subset \mathsf{S}$ is finite, $\mathbf{E}\Psi(K) < \infty$. Then there exists a measure Λ_Ψ, called the *intensity measure* of Ψ, such that

$$\Lambda_\Psi(B) = \mathbf{E}\Psi(B)$$

for all Borel sets B.

For the homogeneous Poisson process Π_λ on $\mathbb{R}^d$ with intensity parameter λ, the intensity measure is $\Lambda_{\Pi_\lambda}(B) = \lambda \nu_d(B)$ for all Borel B, by property 2 of Definition 3.3.

For the general Poisson process Π on $\mathbb{R}^d$, the intensity measure Λ as described in Definition 3.4 coincides with the intensity measure Λ_Π defined above, i.e. $\Lambda_\Pi(B) = \mathbf{E}\Pi(B) = \Lambda(B)$, by property 2 of Definition 3.4.

Example 3.3. For the binomial point process (Exercise 3.1),

$$\begin{aligned}
\mathbf{E}\Psi(B) &= \mathbf{E}\sum_i \mathbf{1}(X_i \in B) \\
&= \sum_i \mathbf{E}\mathbf{1}(X_i \in B) \\
&= \sum_i \mathbf{P}(X_i \in B) \\
&= nP(X_1 \in B) = n\frac{\nu_d(B \cap W)}{\nu_d(W)}
\end{aligned}$$

where ν_d is Lebesgue measure in $\mathbb{R}^d$.

Definition 3.6. A point process Ψ in $\mathbb{R}^d$ has *intensity function* λ if

$$\Lambda_\Psi(B) = \mathbf{E}\Psi(B) = \int_B \lambda(u)\,\mathrm{d}u$$

for all Borel sets $B \subseteq \mathbb{R}^d$.

For example, the homogeneous Poisson process with intensity parameter $\lambda > 0$ has intensity function $\lambda(u) \equiv \lambda$.

Note that a point process need not have an intensity function, since the measure Λ_Ψ need not be absolutely continuous with respect to Lebesgue measure.

Exercise 3.7. Find the intensity function of the binomial process (Example 3.1).

Similarly one may define the *second moment intensity* λ_2, if it exists, to satisfy

$$\mathbf{E}[\Psi(A)\Psi(B)] = \int_A \int_B \lambda_2(u, v)\,\mathrm{d}u\,\mathrm{d}v$$

for disjoint bounded Borel sets $A, B \subset \mathbb{R}^d$.

3.1.4 Poisson-Driven Processes

The Poisson process is a plausible model for many natural processes. It is also easy to analyse and simulate. Hence, the Poisson process is a convenient basis for building new point process models.

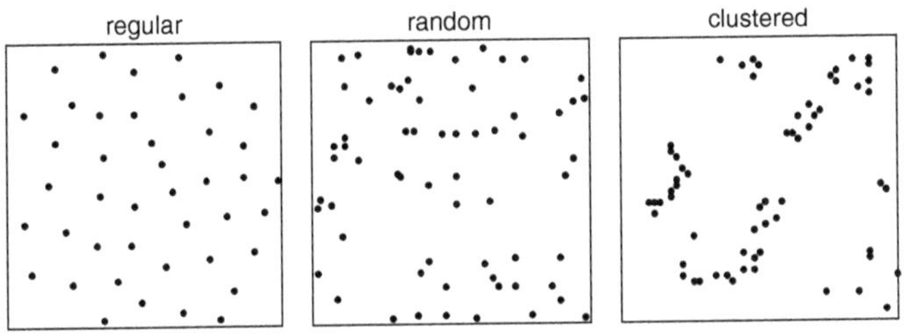

Fig. 3.12 Classical trichotomy between regular (negatively associated), random (Poisson) and clustered (positively associated) point processes

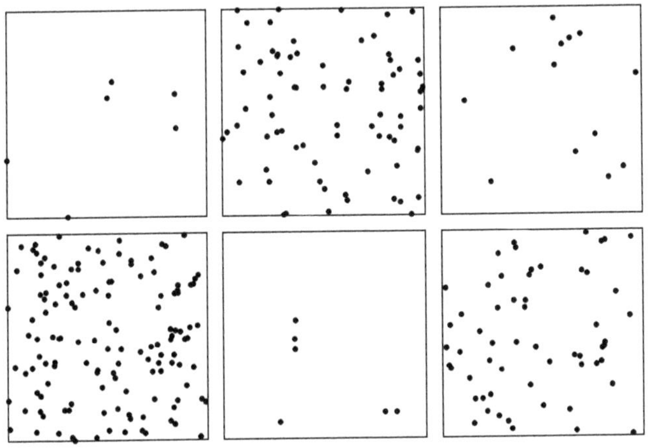

Fig. 3.13 Six realizations of a Cox process with driving measure $\Lambda = \Gamma \nu_d$ where Γ is an exponential random variable with mean 50

A basic objective is to be able to construct models which are either clustered (positively associated) or regular (negatively associated) relative to a Poisson process. See Fig. 3.12.

Cox Process

A *Cox process* is "a Poisson process whose intensity measure is random" (Fig. 3.13).

Definition 3.7. Let Λ be a random locally-finite measure on $\mathbb{R}^d$. Conditional on $\Lambda = \lambda$, let Ψ be a Poisson point process with intensity measure λ. Then Ψ is called a Cox process with driving random measure Λ.

A Cox process Ψ is not ergodic (unless the distribution of Λ is degenerate and Ψ is Poisson). A single realization of Ψ, observed in an arbitrarily large region, cannot be distinguished from a realization of a Poisson process. If multiple realizations can be observed (e.g. multiple point patterns) then the Cox model may be identifiable.

For any Cox process, we obtain by conditioning

$$\mathbf{E}\Psi(B) = \mathbf{E}(\mathbf{E}(\Psi(B) \mid \Lambda)) = \mathbf{E}\Lambda(B) \tag{3.1}$$

$$\begin{aligned}\mathbf{var}\,\Psi(B) &= \mathbf{var}(\mathbf{E}(\Psi(B) \mid \Lambda)) + \mathbf{E}(\mathbf{var}(\Psi(B) \mid \Lambda)) \\ &= \mathbf{var}\,\Lambda(B) + \mathbf{E}\Lambda(B) \end{aligned}\tag{3.2}$$

$$\begin{aligned}\mathbf{P}(\Psi(B) = 0) &= \mathbf{E}(\mathbf{P}(\Psi(B) = 0 \mid \Lambda)) \\ &= \mathbf{E}\exp\{-\Lambda(B)\}\end{aligned} \tag{3.3}$$

Thus a Cox process is always "overdispersed" in the sense that $\mathbf{var}\,\Psi(B) \geq \mathbf{E}\Lambda(B)$. Further progress is limited without a specific model for Λ.

Definition 3.8. Let Γ be a positive real random variable with finite expectation. Conditional on $\Gamma = \gamma$, let Ψ be a homogeneous Poisson process with intensity γ. Then Ψ is called a "mixed Poisson process" with driving intensity Γ.

Exercise 3.8. Find the intensity of the mixed Poisson process with driving intensity Γ that is exponential with mean μ.

Definition 3.9. Let Λ be the measure with density $\lambda(u) = e^{\xi(u)}$, where ξ is a Gaussian random function on $\mathbb{R}^d$. Then Ψ is a "log-Gaussian Cox process".

Moments can be obtained using properties of the lognormal distribution of $e^{\xi(u)}$. If ξ is stationary with mean μ and covariance function

$$\mathbf{cov}(\xi(u), \xi(v)) = c(u - v)$$

then Ξ has the intensity $m_1(u) \equiv \exp\{\mu + c(0)/2\}$ and second moment intensity $m_2(u, v) = \exp\{2\mu + c(0) + c(u - v)\}$.

Exercise 3.9. Verify these calculations using only the characteristic function of the normal distribution.

Poisson Cluster Processes

Definition 3.10. Suppose we can define, for any $x \in \mathbb{R}^d$, the distribution of a point process ζ_x containing a.s. finitely many points, $\mathbf{P}(\zeta_x(\mathbb{R}^d) < \infty) = 1$. Let Π be a Poisson process in $\mathbb{R}^d$. Given Π, let

$$\Psi = \bigcup_{x_i \in \Pi} \Phi_i$$

be the superposition of independent point processes Φ_i where $\Phi_i \sim \zeta_{x_i}$. Then Ψ is called the Poisson cluster process with parent process Π and cluster mechanism $\{\zeta_x, x \in \mathbb{R}^d\}$ (Fig. 3.14).

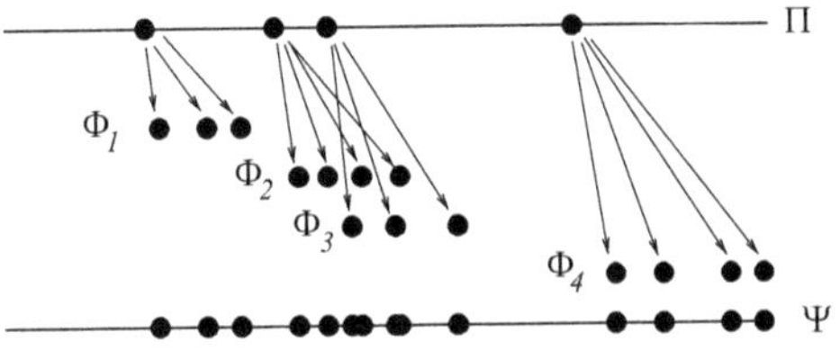

Fig. 3.14 Schematic construction of a Poisson cluster process

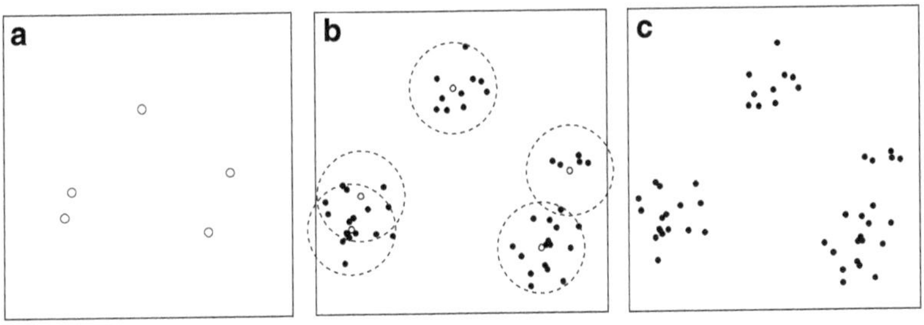

Fig. 3.15 Construction of Matérn cluster process. (**a**) Poisson process of parent points. (**b**) Each parent gives rise to a cluster of offspring lying in a disc of radius r around the parent. (**c**) Offspring points only

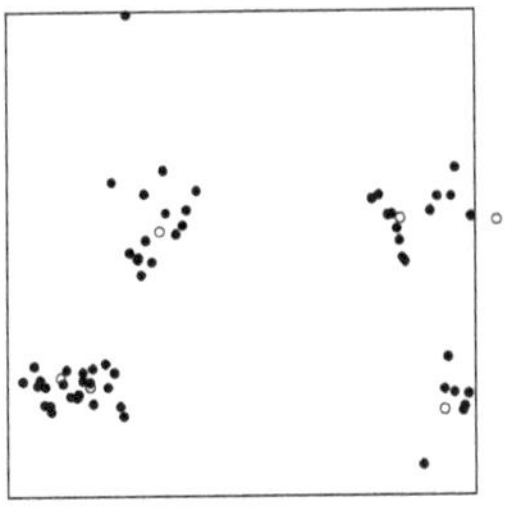

Fig. 3.16 Realization of modified Thomas process

The *Matérn cluster process* is the case where the typical cluster ζ_x consists of $N \sim \text{Pois}(\mu)$ i.i.d. random points uniformly distributed in the disc $B_r(x)$ (Fig. 3.15).

The *modified Thomas process* is the case where the cluster ζ_x is a Poisson process with intensity (Fig. 3.16)

$$\kappa(u) = \mu\varphi(u - x) \tag{3.4}$$

where φ is the probability density of the isotropic Gaussian distribution with mean o and covariance matrix $\Sigma = \text{diag}(\sigma^2, \sigma^2, \ldots, \sigma^2)$. Equivalently there are $N \sim \text{Pois}(\mu)$ points, each point generated as $Y_i = x + E_i$ where $E_1, E_2, \ldots$ are i.i.d. isotropic Gaussian.

Definition 3.11. A Poisson cluster process in which the cluster mechanism ζ_x is a finite Poisson process with intensity μ_x, is called a *Neyman–Scott process*.

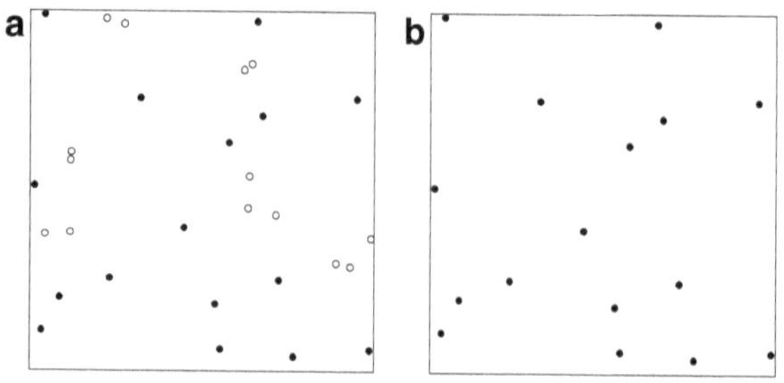

Fig. 3.17 Construction of Matérn thinning Model I. (**a**) Poisson process. (**b**) After deletion of any points that have close neighbours

Matérn's cluster process and the modified Thomas process are both Neyman–Scott processes. A Neyman–Scott process is also a Cox process, with driving random measure $\Lambda = \sum_{x \in \Pi} \mu_x$.

Theorem 3.2 (Cluster formula [11, 349, 489]). *Suppose Ψ is a Poisson cluster process whose cluster mechanism is equivariant under translations, $\zeta_x \equiv x + \zeta_o$. Then the Palm distribution P^x of Ψ is*

$$P^x = P * C^x \tag{3.5}$$

where P is the distribution of Ψ and

$$C_x(A) = \frac{\mathbf{E}\left(\sum_{z_i \in \zeta_o} \mathbf{1}(\zeta_o + (x - z_i) \in A)\right)}{\mathbf{E}(\zeta_o(\mathbb{R}^d))} \tag{3.6}$$

is the finite Palm distribution of the cluster mechanism.

This allows detailed analysis of some properties of Ψ, including its K-function, see Sect. 4.2.1. Thus, Poisson cluster processes are useful in constructing tractable models for clustered (positively associated) point processes.

Dependent Thinning

If the points of a Poisson process are randomly deleted or retained independently of each other, the result is a Poisson process. To get more interesting behaviour we can thin the points in a dependent fashion.

Definition 3.12 (Matérn Model I). Matérn's thinning Model I is constructed by generating a uniform Poisson point process Π_1, then deleting any point of Π_1 that lies closer than r units to another point of Π_1 (Fig. 3.17).

Definition 3.13 (Matérn Model II). Matérn's thinning Model II is constructed by the following steps (Fig. 3.18):

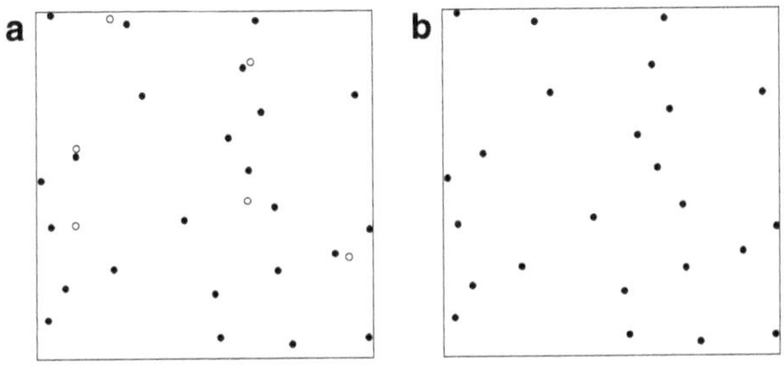

Fig. 3.18 Construction of Matérn thinning Model II. (**a**) Poisson process. (**b**) After deletion of any points that have close neighbours that are older

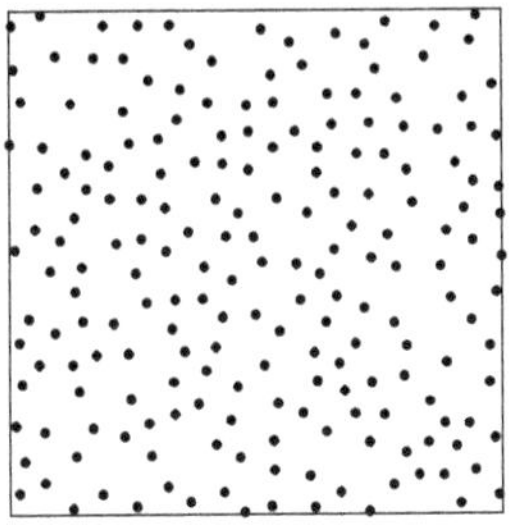

Fig. 3.19 Realization of simple sequential inhibition

1. Generate a uniform Poisson point process Π_1.
2. Associate with each point x_i of Π_1 a random "birth time" t_i.
3. Delete any point x_i that lies closer than r units to another point x_j with earlier birth time $t_j < t_i$.

Iterative Constructions

To obtain regular point patterns with higher densities, one can use iterative constructions.

Definition 3.14. To perform *simple sequential inhibition* in a bounded domain $W \subset \mathbb{R}^d$, we start with an empty configuration $\mathbf{x} = \emptyset$. When the state is $\mathbf{x} = \{x_1, \ldots, x_n\}$, compute

$$A(\mathbf{x}) = \{u \in W : \|u - x_i\| > r \text{ for all } i\}.$$

If $\nu_d(A(\mathbf{x})) = 0$, we terminate and return $\mathbf{x}$. Otherwise, we generate a random point u uniformly distributed in $A(\mathbf{x})$, and add the point u into $\mathbf{x}$. This process is repeated until it terminates (Fig. 3.19).

3.1.5 *Finite Point Processes*

Probability Densities for Point Processes

In most branches of statistical science, we formulate a statistical model by writing down its likelihood (probability density). It would be desirable to be able to formulate models for spatial point pattern data using probability densities. However, it is not possible to handle point processes on the infinite Euclidean space $\mathbb{R}^d$ using probability densities.

Example 3.4. Let Π_λ denote the homogeneous Poisson process with intensity $\lambda > 0$ on $\mathbb{R}^2$. Let A be the event that the limiting average density of points is equal to 5:

$$A = \{\mu \in \mathcal{N} : \frac{\mu(B_R(o))}{\pi R^2} \to 5 \text{ as } R \to \infty\}$$

By the Strong Law of Large Numbers, $\mathbf{P}(\Pi_5 \in A) = 1$ but $\mathbf{P}(\Pi_1 \in A) = 0$. Hence the distributions of Π_5 and Π_1 are mutually singular. Consequently, Π_5 does not have a probability density with respect to Π_1.

To ensure that probability densities are available, we need to avoid point processes with an infinite number of points.

Finite Point Processes

Definition 3.15. A point process Ψ on a space S with $\Psi(\mathsf{S}) < \infty$ a.s. is called a *finite point process*.

One example is the *binomial process* consisting of n i.i.d. random points uniformly distributed in a bounded set $W \subset \mathbb{R}^d$.

Another example is the Poisson process on $\mathbb{R}^d$ with an intensity measure Λ that is totally finite, $\Lambda(\mathbb{R}^d) < \infty$. The total number of points $\Pi(\mathbb{R}^d) \sim \text{Pois}(\Lambda(\mathbb{R}^d))$ is finite a.s.

The distribution of a finite point process can be specified by giving the probability distribution of $N = \Psi(\mathsf{S})$, and given $N = n$, the conditional joint distribution of the n points.

Example 3.5. Consider a Poisson process on $\mathbb{R}^d$ with intensity measure Λ that is totally finite ($\Lambda(\mathbb{R}^d) < \infty$). This is equivalent to choosing a random number $K \sim \text{Pois}(\Lambda(\mathbb{R}^d))$, then given $K = k$, generating k i.i.d. random points with common distribution $Q(B) = \Lambda(B)/\Lambda(\mathbb{R}^d)$.

Space of Realizations

Realizations of a finite point process Ψ belong to the space

$$\mathcal{N}^f = \{\mu \in \mathcal{N} : \mu(\mathsf{S}) < \infty\}$$

of totally finite, simple, counting measures on S, called the (Carter–Prenter) *exponential space* of S. This may be decomposed into subspaces according to the total number of points:

$$\mathcal{N}^f = \mathcal{N}_0 \cup \mathcal{N}_1 \cup \mathcal{N}_2 \cup \ldots$$

where for each $k = 0, 1, 2, \ldots$

$$\mathcal{N}_k = \{\mu \in \mathcal{N} : \mu(\mathsf{S}) = k\}$$

is the set of all counting measures with total mass k, that is, effectively the set of all configurations of k points. The space $\mathcal{N}_k$ can be represented more explicitly by introducing the space of ordered k-tuples

$$\mathsf{S}^{!k} = \{(x_1, \ldots, x_k) : x_i \in \mathsf{S}, \; x_i \neq x_j \text{ for all } i \neq j\}.$$

Define a mapping $I_k : \mathsf{S}^{!k} \to \mathcal{N}_k$ by

$$I_k(x_1, \ldots, x_k) = \delta_{x_1} + \ldots + \delta_{x_k}.$$

This gives

$$\mathcal{N}_k \equiv \mathsf{S}^{!k} / \sim$$

where $\sim$ is the equivalence relation under permutation, i.e.

$$(x_1, \ldots, x_k) \sim (y_1, \ldots, y_k) \quad \Leftrightarrow \quad \{x_1, \ldots, x_k\} = \{y_1, \ldots, y_k\}.$$

Point Process Distributions

Using the exponential space representation, we can give explicit formulae for point process distributions.

Example 3.6 (Distribution of the binomial process). Fix $n > 0$ and let $X_1, \ldots, X_n$ be i.i.d. random points uniformly distributed in $W \subset \mathbb{R}^d$. Set $\Psi = I_n(X_1, \ldots, X_n)$. The distribution of Ψ is the probability measure P_Ψ on $\mathcal{N}$ defined by

$$\begin{aligned} P_\Psi(A) &= \mathbf{P}(I_n(X_1, \ldots, X_n) \in A) \\ &= \frac{1}{\nu_d(W)^n} \int_W \cdots \int_W \mathbf{1}(I_n(x_1, \ldots, x_n) \in A) dx_1 \ldots dx_n. \end{aligned}$$

Example 3.7 (Distribution of the finite Poisson process). Let Π be the Poisson process on $\mathbb{R}^d$ with totally finite intensity measure Λ. We know that $\Pi(\mathbb{R}^d) \sim \text{Pois}(\Lambda(\mathbb{R}^d))$ and that, given $\Pi(\mathbb{R}^d) = n$, the distribution of Π is that of a binomial process of n points i.i.d. with common distribution $Q(B) = \Lambda(B)/\Lambda(\mathbb{R}^d)$. Thus

$$P_{\Pi}(A) = \sum_{n=0}^{n} \mathbf{P}(\Pi(\mathbb{R}^d) = n)\mathbf{P}(I_n(X_1, \ldots, X_n) \in A)$$

$$= \sum_{n=0}^{\infty} \mathrm{e}^{-\Lambda(\mathbb{R}^d)} \frac{\Lambda(\Pi)^n}{n!} \int_{\mathbb{R}^d} \cdots \int_{\mathbb{R}^d} \mathbf{1}(I_n(x_1, \ldots, x_n) \in A) dQ(x_1) \ldots dQ(x_n)$$

$$= \mathrm{e}^{-\Lambda(\mathbb{R}^d)} \sum_{n=0}^{\infty} \frac{1}{n!} \int_{\mathbb{R}^d} \cdots \int_{\mathbb{R}^d} \mathbf{1}(I_n(x_1, \ldots, x_n) \in A) d\Lambda(x_1) \ldots d\Lambda(x_n).$$

The term for $n = 0$ in the sum should be interpreted as $\mathbf{1}(\mathbf{0} \in A)$ (where $\mathbf{0}$ is the zero measure, corresponding to an empty configuration).

Point Process Densities

Henceforth we fix a measure μ on S to serve as the reference measure. Typically μ is Lebesgue measure restricted to a bounded set W in $\mathbb{R}^d$. Let π_μ denote the distribution of the Poisson process with intensity measure μ.

Definition 3.16. Let $f : \mathcal{N}^f \to \mathbb{R}_+$ be a measurable function for which the equality $\int_{\mathcal{N}} f(\mathbf{x}) d\pi_\mu(\mathbf{x}) = 1$ holds. Define

$$P(A) = \int_A f(\mathbf{x}) d\pi_\mu(\mathbf{x}).$$

for any event $A \in \mathfrak{N}$. Then P is a point process distribution. The function f is said to be the *probability density* of the point process with distribution $\mathbf{P}$.

For a point process Ψ with probability density f we have

$$\mathbf{P}(\Psi \in A) = \mathrm{e}^{-\mu(\mathsf{S})} \sum_{n=0}^{\infty} \frac{1}{n!} \int_{\mathsf{S}} \cdots \int_{\mathsf{S}} \mathbf{1}(I_n(x_1, \ldots, x_n) \in A)$$
$$f(I_n(x_1, \ldots, x_n)) d\mu(x_1) \ldots d\mu(x_n)$$

for any event $A \in \mathfrak{N}$, and

$$\mathbf{E}g(\Xi) = \mathrm{e}^{-\mu(\mathsf{S})} \sum_{n=0}^{\infty} \frac{1}{n!} \int_{\mathsf{S}} \cdots \int_{\mathsf{S}} g(I_n(x_1, \ldots, x_n))$$
$$f(I_n(x_1, \ldots, x_n)) d\mu(x_1) \ldots d\mu(x_n)$$

for any integrable function $g : \mathcal{N} \to \mathbb{R}_+$. We can also rewrite these identities as

$$\mathbf{P}(\Psi \in A) = \mathbf{E}(f(\Pi) 1_A(\Pi)), \quad \mathbf{E}g(\Psi) = \mathbf{E}(g(\Pi) f(\Pi))$$

where Π is the Poisson process with intensity measure μ.

For some elementary point processes, it is possible to determine the probability density directly.

Example 3.8 (Density of the uniform Poisson process). Let S be a bounded set $W \subset \mathbb{R}^d$ and take the reference measure μ to be Lebesgue measure. Let $\beta > 0$. Set

$$f(\mathbf{x}) = \alpha\, \beta^{|\mathbf{x}|}$$

where α is a normalizing constant and $|\mathbf{x}| =$ number of points in $\mathbf{x}$. Then for any event A

$$\mathbf{P}(A) = \alpha \mathrm{e}^{-\nu_d(W)} \sum_{n=0}^{\infty} \frac{1}{n!} \int_W \cdots \int_W \mathbf{1}(I_n(x_1,\ldots,x_n) \in A)\beta^n dx_1 \ldots dx_n.$$

But this is the distribution of the Poisson process with intensity β. The normalizing constant must be $\alpha = \mathrm{e}^{(1-\beta)\nu_d(W)}$. Thus, the uniform Poisson process with intensity β has probability density

$$f(\mathbf{x}) = \beta^{|\mathbf{x}|} \mathrm{e}^{(1-\beta)\nu_d(W)}.$$

Exercise 3.10. Find the probability density of the binomial process (Example 3.1) consisting of n i.i.d. random points uniformly distributed in W.

Example 3.9 (Density of an inhomogeneous Poisson process). The finite Poisson point process in W with intensity function $\lambda(u)$, $u \in W$ has probability density

$$f(\mathbf{x}) = \alpha \prod_{i=1}^{|\mathbf{x}|} \lambda(x_i)$$

where

$$\alpha = \exp\left\{ \int_W (1 - \lambda(u)) du \right\}.$$

Exercise 3.11. Verify that (3.9) is indeed the probability density of the Poisson process with intensity function $\lambda(u)$.

Hard Core Process

Fix $r > 0$ and $W \subset \mathbb{R}^d$. Let

$$H_n = \{(x_1,\ldots,x_n) \in W^{!n} : \|x_i - x_j\| \geq r \text{ for all } i \neq j\}$$

and

$$H = \bigcup_{n=0}^{\infty} I_n(H_n).$$

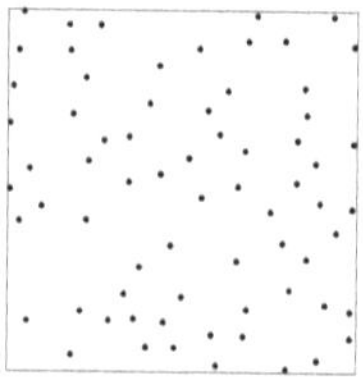

Fig. 3.20 Realization of the hard core process with $\beta = 200$ and $r = 0.07$ in the unit square

Thus H is the subset of $\mathcal{N}$ consisting of all point patterns $\mathbf{x}$ with the property that every pair of distinct points in $\mathbf{x}$ is at least r units apart.

Definition 3.17. Suppose H has positive probability under the unit rate Poisson process. Define the probability density

$$f(\mathbf{x}) = \alpha\beta^n \mathbf{1}(\mathbf{x} \in H)$$

where α is the normalizing constant and $\beta > 0$ is a parameter. A point process with this density is called a *hard core process*.

Lemma 3.4. *A hard core process is equivalent to a Poisson process of rate β conditioned on the event that there are no pairs of points closer than r units apart (Fig. 3.20).*

Proof. First suppose $\beta = 1$. Then the hard core process satisfies, for any event $A \in \mathfrak{N}$,

$$P(A) = \mathbf{E}(f(\Pi_1)\mathbf{1}(\Pi_1 \in A)) = \alpha\mathbf{E}(\mathbf{1}(\Pi_1 \in H)\mathbf{1}(\Pi_1 \in A)) = \alpha\mathbf{P}(\Pi_1 \in H \cap A).$$

It follows that $\alpha = 1/\mathbf{P}(\Pi_1 \in H)$ and hence

$$P(A) = \mathbf{P}(\Pi_1 \in A \mid \Pi_1 \in H),$$

that is, P is the conditional distribution of the unit rate Poisson process Π_1 given that $\Pi_1 \in H$. For general β the result follows by a similar argument. □

Conditional Intensity

Definition 3.18. Consider a finite point process Ψ in a compact set $W \subset \mathbb{R}^d$. The (Papangelou) *conditional intensity* $\beta^*(u, \Psi)$ of Ψ at locations $u \in W$, if it exists, is the stochastic process which satisfies

$$\mathbf{E}\sum_{x\in\Psi} g(x, \Psi \setminus x) = \int_W \mathbf{E}(\beta^*(u, \Psi)g(u, \Psi))du \tag{3.7}$$

for all measurable functions g such that either side exists.

Equation (3.7) is usually known as the *Georgii–Nguyen–Zessin formula* [183, 281, 383].

Suppose Ψ has probability density $f(\mathbf{x})$ (with respect to the uniform Poisson process Π_1 with intensity 1 on W). Then the expectation of any integrable function $h(\Psi)$ may be written explicitly as an integral over $\mathcal{N}^f$. Applying this to both sides of (3.7), we get

$$\mathbf{E}(f(\Pi_1)\sum_{x\in\Pi_1} g(x,\Pi_1\setminus x)) = \int_W \mathbf{E}(\beta^*(u,\Pi_1)f(\Pi_1)g(u,\Pi_1))du.$$

If we write

$$h(x,\Psi) = f(\Psi\cup\{x\})g(x,\Psi),$$

then

$$\mathbf{E}(f(\Pi_1)\sum_{x\in\Pi_1} g(x,\Pi_1\setminus x)) = \mathbf{E}(\sum_{x\in\Pi_1} h(x,\Pi_1\setminus x)) = \int_W \mathbf{E}(h(u,\Psi))du,$$

where the last expression follows since the conditional intensity of Π_1 is identically equal to 1 on W. Thus we get

$$\int_W \mathbf{E}(\beta^*(u,\Pi_1)f(\Pi_1)g(u,\Pi_1))du = \int_W \mathbf{E}(f(\Pi_1\cup\{u\})g(u,\Pi_1))du$$

for all integrable functions g. It follows that

$$\beta^*(u,\Pi_1)f(\Pi_1) = f(\Pi_1\cup u)$$

almost surely, for almost all $u\in W$. Thus we have obtained the following result.

Lemma 3.5. *Let f be the probability density of a finite point process Ψ in a bounded region W of $\mathbb{R}^d$. Assume that*

$$f(\mathbf{x}) > 0 \Longrightarrow f(\mathbf{y}) > 0 \textit{ for all } \mathbf{y}\subset\mathbf{x}.$$

Then the conditional intensity of Ψ exists and equals

$$\beta^*(u,\mathbf{x}) = \frac{f(\mathbf{x}\cup u)}{f(\mathbf{x})}$$

almost everywhere.

Example 3.10 (Conditional intensity of homogeneous Poisson process). The uniform Poisson process on W with intensity β has density

$$f(\mathbf{x}) = \alpha\beta^{|\mathbf{x}|}$$

where α is a certain normalizing constant. Applying Lemma 3.5 we get

$$\beta^*(u, \mathbf{x}) = \beta$$

for $u \in W$.

Example 3.11 (Conditional intensity of Hard Core process). The probability density of the hard core process

$$f(\mathbf{x}) = \alpha \beta^{|\mathbf{x}|} \mathbf{1}(\mathbf{x} \in H)$$

yields

$$\beta^*(u, \mathbf{x}) = \beta \mathbf{1}(\mathbf{x} \cup u \in H).$$

Lemma 3.6. *The probability density of a finite point process is completely determined by its conditional intensity.*

Proof. Invert the relationship, starting with the empty configuration $\emptyset$ and adding one point at a time:

$$\begin{aligned} f(\{x_1, \ldots, x_n\}) &= f(\emptyset) \frac{f(\{x_1\})}{f(\emptyset)} \frac{f(\{x_1, x_2\})}{f(\{x_1\})} \cdot \ldots \cdot \frac{f(\{x_1, \ldots, x_n\})}{f(\{x_1, \ldots, x_{n-1}\})} \\ &= f(\emptyset) \beta^*(x_1, \emptyset) \beta^*(x_2, \{x_1\}) \cdot \ldots \cdot \beta^*(x_n, \{x_1, \ldots, x_{n-1}\}). \end{aligned}$$

If the values of β^* are known, then this determines f up to a constant $f(\emptyset)$, which is then determined by the normalization of f. □

Lemma 3.7. *For a finite point process Ψ in W with conditional intensity $\beta^*(u, \mathbf{x})$, the intensity function is*

$$\lambda(u) = \mathbf{E}(\beta^*(u, \Psi)) \tag{3.8}$$

almost everywhere.

Proof. For $B \subset W$ take $g(u) = \mathbf{1}(u \in B)$ in formula (3.7) to get

$$\mathbf{E}\Psi(B) = \int_B \mathbf{E}(\beta^*(u, \Psi)) du.$$

But the left side is the integral of $\lambda(u)$ over B so the result follows. □

Exercise 3.12. For a hard core process Ψ (Definition 3.17) use Lemma 3.7 to prove that the mean number of points is related to the mean uncovered area:

$$\mathbf{E}|\Psi| = \beta \mathbf{E} A(\Psi)$$

where $A(\mathbf{x}) = \int_W \mathbf{1}(u \cup \mathbf{x} \in H)\, du$.

Modelling with Conditional Intensity

It is often convenient to formulate a point process model in terms of its conditional intensity $\beta^*(u, \mathbf{x})$, rather than its probability density $f(\mathbf{x})$.

The conditional intensity has a natural interpretation (in terms of conditional probability) which may be easier to understand than the density. Using the conditional intensity also eliminates the normalizing constant needed for the probability density.

However, we are not free to choose the functional form of $\beta^*(u, \mathbf{x})$ at will. It must satisfy certain consistency relations.

Finite Gibbs Models

Definition 3.19. A *finite Gibbs process* is a finite point process Ψ with probability density $f(\mathbf{x})$ of the form

$$f(\mathbf{x}) = \exp\{V_0 + \sum_{x \in \mathbf{x}} V_1(x) + \sum_{\{x,y\} \subset \mathbf{x}} V_2(x, y) + \ldots\} \tag{3.9}$$

where $V_k : \mathcal{N}_k \to \mathbb{R} \cup \{-\infty\}$ is called the *potential of order* k.

Gibbs models arise in statistical physics, where $\log f(\mathbf{x})$ may be interpreted as the *potential energy* of the configuration $\mathbf{x}$. The term $-V_1(u)$ can be interpreted as the energy required to create a single point at a location u. The term $-V_2(u, v)$ can be interpreted as the energy required to overcome a force between the points u and v.

Example 3.12 (Hard core process, Gibbs form). Given parameters $\beta, r > 0$, define $V_1(u) = \log \beta$,

$$V_2(u, v) = \begin{cases} 0 & \text{if } \|u - v\| > r \\ -\infty & \text{if } \|u - v\| \leq r \end{cases}$$

and $V_k \equiv 0$ for all $k \geq 3$. Then $\sum_{\{x,y\} \subset \mathbf{x}} V_2(x, y)$ is equal to zero if all pairs of points in $\mathbf{x}$ are at least r units apart, and otherwise this sum is equal to $-\infty$. Taking $\exp\{-\infty\} = 0$, we find

$$f(\mathbf{x}) = \alpha \beta^{|\mathbf{x}|} \mathbf{1}(\mathbf{x} \in H)$$

where H is the hard core constraint set, and $\alpha = \exp\{V_0\}$ is a normalizing constant. This is the probability density of the hard core process.

Lemma 3.8. *Let f be the probability density of a finite point process Ψ in a bounded region W in $\mathbb{R}^d$. Suppose that f is* hereditary, *i.e.*

$$f(\mathbf{x}) > 0 \implies f(\mathbf{y}) > 0 \textit{ for all } \mathbf{y} \subset \mathbf{x}.$$

Then f can be expressed in the Gibbs form (3.9).

Proof. This is a consequence of the Möbius inversion formula (the "inclusion–exclusion principle"), see Lemma 9.2. The functions V_k can be obtained explicitly as

$$\begin{aligned} V_0 &= \log f(\emptyset) \\ V_1(u) &= \log f(\{u\}) - \log f(\emptyset) \\ V_2(u,v) &= \log f(\{u,v\}) - \log f(\{u\}) - \log f(\{v\}) + \log f(\emptyset) \end{aligned}$$

and in general

$$V_k(\mathbf{x}) = \sum_{\mathbf{y}\subseteq\mathbf{x}} (-1)^{|\mathbf{x}|-|\mathbf{y}|} \log f(\mathbf{y}).$$

Then (3.9) can be verified by induction on $|\mathbf{x}|$. □

Exercise 3.13. Complete the proof.

Any process with hereditary density f also has a conditional intensity,

$$\beta^*(u,\mathbf{x}) = \exp\left\{ V_1(u) + \sum_{x\in\Psi} V_2(u,x) + \sum_{\{x,y\}\subset\Psi} V_3(u,x,y) + \dots \right\} \tag{3.10}$$

Hence, the following gives the most general form of a conditional intensity:

Theorem 3.3. *A function $\beta^*(u,\mathbf{x})$ is the conditional intensity of some finite point process Ψ iff it can be expressed in the form* (3.10).

Exercise 3.14. Prove Theorem 3.3.

Example 3.13 (Strauss process). For parameters $\beta > 0$, $0 \le \gamma \le 1$ and $r > 0$, suppose

$$\begin{aligned} V_1(u) &= \log \beta \\ V_2(u,v) &= (\log \gamma)\, \mathbf{1}(\|u-v\| \le r). \end{aligned}$$

This defines a finite point process called the *Strauss process* with conditional intensity

$$\beta^*(u,\mathbf{x}) = \beta\gamma^{t(u,\mathbf{x})}$$

and probability density

$$f(\mathbf{x}) = \alpha\, \beta^{|\mathbf{x}|} \gamma^{s(\mathbf{x})}$$

where

$$t(u,\mathbf{x}) = \sum_{x\in\mathbf{x}} \mathbf{1}(\|u-x\| \le r)$$

is the number of points of $\mathbf{x}$ which are close to u, and

$$s(\mathbf{x}) = \sum_{x,y\in\mathbf{x}} \mathbf{1}(\|x-y\| \le r)$$

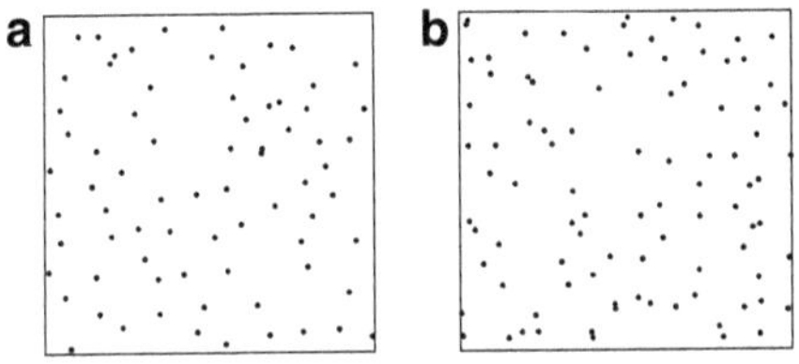

Fig. 3.21 Realizations of the Strauss process with interaction parameter $\gamma = 0.2$ (**a**) and $\gamma = 0.5$ (**b**) in the unit square, both having activity $\beta = 200$ and interaction range $r = 0.07$

is the number of pairs of close points in $\mathbf{x}$. The normalizing constant α is not available in closed form.

When $\gamma = 1$, the Strauss process reduces to the Poisson process with intensity β. When $\gamma = 0$, we have

$$\beta^*(u, \mathbf{x}) = \mathbf{1}(\|u - x\| > r \ \text{ for all } \ x \in \mathbf{x})$$

and

$$f(\mathbf{x}) = \alpha \beta^{|\mathbf{x}|} \mathbf{1}(\mathbf{x} \in H)$$

so we get the hard core process. For $0 < \gamma < 1$, the Strauss process has "soft inhibition" between neighbouring pairs of points (Fig. 3.21).

For $\gamma > 1$ the Strauss density is not integrable, so it does not give a well-defined point process.

The intensity function of the Strauss process is, applying equation (3.8),

$$\beta(u) = \mathbf{E}(\beta^*(u, \Psi)) = \mathbf{E}(\beta\gamma^{\,t(u,\Psi)}) \le \beta$$

It is not easy to evaluate $\beta(u)$ explicitly as a function of β, γ, r.

Exercise 3.15. Verify that the Strauss density is not integrable when $\gamma > 1$.

Pairwise Interaction Processes

More generally we could consider a *pairwise interaction* model of the form

$$f(\mathbf{x}) = \alpha \prod_{i=1}^{|\mathbf{x}|} b(x_i) \prod_{i<j} c(x_i, x_j) \tag{3.11}$$

where $b(u), u \in W$ is the activity function and $c(u, v)$ is the pair interaction. For simplicity, take $c(u, v) = c(\|u - v\|)$ (Fig. 3.22). Pairwise interaction models are very common in statistical physics as models for particle systems.

Pairwise interaction processes usually exhibit "regularity" or "inhibition" between points. For example, the Strauss density is

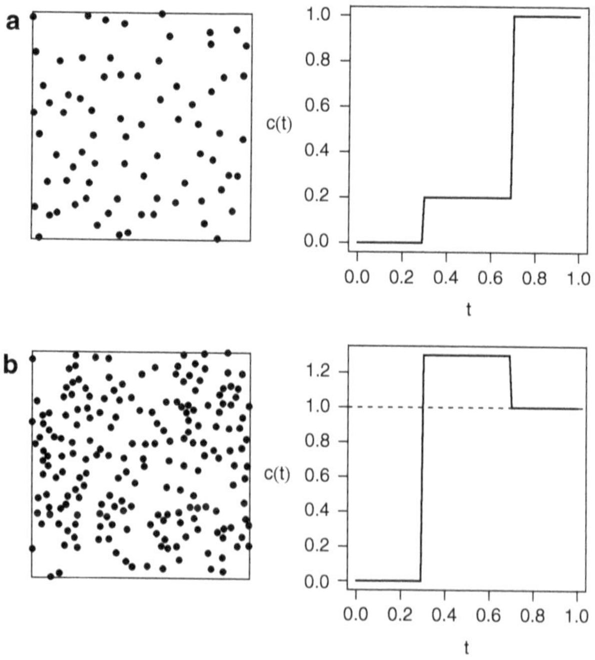

Fig. 3.22 Realizations of two pairwise interaction processes with the pair interaction function c shown at right

$$f(\mathbf{x}) = \alpha\beta^{|\mathbf{x}|}\gamma^{s(\mathbf{x})}$$

where

$$s(\mathbf{x}) = \sum_{i<j} \mathbf{1}(\|x_i - x_j\| < r)$$

is the number of r-close pairs in $\mathbf{x}$. We cannot allow $\gamma > 1$ since f is not integrable in that case. Thus the Strauss process is only a model for inhibition. Note that

$$s(\mathbf{x}) = \frac{1}{2}\sum_i t(x_i, \mathbf{x})$$

where

$$t(x_i, \mathbf{x}) = s(\mathbf{x}) - s(\mathbf{x} \setminus \{x_i\}) = \sum_{j \neq i} \mathbf{1}(\|x_j - x_i\| < r)$$

is the number of r-close neighbours of x_i. Thus the Strauss density can be rewritten

$$f(\mathbf{x}) = \alpha\beta^{|\mathbf{x}|}\prod_{i=1}^{|\mathbf{x}|}\gamma^{\frac{1}{2}t(x_i,\mathbf{x})}.$$

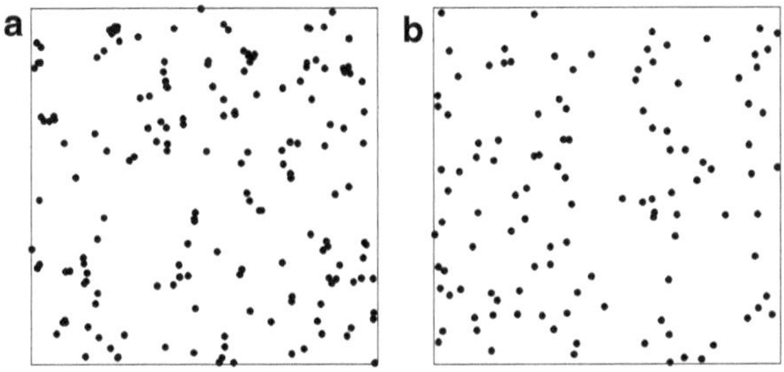

Fig. 3.23 Realizations of the Geyer saturation process with $\gamma = 1.6$ (**a**) and $\gamma = 0.625$ (**b**)

One way to obtain a "clustered" (positively associated) point process is to modify the expression above so that the contribution from each point x_i is bounded.

Definition 3.20 (Geyer saturation process). Define the *saturation process* [186] to have density

$$f(\mathbf{x}) = \alpha\beta^{|\mathbf{x}|}\prod_{i=1}^{|\mathbf{x}|}\gamma^{\min\{s;t(u,\mathbf{x})\}}$$

where α is the normalizing constant, $\beta > 0$ the activity parameter, $\gamma \geq 0$ the interaction parameter, and s the "saturation" parameter.

The Geyer saturation density is integrable for all values of γ. This density has infinite order of interaction. If $\gamma < 1$ the process is inhibited, while if $\gamma > 1$ it is clustered (Fig. 3.23).

Definition 3.21 (Area-interaction or Widom–Rowlinson model). This process [32, 516] has density

$$f(\mathbf{x}) = \alpha\beta^{|\mathbf{x}|}\exp\{\theta V(\mathbf{x})\}$$

where α is the normalizing constant and $V(\mathbf{x}) = \nu_d(U(\mathbf{x}))$ is the area or volume of

$$U(\mathbf{x}) = W \cap \bigcup_{i=1}^{|\mathbf{x}|} B_r(x_i).$$

Since $V(\mathbf{x}) \leq \nu_d(W)$, the density is integrable for all values of $\theta \in \mathbb{R}$. For $\theta = 0$ the process is Poisson. For $\theta < 0$ it is a regular (negatively associated) process, and for $\theta > 0$ a clustered (positively associated) process. The interpoint interactions in this process are very "mild" in the sense that its realizations look very similar to a Poisson process (Fig. 3.24).

By the inclusion–exclusion formula

$$V(\mathbf{x}) = \sum_i V(\{x_i\}) - \sum_{i<j} V(\{x_i, x_j\}) + \ldots + (-1)^n V(\mathbf{x})$$

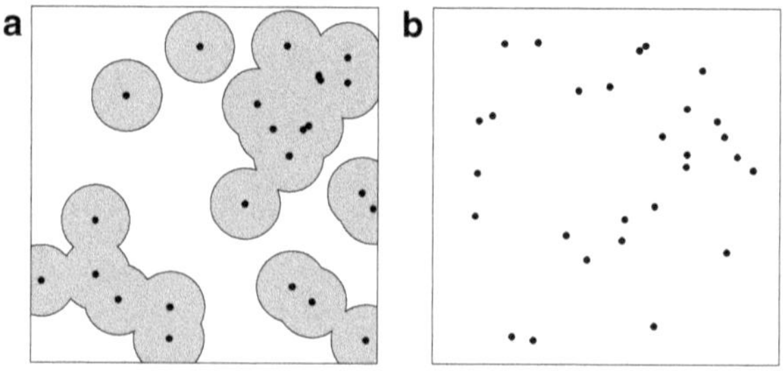

Fig. 3.24 Area-interaction process. (**a**) The dilation set $U(\mathbf{x})$. (**b**) Simulated realization

so that the area-interaction process has interaction potentials of all orders—it has infinite order.

3.2 Simulation

D.G. Kendall told his students that we only really understand a stochastic process when we know how to simulate it. Stochastic simulation also has many practical applications in probability, statistical inference and optimization.

In this section, we cover some basic simulation principles (Sect. 3.2.1), discuss methods for simulating a Poisson process (Sect. 3.2.2) and simulating Poisson-driven processes (Sect. 3.2.3), then discuss elementary Markov Chain Monte Carlo methods (Sect. 3.2.4).

3.2.1 Basic Simulation Principles

Assume we have a supply of independent, identically distributed (i.i.d.) random variables $\eta_1, \eta_2, \ldots$ which are uniformly distributed in [0, 1], written $\eta_i \sim \text{Unif}[0, 1]$.

In practice these would be supplied by a computer's random number generator (RNG). The RNG is a deterministic algorithm designed to imitate i.i.d. uniform random variables. The theory of RNG's will not be discussed here.

Our aim is to generate a random variable ξ (or stochastic process) with a desired probability distribution, using the variables η_i.

Three basic simulation principles are *transformation*, *rejection* and *marginalization*.

Transformation

If $\eta \sim \text{Unif}[0, 1]$ and we set $\xi = a + (b - a)\eta$ where $a < b$, it is intuitively clear that η is uniformly distributed in $[a, b]$.

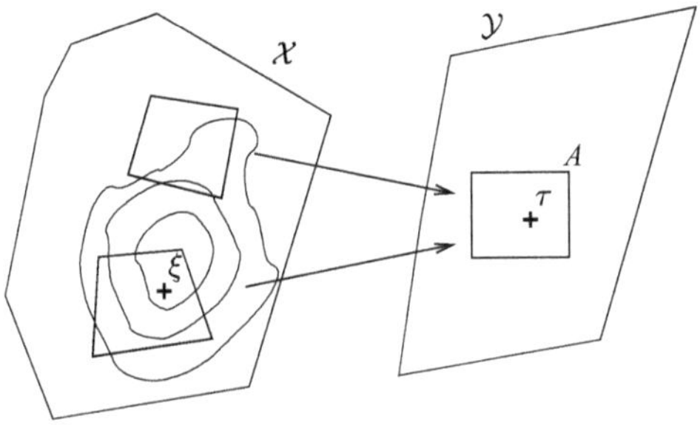

Fig. 3.25 Image of a distribution under a transformation

Exercise 3.16. Prove that $\xi \sim \text{Unif}[a, b]$.

Definition 3.22. Let ξ be a random element in some space $\mathcal{X}$, and $T : \mathcal{X} \to \mathcal{Y}$ a measurable mapping. Let $\tau = T(\xi)$. The distribution of τ is given by

$$\mathbf{P}(\tau \in A) = \mathbf{P}(T(\xi) \in A) = \mathbf{P}(\xi \in T^{-1}(A))$$

where $T^{-1}(A) = \{x \in \mathcal{X} : T(x) \in A\}$ for all measurable $A \subseteq \mathcal{Y}$ (Fig. 3.25).

Lemma 3.9 (Probability integral transformation). *Let ξ be a real random variable with cumulative distribution function (c.d.f.) $F(x) = \mathbf{P}(\xi \leq x)$. Define the right-continuous quantile function*

$$F^{-1}(u) = \min\{x : F(x) \geq u\}.$$

Then:

1. *Let η be uniformly distributed on* $[0, 1]$. *Then $\zeta = F^{-1}(\eta)$ has the same distribution as ξ.*
2. *If F^{-1} is* continuous, *then $\eta = F(\xi)$ is uniformly distributed on* $[0, 1]$.

Typically, property 1 is used to simulate random variables, while property 2 is used to test whether observed data conform to a specified model (Fig. 3.26).

Proof (in absolutely continuous case). Assume $F'(x) = f(x) > 0$ for all x. Then F is a strictly increasing, continuous function, and F^{-1} is its strictly increasing, continuous inverse function: $F(F^{-1}(u)) \equiv u$ and $F^{-1}(F(x)) \equiv x$.

1. Let $\zeta = F^{-1}(\eta)$. Then for $x \in \mathbb{R}$

$$\mathbf{P}(\zeta \leq x) = \mathbf{P}(F^{-1}(\eta) \leq x) = \mathbf{P}(F(F^{-1}(\eta)) \leq F(x)) = \mathbf{P}(\eta \leq F(x)) = F(x).$$

Thus, ζ has the same distribution as ξ.

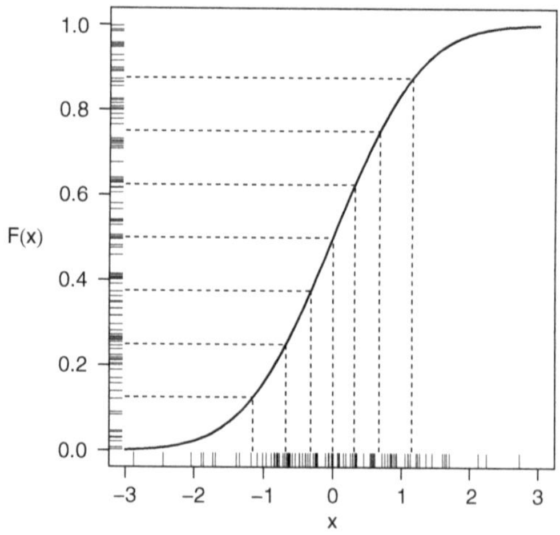

Fig. 3.26 The probability integral transformation for $N(0, 1)$

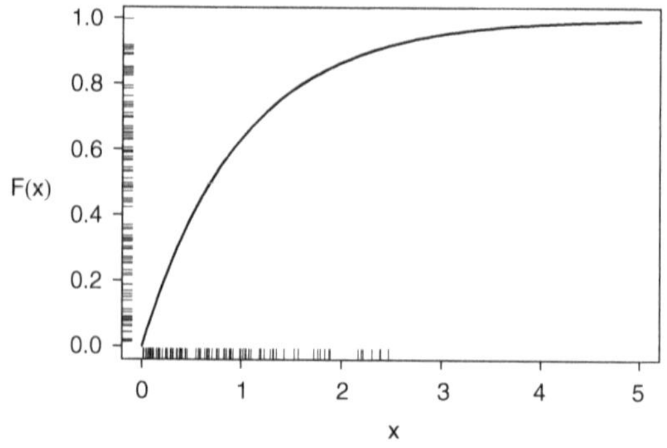

Fig. 3.27 Probability integral transformation for the exponential distribution

2. Let $\eta = F(\xi)$. Then for $v \in (0, 1)$

$$\begin{aligned}\mathbf{P}(\eta \leq v) &= \mathbf{P}(F(\xi) \leq v) = \mathbf{P}(F^{-1}(F(\xi)) \leq F^{-1}(v)) \\ &= \mathbf{P}(\xi \leq F^{-1}(v)) = F(F^{-1}(v)) = v.\end{aligned}$$

Thus, η is uniformly distributed. □

Example 3.14 (Exponential distribution). Let ξ have density $f(x) = \lambda \exp\{-\lambda x\}$ for $x > 0$, and zero otherwise, where $\lambda > 0$ is the parameter. Then $F(x) = 1 - \exp\{-\lambda x\}$ for $x > 0$, and zero otherwise. Hence $F^{-1}(u) = -\log(1 - u)/\lambda$ (Fig. 3.27). If $\eta \sim \text{Unif}[0, 1]$ then $\zeta = -\log(1 - \eta)/\lambda$ has the same distribution as ξ. Since $1 - \eta \sim \text{Unif}[0, 1]$ we could also take $\zeta = -\log(\eta)/\lambda$.

Example 3.15 (Biased coin flip). Suppose we want ξ to take the values 1 and 0 with probabilities p and $1 - p$ respectively. Then

$$F(x) = \begin{cases} 0 & \text{if } x < 0 \\ 1 - p & \text{if } 0 \le x < 1 \\ 1 & \text{if } x \ge 1. \end{cases}$$

The inverse is

$$F^{-1}(u) = \begin{cases} 0 & \text{if } u < 1 - p \\ 1 & \text{if } u \ge 1 - p \end{cases}$$

for $0 < u < 1$. If $\eta \sim \text{Unif}[0, 1]$ then $\zeta = F^{-1}(\eta) = \mathbf{1}(\eta \ge 1 - p) = \mathbf{1}(1 - \eta \le p)$ has the same distribution as ξ. Since $1 - \eta \sim \text{Unif}[0, 1]$ we could also take $\zeta = \mathbf{1}(\eta \le p)$.

Example 3.16 (Poisson random variable). To generate a realization of $N \sim \text{Pois}(\lambda)$, first generate $\eta \sim \text{Unif}[0, 1]$, then find

$$N = \min\{n : \eta \le \sum_{k=0}^{n} \mathrm{e}^{-\lambda} \frac{\lambda^k}{k!}\}$$

Lemma 3.10 (Change-of-variables). *Let $\boldsymbol{\xi}$ be a random element of $\mathbb{R}^d$, $d \ge 1$ with probability density function $f(\mathbf{x})$, $\mathbf{x} \in \mathbb{R}^d$. Let $T : \mathbb{R}^d \to \mathbb{R}^d$ be a differentiable transformation such that, at any $\mathbf{x} \in \mathbb{R}^d$, the derivative $DT(\mathbf{x})$ is nonsingular. Then the random vector $\boldsymbol{\zeta} = T(\boldsymbol{\xi})$ has probability density*

$$g(\mathbf{y}) = \sum_{\mathbf{x} \in T^{-1}(\mathbf{y})} \frac{f(\mathbf{x})}{\det DT(\mathbf{x})}.$$

Example 3.17 (Box–Muller device). Let $\boldsymbol{\xi} = (\xi_1, \xi_2)^{\top}$ where ξ_1, ξ_2 are i.i.d. normal $N(0, 1)$, with joint density

$$f(x_1, x_2) = \frac{1}{2\pi} \exp\left\{-\frac{1}{2}(x_1^2 + x_2^2)\right\}.$$

Since the density is invariant under rotation, consider the polar transformation $T(x_1, x_2) = (x_1^2 + x_2^2, \arctan(x_2/x_1))$, which is one-to-one and has the Jacobian $\det DT(\mathbf{x}) \equiv 2$. The transformed variables $\tau = \xi_1^2 + \xi_2^2$ and $\theta = \arctan(\xi_2/\xi_1)$ have joint density

$$g(t, y) = \frac{f(T^{-1}(t, y))}{2} = \frac{1}{2} f(\sqrt{t} \cos y, \sqrt{t} \sin y) = \frac{1}{2\pi} \frac{1}{2} \exp\{-t/2\}$$

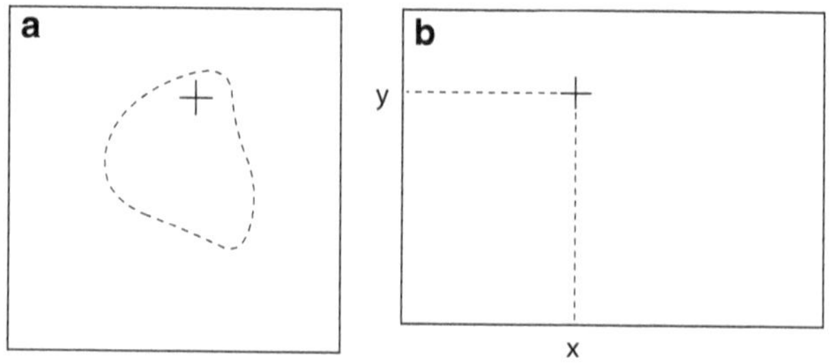

Fig. 3.28 (**a**) Uniformly random point. (**b**) Generating a uniformly random point in a box

Thus τ and θ are independent; τ has an exponential distribution with parameter $1/2$, and θ is uniform on $[0, 2\pi)$. Applying the inverse transformation, when η_1, η_2 are i.i.d. uniform $[0, 1]$

$$\xi = \cos(2\pi\eta_1)\sqrt{-2\log(\eta_2)}$$

has a standard normal $N(0, 1)$ distribution (*Box–Muller device* [82]).

Uniform Random Points

Definition 3.23. Let $A \subset \mathbb{R}^d$ be a measurable set with volume $0 < \nu_d(A) < \infty$. A *uniformly random (UR)* point in A is a random point $\mathbf{X} \in \mathbb{R}^d$ with probability density

$$f(\mathbf{x}) = \frac{\mathbf{1}(\mathbf{x} \in A)}{\nu_d(A)}.$$

Equivalently, for any measurable $B \subset \mathbb{R}^d$

$$\mathbf{P}(\mathbf{X} \in B) = \int_B f(\mathbf{x})d\mathbf{x} = \frac{\nu_d(A \cap B)}{\nu_d(A)}.$$

Example 3.18 (Uniform random point in a box). If $X_1, \ldots, X_d$ are independent random variables such that $X_i \sim \text{Unif}[a_i, b_i]$, then the random point

$$\mathbf{X} = (X_1, \ldots, X_d)$$

is a uniformly random point in the parallelepiped (Fig. 3.28)

$$A = \prod_{i=1}^{d} [a_i, b_i].$$

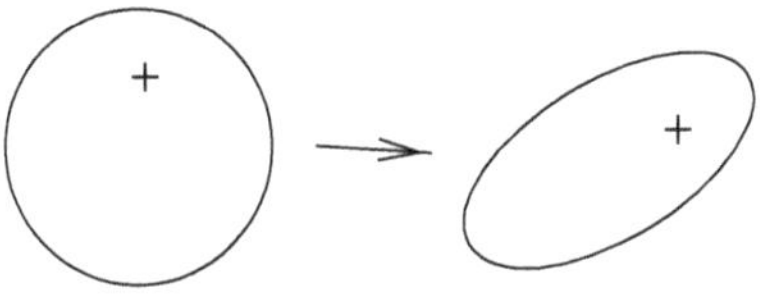

Fig. 3.29 Affine maps preserve the uniform distribution

Example 3.19 (Uniform random point in disc). Let $\mathbf{X}$ be a uniformly random point in the disc $B_r(o)$ of radius r centred at the origin in $\mathbb{R}^2$. Consider the polar coordinates

$$R = \sqrt{X_1^2 + X_2^2}, \quad \theta = \arctan(X_2/X_1).$$

By elementary geometry, R^2 and Θ are independent and uniformly distributed on $[0, r^2]$ and $[0, 2\pi]$ respectively. Thus, if η_1, η_2 are i.i.d. Unif[0, 1] and we set

$$X_1 = r\sqrt{\eta_1}\cos(2\pi\eta_2) \quad X_2 = r\sqrt{\eta_1}\sin(2\pi\eta_2)$$

then $\mathbf{X} = (X_1, X_2)^\top$ is a uniformly random point in $B_r(o)$.

Lemma 3.11 (Uniformity under affine maps). *Let $T : \mathbb{R}^d \to \mathbb{R}^d$ be a linear transformation with nonzero determinant* $\det(T) = \delta$. *Then $\nu_d(T(B)) = \delta\nu_d(B)$ for all compact $B \subset \mathbb{R}^d$. If $\mathbf{X}$ is a uniformly random point in $A \subset \mathbb{R}^d$, then $T(\mathbf{X})$ is a uniformly random point in $T(A)$ (Fig. 3.29).*

Exercise 3.17. Write an algorithm to generate uniformly random points inside an ellipse in $\mathbb{R}^2$.

Uniformity in Non-Euclidean Spaces

The following is how *not* to choose your next holiday destination:

1. Choose a random longitude $\theta \sim$ Unif[−180, 180]
2. Independently choose a random latitude $\varphi \sim$ Unif[−90, 90]

When the results are plotted on the globe (Fig. 3.30), they show a clear preference for locations near the poles.

This procedure is equivalent to projecting the globe onto a flat map in which the latitude lines are equally spaced (Fig. 3.31) and selecting points uniformly at random on the flat atlas. There is a higher probability of selecting a destination in Greenland than in Australia, although Australia is five times larger than Greenland.

This paradox arises because, in a general space $\mathcal{S}$, the probability density of a random element $\mathbf{X}$ must always be defined relative to an agreed reference measure μ, through

$$\mathbf{P}(\mathbf{X} \in A) = \int_A f(\mathbf{x})d\mu(\mathbf{x}).$$

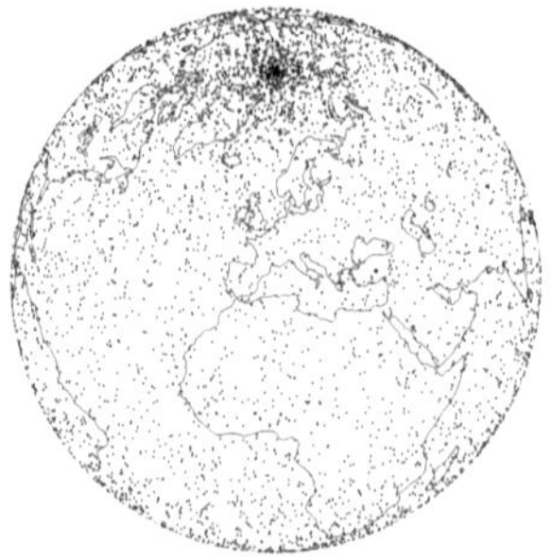

Fig. 3.30 How not to choose the next holiday destination

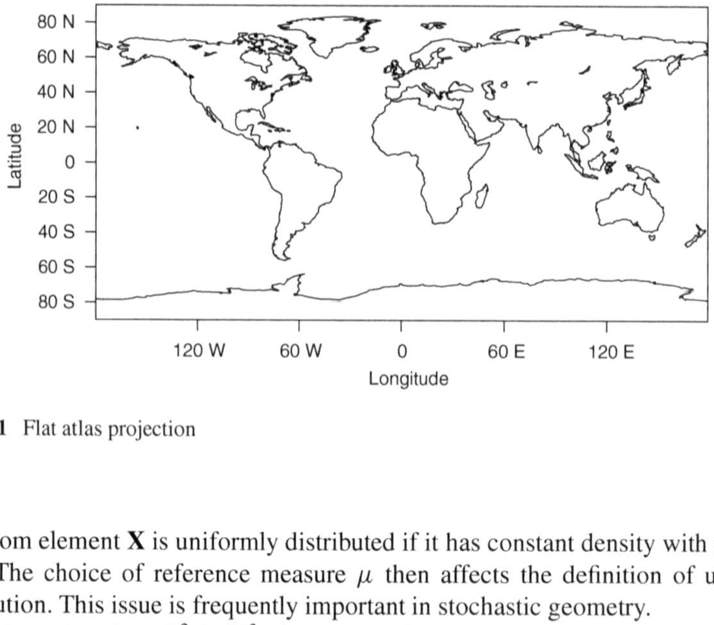

Fig. 3.31 Flat atlas projection

A random element **X** is uniformly distributed if it has constant density with respect to μ. The choice of reference measure μ then affects the definition of uniform distribution. This issue is frequently important in stochastic geometry.

On the unit sphere $\mathbb{S}^2$ in $\mathbb{R}^3$, the usual reference measure μ is spherical area. For the (longitude, latitude) coordinate system $T : [-\pi, \pi) \times [-\pi/2, \pi/2] \to \mathbb{S}^2$ defined by

$$T(\theta, \varphi) = (\cos\theta\cos\varphi, \sin\theta\cos\varphi, \sin\varphi)$$

we have

$$\int_A h(\mathbf{x}) d\mu(\mathbf{x}) = \int_{T^{-1}(A)} h(T(\theta, \varphi)) \cos\varphi d\theta d\varphi.$$

or in terms of differential elements, "$d\mu = \cos\varphi d\theta d\varphi$". Hence the following algorithm generates uniform random points on the globe in the usual sense.

Algorithm 3.1 (Uniform random point on a sphere). To generate a uniform random point on the earth (Fig. 3.32),

Fig. 3.32 Uniformly random points on the earth

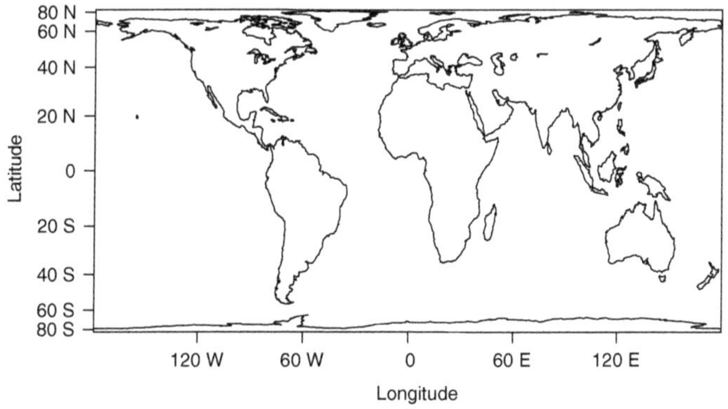

Fig. 3.33 Equal-area (cylindrical) projection

1. Choose a random longitude $\theta \sim \text{Unif}[-180, 180]$
2. Independently choose a random latitude φ with probability density proportional to $|\cos(\varphi)|$

To achieve the second step we can take $\varphi = \arcsin(\eta)$ where $\eta \sim \text{Unif}[-1, 1]$.

This procedure is equivalent to projecting the globe using an equal-area projection (Fig. 3.33) and selecting a uniformly random point in the projected atlas.

Exercise 3.18. Let $\xi = \arcsin(\eta)$ where $\eta \sim \text{Unif}[-1, 1]$. Prove that ξ has probability density proportional to $\cos(x)$ on $(-\pi/2, \pi/2)$.

Rejection

Algorithm 3.2 (Rejection). Suppose we wish to generate a realization of a random variable $\mathbf{X}$ (in some space) conditional on $\mathbf{X} \in A$, where A is a subset of the possible outcomes of $\mathbf{X}$. Assume $\mathbf{P}(\mathbf{X} \in A) > 0$.

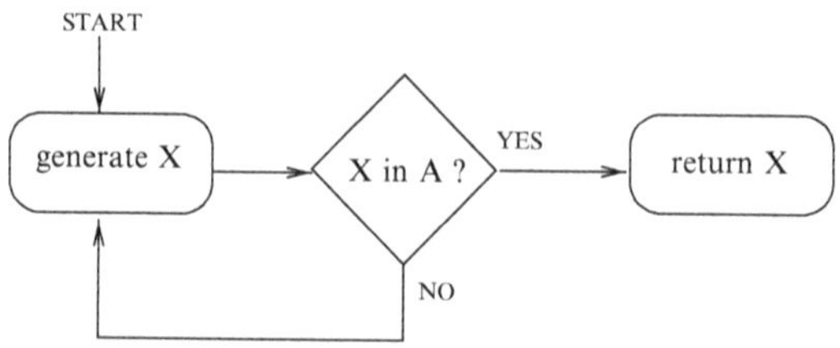

Fig. 3.34 Flowchart for the rejection algorithm

1. Generate a realization X of $\mathbf{X}$.
2. If $X \in A$, terminate and return X.
3. Otherwise go to step 1.

To understand the validity of the rejection algorithm (Fig. 3.34), let $\mathbf{X}_1, \mathbf{X}_2, \ldots$ be i.i.d. with the same distribution as $\mathbf{X}$. The events $B_n = \{\mathbf{X}_n \in A\}$ are independent and have probability $p = \mathbf{P}(\mathbf{X} \in A) > 0$. Hence the algorithm termination time

$$N = \min\{n : \mathbf{X}_n \in A\}$$

has a geometric (p) distribution $\mathbf{P}(N = n) = (1 - p)^{n-1} p$. The algorithm output $\mathbf{X}_N$ is well defined and has distribution

$$\begin{aligned}
\mathbf{P}(\mathbf{X}_N \in C) &= \sum_n \mathbf{P}(\mathbf{X}_N \in C \mid N = n)\mathbf{P}(N = n) \\
&= \sum_n \mathbf{P}(\mathbf{X}_n \in C \mid N = n)\mathbf{P}(N = n) \\
&= \sum_n \mathbf{P}(\mathbf{X}_n \in C \mid \mathbf{X}_n \in A;\ X_i \notin A, i < n)\mathbf{P}(N = n) \\
&= \sum_n \mathbf{P}(\mathbf{X}_n \in C \mid \mathbf{X}_n \in A)\mathbf{P}(N = n) \quad \text{by independence} \\
&= \sum_n \mathbf{P}(\mathbf{X} \in C \mid \mathbf{X} \in A)\mathbf{P}(N = n) \\
&= \mathbf{P}(\mathbf{X} \in C \mid \mathbf{X} \in A),
\end{aligned}$$

the desired conditional distribution.

Example 3.20 (Uniform random point in any region). To generate a random point $\mathbf{X}$ uniformly distributed inside an irregular region $B \subset \mathbb{R}^d$,

1. Enclose B in a simpler set $C \supset B$.
2. Using the rejection method, generate i.i.d. random points uniformly in C until a point falls in B.

See Fig. 3.35.

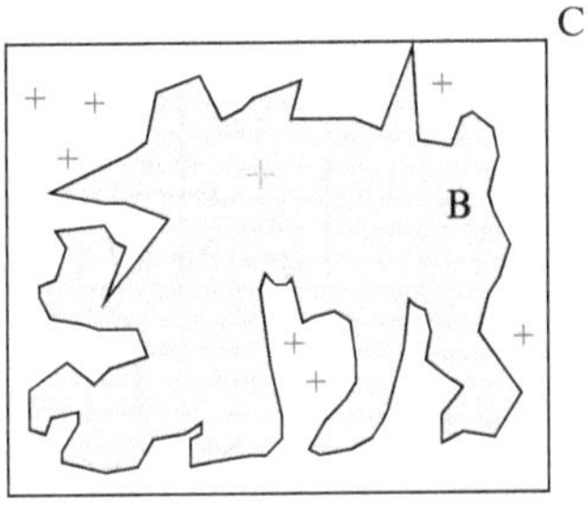

Fig. 3.35 Rejection method for generating uniformly random points in an irregular region B

Lemma 3.12 (Conditional property of uniform distribution). *Suppose* $\mathbf{X}$ *is uniformly distributed in* $A \subset \mathbb{R}^d$ *with* $\nu_d(A) < \infty$. *Let* $B \subset A$. *The conditional distribution of* $\mathbf{X}$ *given* $\mathbf{X} \in B$ *is uniform in* B.

Proof. For any measurable $C \subset \mathbb{R}^d$

$$\mathbf{P}(\mathbf{X} \in C \mid \mathbf{X} \in B) = \frac{\mathbf{P}(X \in C \cap B)}{\mathbf{P}(X \in B)} = \frac{\nu_d(C \cap B \cap A)/\nu_d(A)}{\nu_d(B \cap A)/\nu_d(A)} = \frac{\nu_d(C \cap B)}{\nu_d(B)}.$$

The proof is complete. □

In summary, the rejection algorithm is simple, adaptable and generic, but may be slow. The transformation technique is fast, and has a fixed computation time, but may be complicated to implement, and is specific to one model.

Marginalization

Let ξ be a real random variable with probability density f. Suppose $f(x) \leq M$ for all x. Consider the subgraph

$$A = \{(x, y) : 0 \leq y \leq f(x)\}.$$

Let $(X_1, X_2)^\top$ be a uniformly random point in A. The joint density of $(X_1, X_2)^\top$ is $g(x_1, x_2) = \mathbf{1}(0 \leq x_2 \leq f(x_1))$ since A has unit area. The marginal density of X_1 is

$$h(x_1) = \int_0^M \mathbf{1}(x_2 \leq f(x_1))dx_2 = f(x_1),$$

that is, X_1 has probability density f (Fig. 3.36).

Algorithm 3.3 (Marginalization). Suppose f is a probability density on $[a, b]$ with $\sup_{x \in [a,b]} f(x) < M$.

1. Generate $\xi \sim \text{Unif}[a, b]$.
2. Independently generate $\eta \sim [0, M]$.
3. If $\eta < f(\xi)/M$, terminate and return the value ξ. Otherwise, go to step 1.

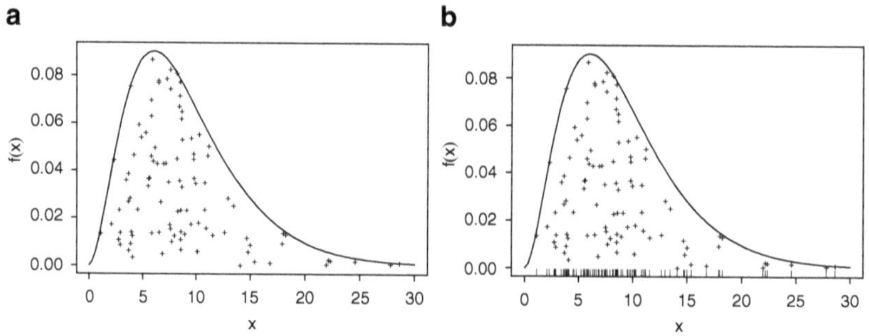

Fig. 3.36 Marginalization principle. (**a**) Uniformly random points in the subgraph of a density f. (**b**) Projections onto the x-axis have density f

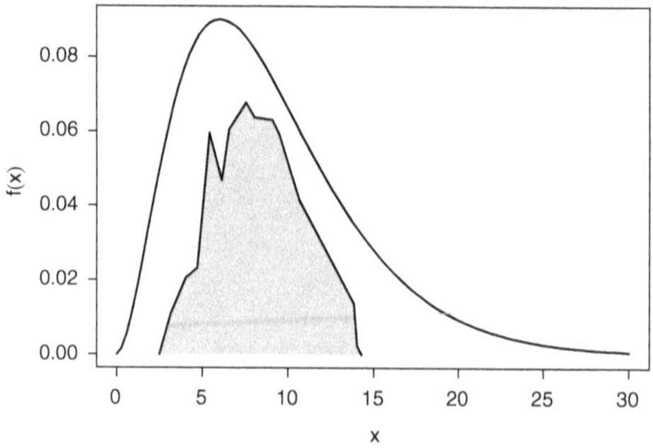

Fig. 3.37 Importance sampling

It is easy to check (following previous proofs) that this algorithm terminates in finite time and returns a random variable with density f.

Importance Sampling

We wish to generate $\mathbf{X}$ with a complicated probability density f. Let g be another probability density, that is easier to simulate, such that $f(\mathbf{x}) \leq Mg(\mathbf{x})$ for all $\mathbf{x}$ (Fig. 3.37).

Algorithm 3.4 (Importance sampling). Let f and g be probability densities and $M < \infty$ such that $f(\mathbf{x}) \leq M g(\mathbf{x})$ for all $\mathbf{x}$.

1. Generate $\mathbf{Y}$ with density g.
2. Independently generate $\eta \sim \text{Unif}[0, M]$.
3. If $\eta \leq f(\mathbf{Y})/g(\mathbf{Y})$, set $\mathbf{X} = \mathbf{Y}$ and exit. Otherwise, go to step 1.

Given $\mathbf{Y}$ let $\xi = \eta g(\mathbf{Y})$. The pair $(\mathbf{Y}, \xi)$ is uniformly distributed in the subgraph of Mg. This point falls in the subgraph of f if $\xi \leq f(\mathbf{Y})$, equivalently if $\eta \leq f(\mathbf{Y})/g(\mathbf{Y})$. This is the rejection algorithm.

3.2.2 *Simulating the Poisson Process*

To simulate the homogeneous Poisson process in one dimension, we may use any of the properties described in Sect. 3.1.

Algorithm 3.5 (Poisson simulation using waiting times). To simulate a realization of a homogeneous Poisson process in $[0, a]$:

1. Generate i.i.d. waiting times $S_1, S_2, \ldots$ from the Exp(λ)-distribution.
2. Compute arrival times $T_n = \sum_{i=1}^{n} S_i$.
3. Return the set of arrival times T_n that satisfy $T_n < a$.

The homogeneous Poisson process in $[0, \infty)$ has *conditionally uniform arrivals*: given $N_t = n$, the arrival times in $[0, t]$

$$T_1 < T_2 < \ldots < T_n < t$$

are (the order statistics of) n i.i.d. uniform random variables in $[0, t]$.

Algorithm 3.6 (Poisson simulation using conditional property). To simulate a realization of a homogeneous Poisson process in $[0, a]$:

1. Generate $N \sim \text{Pois}(\lambda a)$.
2. Given $N = n$, generate n i.i.d. variables $\eta_1, \ldots, \eta_n \sim \text{Unif}[0, a]$.
3. Sort $\{\eta_1, \ldots, \eta_n\}$ to obtain the arrival times.

For the homogeneous Poisson process on $\mathbb{R}^d$, given $\Pi_\lambda(B) = n$, the restriction of Π_λ to B has the same distribution as a binomial process (n i.i.d. random points uniformly distributed in B).

These properties can be used directly to simulate the Poisson process in $\mathbb{R}^d$ with constant intensity $\lambda > 0$ (Fig. 3.38).

Algorithm 3.7 (Poisson simulation in $\mathbb{R}^d$ using conditional property). To simulate a realization of a homogeneous Poisson process in $\mathbb{R}^d$:

1. Divide $\mathbb{R}^d$ into unit hypercubes $Q_k, k = 1, 2, \ldots$
2. Generate i.i.d. random variables $N_k \sim \text{Pois}(\lambda)$.
3. Given $N_k = n_k$, generate n_k i.i.d. random points uniformly distributed in Q_k.

To appreciate the validity of the latter algorithm, see the explicit construction of Poisson processes in Sect. 4.1.1.

Remark 3.1. To generate a realization of a *Poisson process* Π_λ with constant intensity λ inside an irregular region $B \subset \mathbb{R}^d$, it is easiest to generate a Poisson

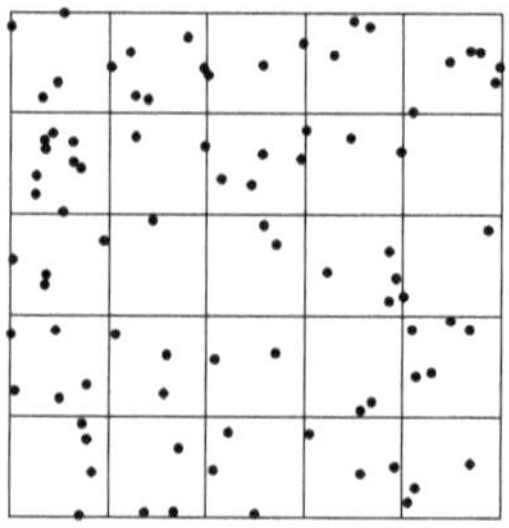

Fig. 3.38 Direct simulation of Poisson point process

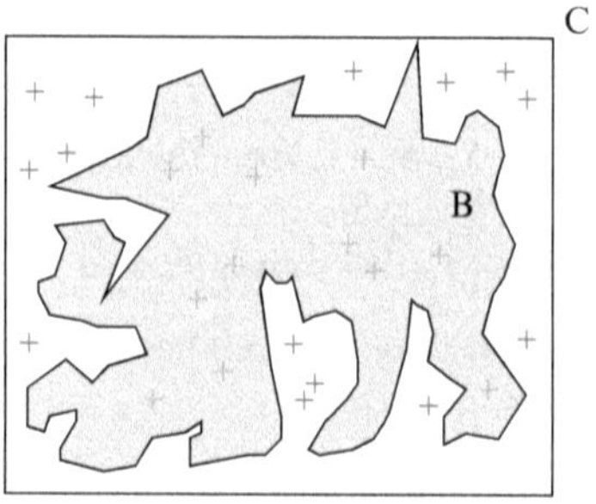

Fig. 3.39 Rejection method for simulating Poisson points in an arbitrary domain B

process Ψ with intensity λ in a simpler set $C \supset B$, and take $\Pi_\lambda = \Psi \cap B$, i.e. retain only the points of Ψ that fall in B. Thus it is computationally easier to generate a Poisson random number of random points, than to generate a single random point (Fig. 3.39).

Transformations of Poisson Process

Lemma 3.13. *Let Π be a Poisson point process in $\mathbb{R}^d$, $d \geq 1$ with intensity function $\lambda(u)$, $u \in \mathbb{R}^d$. Let $T : \mathbb{R}^d \to \mathbb{R}^d$ be a differentiable transformation with nonsingular derivative. Then $\Psi = T(\Pi)$ is a* Poisson process *with intensity*

$$\nu(\mathbf{y}) = \sum_{\mathbf{x} \in T^{-1}(\mathbf{y})} \frac{\lambda(\mathbf{x})}{|\det DT(\mathbf{x})|}.$$

Here DT denotes the differential of T, and $|\det DT|$ is the Jacobian (Fig. 3.40).

Example 3.21 (Polar coordinates). Let Π_λ be a homogeneous Poisson process in $\mathbb{R}^2$ with intensity λ. The transformation

$$T(x_1, x_2) = (x_1^2 + x_2^2, \arctan(x_2/x_1))$$

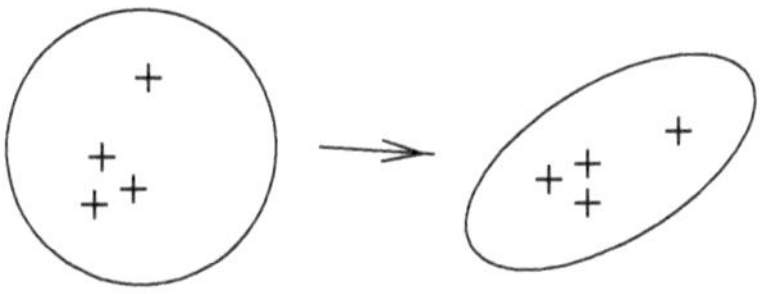

Fig. 3.40 Affine transformations preserve the homogeneous Poisson process up to a constant factor

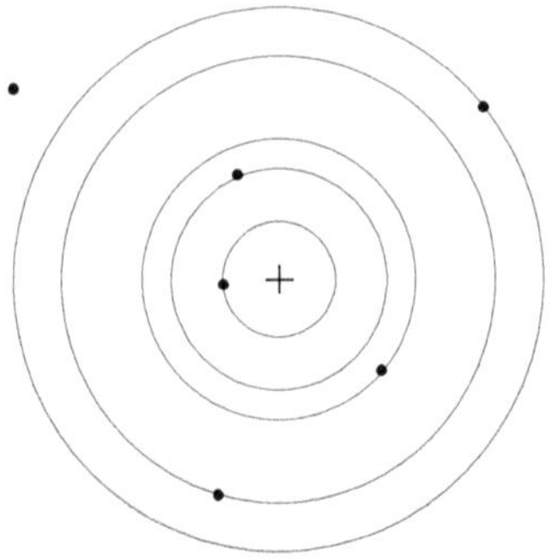

Fig. 3.41 Polar simulation of a Poisson process

has Jacobian 2. So $\Psi = T(\Pi_\lambda)$ is a homogeneous Poisson process in $(0, \infty) \times [0, 2\pi)$ with intensity $\lambda/2$. Projecting onto the first coordinate gives a homogeneous Poisson process with intensity $\lambda/2 \times 2\pi = \lambda\pi$.

Algorithm 3.8 (Polar simulation of a Poisson process). We may simulate a Poisson process by taking the points $X_n = (\sqrt{T_n} \cos \eta_n, \sqrt{T_n} \sin \eta_n)$ where $T_1, T_2, \ldots$ are the arrival times of a homogeneous Poisson process in $(0, \infty)$ with intensity $\lambda\pi$, and $\eta_1, \eta_2, \ldots$ are i.i.d. Unif$[0, 2\pi)$ (Fig. 3.41).

Example 3.22 (Transformation to uniformity). In one dimension, let Π be a Poisson point process in $[0, a]$ with rate (intensity) function $\lambda(u)$, $0 \leq u \leq a$. Let

$$T(v) = \int_0^v \lambda(u) du.$$

Then $T(\Pi)$ is a *Poisson process with rate 1* on $[0, T(a)]$. Conversely if Ψ is a unit rate Poisson process on $[0, T(a)]$ then $T^{-1}(\Psi)$ is a Poisson process with intensity $\lambda(u)$ on $[0, a]$, where $T^{-1}(t) = \min\{v : T(v) \geq t\}$. This is often used for checking goodness-of-fit.

Thinning

Let Π_λ be a homogeneous Poisson point process in $\mathbb{R}^d$ with intensity λ. Suppose we randomly delete or retain each point of Π_λ, with retention probability p for each

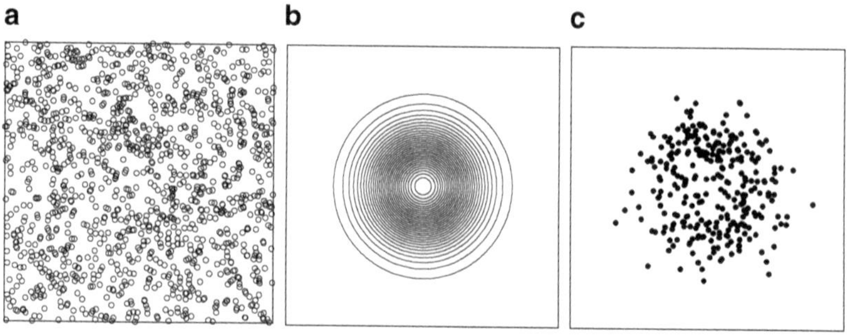

Fig. 3.42 Lewis–Shedler simulation of inhomogeneous Poisson process. (**a**) Uniform Poisson points. (**b**) Contours of desired intensity. (**c**) thinned points constituting the inhomogeneous Poisson process

point, independently of other points. The process Ψ of retained points is a Poisson process with intensity measure $p\lambda$.

To define the thinning procedure, we construct the marked point process Ψ obtained by attaching to each point x_i of Π_λ a random mark $\eta_i \sim \text{Unif}[0, 1]$ independently of other points and other marks. A point x_i is then retained if $\eta_i < p$ and thinned otherwise. When Π_λ is a homogeneous Poisson process of intensity λ in $\mathbb{R}^2$, the marked process Ψ is homogeneous Poisson with intensity λ in $\mathbb{R}^2 \times [0, 1]$. Consider the restriction $\Psi \cap A$ of Ψ to the set $A = \mathbb{R}^2 \times [0, p]$. Project $\Psi \cap A$ onto $\mathbb{R}^2$. This is the process of retained points. It is Poisson with intensity $p\lambda$.

Example 3.23 (Independent thinning). Let Π be a Poisson point process in $\mathbb{R}^d$ with intensity function $\lambda(u)$, $u \in \mathbb{R}^d$. Suppose we randomly delete or retain each point of Π, a point x_i being retained with probability $p(x_i)$ independently of other points, where p is a measurable function. The process Ψ of retained points is a Poisson process with intensity function $\kappa(u) = p(u)\lambda(u)$.

Exercise 3.19. In Example 3.23, prove that Ψ is Poisson with intensity function κ.

Algorithm 3.9 (Lewis–Shedler algorithm for Poisson process). We want to generate a realization of the Poisson process with intensity function λ in $\mathbb{R}^d$. Assume $\lambda(\mathbf{x}) \leq M$ for all $\mathbf{x} \in \mathbb{R}^d$ (Fig. 3.42).

1. Generate a homogeneous Poisson process with intensity M.
2. Apply independent thinning, with retention probability function

$$p(\mathbf{x}) = \lambda(\mathbf{x})/M.$$

3.2.3 *Simulating Poisson-Driven Processes*

Point processes that are defined by modifying the Poisson process are relatively straightforward to simulate.

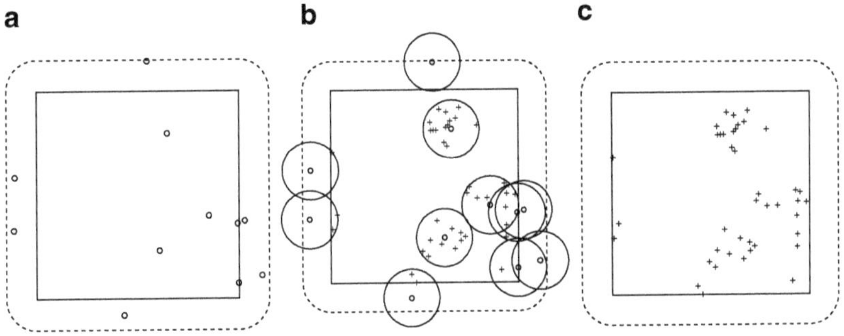

Fig. 3.43 Simulation of Matérn cluster process when cluster size is bounded. (**a**) Poisson parent points in dilated window. (**b**) Generation of offspring. (**c**) Offspring restricted to original window

Algorithm 3.10 (Cox process). To simulate a Cox process

1. Generate a realization of the random intensity measure Λ.
2. Given Λ, generate Ψ according to a Poisson process with intensity Λ.

Let Ψ be a *stationary* Poisson cluster process formed by a homogeneous Poisson parent process Π_κ of intensity κ and a translation-equivariant cluster mechanism $\zeta_x \equiv \zeta_o + x$. To simulate a realization of Ψ inside a bounded window W, we may need to consider parents $x_i \in \Pi_\kappa$ that lie outside W as well as inside.

Algorithm 3.11 (Poisson cluster process with bounded clusters). If $\zeta_o \subset B_r(o)$ a.s. where r is fixed, then

1. Generate a realization $\mathbf{x}$ of the homogeneous Poisson process of intensity κ in $W \oplus B_r(o)$.
2. For each $x_i \in \mathbf{x}$ generate a realization of ζ_{x_i}.
3. Set
$$\mathbf{y} = W \cap \bigcup_i \zeta_{x_i}.$$

See Fig. 3.43.

If the clusters are unbounded, consider the marked point process consisting of parent points x_i marked by ζ_{x_i}. We need to thin this process, retaining only those (x_i, ζ_{x_i}) such that $\zeta_{x_i} \cap W \neq \emptyset$, to obtain a marked process Ψ. This is the process of clusters that intersect W. It must be a finite marked point process. We then use any of the preceding tricks to generate the finite process Ψ.

For example, consider a modification of Matérn's cluster model in which the cluster consists of $N \sim \text{Pois}(\mu)$ i.i.d. random points uniformly distributed in a disc of *random* radius R where R has probability density ρ.

To simulate this process inside a bounded W, first assume without loss of generality that $W = B_s(o)$. Construct the marked point process consisting of parent points x_i marked by cluster radii R_{x_i}. This is a homogeneous Poisson marked point process with point intensity κ and i.i.d. marks with probability density ρ.

For a given r, we have $B_r(x) \cap W \neq \emptyset$ iff $x \in B_{s+r}(o)$. The thinned marked process (obtained by retaining only those (x, r) such that $B_r(x) \cap W \neq \emptyset$) is Poisson with intensity $\lambda(x, r) = \kappa\rho(r)$ if $x \in B_{s+r}(o)$ and 0 otherwise.

The expected number of retained points is

$$\mathbf{E}N = \int_{\mathbb{R}^2} \int_0^\infty \lambda(x, r)dxdr = \kappa\mathbf{E}(\pi(s + R)^2) = \kappa\pi(s^2 + 2s\mathbf{E}R + \mathbf{E}R^2).$$

Algorithm 3.12 (Unbounded Matérn cluster process).

1. Generate $N \sim \text{Pois}(\nu)$ where $\nu = \kappa\pi(s^2 + 2s\mathbf{E}R + \mathbf{E}R^2)$.
2. Given $N = n$, generate n i.i.d. values R_i from the distribution with density $g(r) = \kappa\pi(s + r)^2/\nu$.
3. Given $R_1, \ldots, R_n$, generate independent random points $x_1, \ldots, x_n$ such that x_i is uniformly random in $B_{s+R_i}(o)$.
4. Generate $N_i \sim \text{Pois}(\mu)$ and given $N_i = n_i$, generate n_i i.i.d. uniform random points in $B_{R_i}(x_i)$.
5. Combine all random points generated in step 4 and retain those which fall in W.

An alternative technique is described in [86].

3.2.4 *Markov Chain Monte Carlo*

Next we want to generate simulated realizations of Gibbs point processes, such as the hard-core process.

Algorithm 3.13 (Hard core process, rejection method).

1. Generate a realization $\mathbf{x}$ of the Poisson point process with intensity β in W.
2. If $\|x_i - x_j\| > r$ for all $i \neq j$, exit and return the point pattern $\mathbf{x}$. Otherwise, go to step 1.

This brute force method will take a very long time. An alternative approach is *Markov Chain* simulation. This may be explained by an analogy with card-shuffling. The easiest way to randomize a deck of playing cards is to shuffle them: starting from any ordering of the cards (effectively a permutation of the integers 1–52), we apply a sequence of random shuffles. The successive states of the deck after each shuffle, $X_0, X_1, \ldots$ constitute a *Markov chain*. After enough random shuffles, the deck is randomised.

Definition 3.24. Recall that a (discrete-time) *Markov chain* is a sequence of random elements $X_0, X_1, X_2, \ldots$ in a finite or countable set $\mathcal{X}$ such that

$$\begin{aligned}\mathbf{P}(X_{n+1} = x_{n+1} \mid X_n = x_n, X_{n-1} = x_{n-1}, \ldots, X_0 = x_0)\\ = \mathbf{P}(X_{n+1} = x_{n+1} \mid X_n = x_n)\end{aligned}$$

for all $x_{n+1}, \ldots, x_0 \in \mathcal{X}$. It is *time-homogeneous* if

$$\mathbf{P}(X_{n+1} = \mathbf{y} \mid X_n = \mathbf{x}) = \mathbf{P}(X_1 = \mathbf{y} \mid X_0 = \mathbf{x}) = p(\mathbf{x}, \mathbf{y}), \quad \mathbf{x}, \mathbf{y} \in \mathcal{X}.$$

An *equilibrium distribution* for the chain is a probability distribution $\boldsymbol{\pi} = (\pi(\mathbf{x}), \mathbf{x} \in \mathcal{X})$ such that

$$\sum_{\mathbf{x}} \pi(\mathbf{x}) p(\mathbf{x}, \mathbf{y}) = \pi(\mathbf{y}) \text{ for all } \mathbf{y} \in \mathcal{X}.$$

If the chain is "irreducible" and "aperiodic", then the chain converges in distribution to $\boldsymbol{\pi}$ from any initial state. See [168, 203, 291].

Definition 3.25 (Detailed balance). A chain is in *detailed balance* [291] with a probability distribution $\boldsymbol{\pi}$ if

$$\pi(\mathbf{x}) p(\mathbf{x}, \mathbf{y}) = \pi(\mathbf{y}) p(\mathbf{y}, \mathbf{x}) \text{ for all } \mathbf{x}, \mathbf{y} \in \mathcal{X}.$$

Detailed balance implies that $\boldsymbol{\pi}$ is an equilibrium distribution of the chain, since

$$\sum_{\mathbf{x}} \pi(\mathbf{x}) p(\mathbf{x}, \mathbf{y}) = \sum_{\mathbf{x}} \pi(\mathbf{y}) p(\mathbf{y}, \mathbf{x}) = \pi(\mathbf{y}) \sum_{\mathbf{x}} p(\mathbf{y}, \mathbf{x}) = \pi(\mathbf{y}).$$

Now suppose we want to construct a Markov chain $\{X_n\}$ which has a given equilibrium distribution $\boldsymbol{\pi}$. This can be achieved by constructing $(p(\mathbf{x}, \mathbf{y}))$ to satisfy detailed balance with $\boldsymbol{\pi}$.

Example 3.24. On the state space $\mathcal{X} = \{0, 1\}$ consider the distribution $\pi(0) = \frac{1}{3}$, $\pi(1) = \frac{2}{3}$. For detailed balance we require

$$\pi(0) p(0, 1) = \pi(1) p(1, 0)$$

implying

$$p(0, 1) = 2p(1, 0).$$

The transition probability matrix can be

$$\begin{bmatrix} 1 - 2s & 2s \\ s & 1 - s \end{bmatrix}$$

for any $0 < s < \frac{1}{2}$.

Example 3.25 (Poisson distribution). Let $\pi_n = e^{-\lambda} \lambda^n / n!$. Let us try to construct a chain with steps of ± 1 only (called a *birth-and-death process*). Thus we assume $p_{n,m} = 0$ unless $|m - n| = 1$. Detailed balance requires $\pi_n p_{n,n+1} = \pi_{n+1} p_{n+1,n}$ so that

$$\frac{p_{n,n+1}}{p_{n+1,n}} = \frac{e^{-\lambda} \lambda^{n+1} / (n + 1)!}{e^{-\lambda} \lambda^n / n!} = \frac{\lambda}{n + 1}.$$

To achieve this, we set

$$p_{n,n+1} = \frac{1}{2}\min\left\{1, \frac{\lambda}{n+1}\right\}, \quad p_{n+1,n} = \frac{1}{2}\min\left\{1, \frac{n+1}{\lambda}\right\}$$

i.e.

$$p_{n,n+1} = \frac{1}{2}\min\left\{1, \frac{\lambda}{n+1}\right\}, \quad p_{n,n-1} = \frac{1}{2}\min\left\{1, \frac{n}{\lambda}\right\}, \quad p_{n,n} = 1-p_{n,n+1}-p_{n,n-1}.$$

Then detailed balance is satisfied.

On countable state spaces there is the possibility of "escaping to infinity", $\mathbf{P}(X_n = \mathbf{x}) \to 0$ as $n \to \infty$ for each fixed $\mathbf{x}$. This does not occur if the sufficient condition

$$\sum_{n=1}^{\infty} \frac{p_{01}\, p_{12} \cdots p_{n-1,n}}{p_{n,n-1} \cdots p_{21}\, p_{10}} < \infty$$

is satisfied. To verify this condition in the present case we note that $p_{n,n+1}/p_{n+1,n} = \lambda/(n+1) < 1/2$ when $n > \lambda/2$. Thus the series above is dominated by a geometric series. The Markov chain with these transition probabilities converges in distribution to $\text{Pois}(\lambda)$ from any initial state.

Preserving Detailed Balance

Note that detailed balance is an equation relating the transition probabilities of *pairs of mutually inverse transitions*. We can modify the transition probabilities and still preserve detailed balance, so long as we modify *pairs of transition probabilities*.

Example 3.26. We may modify Example 3.25 to add the possibility of visiting 0 from any state. We re-set the values of $p_{n,0}$ and $p_{0,n}$ (where $n > 1$) so that

$$\frac{p_{0,n}}{p_{n,0}} = \frac{\pi_n}{\pi_0} = \frac{\lambda^n}{n!}$$

Then the chain is still in detailed balance with $\text{Pois}(\lambda)$.

Definition 3.26 (Proposal-acceptance). In a proposal-acceptance algorithm, when the current state is $X_n = \mathbf{x}$,

1. we generate a random state $\mathbf{y}$ with probability $q(\mathbf{x}, \mathbf{y})$ and "propose" jumping to state $\mathbf{y}$.
2. with probability $a(\mathbf{x}, \mathbf{y})$ the proposal is "accepted" and $X_{n+1} = \mathbf{y}$. Otherwise the proposal is "rejected" and $X_{n+1} = \mathbf{x}$.

This is a Markov chain with transition probabilities $p(\mathbf{x}, \mathbf{y}) = q(\mathbf{x}, \mathbf{y})a(\mathbf{x}, \mathbf{y})$.

Metropolis-Hastings

A proposal-acceptance algorithm is in detailed balance with a distribution π if

$$\pi(\mathbf{x})q(\mathbf{x},\mathbf{y})a(\mathbf{x},\mathbf{y}) = \pi(\mathbf{y})q(\mathbf{y},\mathbf{x})a(\mathbf{y},\mathbf{x})$$

Equivalently

$$\frac{a(\mathbf{x},\mathbf{y})}{a(\mathbf{y},\mathbf{x})} = \frac{\pi(\mathbf{y})q(\mathbf{y},\mathbf{x})}{\pi(\mathbf{x})q(\mathbf{x},\mathbf{y})}$$

One way to achieve this is to choose

$$a(\mathbf{x},\mathbf{y}) = \min\left\{1, \frac{\pi(\mathbf{y})q(\mathbf{y},\mathbf{x})}{\pi(\mathbf{x})q(\mathbf{x},\mathbf{y})}\right\}$$

With these acceptance probabilities, the algorithm is called a *Metropolis-Hastings algorithm* [355].

Exercise 3.20. Derive the Metropolis-Hastings acceptance probabilities when the target distribution π is the Poisson distribution with mean μ and the proposal mechanism is the symmetric random walk on the nonnegative integers, i.e. $p_{n,n+1} = p_{n,n-1} = 1/2$ for $n > 0$ and $p_{0,1} = 1$.

Spatial Birth-and-Death Processes

Definition 3.27. A *spatial birth-and-death process* (in discrete time) is a time-homogeneous Markov chain X_n, $n = 0, 1, 2, \ldots$ in $\mathcal{N}$ in which the only possible transitions are

1. "births" $\mathbf{x} \mapsto \mathbf{x} \cup \{u\}$ where $u \in \mathsf{S}$;
2. "deaths" $\mathbf{x} \mapsto \mathbf{x} \setminus \{x_i\}$ for some i.

A (discrete time) spatial birth-and-death process is characterized by its *death probabilities*

$$D(\mathbf{x}, x_i) = \mathbf{P}(X_1 = \mathbf{x} \setminus \{x_i\} \mid X_0 = \mathbf{x})$$

and *birth measure*

$$B(\mathbf{x}, A) = \mathbf{P}(X_1 = \mathbf{x} \cup \{u\},\ u \in A \mid X_0 = \mathbf{x}), \quad A \in \mathfrak{N}.$$

Often we represent the birth measure by its *birth density* $b(\mathbf{x}, u)$ such that

$$B(\mathbf{x}, A) = \int_W b(\mathbf{x}, u)\mathbf{1}(\mathbf{x} \cup \{u\} \in A)du.$$

For a spatial birth-and-death process with death probabilities $D(\mathbf{x}, x_i)$ and birth density $b(\mathbf{x}, u)$, the transition kernel is

$$
\begin{aligned}
P_{\mathbf{x}}(A) &= \mathbf{P}(X_1 \in A \mid X_0 = x) \\
&= \sum_{i=1}^{|\mathbf{x}|} D(\mathbf{x}, x_i)\mathbf{1}(\mathbf{x} \setminus \{x_i\} \in A) + \int_W b(\mathbf{x}, u)\mathbf{1}(\mathbf{x} \cup \{u\} \in A)\, du
\end{aligned}
$$

for $A \in \mathfrak{N}$ with $\mathbf{x} \notin A$, where $|\mathbf{x}|$ = number of points in $\mathbf{x}$.

Example 3.27. Consider the following algorithm:

1. Start with an empty point pattern $\mathbf{x} = \emptyset$.
2. When the current state $\mathbf{x}$ contains m points,
   (a) With probability d_m choose one of the existing points at random with equal probability $1/m$ and delete it from $\mathbf{x}$ ("death").
   (b) With probability b_m generate a new random point U uniformly distributed in W and add it to $\mathbf{x}$ ("birth").
   (c) Otherwise (probability $1 - d_m - b_m$) do not change state.
3. Go to step 2.

Let X_n be the state of the algorithm at time $n = 0, 1, 2, \ldots$. Then $\{X_n\}$ is a spatial birth-and-death process with death probabilities $D(\mathbf{x}, x_i) = d_m/m$ and birth density $b(\mathbf{x}, u) = b_m/\nu_d(W)$ where $m = |\mathbf{x}|$.

Lemma 3.14. *In Example 3.27, assume $(d_m), (b_m)$ satisfy $b_m/d_m = \mu/(m+1)$ where $\mu = \beta\nu_d(W)$. Then X_n converges in distribution as $n \to \infty$ to a Poisson point process with intensity β.*

Proof. The number of points $Y_n = |X_n|$ follows a time-homogeneous Markov chain on the nonnegative integers, with transition probabilities

$$
\begin{aligned}
p_{m,m+1} &= b_m \\
p_{m,m-1} &= d_m \\
p_{m,m} &= 1 - b_m - d_m
\end{aligned}
$$

which converges in distribution to Pois(μ). Given $Y_n = m$, the state $X^{(t)}$ consists of m i.i.d. uniform random points. By the conditional property of the Poisson process, X_n converges in distribution as $n \to \infty$ to a Poisson point process with intensity β. □

Definition 3.28 (Detailed balance in continuous state space). For a chain in a continuous state space $\mathcal{X}$, the detailed balance condition is

$$
\int_A P_{\mathbf{x}}(B)\, d\pi(\mathbf{x}) = \int_B P_{\mathbf{x}}(A)\, d\pi(\mathbf{x}) \tag{3.12}
$$

for all measurable $A, B \subset \mathcal{X}$. It suffices to assume $A \cap B = \emptyset$.

Theorem 3.4 (Detailed balance for spatial birth–death process). *Let* $\{X_n : n = 0, 1, 2, \ldots\}$ *be a spatial birth-and-death process on a compact* $W \subset \mathbb{R}^d$ *with death probabilities* $D(\mathbf{x}, x_i)$ *and birth density* $b(\mathbf{x}, u)$. *Let* $\boldsymbol{\pi}$ *be the distribution of a finite point process on* W *with density* $f(\mathbf{x})$ *with respect to the unit rate Poisson process. Then* $\{X_n\}_n$ *is in detailed balance with* $\boldsymbol{\pi}$ *iff*

$$f(\mathbf{x})\, b(\mathbf{x}, u) = f(\mathbf{x} \cup \{u\})\, D(\mathbf{x} \cup \{u\}, u) \tag{3.13}$$

for almost all $\mathbf{x} \in \mathcal{N}^f$ *and* $u \in W$.

Proof. Assume (3.13) holds. Define for $A, B \in \mathfrak{N}$ with $A \cap B = \emptyset$

$$\begin{aligned} h_{A,B}(\mathbf{x}, u) &= f(\mathbf{x} \cup \{u\})\, D(\mathbf{x} \cup \{u\}, u)\mathbf{1}(\mathbf{x} \cup \{u\} \in A,\ \mathbf{x} \in B) \\ &= f(\mathbf{x})\, b(\mathbf{x}, u)\mathbf{1}(\mathbf{x} \cup \{u\} \in A,\ \mathbf{x} \in B). \end{aligned}$$

Observe that

$$f(\mathbf{x})\mathbf{1}(\mathbf{x} \in A) \sum_{i=1}^{|\mathbf{x}|} D(\mathbf{x}, x_i)\mathbf{1}(\mathbf{x} \setminus \{x_i\} \in B) = \sum_{i=1}^{|\mathbf{x}|} h_{A,B}(\mathbf{x} \setminus \{x_i\}, x_i)$$

$$f(\mathbf{x})\mathbf{1}(\mathbf{x} \in A) \int_W b(\mathbf{x}, u)\mathbf{1}(\mathbf{x} \cup \{u\} \in B) du = \int_W h_{B,A}(\mathbf{x}, u) du.$$

Thus we have

$$\begin{aligned} \int_A P_{\mathbf{x}}(B) d\boldsymbol{\pi}(\mathbf{x}) &= \mathbf{E}(f(\Pi_1)\mathbf{1}(\Pi_1 \in A) P_{\Pi_1}(B)) \\ &= \mathbf{E}\left(\sum_{i=1}^{|\Pi_1|} h_{A,B}(\Pi_1 \setminus \{z_i\}, z_i) + \int_W h_{B,A}(\Pi_1, u) du \right) \end{aligned}$$

where $\Pi_1 = \{z_i\}$ denotes the unit rate Poisson process.
But by the GNZ formula for Π_1,

$$\mathbf{E}\left(\sum_{i=1}^{|\Pi_1|} h(\Pi_1 \setminus \{z_i\}, z_i) \right) = \mathbf{E}\left(\int_W h(\Pi_1, u) du \right)$$

Hence

$$\begin{aligned} \int_A P_{\mathbf{x}}(B) d\boldsymbol{\pi}(\mathbf{x}) &= \mathbf{E}\left(\sum_{i=1}^{|\Pi_1|} h_{A,B}(\Pi_1 \setminus \{z_i\}, z_i) + \int_W h_{B,A}(\Pi_1, u) du \right) \\ &= \mathbf{E}\left(\int_W h_{A,B}(\Pi_1, u) du \right) + \int_W h_{B,A}(\Pi_1, u) du] \end{aligned}$$

is symmetric in A, B. So detailed balance (3.12) holds.

Conversely, suppose (3.12) holds. Let $A, B \in \mathfrak{N}$ be such that $A \subset \mathcal{N}_n$ and $B \subset \mathcal{N}_{n+1}$ for some $n \geq 0$. The only transitions from A to B are births, so for $\mathbf{x} \in A$

$$P_{\mathbf{x}}(B) = \int_W b(\mathbf{x}, u) \mathbf{1}(\mathbf{x} \cup \{u\} \in B) \, du$$

while for $\mathbf{x} \in B$

$$P_{\mathbf{x}}(A) = \sum_{x_i \in \mathbf{x}} D(\mathbf{x}, x_i) \mathbf{1}(\mathbf{x} \setminus \{x_i\} \in A).$$

Substituting into (3.12) and applying the Radon–Nikodým theorem yields (3.13). □

In Example 3.27 and Lemma 3.14, $D(\mathbf{x}, x_i) = d_m/m$ and $b(\mathbf{x}, u) = b_m/\nu_d(W)$ satisfy

$$\frac{b(\mathbf{x}, u)}{D(\mathbf{x} \cup \{u\}, u)} = \frac{b_m/\nu_d(W)}{d_{m+1}/(m+1)} = \frac{\beta \nu_d(W)(m+1)}{(m+1)\nu_d(W)} = \beta$$

The probability density of the Poisson process of rate β is $f(\mathbf{x}) = \alpha \beta^{|\mathbf{x}|}$ where $\alpha = \exp\{(1-\beta)\nu_d(W)\}$. Thus $f(\mathbf{x} \cup \{u\})/f(\mathbf{x}) = \beta$. Hence detailed balance applies.

Simulation of Hard-Core Process

Suppose we want to simulate a hard-core process with hard core diameter r in a region W. The hard core process is a Poisson process conditioned on the event

$$H = \{\mathbf{x} \in \mathcal{N}^f : \|x_i - x_j\| > r \text{ for all } i \neq j\}.$$

To construct a spatial birth-and-death process that is in detailed balance with the hard core process, we simply modify the previous process by forbidding transitions out of H.

Algorithm 3.14 (Hard-core spatial-birth-and-death).

1. Start with an empty point pattern $\mathbf{x} = \emptyset$.
2. When the current state $\mathbf{x}$ contains m points,
   (a) With probability $1 - d_m - b_m$ do nothing.
   (b) With probability d_m choose one of the existing points at random and delete it from $\mathbf{x}$.
   (c) With probability b_m generate a new random point U uniformly distributed in W. If $\|U - x_i\| > r$ for all i, then add U to the configuration, $\mathbf{x} := \mathbf{x} \cup \{U\}$. Otherwise leave the current configuration unchanged.
3. Go to step 2.

Algorithm 3.15 (Metropolis–Hastings for Gibbs point process). For any finite Gibbs process with density $f(\mathbf{x})$ the following algorithm is in detailed balance:

1. Start in any initial state $\mathbf{x}$ for which $f(\mathbf{x}) > 0$
2. When the current state is $\mathbf{x}$,

   (a) With probability p, propose a death: select one of the points $x_i \in \mathbf{x}$ with equal probability, and propose deleting x_i. Accept with probability

$$a(\mathbf{x}, \mathbf{x} \setminus \{x_i\}) = \min\left\{1, \frac{(1-p)|\mathbf{x}| f(\mathbf{x} \setminus \{x_i\})}{p\nu_d(W) f(\mathbf{x})}\right\}$$

   (b) Otherwise (with probability $1 - p$) propose a birth: generate a uniformly random point u in W and propose adding it. Accept with probability

$$a(\mathbf{x}, \mathbf{x} \cup \{u\}) = \min\left\{1, \frac{p\nu_d(W) f(\mathbf{x} \cup \{u\})}{(1-p)(|\mathbf{x}|+1) f(\mathbf{x})}\right\}$$

Under additional conditions (e.g. that $f(\mathbf{x} \cup \{u\})/f(\mathbf{x}) < B < \infty$ for all $\mathbf{x}, u$) the algorithm converges in distribution to f from any initial state. See [369, Chap. 7].

Continuous Time

This may be easier to understand in continuous time. A spatial birth–death process in continuous time is a Markov process $\{X(t),\ t \geq 0\}$ whose trajectories are piecewise constant ("pure jump process") with

$$\mathbf{P}(X(t+h) \in A \mid X(t) = \mathbf{x}) = \mathbf{1}(\mathbf{x} \in A) + Q_{\mathbf{x}}(A)h + o(h), \quad h \downarrow 0$$

where the *rate measure* $Q_{\mathbf{x}}$ is

$$Q_{\mathbf{x}}(A) = \sum_{i=1}^{|\mathbf{x}|} D(\mathbf{x}, x_i)\mathbf{1}(\mathbf{x} \setminus \{x_i\} \in A) + \int_W b(\mathbf{x}, u)\mathbf{1}(\mathbf{x} \cup \{u\} \in A)du.$$

The quantities $D(\mathbf{x}, x_i)$ and $b(\mathbf{x}, u)$ are rates which may take any nonnegative value.

Definition 3.29 (Detailed balance for spatial birth–death process). A continuous time spatial birth–death process is in detailed equilibrium with a point process density f iff

$$f(\mathbf{x})b(\mathbf{x}, u) = f(\mathbf{x} \cup \{u\})D(\mathbf{x} \cup \{u\}, u).$$

This is now easier to satisfy, because $D(\mathbf{x}, x_i)$ is not required to be a probability.

Example 3.28 (Gibbs sampler). We may take

$$D(\mathbf{x}, x_i) = 1$$

$$b(\mathbf{x}, u) = \lambda(u, \mathbf{x}) = \frac{f(\mathbf{x} \cup \{u\})}{f(\mathbf{x})}$$

A simple interpretation is that each existing point has an exponentially distributed lifetime with mean 1, independent of other points, after which it is deleted; and new points u are added at rate $\lambda(u, \mathbf{x})$ where $\mathbf{x}$ is the current state.

3.3 Inference

This section discusses statistical inference for point process models. It covers maximum likelihood (Sect. 3.3.1), Markov Chain Monte Carlo maximum likelihood (Sect. 3.3.2), fitting models using summary statistics (Sect. 3.3.3) and estimating equations and maximum pseudolikelihood (Sect. 3.3.4).

3.3.1 Maximum Likelihood

General Definitions

Suppose that we observe a point pattern $\mathbf{x}$ in a window W. We wish to fit a finite point process model which has probability density $f(\mathbf{x}) = f(\mathbf{x}; \theta)$ depending on a vector parameter $\theta \in \Theta \subseteq \mathbb{R}^k$. Define the *likelihood*

$$L(\theta) = f(\mathbf{x}; \theta).$$

The maximum likelihood estimate (MLE) of θ is

$$\hat{\theta} = \operatorname{argmax}_\theta L(\theta).$$

Under regularity conditions, $\hat{\theta}$ is the root of the *score*

$$U(\theta) = \frac{\partial}{\partial \theta} \log L(\theta).$$

If θ_0 denotes the true value, then [134, p. 107 ff.] the score satisfies

$$\mathbf{E}_{\theta_0}(U(\theta_0)) = 0, \quad I(\theta_0) = \mathbf{var}_{\theta_0}(U(\theta_0)) = \mathbf{E}\left(-\frac{\partial}{\partial \theta} U(\theta_0)\right),$$

where $I(\theta)$ is the *Fisher information*. Note that the classical theory under which the MLE is optimal, is based on i.i.d. observations, and does not automatically apply here.

Maximum Likelihood for Poisson Processes

For Poisson point processes, maximum likelihood estimation is outlined in [317].

Example 3.29 (MLE for homogeneous Poisson). For the homogeneous Poisson process in W with intensity λ,

$$f(\mathbf{x};\lambda) = \lambda^{|\mathbf{x}|}\exp\{(1-\lambda)\nu_d(W)\}.$$

The score is

$$U(\lambda) = \frac{|\mathbf{x}|}{\lambda} - \nu_d(W)$$

so the MLE is

$$\hat{\lambda} = \frac{|\mathbf{x}|}{\nu_d(W)}.$$

Example 3.30 (MLE for inhomogeneous Poisson). For the inhomogeneous Poisson process in W with intensity function $\lambda_\theta(u)$, $u \in W$ depending on θ, the density is

$$f(\mathbf{x};\theta) = \left[\prod_{i=1}^{|\mathbf{x}|}\lambda_\theta(x_i)\right]\exp\left\{\int_W (1-\lambda_\theta(u))d\right\}.$$

Assuming $\lambda_\theta(u)$ is differentiable with respect to θ, the score is

$$U(\theta) = \sum_{i=1}^{|\mathbf{x}|}\frac{\partial}{\partial\theta}\log\lambda_\theta(x_i) - \int_W \frac{\partial}{\partial\theta}\lambda_\theta(u)du.$$

Exercise 3.21. Let $V \subset W$ and consider the inhomogeneous Poisson process with different constant intensities in V and $W \setminus V$, i.e. with intensity function

$$\lambda(u) = \begin{cases}\lambda_1 \text{ if } u \in V\\ \lambda_2 \text{ if } u \in W \setminus V\end{cases}$$

Find the MLE of (λ_1, λ_2).

Example 3.31. Assume $\lambda(u) = \theta Z(u)$ where $Z(u)$ is a nonnegative real-valued covariate. Then

$$f(\mathbf{x};\theta) = \theta^{|\mathbf{x}|}\prod_i Z(x_i)\exp\left\{\int_W (1-\theta Z(u))du\right\}$$

and

$$U(\theta) = |\mathbf{x}|\log\theta - \int_W Z(u)du$$

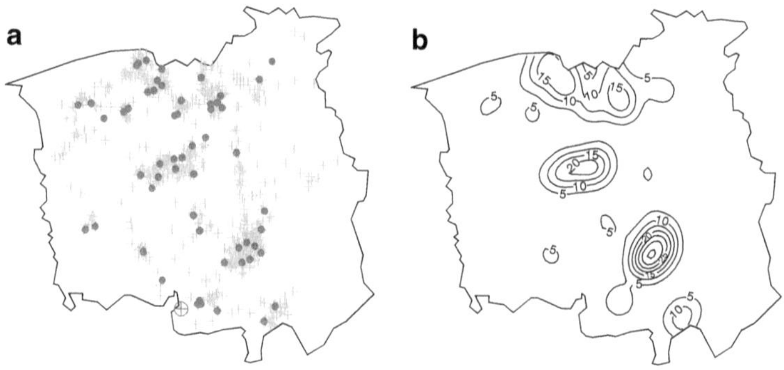

Fig. 3.44 Analysis of Chorley–Ribble data. (**a**) Original data. (**b**) Kernel smoothed estimate of lung cancer incidence

so

$$\hat{\theta} = \frac{|\mathbf{x}|}{\int_W Z(u)du}.$$

The model above is particularly relevant to epidemiological studies where it represents the hypothesis of constant relative risk. Figure 3.44 shows the original data from the Chorley–Ribble dataset (Fig. 3.2 and [153]), and a kernel-smoothed estimate of lung cancer incidence which serves as a surrogate for the spatially-varying intensity of the susceptible population.

Example 3.32 (Loglinear Poisson). Consider the *loglinear* intensity model

$$\lambda_\theta(u) = \exp\{\theta^{\mathrm{T}} Z(u)\} \tag{3.14}$$

where $Z : W \to \mathbb{R}^k$ is a vector-valued spatial covariate function. Then the log-likelihood is concave, and the score

$$U(\theta) = \sum_{i=1}^{|\mathbf{x}|} Z(x_i) - \int_W Z(u) \exp\{\theta^{\mathrm{T}} Z(u)\} du$$

has a unique root provided $\sum_i Z(x_i) \neq 0$. The Fisher information matrix is

$$I(\theta) = \int_W Z(u)\, Z(u)^{\mathrm{T}} \exp\{\theta^{\mathrm{T}} Z(u)\} du.$$

Under regularity and positivity conditions, $\hat{\theta}$ is consistent and asymptotically normal with variance $I(\theta)^{-1}$ in a limiting scenario where $W \nearrow \mathbb{R}^d$. See [317].

The maximum likelihood estimator $\hat{\theta}$ for (3.14) does not have a simple analytic form. Several numerical algorithms exist for computing $\hat{\theta}$ approximately [5,25,64]. In each technique, the domain W is divided into disjoint subsets $Q_1, \ldots, Q_m$

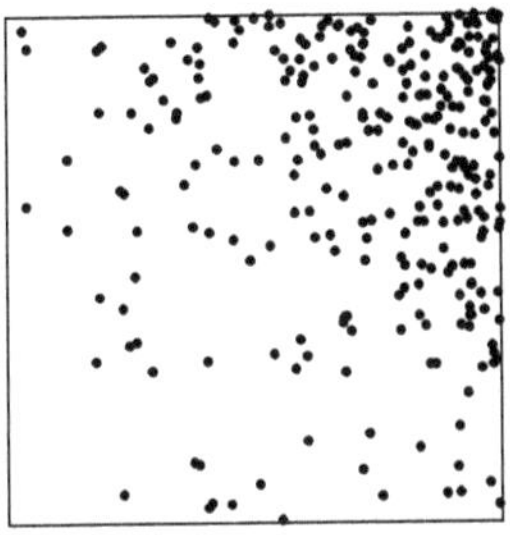

Fig. 3.45 Simulated inhomogeneous Poisson data

and the numbers $n_j = N(\mathbf{x} \cap Q_j)$ of points in each subset are counted. Then the point process likelihood is approximated by the likelihood of a Poisson loglinear regression, which can be maximised using standard software.

Figure 3.45 shows a simulated realization of the inhomogeneous Poisson process with intensity function

$$\lambda(x, y) = \mathrm{e}^{\beta_0+\beta_1 x+\beta_2 y^2}$$

where $(\beta_0, \beta_1, \beta_2) = (3, 3, 2)$. The algorithm above was applied using a 100×100 grid of squares, yielding the approximate MLE of $(2.7, 3.2, 2.1)$.

Maximum Likelihood for Gibbs Processes

For Gibbs point processes, maximum likelihood estimation is generally intractable.

Example 3.33. Consider a Gibbs process with density of the loglinear form

$$f(\mathbf{x}; \theta) = \alpha(\theta) \exp(\theta^{\mathrm{T}} V(\mathbf{x})) \tag{3.15}$$

where $V(\mathbf{x})$ is a statistic and $\alpha(\theta)$ is the normalizing constant given by

$$\alpha(\theta)^{-1} = \mathbf{E}_0(\exp\{\theta^{\mathrm{T}} V(\mathbf{X})\})$$

where $\mathbf{E}_0$ denotes expectation with respect to the unit rate Poisson process on W.

In Example 3.33, the score is

$$\begin{aligned} U(\theta) &= V(\mathbf{x}) + \frac{\partial}{\partial\theta} \log \alpha(\theta) = V(\mathbf{x}) - \frac{\frac{\partial}{\partial\theta}\mathbf{E}_0(\exp\{\theta^{\mathrm{T}} V(\mathbf{X})\})}{\mathbf{E}_0(\exp\{\theta^{\mathrm{T}} V(\mathbf{X})\})} \\ &= V(\mathbf{x}) - \frac{\mathbf{E}_0(V(\mathbf{X}) \exp\{\theta^{\mathrm{T}} V(\mathbf{X})\})}{\mathbf{E}_0(\exp\{\theta^{\mathrm{T}} V(\mathbf{X})\})} = V(\mathbf{x}) - \mathbf{E}_\theta(V(\mathbf{X})) \end{aligned}$$

and Fisher information

$$I(\theta) = \mathbf{var}_\theta(V(\mathbf{X}))$$

where $\mathbf{E}_\theta$, $\mathbf{var}_\theta$ denote expectation and variance with respect to $f(\cdot;\theta)$.

Unfortunately, $\alpha(\theta)$, $M(\theta) = \mathbf{E}_\theta(V(\mathbf{X}))$ and $I(\theta)$ are usually intractable functions of θ.

Example 3.34 (Strauss process). The Strauss process has density

$$f(\mathbf{x};\beta,\gamma) = \alpha\beta^{|\mathbf{x}|}\gamma^{s(\mathbf{x})} = \alpha\exp\{|\mathbf{x}|\log\beta + s(\mathbf{x})\log\gamma\}$$

where $s(\mathbf{x}) = \sum_{i<j} \mathbf{1}(\|x_i - x_j\| < r)$ is the number of pairs of points that are closer than r units apart.

We see that evaluating α is equivalent to evaluating the characteristic function of $(|\mathbf{X}|, s(\mathbf{X}))$ for a unit rate Poisson process. Evaluating $M(\theta)$ and $I(\theta)$ requires the mean and variance of $(|\mathbf{X}|, s(\mathbf{X}))$ for the Strauss process.

3.3.2 MCMC Maximum Likelihood

Continuing with the loglinear Gibbs model (Example 3.33), another strategy is to estimate the functions $\alpha(\theta)^{-1}$, $M(\theta) = \mathbf{E}_\theta(V(\mathbf{X}))$ and $I(\theta) = \mathbf{var}_\theta(V(\mathbf{X}))$ by simulation. Three simple strategies are described below.

Simulation from Poisson

Using

$$\alpha(\theta)^{-1} = \mathbf{E}_0(\exp\{\theta^{\mathsf{T}} V(\mathbf{X})\}), \quad M(\theta) = \frac{\mathbf{E}_0(V(\mathbf{X})\exp\{\theta^{\mathsf{T}} V(\mathbf{X})\})}{\mathbf{E}_0(\exp\{\theta^{\mathsf{T}} V(\mathbf{X})\})}$$

we could generate K simulated point patterns $X^{(1)}, \ldots, X^{(K)}$ from the unit rate Poisson process, and take the sample moments

$$\hat{\alpha}(\theta)^{-1} = \frac{1}{K}\sum_{k=1}^{K}\exp\{\theta^{\mathsf{T}} V(X^{(k)})\}, \quad \hat{M}(\theta) = \frac{\sum_k V(X^{(k)})\exp\{\theta^{\mathsf{T}} V(X^{(k)})\}}{\sum_k \exp\{\theta^{\mathsf{T}} V(X^{(k)})\}} \tag{3.16}$$

and take $\hat{I}(\theta)$ to be the corresponding sample variance.

Algorithm 3.16 (MLE using massive simulation from Poisson).

1. Generate a huge number of simulations from the unit rate Poisson process;
2. Compute estimates of the functions $\alpha(\theta)$, $M(\theta)$ and $I(\theta)$ using (3.16);

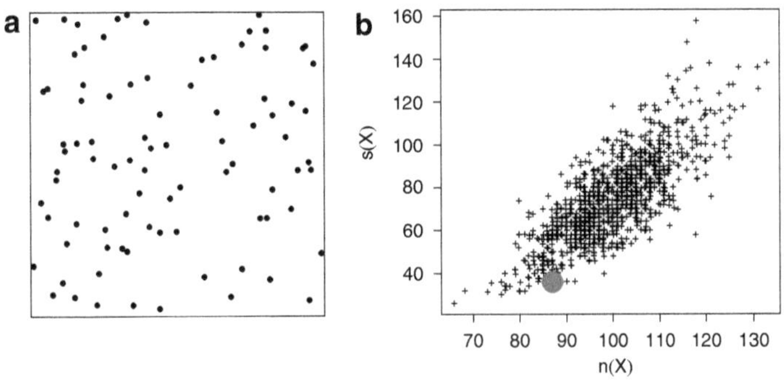

Fig. 3.46 Approximate MLE by simulation from Poisson. (**a**) Data pattern: Realization of Strauss process. (**b**) Bivariate scatter plot of $(|\mathbf{X}|, s(\mathbf{X}))$ for 1,000 simulated realizations of Poisson process. Large dot shows observed value of $(|\mathbf{X}|, s(\mathbf{X}))$ for data

3. Approximate the likelihood (3.15) using the estimate of $\alpha(\theta)$, and/or approximate the score function using the estimate of $M(\theta)$, and maximize the approximate likelihood numerically.

Example 3.35. The left panel of Fig. 3.46 shows synthetic data, a simulated realization of the Strauss process with $\beta = 100$, $\gamma = 0.7$, $r = 0.05$ in unit square.

The right panel shows a bivariate scatter plot of $(|\mathbf{X}|, s(\mathbf{X}))$ for 1,000 simulated realizations of the Poisson process with intensity $\lambda = 100$ in the unit square. The large dot shows the observed value of $(|\mathbf{X}|, s(\mathbf{X}))$ for the data.

An approximation to $\alpha(\theta)$ was obtained using (3.16). This was substituted into (3.15) to obtain an approximate likelihood function, which was then maximised to obtain the approximate MLE $\hat{\beta} = 132$, $\hat{\gamma} = 0.60$.

The main difficulty with this method is that it becomes very inaccurate when the observed value $V(\mathbf{x})$ is not near the centre of the cloud of simulated values $V(X^{(k)})$.

Simulation from Target Distribution

Alternatively, we could simulate K times from the target distribution $f(\cdot;\theta)$ using a current estimate of θ, use the estimator

$$\hat{M}(\theta) = \frac{1}{K} \sum_k V(X^{(k)})$$

and estimate $I(\theta)$ by the analogous sample variance.

Algorithm 3.17. MLE using simulation from current estimate

1. Choose a sensible initial estimate θ_0.
2. When the current estimate is θ_n, generate simulations from $f(\cdot,\theta_n)$ and compute the sample estimates $\hat{M}(\theta_n)$ and $\hat{I}(\theta_n)$ of $M(\theta_n)$ and $I(\theta_n)$, respectively.

3. Take one step of the Newton–Raphson algorithm [290],

$$\theta_{n+1} = \theta_n - \hat{I}(\theta_n)^{-1}\hat{M}(\theta_n).$$

4. Go to step 2 (unless convergence has occurred).

Importance Sampling

For two values θ, η the corresponding densities $f(\cdot,\theta)$, $f(\cdot,\eta)$ are related by

$$f(\mathbf{x},\theta) = \frac{\mathrm{e}^{(\theta-\eta)^\top V(\mathbf{x})} f(\mathbf{x},\eta)}{\mathbf{E}_\eta(\mathrm{e}^{(\theta-\eta)^\top V(\mathbf{X})})} \tag{3.17}$$

For a reference value η we could generate K simulated point patterns $X^{(1)},\ldots,X^{(K)}$ from $f(\cdot,\eta)$, and take

$$\hat{\alpha}(\theta)^{-1} = \frac{1}{K}\sum_{k=1}^{K} \exp\{(\theta-\eta)^{\mathrm{T}} V(X^{(k)})\}$$

$$\hat{M}(\theta) = \frac{\sum_k V(X^{(k)}) \exp\{(\theta-\eta)^{\mathrm{T}} V(X^{(k)})\}}{\sum_k \exp\{(\theta-\eta)^{\mathrm{T}} V(X^{(k)})\}}$$

Similarly for $I(\theta)$. These are known as the "importance sampling" estimates.

Algorithm 3.18 (MLE using importance sampling). Set up a grid consisting of values $\eta_1,\ldots,\eta_M$ in the parameter space. For each m generate a substantial number of simulated realizations from $f(\cdot,\eta_m)$.

1. Choose a sensible initial estimate θ_0.
2. When the current estimate is θ_n, find the grid value η_m that is closest to θ_n, and use it to compute the importance-sampling estimates of $M(\theta_n)$ and $I(\theta_n)$.
3. Take one step of the Newton–Raphson algorithm,

$$\theta_{n+1} = \theta_n - \hat{I}(\theta_n)^{-1}\hat{M}(\theta_n).$$

4. Go to step 2 (unless convergence has occurred).

3.3.3 *Fitting Models Using Summary Statistics*

Summary Statistics

When investigating a spatial point pattern dataset $\mathbf{x} = \{x_1,\ldots,x_n\}$ in a domain W, it is often useful to begin by computing various summary statistics.

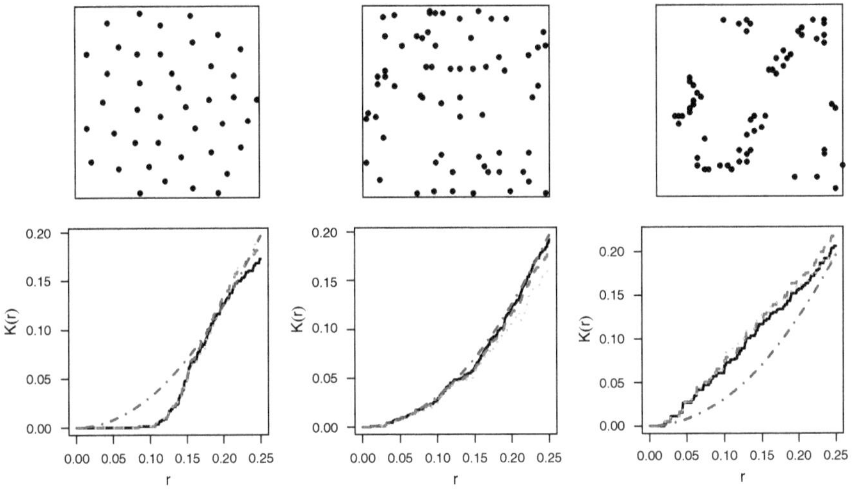

Fig. 3.47 Summarizing interpoint interaction with the K function. *Top row*: examples of point pattern datasets showing (from *left* to *right*) regular, random, and clustered patterns. *Bottom row*: corresponding empirical K functions

One important example is the empirical K-function, essentially a renormalised empirical distribution of the distances between all pairs of points in $\mathbf{x}$:

$$\hat{K}(r) = \frac{1}{\widehat{\lambda^2}\nu(W)} \sum_i \sum_{j \neq i} e(x_i, x_j, r)\mathbf{1}(\|x_i - x_j\| \leq r), \quad r \geq 0, \tag{3.18}$$

where typically $\widehat{\lambda^2} = n(n-1)/\nu(W)^2$, and $e(u, v, r)$ is an edge correction term, depending on the choice of technique.

The empirical K-function is a useful exploratory tool for distinguishing between different types of spatial pattern. See Fig. 3.47.

If $\mathbf{x}$ is a realisation of a stationary point process Ψ, then $\hat{K}(r)$ is an approximately unbiased estimator of the ("theoretical") K-function of Ψ, defined in Sect. 4.2.1, formula (4.14).

Model-Fitting

Summary statistics such as the K-function, are normally used for exploratory data analysis. Summary statistics can also be used for inference, especially for parameter estimation:

1. From the point pattern data, compute the estimate $\hat{K}(r)$ of the K function.
2. For each possible value of θ, determine the K-function of the model, $K_\theta(r)$.
3. Select the value of θ for which K_θ is closest to $\hat{K}$. For example, choose

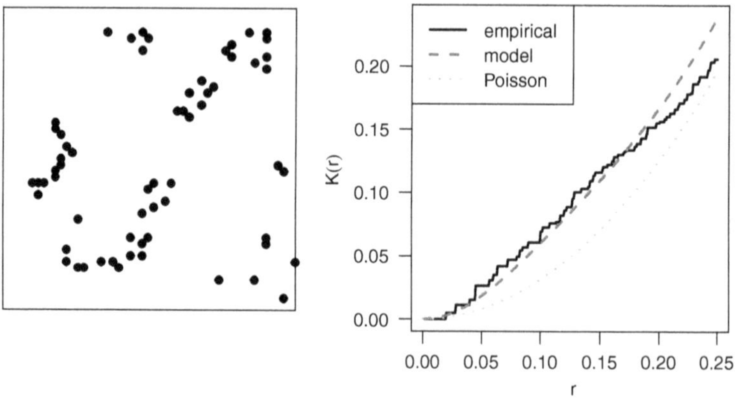

Fig. 3.48 Minimum contrast estimation. (**a**) Subset of Strauss's redwood data. (**b**) Minimum contrast fit of Thomas model using K function

$$\hat{\theta} = \text{argmin} \int_0^R |\hat{K}(r) - K_\theta(r)|^p dr$$

This is called the "method of minimum contrast" [152, 406].

Minimum Contrast for Poisson Cluster Processes

For a Poisson cluster process, the "cluster formula" (3.5) yields an expression for the K-function.

Example 3.36 (Minimum contrast for Thomas process). For the modified Thomas process with parent intensity κ, with a Pois(μ) number of offspring at Gaussian locations with standard deviation σ, the intensity is $\lambda = \kappa\mu$ and the K-function is [265, p. 375 ff.]

$$K(r) = \pi r^2 + \frac{1}{\kappa}\left(1 - \exp\left\{-\frac{r^2}{4\sigma^2}\right)\right\}.$$

We can estimate the parameters κ, σ by minimum contrast (Fig. 3.48), then estimate μ by $\hat{\mu} = \hat{\lambda}/\hat{\kappa}$ where $\hat{\lambda}$ is the usual estimate of intensity, see e.g. Example 3.29.

Monte Carlo Minimum Contrast

For a general point process model with parameter θ, we do not know K_θ in closed form. It can usually be estimated by simulation. Hence the discrepancy

$$D(\theta) = \int_0^R |\hat{K}(r) - K_\theta(r)|^p dr$$

can be estimated by simulation, requiring a separate suite of simulations for each θ.

3.3.4 Estimating Equations and Maximum Pseudolikelihood

Score Equation

Under regularity conditions, the MLE is the solution of the *score equation*

$$U(\theta; \mathbf{x}) = 0$$

where $U(\theta; \mathbf{x}) = \frac{\partial}{\partial\theta} \log f(\mathbf{x}; \theta)$. Study of the asymptotic behaviour of $\hat{\theta}$ depends on properties of the score function U. In particular the special property

$$\mathbf{E}_\theta(U(\theta; \mathbf{X})) = 0$$

is essential in proving that the MLE is consistent and asymptotically normal [134, p. 107ff.].

Estimating Equations

More generally we may consider an estimator $\tilde{\theta}$ that is the solution of an estimating equation

$$g(\theta; \mathbf{x}) = 0$$

where g is some estimating function that we choose. If g has the analogous property

$$\mathbf{E}_\theta(g(\theta; \mathbf{X})) = 0$$

we say that g is an unbiased estimating function. Note that this terminology does *not* mean that $\tilde{\theta}$ is unbiased.

Takacs–Fiksel Estimators

For a point process model depending on parameter θ, let $\lambda_\theta(u, \mathbf{x})$ be its conditional intensity. The GNZ formula states

$$\mathbf{E}\left(\sum_i h(x_i, \mathbf{x} \setminus \{x_i\})\right) = \mathbf{E}\left(\int h(u, \mathbf{x})\lambda_\theta(u, \mathbf{x})du\right)$$

for "any" function h. Hence

$$g(\theta; \mathbf{x}) = \sum_i h(x_i, \mathbf{x} \setminus \{x_i\}) - \int h(u, \mathbf{x})\lambda_\theta(u, \mathbf{x})du$$

is an unbiased estimating function, for any choice of h. This approach to estimation is called the *Takacs–Fiksel method*.

Example 3.37. For $h \equiv 1$ we get the unbiased estimating function

$$g(\theta;\mathbf{x}) = |\mathbf{x}| - \int_W \lambda_\theta(u,\mathbf{x})du$$

which matches the observed number of points $|\mathbf{x}|$ to a quantity that estimates the expected number of points if the model is true.

Example 3.38. For $h(u,\mathbf{x}) = t(u,\mathbf{x}) = \sum_i \mathbf{1}(\|u - x_i\| < r)$, the estimating function is

$$g(\theta;\mathbf{x}) = 2s(\mathbf{x}) - \int_W t(u,\mathbf{x})\lambda_\theta(u,\mathbf{x})du.$$

Note that $s(\mathbf{x})$ is closely related to $\hat{K}(r)$, so this is analogous to the minimum contrast method.

If $\lambda_\theta(u;\mathbf{x})$ is easily computable (e.g. pairwise interaction models), then the Takacs–Fiksel estimating function

$$g(\theta;\mathbf{x}) = \sum_i h(x_i, \mathbf{x} \setminus \{x_i\}) - \int h(u,\mathbf{x})\lambda_\theta(u,\mathbf{x})du$$

will be easily computable, for suitable choices of h. Hence it will be relatively easy to find $\tilde{\theta}$. However it is unclear which functions h we should choose.

Maximum Pseudolikelihood

Recall that the likelihood of a Poisson process with intensity function $\lambda_\theta(u)$ is

$$f(\mathbf{x};\theta) = \left[\prod_{i=1}^{|\mathbf{x}|} \lambda_\theta(x_i)\right] \exp\left\{\int_W (1 - \lambda_\theta(u))du\right\}.$$

For non-Poisson processes, Besag [65, 66] proposed replacing the likelihood by the following.

Definition 3.30. For a point process model with (Papangelou) conditional intensity $\lambda_\theta(u;\mathbf{x})$ depending on a parameter θ, the *pseudolikelihood* function is

$$\mathrm{PL}(\theta) = \prod_{i=1}^{|\mathbf{x}|} \lambda_\theta(x_i;\mathbf{x}) \exp\left\{-\int_W \lambda_\theta(u;\mathbf{x})du\right\}.$$

The *maximum pseudolikelihood estimate (MPLE)* is

$$\check{\theta} = \mathrm{argmax}\mathrm{PL}(\theta)$$

Up to a constant factor, the pseudolikelihood is identical to the likelihood if the model is Poisson, and approximately equal to the likelihood if the model is close to

Poisson. The pseudolikelihood is not a likelihood, in general (and is used only as a function of θ) but is supported by "deep statistical intuition".

Example 3.39. Consider a loglinear Gibbs model

$$f(\mathbf{x};\theta) = \alpha(\theta)\exp\{\theta^{\mathrm{T}} V(\mathbf{x})\}.$$

The conditional intensity is

$$\lambda_\theta(u,\mathbf{x}) = \frac{f(\mathbf{x}\cup\{u\})}{f(\mathbf{x})} = \exp\{\theta^{\mathrm{T}} T(u,\mathbf{x})\}$$

where $T(u,\mathbf{x}) = V(\mathbf{x}\cup\{u\}) - V(\mathbf{x})$. Then

$$\log \mathrm{PL}(\theta) = \sum_i \theta^{\mathrm{T}} T(x_i,\mathbf{x}) - \int_W \exp\{\theta^{\mathrm{T}} T(u,\mathbf{x})\} du.$$

The "pseudoscore" is

$$\begin{aligned}\mathrm{PU}(\theta) &= \frac{\partial \log \mathrm{PL}(\theta)}{\partial\theta}\\ &= \sum_i T(x_i,\mathbf{x}) - \int_W T(u,\mathbf{x})\exp\{\theta^{\mathrm{T}} T(u,\mathbf{x})\} du\\ &= \sum_i T(x_i,\mathbf{x}) - \int_W T(u,\mathbf{x})\lambda_\theta(u,\mathbf{x}) du.\end{aligned}$$

This is also the Takacs–Fiksel estimating function based on $T(u,\mathbf{x})$.

Exercise 3.22. Write down the pseudolikelihood and pseudoscore of the Strauss process (Example 3.34). Simplify them as far as possible.

Clyde and Strauss [127] and Baddeley and Turner [26] proposed algorithms for maximising the pseudolikelihood, in which the window W is divided into disjoint subsets $Q_1,\dots,Q_m$, the numbers $n_j = N(\mathbf{x}\cap Q_j)$ of points in each subset are counted, and the pseudolikelihood is approximated by the likelihood of a Poisson loglinear regression, which can be maximised using standard software.

Example 3.40. Figure 3.49 is a simulated realization of the Strauss process with $\beta = 100$, $\gamma = 0.7$, $r = 0.05$ in unit square. The Strauss parameters were fitted by MLE and MPLE yielding the following results.

Method	β	γ	Time taken
True	100	0.70	
MLE	132	0.60	5 s per 1,000 samples
MPLE	102	0.68	0.1 s

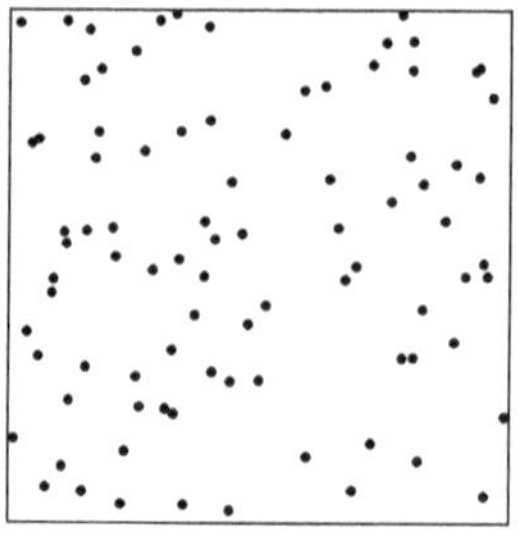

Fig. 3.49 Simulated realization of Strauss process

Time-Invariance Estimating Equations

It remains to understand why (in a more fundamental sense) maximum pseudolikelihood works. For this, we recall that in Sect. 3.2, we saw that a Gibbs point process $\mathbf{X}$ can be represented as the equilibrium distribution of a spatial birth-and-death process $\{\mathbf{Y}_n, n = 0, 1, 2, \ldots\}$.

Consider a time-homogeneous Markov chain $\{Y_n, n = 0, 1, 2, \ldots\}$ on a finite state space, with transition probability matrix P and equilibrium distribution vector $\boldsymbol{\pi}$. The *generator* is the matrix $A = P - I$ where I is the identity matrix. An important fact is

$$\pi A = 0$$

because

$$\pi A = \pi(P - I) = \pi P - \pi I = \pi - \pi = 0.$$

Definition 3.31. Let $\{\mathbf{Y}_n, n = 0, 1, 2, \ldots\}$ be a time-homogeneous Markov chain with states in $\mathcal{X}$. Let $\mathfrak{F}$ be a suitable class of functions $S : \mathcal{X} \to \mathbb{R}^k$. The *generator* of the chain is the operator $\mathcal{A}$ on $\mathfrak{F}$ defined by

$$(\mathcal{A}S)(\mathbf{x}) = \mathbf{E}(S(\mathbf{Y}_1) \mid \mathbf{Y}_0 = \mathbf{x}) - S(\mathbf{x})$$

for all $\mathbf{x} \in \mathcal{X}$.

Thus $(\mathcal{A}S)(\mathbf{x})$ is the expected change in the value of the function S occurring in one step of the chain, starting from state $\mathbf{x}$.

Lemma 3.15. *Let $\{\mathbf{Y}_n, n = 0, 1, 2, \ldots\}$ be a time-homogeneous Markov chain with states in $\mathcal{X}$ which has equilibrium distribution π. Let $\mathcal{A}$ be its generator. If $\mathbf{X}$ is a random element of $\mathcal{X}$ with distribution $\boldsymbol{\pi}$, then for any $S \in \mathfrak{F}$*

$$\mathbf{E}((\mathcal{A}S)(\mathbf{X})) = 0. \tag{3.19}$$

Proof. Let $\{\mathbf{Y}_n^*\}$ be a version of the chain which starts in state $\mathbf{X}$, i.e. $\mathbf{Y}_0^* = \mathbf{X}$. Since $\mathbf{X}$ is drawn from the equilibrium distribution $\boldsymbol{\pi}$, the distribution of $\mathbf{Y}_1^*$ is also $\boldsymbol{\pi}$. Hence

$$\mathbf{E}((\mathcal{A}S)(\mathbf{X})) = \mathbf{E}(\mathbf{E}(S(\mathbf{Y}_1^*) \mid \mathbf{Y}_0 = \mathbf{X}) - S(\mathbf{X})) = \mathbf{E}(S(\mathbf{Y}_1^*)) - \mathbf{E}(S(\mathbf{X})) = 0.$$

The proof is complete. □

Definition 3.32. Consider a model for a point process $\mathbf{X}$ with distribution π_θ governed by a parameter θ. For each θ let $(\mathbf{Y}_n^{(\theta)}, n = 0, 1, 2, \ldots)$ be a spatial birth-and-death process whose equilibrium distribution is π_θ. Let $\mathcal{A}_\theta$ be the generator. Then for any $S \in \mathfrak{F}$

$$g(\theta, \mathbf{x}) = (\mathcal{A}_\theta S)(\mathbf{x}) \tag{3.20}$$

is an unbiased estimating equation for θ called the *time-invariance estimating equation* [29].

More specifically, let the birth–death process have death probabilities $D(\mathbf{x}, x_i)$ and birth density $b(\mathbf{x}, u)$. Then the generator is

$$(\mathcal{A}S)(\mathbf{x}) = \sum_i D(\mathbf{x}, x_i)T(x_i, \mathbf{x}) - \int_W b(\mathbf{x}, u)T(u, \mathbf{x})du$$

where for any integrable function of point patterns $S(\mathbf{x})$ we define

$$T(u, \mathbf{x}) = S(\mathbf{x} \cup \{u\}) - S(\mathbf{x} \setminus \{u\}).$$

In particular if $\mathbf{X}$ is the point process with density $f(\mathbf{x}; \theta) \propto \exp\{\theta S(\mathbf{x})\}$ and we choose $(\mathbf{Y}_t)$ to be the Gibbs sampler

$$D(\mathbf{x}, x_i) = 1, \quad b(\mathbf{x}, u) = \lambda_\theta(u, \mathbf{x})$$

we obtain

$$(\mathcal{A}S)(\mathbf{x}) = \sum_i T(x_i, \mathbf{x}) - \int_W T(u, \mathbf{x})\lambda_\theta(\mathbf{x}, u)du \tag{3.21}$$

equal to the pseudoscore.

That is, maximum pseudolikelihood and other estimation methods are (mathematically) associated with MCMC samplers through time-invariance estimation. The following table summarises some of the main examples, where "canonical" refers to the canonical sufficient statistic V in an exponential family (Example 3.39).

Sampler $\mathbf{Y}$	Statistic S	Time-invariance estimator
Gibbs sampler	Canonical	Maximum pseudolikelihood
i.i.d. sampler	Canonical	Maximum likelihood
i.i.d. sampler	K-function	Minimum contrast

See [29] for details.

Literature and Software

For further technical background, we recommend [140, 154, 369]. Recent surveys of statistical methodology for point patterns are [180, 265]. Software for performing the simulations and data analysis is available [27] along with detailed notes [23].

Chapter 4
Asymptotic Methods in Statistics of Random Point Processes

Lothar Heinrich

Abstract First we put together basic definitions and fundamental facts and results from the theory of (un)marked point processes defined on Euclidean spaces $\mathbb{R}^d$. We introduce the notion *random marked point process* together with the concept of Palm distributions in a rigorous way followed by the definitions of factorial moment and cumulant measures and characteristics related with them. In the second part we define a variety of estimators of second-order characteristics and other so-called *summary statistics* of stationary point processes based on observations on a "convex averaging sequence" of windows $\{W_n,\ n \in \mathbb{N}\}$. Although all these (mostly edge-corrected) estimators make sense for fixed bounded windows our main issue is to study their behaviour when W_n grows unboundedly as $n \to \infty$. The first problem of *large-domain statistics* is to find conditions ensuring strong or at least mean-square consistency as $n \to \infty$ under ergodicity or other mild mixing conditions put on the underlying point process. The third part contains weak convergence results obtained by exhausting strong mixing conditions or even m-dependence of spatial random fields generated by Poisson-based point processes. To illustrate the usefulness of asymptotic methods we give two Kolmogorov–Smirnov-type tests based on K-functions to check *complete spatial randomness* of a given point pattern in $\mathbb{R}^d$.

4.1 Marked Point Processes: An Introduction

First we present a rigorous definition of the marked point process on Euclidean spaces with marks in some Polish space and formulate an existence theorem for marked point processes based on their finite-dimensional distributions. Further, all essential notions and tools of point process theory such as factorial moment

L. Heinrich (✉)
Augsburg University, Augsburg, Germany
e-mail: heinrich@math.uni-augsburg.de

E. Spodarev (ed.), *Stochastic Geometry, Spatial Statistics and Random Fields*, Lecture Notes in Mathematics 2068, DOI 10.1007/978-3-642-33305-7_4,

and cumulant measures with their densities and reduced versions as well as the machinery of Palm distributions in the marked and unmarked case are considered in detail.

4.1.1 Marked Point Processes: Definitions and Basic Facts

Point processes are mathematical models for irregular point patterns formed by randomly scattered points in some locally compact Hausdorff space. Throughout this chapter, this space will be the Euclidean space $\mathbb{R}^d$ of dimension $d \in \mathbb{N}$. In many applications to each point X_i of the pattern a further random element M_i, called *mark*, can be assigned which carries additional information and may take values in a rather general *mark space* M equipped with an appropriate σ-algebra $\mathcal{M}$. For example, for $d = 1$, the X_i's could be arrival times of customers and the M_i's their sojourn times in a queueing system and, for $d = 2$, one can interpret the X_i's as locations of trees in a forest with the associated random vectors M_i of stem diameter, stem height and distance to the nearest-neighbour tree. In this way we are led to the notion of a (random) *marked point process* which can be adequately modeled as random counting measure $\Psi_{\mathsf{M}}(\cdot)$ on Cartesian products $B \times L$, which gives the total number of points in a bounded subset B of $\mathbb{R}^d$ whose marks belong to a set of marks $L \subseteq \mathsf{M}$. To be precise, we need some further notation. Let $\mathcal{N}_{\mathsf{M}}$ denote the set of *locally finite counting measures* $\psi(\cdot)$ on the measurable product space $(\mathbb{R}^d \times \mathsf{M}, \mathcal{B}(\mathbb{R}^d) \otimes \mathcal{M})$, i.e. $\psi \in \mathcal{N}_{\mathsf{M}}$ is a σ-additive set function on $\mathcal{B}(\mathbb{R}^d) \otimes \mathcal{M}$ taking on non-negative integer values such that $\psi(B \times \mathsf{M}) < \infty$ for any bounded Borel set $B \in \mathcal{B}(\mathbb{R}^d)$. We then define $\mathfrak{N}_{\mathsf{M}}$ to be the smallest σ-algebra containing all sets $\{\psi \in \mathcal{N}_{\mathsf{M}} : \psi(B \times L) = n\}$ for $n \in \mathbb{N} \cup \{0\}$, any bounded $B \in \mathcal{B}(\mathbb{R}^d)$ and $L \in \mathcal{M}$. Finally, let $(\Omega, \mathcal{F}, \mathbf{P})$ be a hypothetical probability space on which all subsequent random elements will be defined.

Definition 4.1. A $(\mathcal{F}, \mathfrak{N}_{\mathsf{M}})$-measurable mapping

$$\Psi_{\mathsf{M}} \mid (\Omega, \mathcal{F}, \mathbf{P}) \mapsto (\mathcal{N}_{\mathsf{M}}, \mathfrak{N}_{\mathsf{M}}), \qquad \Omega \ni \omega \mapsto \Psi_{\mathsf{M}}(\omega, \cdot) \in \mathcal{N}_{\mathsf{M}}$$

is said to be a (random) marked point process (briefly: MPP) on $\mathbb{R}^d$ with mark space $(\mathsf{M}, \mathcal{M})$. In other words, a MPP $\Psi_{\mathsf{M}}(\cdot)$ (ω will be mostly suppressed) is a random locally finite counting measure on $(\mathbb{R}^d \times \mathsf{M}, \mathcal{B}(\mathbb{R}^d) \otimes \mathcal{M})$.

The probability measure $P_{\mathsf{M}}(A) = \mathbf{P}(\{\omega \in \Omega : \Psi_{\mathsf{M}}(\omega, \cdot) \in A\})$ for $A \in \mathfrak{N}_{\mathsf{M}}$ induced on $(\mathcal{N}_{\mathsf{M}}, \mathfrak{N}_{\mathsf{M}})$ is called the *distribution* of Ψ_{M}—briefly expressed by $\Psi_{\mathsf{M}} \sim P_{\mathsf{M}}$. Here and in what follows we put $\Psi(\cdot) = \Psi_{\mathsf{M}}(\cdot \times \mathsf{M})$ to denote the corresponding unmarked point process and write in general $\Psi \sim P$ to indicate point processes without marks. One often uses the notation

$$\Psi_{\mathsf{M}} = \sum_{i \geq 1} \delta_{(X_i, M_i)} \text{ or } \Psi = \sum_{i \geq 1} \delta_{X_i}, \tag{4.1}$$

where $\delta_x(A) = 1$ for $x \in A$ and $\delta_x(A) = 0$ for $x \notin A$ (Dirac measure). Note that due to the local finiteness of Ψ_{M} there are at most countably many atoms but each atom occurs with random (**P**-a.s. finite) multiplicity. The indexing in (4.1) does not need to be unique and the X_i's occur in the sums according to their multiplicity. In accordance with the general theory of random processes our next result formulates an analogue to Kolmogorov's extension theorem stating the existence of a probability space $(\Omega, \mathcal{F}, \mathbf{P})$ in Definition 4.1 in case of Polish mark spaces M given the family of *finite-dimensional distributions* $\mathbf{P}(\Psi_{\mathsf{M}}(B_1 \times L_1) = n_1, \ldots, \Psi_{\mathsf{M}}(B_k \times L_k) = n_k)$.

Theorem 4.1. *Let* M *be a Polish space equipped with the corresponding Borel-*σ*-algebra* $\mathcal{B}(\mathsf{M})$ *generated by a family* $\mathcal{M}_0$ *of subsets in* M *such that* $\mathcal{S} = \{ \times_{i=1}^{d} [a_i, b_i) \times L : -\infty < a_i \le b_i < \infty\,,\ L \in \mathcal{M}_0 \}$ *is a semi-ring which generates the ring of bounded Borel sets in* $\mathbb{R}^d \times \mathsf{M}$ *and each bounded Borel set* $X \subset \mathbb{R}^d \times \mathsf{M}$ *can be covered by a finite sequence* $S_1, \ldots, S_m \in \mathcal{S}$.

For any finite sequence of pairwise disjoint $B_1 \times L_1, \ldots, B_k \times L_k \in \mathcal{S}$ *define the distribution*

$$p_{n_1,\ldots,n_k}(B_1 \times L_1, \ldots, B_k \times L_k) \text{ for } n_1, \ldots, n_k = 0, 1, 2, \ldots$$

of a k*-dimensional random vector with non-negative integer-valued components.*

Then there exists a unique probability measure P_{M} *on the measurable space* $(\mathcal{N}_{\mathsf{M}}, \mathfrak{N}_{\mathsf{M}})$ *with finite-dimensional distributions*

$$P_{\mathsf{M}}(\{\psi \in \mathcal{N}_{\mathsf{M}} : \psi(B_j \times L_j) = n_j,\ j = 1, \ldots, k\}) = p_{n_1,\ldots,n_k}(B_1 \times L_1, \ldots, B_k \times L_k)$$

for $n_1, \ldots, n_k \in \mathbb{N} \cup \{0\}$ *and any* $k \in \mathbb{N}$, *if the following conditions for the family of probabilities* $p_{n_1,\ldots,n_k}(B_1 \times L_1, \ldots, B_k \times L_k)$ *are satisfied:*

1. *Symmetry:*

$$p_{n_1,\ldots,n_k}(B_1 \times L_1, \ldots, B_k \times L_k) = p_{n_{\pi(1)},\ldots,n_{\pi(k)}}(B_{\pi(1)} \times L_{\pi(1)}, \ldots, B_{\pi(k)} \times L_{\pi(k)})$$

for any permutation $\pi : \{1, \ldots, k\} \mapsto \{1, \ldots, k\}$,

2. *Consistency:*

$$\sum_{n=0}^{\infty} p_{n_1,\ldots,n_{k-1},n}(B_1 \times L_1, \ldots, B_{k-1} \times L_{k-1}, B_k \times L_k) = p_{n_1,\ldots,n_{k-1}}(B_1 \times L_1, \ldots, B_{k-1} \times L_{k-1}),$$

3. *Additivity: If* $B_j \times L_j \cup \cdots \cup B_k \times L_k \in \mathcal{S}$, *then*

$$\sum_{\substack{n_j+\cdots+n_k=n\\ n_j,\ldots,n_k\geq 0}} p_{n_1,\ldots n_{j-1},n_j,\ldots,n_k}(B_1\times L_1,\ldots,B_{j-1}\times L_{j-1},B_j\times L_j,\ldots,B_k\times L_k)$$
$$= p_{n_1,\ldots,n_{j-1},n}(B_1\times L_1,\ldots,B_{j-1}\times L_{j-1},(B_j\times L_j\cup\cdots\cup B_k\times L_k))$$

for $j=1,\ldots,k$.

4. *Continuity: For any sequence of pairwise disjoint sets* $B_j^{(n)}\times L_j^{(n)}\in\mathcal{S}$ *for* $j=1,\ldots,k_n$ *with* $k_n\uparrow\infty$ *satisfying* $\bigcup_{j=1}^{k_n}(B_j^{(n)}\times L_j^{(n)})\downarrow\emptyset$, *it holds* $p_{0,\ldots,0}(B_1^{(n)}\times L_1^{(n)},\ldots,B_{k_n}^{(n)}\times L_{k_n}^{(n)})\uparrow 1$ *as* $n\to\infty$.
If $\mathsf{M}=\{m\}$ *consists of a single (or at most finitely many) element(s), then the latter condition can be replaced by*

$$\lim_{n\to\infty} p_0(\mathop{\times}_{i=1}^{j-1}[a_i,b_i)\times[x_j-\frac{1}{n},x_j)\times\mathop{\times}_{i=j+1}^{d}[a_i,b_i)\times\{m\})=1$$

for all $(x_1,\ldots,x_d)^\top\in\mathbb{R}^d$ *and* $-\infty<a_i<b_i<\infty$, $i=1,\ldots,d$.

This allows a canonical choice of the probability space $(\Omega,\mathcal{F},\mathbf{P})$ mapping Ψ_{M} :

$$\Omega:=\mathcal{N}_{\mathsf{M}},\ \ \mathcal{F}:=\mathfrak{N}_{\mathsf{M}},\ \ \mathbf{P}:=P_{\mathsf{M}},\ \ \psi\mapsto\Psi_{\mathsf{M}}(\psi,\cdot):=\psi(\cdot)\ \ \text{(identical mapping)}.$$

Remark 4.1. There exists a metric τ in the set $\mathcal{N}_{\mathsf{M}}$ such that the metric space $(\mathcal{N}_{\mathsf{M}},\tau)$ is separable and complete whose Borel-σ-algebra $\mathcal{B}(\mathcal{N}_{\mathsf{M}})$ coincides with $\mathfrak{N}_{\mathsf{M}}$. This allows to introduce the notion of *weak convergence* of MPP's in the usual way, see [69].

To prove Theorem 4.1 one has only to reformulate the proof of a corresponding result for (unmarked) PP's on Polish spaces in [346]. Readers interested in more background of and a rigouros introduction to the theory of marked and unmarked PP's are referred to the two-volume monograph [140]. Less technical approaches combined with statistics of point processes are presented in [265, 489] and in the survey papers [22, 488]. In the following we reduce the rigour and in particular measurability questions will be not considered.

An advantage of the counting measure approach to point processes (in contrast to modelling with discrete random closed sets) consists in catching random multiplicities of the point or atoms which is important in several fields of application. For example, the description of batch arrivals in queueing systems or of end-points of edges in planar random tessellations requires multiplicities of points. On the other hand, for quite a few point processes in particular for the upmost occurring in stochastic geometry it is natural to assume that at most one point occurs in any $x\in\mathbb{R}^d$, more precisely

$$P_{\mathsf{M}}(\{\psi\in\mathcal{N}_{\mathsf{M}}:\psi(\{x\}\times\mathsf{M})\leq 1,\ \forall\,x\in\mathbb{R}^d\})=1\,.$$

MPP's $\Psi_{\mathsf{M}} \sim P_{\mathsf{M}}$ satisfying the later condition are called *simple*. In view of (4.1) we get $\text{supp}(\Psi_{\mathsf{M}}) = \{(X_i, M_i) : i \in \mathbb{N}\}$ for the support set of a simple MPP Ψ_{M} which motivates the somewhat loose writing $\Psi_{\mathsf{M}} = \{(X_i, M_i) : i \in \mathbb{N}\}$. By the **P**-a.s. local boundedness of the counting measure Ψ_{M} its support set has no accumulation point in $\mathbb{R}^d$ and can therefore be considered as a discrete *random closed set*. The characterization of distributions of random closed sets Ξ by the family of hitting probabilities $\mathbf{P}(\Xi \cap K \neq \emptyset)$ for any compact set $K \subset \mathbb{R}^d$ leads to the following pendant to Theorem 4.1:

Theorem 4.2. *Let $\Psi_{\mathsf{M}} \sim P_{\mathsf{M}}$ be a simple MPP on $\mathbb{R}^d$ with Polish mark space M. Then the distribution P_{M} is completely determined by the void probabilities*

$$\mathbf{P}(\Psi_{\mathsf{M}}(X) = 0) = P_{\mathsf{M}}(\{\psi \in \mathcal{N}_{\mathsf{M}} : \psi(X) = 0\})$$

for all compact $X \subset \mathbb{R}^d \times \mathsf{M}$.

For the sake of simplicity we shall consider only simple MPP's after the introductory Sect. 4.1.1.

The simplest numerical characteristic of a MPP $\Psi_{\mathsf{M}} \sim P_{\mathsf{M}}$ describes the mean number of points in bounded sets $B \in \mathcal{B}(\mathbb{R}^d)$ having marks in an arbitrary set $L \in \mathcal{B}(\mathsf{M})$. In this way we obtain the *intensity measure* Λ_{M} (on $\mathcal{B}(\mathbb{R}^d) \otimes \mathcal{B}(\mathsf{M})$) defined by

$$\Lambda_{\mathsf{M}}(B \times L) = \mathbf{E}\Big(\sum_{i \geq 1} \mathbf{1}((X_i, M_i) \in B \times L)\Big) = \int_{\mathcal{N}_{\mathsf{M}}} \psi(B \times L)\, P_{\mathsf{M}}(d\psi) \quad (4.2)$$

provided that $\Lambda(B) := \Lambda_{\mathsf{M}}(B \times \mathsf{M}) < \infty$ for any bounded $B \in \mathcal{B}(\mathbb{R}^d)$ expressing the local finiteness of Λ_{M}. By Theorem 4.1 we are now in a position to define the *marked Poisson process* $\Psi_{\mathsf{M}} \sim \Pi_{\Lambda_{\mathsf{M}}}$ with a given locally finite intensity measure Λ_{M}:

Definition 4.2. A marked Poisson process $\Psi_{\mathsf{M}} \sim \Pi_{\Lambda_{\mathsf{M}}}$ (more precisely its distribution) is completely determined by the following two conditions:

1. $\Psi_{\mathsf{M}}(B_1 \times L_1), \ldots, \Psi_{\mathsf{M}}(B_k \times L_k)$ are mutually independent random variables for pairwise disjoint $B_j \times L_j \in \mathcal{B}(\mathbb{R}^d) \times \mathcal{B}(\mathsf{M})$ with bounded B_j for $j = 1, \ldots, k$ and $k \in \mathbb{N}$.
2. $\Psi_{\mathsf{M}}(B \times L)$ is Poisson distributed with mean $\Lambda_{\mathsf{M}}(B \times L)$ for any $B \times L \in \mathcal{B}(\mathbb{R}^d) \times \mathcal{B}(\mathsf{M})$ with bounded B.

Remark 4.2. Since $\Lambda_{\mathsf{M}}(B \times L) \leq \Lambda(B)$ there exists (by the Radon–Nikodym theorem and disintegration arguments) a family $\{Q_{\mathsf{M}}^x, x \in \mathbb{R}^d\}$ of (regular) conditional distributions on $(\mathsf{M}, \mathcal{B}(\mathsf{M}))$ such that $\Lambda_{\mathsf{M}}(B \times L) = \int_{\mathbb{R}^d} Q_{\mathsf{M}}^x(L)\, \Lambda(dx)$, which justifies the interpretation $Q_{\mathsf{M}}^x(L) = \mathbf{P}(M_i \in L \mid X_i = x)$. It turns out that $\Psi_{\mathsf{M}} \sim \Pi_{\Lambda_{\mathsf{M}}}$ can be obtained from an *unmarked Poisson process* $\Psi \sim \Pi_\Lambda$ with *intensity measure* Λ by *location-dependent, independent marking*, that is, to each atom X_i of Ψ located in $x \in \mathbb{R}^d$ the mark M_i is assigned according to the probability law Q_{M}^x independent of the location of all other atoms of Ψ and also independent of all other marks even those assigned to further atoms located in x (if $\Psi(\{x\}) > 1$).

Remark 4.3. A marked Poisson process $\Psi_{\mathsf{M}} \sim \Pi_{\Lambda_{\mathsf{M}}}$ is simple iff the intensity measure $\Lambda(\cdot)$ is *diffuse*, i.e. $\Lambda_{\mathsf{M}}(\{x\} \times \mathsf{M}) = 0$ for all $x \in \mathbb{R}^d$, see [346].

Remark 4.4. The conditions (1) and (2) in the above definition of $\Psi_{\mathsf{M}} \sim \Pi_{\Lambda_{\mathsf{M}}}$ can be expressed equivalently by means of the characteristic function of the random vector $(\Psi_{\mathsf{M}}(B_1 \times L_1), \ldots, \Psi_{\mathsf{M}}(B_k \times L_k))$ as follows: For any $k \in \mathbb{N}$, any pairwise disjoint $B_1 \times L_1, \ldots, B_k \times L_k \in \mathcal{B}(\mathbb{R}^d) \times \mathcal{B}(\mathsf{M})$ and all $u_1, \ldots, u_k \in \mathbb{R}^d$

$$\mathbf{E}\exp\{\, i \sum_{j=1}^{k} u_j\, \Psi_{\mathsf{M}}(B_j \times L_j)\,\} = \prod_{j=1}^{k} \exp\{\Lambda_{\mathsf{M}}(B_j \times L_j)\,(\mathrm{e}^{i\, u_j} - 1\,)\}\,. \tag{4.3}$$

Next, we give an elementary explicit construction of an unmarked Poisson process $\Psi \sim \Pi_\Lambda$ with locally finite intensity measure $\Lambda(\cdot)$, see [102]. This construction is also the basis for simulations of Poisson processes in bounded Borel sets.

Let $\{K_m,\ m \in \mathbb{N}\}$ be a partition of $\mathbb{R}^d$ into bounded Borel sets. Consider an array of independent random variables $\{\tau_m, X_{mj}\}_{m,j\in\mathbb{N}}$ defined on $(\Omega, \mathcal{F}, \mathbf{P})$ such that τ_m and X_{mj} take values in $\mathbb{Z}_+$ and $\mathbb{R}^d$ respectively, namely,

1. τ_m is Poisson distributed with mean $\Lambda(K_m)$ (briefly $\tau_m \sim \mathrm{Pois}(\Lambda(K_m))$).
2. $\mathbf{P}(X_{mj} \in C) = \begin{cases} \Lambda(C \cap K_m)/\Lambda(K_m), & \Lambda(K_m) \neq 0, \\ 0, & \text{otherwise} \end{cases}$ for any $C \in \mathcal{B}(\mathbb{R}^d)$.

Here $Y \sim \mathrm{Pois}(0)$ means that $Y \equiv 0$. Clearly, $\mathbf{P}(X_{mj} \in C) = \mathbf{P}(X_{mj} \in C \cap K_m)$ for $m, j \in \mathbb{N}$. Note that the random variables X_{mj} are uniformly distributed on K_m if $\Lambda(K_m)$ is a multiple of $\nu_d(K_m)$.

For any $B \in \mathcal{B}(\mathbb{R}^d)$ and $m \in \mathbb{N}$ put

$$\Psi_m(B) := \sum_{j=1}^{\tau_m} \mathbf{1}(X_{mj} \in B) \quad \text{and} \quad \Psi(B) := \sum_{m=1}^{\infty} \Psi_m(B)\,. \tag{4.4}$$

Obviously, $\Psi_m(B)$ is a random variable for each $m \in \mathbb{N}$ and any $B \in \mathcal{B}(\mathbb{R}^d)$ such that

$$\mathbf{P}(\,0 \leq \Psi_m(B) \leq \tau_m < \infty\,) = 1 \quad \text{for any} \quad B \in \mathcal{B}(\mathbb{R}^d)\,,\ m \in \mathbb{N}\,.$$

Moreover, it turns out that $\Psi(\cdot)$ is a locally finite random counting measure.

Theorem 4.3. *Let Λ be a locally finite measure on $(\mathbb{R}^d, \mathcal{B}(\mathbb{R}^d))$. For any partition of $\mathbb{R}^d$ into bounded Borel sets K_m, $m \in \mathbb{N}$, the family of non-negative,* *integer-valued random variables* $\{\Psi(B),\, B \in \mathcal{B}(\mathbb{R}^d)\}$ *introduced in* (4.4) *defines an unmarked Poisson process with intensity measure* $\Lambda(\cdot)$.

Proof. For any $m, k \in \mathbb{N}$ consider pairwise disjoint Borel sets $B_1, \ldots, B_k \subset K_m$. Then $\Psi(B_r) = \Psi_m(B_r)$ for $r = 1, \ldots, k$ and it follows that

$$\begin{aligned}
&\mathbf{E}\exp\{i\,(u_1\,\Psi_m(B_1) + \ldots + u_k\,\Psi_m(B_k))\}\\
&= \sum_{n=0}^{\infty} \mathbf{E}\Big(\exp\Big\{i\,\sum_{j=1}^{n}\big(u_1\,\mathbf{1}(X_{mj} \in B_1) + \ldots + u_k\,\mathbf{1}(X_{mj} \in B_k)\big)\Big\}\mathbf{1}(\tau_m = n)\Big)\\
&= \sum_{n=0}^{\infty} \varphi_n(u_1, \ldots, u_k)\frac{\Lambda(K_m)^n}{n!}\mathrm{e}^{-\Lambda(K_m)}, \qquad (4.5)
\end{aligned}$$

where $u_1, \ldots, u_k \in \mathbb{R}^1$ and $\varphi_n(u_1, \ldots, u_k)$ is the characteristic function of the random vector $Y_n = (Y_{n1}, \ldots, Y_{nk})$ with components $Y_{nr} = \sum_{j=1}^{n} \mathbf{1}(X_{mj} \in B_r)$ for $r = 1, \ldots, k$.

By setting $Y_{n0} = \sum_{j=1}^{n} \mathbf{1}(X_{mj} \in B_0)$ for $B_0 = K_m \setminus \cup_{r=1}^{n} B_r$ we get a multinomially distributed random vector $(Y_{n0}, Y_{n1}, \ldots, Y_{nk})$ with success probabilities $p_r = \mathbf{P}(X_{mj} \in B_r) = \Lambda(B_r)/\Lambda(K_m)$, $r = 0, 1, \ldots, k$, i.e.,

$$\mathbf{P}(Y_{n0} = l_0, Y_{n1} = l_1, \ldots, Y_{nk} = l_k) = \frac{n!}{l_0!\, l_1! \ldots l_k!}\, p_0^{l_0}\, p_1^{l_1} \ldots p_k^{l_k}$$

for $l_0, l_1, \ldots, l_k \geq 0$ and $\sum_{r=0}^{k} l_r = n$. Hence, $\zeta_n(u_0, u_1, \ldots, u_k) = (\,p_0\,\mathrm{e}^{i\,u_0} + p_1\,\mathrm{e}^{i\,u_1} + \ldots + p_k\,\mathrm{e}^{i\,u_k}\,)^n$, $u_0, \ldots, u_k \in \mathbb{R}^1$ is the characteristic function of $(Y_{n0}, Y_{n1}, \ldots, Y_{nk})$ and we obtain

$$\varphi_n(u_1, \ldots, u_k) = \zeta_n(0, u_1, \ldots, u_k) = (\,p_0 + p_1\,\mathrm{e}^{i\,u_1} + \ldots + p_k\,\mathrm{e}^{i\,u_k}\,)^n.$$

Using (4.5) and the latter relation together with $p_0 = 1 - p_1 - \ldots - p_k$ we may write

$$\begin{aligned}
\mathbf{E}\exp\{i\,\sum_{r=1}^{k} u_r\,\Psi_m(B_r)\} &= \mathrm{e}^{-\Lambda(K_m)}\sum_{n=0}^{\infty}\frac{1}{n!}\,[\,\Lambda(K_m)(p_0 + p_1\,\mathrm{e}^{i\,u_1} + \ldots + p_k\,\mathrm{e}^{i\,u_k})\,]^n\\
&= \exp\{\Lambda(K_m)(p_1(\mathrm{e}^{i\,u_1} - 1) + \ldots + p_k(\mathrm{e}^{i\,u_k} - 1))\}\\
&= \prod_{r=1}^{k}\exp\{\Lambda(B_r)(\mathrm{e}^{i\,u_r} - 1)\}\,.
\end{aligned}$$

This, however, is just (4.3) (rewritten for the unmarked case) for pairwise disjoint Borel sets $B_1, \ldots, B_k \subset K_m$.

Now, consider pairwise disjoint bounded $B_1, \ldots, B_k \in \mathcal{B}(\mathbb{R}^d)$ for $k \in \mathbb{N}$ so that

$$(\Psi(B_1), \ldots, \Psi(B_k)) = \sum_{m=1}^{\infty} (\Psi_m(B_1 \cap K_m), \ldots, \Psi_m(B_k \cap K_m)) .$$

Obviously, $(\Psi_m(B_1 \cap K_m), \ldots, \Psi_m(B_k \cap K_m))_{m \in \mathbb{N}}$ forms a sequence of independent random vectors with independent components as we have proved above. Since $\Psi(B_r) = \sum_{m=1}^{\infty} \Psi_m(B_r \cap K_m)$ and the summands are nonnegative, it follows that

$$\mathbf{E}\Psi(B_r) = \sum_{m=1}^{\infty} \mathbf{E}\Psi_m(B_r \cap K_m) = \sum_{m=1}^{\infty} \Lambda(B_r \cap K_m) = \Lambda(B_r) < \infty$$

implying $\mathbf{P}(\,\Psi(B_r) < \infty\,) = 1$ for $r = 1, \ldots, k$, and

$$\Big(\sum_{m=1}^{n} \Psi_m(B_1 \cap K_m), \ldots, \sum_{m=1}^{n} \Psi_m(B_k \cap K_m)\Big) \underset{n \to \infty}{\overset{\mathbf{P}-\text{a.s.}}{\longrightarrow}} (\Psi(B_1), \ldots, \Psi(B_k)) .$$

Since, for each fixed $n \in \mathbb{N}$, the sums at the left-hand side of the latter relation are mutually independent, it follows that the limits $\Psi(B_1), \ldots, \Psi(B_k)$ are independent as well. Finally, in view of the fact that

$$\sum_{m=1}^{n} \Psi_m(B_r \cap K_m) \sim \text{Pois}\Big(\sum_{m=1}^{n} \Lambda(B_r \cap K_m)\Big)$$

we conclude that $\Psi(B_r) \sim \text{Pois}(\Lambda(B_r))$ for $r = 1, \ldots, k$, which completes the proof of Theorem 4.3. $\square$

Next, we introduce the translation- and rotation operator T_x resp. $R_O \mid \mathcal{N}_{\mathsf{M}} \mapsto \mathcal{N}_{\mathsf{M}}$ by

$$(T_x\psi)(\,B \times L\,) := \psi(\,(B + x) \times L\,) \quad \text{resp.} \quad (R_O\psi)(\,B \times L\,) := \psi(\,O\,B \times L\,)$$

for $\psi \in \mathcal{N}_{\mathsf{M}}$, $x \in \mathbb{R}^d$ and $O \in \mathrm{SO}_d$ = group of orthogonal $d \times d$-matrices with determinant equal to 1. Then the MPP $\Psi_{\mathsf{M}} \sim P_{\mathsf{M}}$ is said to be (strictly) *stationary* or *homogeneous* if, for all $x \in \mathbb{R}^d$ and bounded Borel sets $B_1, \ldots, B_k$,

$$T_x\,\Psi_{\mathsf{M}} \overset{d}{=} \Psi_{\mathsf{M}} \quad \Longleftrightarrow \quad (\Psi_{\mathsf{M}}((B_j + x) \times L_j))_{j=1}^k \overset{d}{=} (\Psi_{\mathsf{M}}(B_j \times L_j))_{j=1}^k\,,\ k \in \mathbb{N}\,,$$

and *isotropic* if, for all $O \in \mathrm{SO}_d$ and bounded Borel sets $B_1, \ldots, B_k$,

$$R_O\,\Psi_{\mathsf{M}} \overset{d}{=} \Psi_{\mathsf{M}} \quad \Longleftrightarrow \quad (\Psi_{\mathsf{M}}(OB_j \times L_j))_{j=1}^k \overset{d}{=} (\Psi_{\mathsf{M}}(B_j \times L_j))_{j=1}^k\,,\ k \in \mathbb{N}\,.$$

A MPP is said to be *motion-invariant* if it is both stationary and isotropic.

The stationarity of $\Psi_{\mathsf{M}} \sim P_{\mathsf{M}}$ implies the shift-invariance of the locally finite measure $\Lambda_{\mathsf{M}}((\cdot) \times L)$ for any fixed $L \in \mathcal{B}(\mathsf{M})$ provided the *intensity*

$$\lambda := \mathbf{E}\Psi_{\mathsf{M}}([0,1]^d \times \mathsf{M})$$

of the unmarked point process $\Psi \sim P$ exists. This entails that the intensity measure $\Lambda_{\mathsf{M}}(B \times L)$ is a multiple of the Lebesgue-measure $\nu_d(B)$ which can be rewritten as

$$\Lambda_{\mathsf{M}}(B \times L) = \lambda\, \nu_d(B)\, Q_{\mathsf{M}}^o(L) \text{ for } B \in \mathcal{B}(\mathbb{R}^d),\ L \in \mathcal{B}(\mathsf{M})\,, \tag{4.6}$$

where Q_{M}^o is called the *distribution of the typical mark* or plainly the *mark distribution*.

4.1.2 Higher-Order Moment Measures and Palm Distributions

From now on we suppose that the MPP $\Psi_{\mathsf{M}} = \sum_{i\geq 1} \delta_{(X_i, M_i)} \sim P_{\mathsf{M}}$ is simple. Under the additional assumption $\mathbf{E}((\Psi_{\mathsf{M}}(B \times \mathsf{M}))^k) < \infty$ for some integer $k \geq 2$ and any bounded Borel set $B \subset \mathbb{R}^d$ we define the *factorial moment measure* $\alpha_{\mathsf{M}}^{(k)}$ on $\mathcal{B}((\mathbb{R}^d \times \mathsf{M})^k)$ by

$$\alpha_{\mathsf{M}}^{(k)}\big(\mathop{\times}_{j=1}^{k} (B_j \times L_j)\big) := \mathbf{E}\Big(\sum_{i_1,\dots,i_k \geq 1}^{\neq} \prod_{j=1}^{k} \mathbf{1}((X_{i_j}, M_{i_j}) \in B_j \times L_j)\Big)$$

which is dominated by the kth-order factorial moment measure

$$\alpha^{(k)}\big(\mathop{\times}_{j=1}^{k} B_j\big) = \alpha_{\mathsf{M}}^{(k)}\big(\mathop{\times}_{j=1}^{k} (B_j \times \mathsf{M})\big)$$

of the unmarked simple point process $\Psi = \sum_{i \geq 1} \delta_{X_i} \sim P$. Note that the sum $\sum^{\neq}$ stretches over pairwise distinct indices indicated under the sum sign. If the sum is taken over all k-tuples of indices we get the (ordinary) kth-order moment measure.

For any fixed $L_1, \dots, L_k \in \mathcal{B}(\mathsf{M})$ we obtain as Radon–Nikodym derivative a family of distributions $Q_{\mathsf{M}}^{x_1,\dots,x_k}(L_1 \times \cdots \times L_k)$ (for $\alpha^{(k)}$-almost every $(x_1, \dots, x_k)^\top \in (\mathbb{R}^d)^k$) satisfying

$$\alpha_{\mathsf{M}}^{(k)}\big(\mathop{\times}_{j=1}^{k} (B_j \times L_j)\big) = \int_{B_1} \cdots \int_{B_k} Q_{\mathsf{M}}^{x_1,\dots,x_k}\big(\mathop{\times}_{j=1}^{k} L_j\big)\, \alpha^{(k)}(d(x_1, \dots, x_k))\,,$$

where the integrand is interpreted as (regular) conditional distribution

$$Q_{\mathsf{M}}^{x_1,\dots,x_k}\big(\mathop{\times}_{j=1}^{k} L_j\big) = \mathbf{P}(M_1 \in L_1, \dots, M_k \in L_k \mid X_1 = x_1, \dots, X_k = x_k)\,.$$

Definition 4.3. The stochastic kernel $Q_{\mathsf{M}}^{x_1,\dots,x_k}(A)$ (which is defined to be $\mathcal{B}(\mathbb{R}^{dk})$-measurable in $(x_1,\dots,x_k)^\top \in (\mathbb{R}^d)^k$ and a probability measure in $A \in \mathcal{B}(\mathsf{M}^k)$) is called *kth-order* or *k-point Palm mark distribution* of the MPP $\Psi_{\mathsf{M}} \sim P_{\mathsf{M}}$.

If $\Psi_{\mathsf{M}} \sim P_{\mathsf{M}}$ is stationary then the factorial moment measures $\alpha_{\mathsf{M}}^{(k)}$ as well as $\alpha^{(k)}$ are invariant under diagonal shifts, i.e.,

$$\alpha_{\mathsf{M}}^{(k)}\Big(\mathop{\times}_{j=1}^{k}((B_j + x) \times L_j)\Big) = \alpha_{\mathsf{M}}^{(k)}\Big(\mathop{\times}_{j=1}^{k}(B_j \times L_j)\Big) \text{ for all } x \in \mathbb{R}^d ,$$

which in turn implies $Q_{\mathsf{M}}^{x_1+x,\dots,x_k+x}(A) = Q_{\mathsf{M}}^{x_1,\dots,x_k}(A)$ for all $x \in \mathbb{R}^d$. By disintegration with respect to the Lebesgue-measure ν_d (see [140, Vol. II] for the more details) we can introduce so-called *reduced kth-order factorial moment measures* $\alpha_{\mathsf{M},red}^{(k)}$ and $\alpha_{red}^{(k)}$ by

$$\alpha_{\mathsf{M}}^{(k)}\Big(\mathop{\times}_{j=1}^{k}(B_j \times L_j)\Big) = \lambda \int_{B_1} \alpha_{\mathsf{M},red}^{(k)}\Big(L_1 \times \mathop{\times}_{j=2}^{k}((B_j - x) \times L_j)\Big)\, dx ,$$

where $\lambda > 0$ stands for the intensity of $\Psi \sim P$ which already occurred in (4.6). Putting $\alpha_{red}^{(k)}\Big(\mathop{\times}_{j=2}^{k} B_j\Big) = \alpha_{\mathsf{M},red}^{(k)}\Big(\mathsf{M} \times \mathop{\times}_{j=2}^{k}((B_j - x) \times \mathsf{M})\Big)$ we obtain an analogous relation between $\alpha^{(k)}$ and $\alpha_{red}^{(k)}$ for the unmarked PP $\Psi \sim P$. Rewriting this relation as integrals over indicator functions we are led by algebraic induction to

$$\int_{(\mathbb{R}^d)^k} f(x_1, x_2, \dots, x_k)\alpha^{(k)}\big(d(x_1, x_2, \dots, x_k)\big)$$
$$= \lambda \int_{\mathbb{R}^d} \int_{(\mathbb{R}^d)^{k-1}} f(x_1, x_2 + x_1, \dots, x_k + x_1)\alpha_{red}^{(k)}\big(d(x_2, \dots, x_k)\big)\, dx_1 .$$

for any non-negative $\mathcal{B}(\mathbb{R}^{dk})$-measurable function $f : (\mathbb{R}^d)^k \mapsto \mathbb{R}^1$. Setting $f(x_1, x_2, \dots, x_k) = \mathbf{1}(x_1 \in B)\,\mathbf{1}((x_2 - x_1, \dots, x_k - x_1) \in C)$ for an arbitrary bounded set $C \in \mathcal{B}(\mathbb{R}^{d(k-1)})$ and any $B \in \mathcal{B}(\mathbb{R}^d)$ with $\nu_d(B) = 1$, for example $B = [0,1]^d$, we arrive at the formula

$$\alpha_{red}^{(k)}(C)$$
$$= \frac{1}{\lambda} \int_{(\mathbb{R}^d)^k} \mathbf{1}(x_1 \in B)\,\mathbf{1}((x_2 - x_1, \dots, x_k - x_1) \in C)\,\alpha^{(k)}\big(d(x_1, x_2, \dots, x_k)\big) \qquad (4.7)$$

which again confirms that $\alpha_{red}^{(k)}$ is a locally-finite measure on $(\mathbb{R}^{d(k-1)}, \mathcal{B}(\mathbb{R}^{d(k-1)}))$. Below it will be shown that $\alpha_{red}^{(k)}$ coincides with the $(k-1)$st-order factorial moment measure with respect to the reduced Palm distribution. It should be mentioned that both factorial moment measures $\alpha^{(k)}$ and $\alpha^{(k)}_{red}$ are symmetric in their components and the reduction is possible in any component of $\alpha^{(k)}$ yielding $\alpha^{(k)}_{red}(-C) = \alpha^{(k)}_{red}(C)$. Furthermore, the above reduction and therefore the definition of $\alpha^{(k)}_{red}$ needs only the diagonal shift-invariance of $\alpha^{(k)}$ (and of the intensity measure Λ) and not the shift-invariance of the finite-dimensional distributions of all orders and even not of the kth order.

This fact gives rise to consider weaker notions of kth-order (moment) stationarity, see Sect. 4.2.1 for $k = 2$.

A rather technical, but useful tool in expressing and studying dependences between distant parts of a stationary point pattern is based on so-called (factorial) cumulant measures of both marked and unmarked PP's. The origins of these characteristics can be traced back up to the beginning of the systematic study of random processes, random fields and particle configurations in statistical mechanics. In probability theory cumulants of random variables or vectors are defined by logarithmic derivatives of the corresponding moment-generating or characteristic functions. Along this line cumulant measures of point processes are defined by mixed higher-order partial derivatives of the logarithm of the probability generating functional of the point process, see [140, Chap. 9]. This approach is the background of the following

Definition 4.4. For any fixed $L_1,\ldots,L_k \in \mathcal{B}(\mathsf{M})$ and bounded $B_1,\ldots,B_k \in \mathcal{B}(\mathbb{R}^d)$ we define the *kth-order factorial cumulant measure* $\gamma^{(k)}_{\mathsf{M}}$ of the MPP $\Psi_{\mathsf{M}} \sim P_{\mathsf{M}}$ by

$$\gamma^{(k)}_{\mathsf{M}}\Big(\mathop{\times}_{j=1}^{k}(B_j \times L_j)\Big) := \sum_{j=1}^{k}(-1)^{j-1}(j-1)! \sum_{\substack{K_1\cup\cdots\cup K_j\\ =\{1,\ldots,k\}}} \prod_{r=1}^{j} \alpha^{(|K_r|)}_{\mathsf{M}}\Big(\mathop{\times}_{s\in K_r}(B_s \times L_s)\Big),$$

where the sum $\sum_{K_1\cup\cdots\cup K_j=\{1,\ldots,k\}}$ is taken over all partitions of the set $\{1,\ldots,k\}$ into j non-empty sets $K_1,\ldots,K_j$ and $|K_r|$ denotes the cardinality of K_r.

In general, $\gamma^{(k)}_{\mathsf{M}}$ is a locally finite *signed measure* on $\mathcal{B}((\mathbb{R}^d \times \mathsf{M})^k)$ which in case of a stationary MPP $\Psi_{\mathsf{M}} \sim P_{\mathsf{M}}$ can also be reduced in analogy to $\alpha^{(k)}_{\mathsf{M}}$ which leads to

$$\gamma^{(k)}_{\mathsf{M}}\Big(\mathop{\times}_{j=1}^{k}(B_j \times L_j)\Big) = \lambda \int_{B_1} \gamma^{(k)}_{\mathsf{M},red}\Big(L_1 \times \mathop{\times}_{j=2}^{k}((B_j - x) \times L_j)\Big)\,dx$$

By setting in the latter formula $L_1 = \ldots = L_k = \mathsf{M}$ we obtain the corresponding relationship between the kth-order factorial cumulant measure $\gamma^{(k)}$ and the reduced kth-order factorial cumulant measure $\gamma^{(k)}_{red}$ of the unmarked point process $\Psi \sim P$. In the special case $k = 2$ we have

$$\gamma_{\mathsf{M}}^{(2)}(B_1 \times L_1 \times B_2 \times L_2) = \alpha_{\mathsf{M}}^{(2)}(B_1 \times L_1 \times B_2 \times L_2) - \Lambda_{\mathsf{M}}(B_1 \times L_1)\Lambda_{\mathsf{M}}(B_2 \times L_2)$$

$$\gamma_{\mathsf{M},red}^{(2)}(L_1 \times B_2 \times L_2) = \alpha_{\mathsf{M},red}^{(2)}(L_1 \times B_2 \times L_2) - \lambda\, Q_{\mathsf{M}}^{o}(L_1)\, Q_{\mathsf{M}}^{o}(L_2)\, \nu_d(B_2)\,.$$

Finally, if in addition $\alpha^{(k)}$ is absolutely continuous with respect to the Lebesgue measure ν_{dk} then the *kth-order product density* $\rho^{(k)} : \mathbb{R}^{d(k-1)} \mapsto [0,\infty]$ and the *kth-order cumulant density* $c^{(k)} : \mathbb{R}^{d(k-1)} \mapsto [-\infty,\infty]$ exist such that

$$\alpha_{red}^{(k)}(C) = \int_C \rho^{(k)}(x)\, dx \quad\text{and}\quad \gamma_{red}^{(k)}(C) = \int_C c^{(k)}(x)\, dx \quad\text{for } C \in \mathcal{B}(\mathbb{R}^{d(k-1)})\,.$$

The interpretation of $\rho^{(k)}$ as density of k-point subsets of stationary point configurations is as follows:

$$\mathbf{P}(\Psi(dx_1) = 1, \ldots, \Psi(dx_{k-1}) = 1 \mid \Psi(\{o\}) = 1) = \rho^{(k)}(x_1, \ldots, x_{k-1})\, dx_1 \cdots dx_{k-1}\,.$$

and in a similar way $c^{(2)}(x)\, dx = \mathbf{P}(\Psi(dx) = 1 \mid \Psi(\{o\}) = 1) - \mathbf{P}(\Psi(dx) = 1)$ for any $x \neq o$.

Note that in statistical physics $\rho^{(k)}(x_1, \ldots, x_{k-1})$ and $c^{(k)}(x_1, \ldots, x_{k-1})$ are frequently used under the name *kth-order correlation function* resp. *kth-order truncated correlation function*.

In some cases under slight additional assumptions the knowledge of the (factorial) moment—or cumulant measures and their densities of any order determine the distribution of point processes uniquely. So far this moment problem for point processes is not completely solved. Another longstanding question which is still unanswered to the best of the author's knowledge is: Which properties of a locally-finite measure on $(\mathbb{R}^{dk}, \mathcal{B}(\mathbb{R}^{dk}))$ are sufficient and necessary for being a kth-order (factorial) moment measure of some unmarked point process $\Psi \sim P$?

In the simplest case of a Poisson process with given intensity measure we have the following characterization:

Theorem 4.4. *A MPP* $\Psi_{\mathsf{M}} \sim P_{\mathsf{M}}$ *with intensity measure* Λ_{M} *is a marked Poisson process, i.e.* $P_{\mathsf{M}} = \Pi_{\Lambda_{\mathsf{M}}}$, *iff*

$$\alpha_{\mathsf{M}}^{(k)} = \Lambda_{\mathsf{M}} \times \cdots \times \Lambda_{\mathsf{M}} \;\textit{or equivalently}\; \gamma_{\mathsf{M}}^{(k)} \equiv 0 \;\textit{for any}\; k \geq 2\,.$$

For stationary unmarked PP's with intensity λ this means $\rho^{(k)}(x) = \lambda^{k-1}$ or equivalently $c^{(k)}(x) \equiv 0$ for all $x \in \mathbb{R}^{d(k-1)}$ and any $k \geq 2$.

If the marks are real-valued it is natural to consider the higher-order mixed moments and mixed cumulants between marks conditional on their locations. For a MPP with mark space $\mathsf{M} = \mathbb{R}$ and k-point Palm mark distribution $Q_{\mathsf{M}}^{x_1,\ldots,x_k}$ let us define

$$m_{p_1,\ldots,p_k}(x_1,\ldots,x_k) = \int_{\mathbb{R}^k} m_1^{p_1}\cdots m_k^{p_k}\, Q_{\mathsf{M}}^{x_1,\ldots,x_k}(d(m_1,\ldots,m_k)) \text{ for } p_1,\ldots,p_k \in \mathbb{N}\,.$$

For stationary MPP the function $c_{mm}(x) = m_{1,1}(o,x)$ for $x \neq o$ has been introduced by D. Stoyan in 1984 in order to describe spatial correlations of marks by means of the *mark correlation function* $k_{mm}(r) = c_{mm}(x)/\mu^2$ for $\|x\| = r > 0$, where $\mu = \int_{\mathbb{R}^1} m Q_{\mathsf{M}}^o(dm)$ denotes the mean value of the typical mark, see [265, Chap. 5.3], for more details on the use and [438] for a thorough discussion of this function. Kernel-type estimators of the function $c_{mm}(x)$ and their asymptotic properties including consistency and asymptotic normality have been studied in [233] by imposing total variation conditions on higher-order reduced cumulant measures $\gamma_{\mathsf{M},red}^{(k)}$.

Finally, we give a short introduction to general Palm distributions of (marked) point processes. We first consider a simple stationary unmarked point process $\Psi \sim P$ with positive and finite intensity $\lambda = \mathbf{E}\Psi([0,1]^d)$. Let us define the product measure $\mu^!$ on $(\mathbb{R}^d \times \mathcal{N}, \mathcal{B}(\mathbb{R}^d) \otimes \mathfrak{N})$ by

$$\mu^!(B \times A) = \frac{1}{\lambda} \int_{\mathcal{N}} \sum_{x \in s(\psi)} \mathbf{1}(x \in B)\, \mathbf{1}(T_x\psi - \delta_o \in A)\; P(d\psi)$$

for bounded $B \in \mathcal{B}(\mathbb{R}^d)$ and $A \in \mathfrak{N}$, where the exclamation mark indicates that the atom of $T_x\psi$ in the origin, i.e., the atom of ψ in $x \in \mathbb{R}^d$, is removed from each counting measure $\psi \in \mathcal{N}$; $s(\psi)$ is shorthand for the support
$\mathrm{supp}(\psi) = \{x \in \mathbb{R}^d : \psi(\{x\}) > 0\}$.

By the stationarity of Ψ, that is $P \circ T_x = P$ for any $x \in \mathbb{R}^d$ combined with standard arguments from measure theory it is easily seen that

1. $\mu^!((B + x) \times A) = \mu^!(B \times A)$ for any $x \in \mathbb{R}^d$.
2. $P^{o\,!}(A) := \mu^!([0,1]^d \times A)$ for $A \in \mathfrak{N}$ is a probability measure on $(\mathcal{N}, \mathfrak{N})$.

which is concentrated on the subset $\mathcal{N}^o = \{\psi \in \mathcal{N} : \psi(\{o\}) = 0\}$ of counting measures having no atom in the origin o and called the *reduced Palm distribution* of $\Psi \sim P$. As an immediate consequence of (1) and (2) we obtain the factorization

$$\mu^!(B \times A) = \nu_d(B)\; P^{o\,!}(A) \;\text{ for any fixed }\; B \times A \in \mathcal{B}(\mathbb{R}^d) \times \mathfrak{N}$$

which in turn implies, by algebraic induction, the *Campbell–Mecke formula*—also known as *refined Campbell theorem*

$$\int_{\mathcal{N}} \int_{\mathbb{R}^d} f(x, T_x\psi - \delta_o)\, \psi(dx)\, P(d\psi) = \lambda \int_{\mathbb{R}^d} \int_{\mathcal{N}^o} f(x,\psi)\, P^{o\,!}(d\psi)\, dx \quad (4.8)$$

for any non-negative $\mathcal{B}(\mathbb{R}^d) \otimes \mathfrak{N}$-measurable function $f : \mathbb{R}^d \times \mathcal{N} \mapsto \mathbb{R}^1$. This formula connects the stationary distribution P and the reduced Palm distribution

$P^{o!}$ in a one-to-one correspondence. $P^{o!}(A)$ can be interpreted (justified in a rigorous sense by limit theorems, see [346]) as probability that $\Psi - \delta_o \in A$ conditional on the null event $\Psi(\{o\}) > 0$. Loosely speaking, $P^{o!}$ describes the stationary point pattern by an observer sitting in a "typical atom" shifted in the origin.

To describe the distributional properties of stationary PP's it is often more effective to use $P^{o!}$ rather than P, for example in case of *recurrent*, *semi-Markov-* or *infinitely divisible* PP's, see for example [346].

A crucial result in this direction is Slivnyak's characterization of homogeneous Poisson processes:

Theorem 4.5. *A stationary (unmarked) PP $\Psi \sim P$ on $\mathbb{R}^d$ with intensity $0 < \lambda < \infty$ is a Poisson process, i.e. $P = \Pi_\lambda$ iff $P = P^{o!}$.*

As announced above we apply (4.8) to prove that, for any $k \geq 2$, the kth-order reduced factorial moment measure $\alpha^{(k)}_{red}$ is nothing else but the $(k-1)$st-order factorial moment measure w.r.t the reduced Palm distribution, formally written:

$$\int_{(\mathbb{R}^d)^{k-1}} f(x_2,\ldots,x_k)\,\alpha^{(k)}_{red}(d(x_2,\ldots,x_k)) = \int_{\mathcal{N}^o} \sum_{x_2,\ldots,x_k \in s(\psi)}^{\neq} f(x_2,\ldots,x_k)\,P^{o!}(d\psi)$$

for any non-negative Borel-measurable function f on $\mathbb{R}^{d(k-1)}$. For notational ease we check this only for $k = 2$. From (4.7) and the very definition of $\alpha^{(2)}$ we get for bounded $B, C \in \mathcal{B}(\mathbb{R}^d)$ with $\nu_d(B) = 1$ that

$$\alpha^{(2)}_{red}(C) = \frac{1}{\lambda}\int_{\mathcal{N}} \sum_{x,y \in s(\psi)}^{\neq} \mathbf{1}(x \in B)\,\mathbf{1}(y - x \in C)\,P(d\psi) \tag{4.9}$$

$$= \frac{1}{\lambda}\int_{\mathcal{N}} \sum_{x \in s(\psi)} \mathbf{1}(x \in B)\,(T_x\psi - \delta_o)(C)\,P(d\psi) = \int_{\mathcal{N}^o} \psi(C)\,P^{o!}(d\psi)\,.$$

Quite similarly, we can define reduced Palm distributions $P^{o!}_L$ for simple stationary MPP's with respect to any fixed mark set $L \in \mathsf{M}$ with $Q^o_{\mathsf{M}}(L) > 0$. For this we have to replace λ by $\Lambda_{\mathsf{M}}([0,1]^d \times L) = \lambda\, Q^o_{\mathsf{M}}(L)$ which leads to the following extension of (4.8):

$$\lambda\, Q^o_{\mathsf{M}}(L) \int_{\mathbb{R}^d}\int_{\mathcal{N}^o_L} f(x,\psi)P^{o!}_L(d\psi)dx = \int_{\mathcal{N}_{\mathsf{M}}}\int_{\mathbb{R}^d \times L} f(x, T_x\psi - \delta_{(o,m)})\psi(d(x,m))P_{\mathsf{M}}(d\psi)$$

for any non-negative, measurable function f on $\mathbb{R}^d \times \mathcal{N}$, where $\mathcal{N}^o_L = \{\psi \in \mathcal{N}_{\mathsf{M}} : \psi(\{o\} \times L) = 0\}$.

To include mark sets L and, in particular single marks m, with Q^o_{M}-measure zero, we make use of the Radon–Nikodym derivative of the so-called *reduced Campbell measure* $C^!_{\mathsf{M}}$ defined by

$$C^!_{\mathsf{M}}(B \times L \times A) = \int_{\mathcal{N}_{\mathsf{M}}} \int_{\mathbb{R}^d \times \mathsf{M}} \mathbf{1}((x,m) \in B \times L)\mathbf{1}(\psi - \delta_{(x,m)} \in A)\psi(d(x,m))P_{\mathsf{M}}(d\psi)$$

with respect to the intensity measure $\Lambda_{\mathsf{M}}(B \times L) = \lambda\, \nu_d(B)\, Q^o_{\mathsf{M}}(L)$. The corresponding Radon–Nikodym density $P^{x\,!}_m(A)$ is called the *reduced Palm distribution* of $\Psi_{\mathsf{M}} \sim P_{\mathsf{M}}$ with respect to (x,m) and can be heuristically interpreted as conditional probability that $\Psi - \delta_{(x,m)} \in A$ given a marked point at x with mark m. This interpretation remains also valid for non-stationary MPP's and can even be generalized in an appropriate way to *k-point reduced Palm distributions* $P^{x_1,\dots,x_k\,!}_{m_1,\dots,m_k}(A)$ of $A \in \mathfrak{N}_{\mathsf{M}}$ with respect to $(x_1,m_1),\dots,(x_k,m_k)$ with pairwise distinct $x_1,\dots,x_k$.

In the stationary case we get $P^{x_1,x_2\dots,x_k\,!}_{m_1,\dots,m_k}(A) = P^{x_1-x_j,\dots,o,\dots,x_k-x_j\,!}_{m_1,\dots,m_j,\dots,m_k}(T_{-x_j}A)$ for each $j = 1,\dots,k$ (due to the intrinsic symmetry), which, for $k = 1$, yields the Campbell–Mecke-type formula

$$\mathbf{E}\Big(\sum_{i\geq 1} f(X_i, M_i, T_{X_i}\Psi - \delta_{(o,M_i)})\Big) = \int_{\mathcal{N}_{\mathsf{M}}} \int_{\mathbb{R}^d \times \mathsf{M}} f(x,m,T_x\psi - \delta_{(o,m)})\psi(d(x,m))P_{\mathsf{M}}(d\psi)$$

$$= \lambda \int_{\mathbb{R}^d} \int_{\mathsf{M}} \int_{\mathcal{N}_{\mathsf{M}}} f(x,m,\psi) P^{o\,!}_m(d\psi) Q^o_{\mathsf{M}}(dm)dx \qquad (4.10)$$

for any non-negative measurable function f on $\mathbb{R}^d \times \mathsf{M} \times \mathcal{N}_{\mathsf{M}}$. Furthermore, this formula can be extended for $k \geq 2$ to the following relationship involving the k-point reduced Palm distribution, k-point Palm mark distribution and the k-order reduced factorial moment measure introduced at the beginning of Sect. 4.1.2:

$$\mathbf{E}\Big(\sum^{\neq}_{i_1,i_2,\dots,i_k\geq 1} f_k(X_{i_1}, M_{i_1}, X_{i_2}, M_{i_2}, \dots, X_{i_k}, M_{i_k}, T_{X_{i_1}}\Psi - \delta_{(o,M_{i_1})} - \sum_{j=2}^{k} \delta_{(X_{i_j},M_{i_j})})\Big)$$

$$= \lambda \int_{(\mathbb{R}^d \times \mathsf{M})^k} \int_{\mathcal{N}_{\mathsf{M}}} f_k(x_1,m_1,x_2,m_2,\dots,x_k,m_k,\psi)\, P^{o,x_2,\dots,x_k\,!}_{m_1,m_2,\dots,m_k}(d\psi)$$

$$\times\, \alpha^{(k)}_{\mathsf{M},red}(d(m_1,x_2,m_2,\dots,x_k,m_k))\, dx_1$$

$$= \lambda \int_{\mathbb{R}^d} \int_{\mathbb{R}^{d(k-1)}} \int_{\mathsf{M}^k} \int_{\mathcal{N}_{\mathsf{M}}} f_k(x_1,m_1,x_2,m_2,\dots,x_k,m_k,\psi)\, P^{o,x_2,\dots,x_k\,!}_{m_1,m_2,\dots,m_k}(d\psi)$$

$$\times\, Q^{o,x_2,\dots,x_k}_{\mathsf{M}}(d(m_1,m_2,\dots,m_k))\, \alpha^{(k)}_{red}(d(x_2,\dots,x_k))\, dx_1$$

for any non-negative measurable function f_k on $(\mathbb{R}^d \times \mathsf{M})^k \times \mathcal{N}_{\mathsf{M}}$.

4.1.3 Different Types of Marking and Some Examples

In the following we distinguish three types of MPP's $\Psi_{\mathsf{M}} \sim P_{\mathsf{M}}$ by the dependences between marks at distinct points in $\mathbb{R}^d$, by interactions between separated parts of the point pattern and, last but not least, by cross correlations between the whole point pattern and the whole mark field. In the most general case, only the family of k-point Palm mark distributions seems to be appropriate to describe such complicated structure of dependences.

1. Independently Marked (Stationary) Point Processes

Given an unmarked (not necessarily stationary) PP $\Psi = \sum_{i\geq 1} \delta_{X_i} \sim P$ on $\mathbb{R}^d$ and a stochastic kernel $Q(x, L)$, $x \in \mathbb{R}^d, L \in \mathcal{B}(\mathsf{M})$ we assign to an atom X_i located at x the mark $M_i \sim Q(x, \cdot)$ independently of $\Psi - \delta_{X_i}$ and of any other mark M_j, $j \neq i$. The resulting MPP $\Psi_{\mathsf{M}} = \sum_{i\geq 1} \delta_{(X_i, M_i)}$ is said to be derived from Ψ by *location-dependent independent marking*. We obtain for the intensity measure and the k-point Palm mark distribution

$$\Lambda_{\mathsf{M}}(B \times L) = \int_B Q(x, L)\, \Lambda(dx) \;\; \text{resp.} \;\; Q_{\mathsf{M}}^{x_1,\ldots,x_k}\big(\mathop{\times}_{i=1}^{k} L_i\big) = \prod_{i=1}^{k} Q(x_i, L_i), \;\; k \geq 1,$$

where Λ denotes the intensity measure of Ψ.

Note that the MPP $\Psi_{\mathsf{M}} \sim P_{\mathsf{M}}$ is stationary iff $\Psi \sim P$ is stationary and independent of an i.i.d. sequence of marks $\{M_i, \, i \geq 1\}$ with a common distribution $Q(\cdot)$—the (mark) distribution of the typical mark M_0.

2. Geostatistically or Weakly Independently Marked Point Processes

Let unmarked PP $\Psi = \sum_{i\geq 1} \delta_{X_i} \sim P$ on $\mathbb{R}^d$ be stochastically independent of a random field $\{M(x), \, x \in \mathbb{R}^d\}$ taking values in the measurable mark space $(\mathsf{M}, \mathcal{B}(\mathsf{M}))$. To each atom X_i we assign the mark $M_i = M(X_i)$ for $i \geq 1$. In this way the k-point Palm mark distribution coincide with the finite-dimensional distributions of the mark field, that is,

$$Q_{\mathsf{M}}^{x_1,\ldots,x_k}(L_1 \times \cdots \times L_k) = \mathbf{P}(M(x_1) \in L_1, \ldots, M(x_k) \in L_k) \;\; \text{for all} \;\; k \in \mathbb{N}\,.$$

Note that the MPP $\Psi_{\mathsf{M}} \sim P_{\mathsf{M}}$ is stationary iff both the point process $\Psi \sim P$ and the random field are (strictly) stationary. In case of real-valued marks (stationary) Gaussian random fields $M(x)$ with some covariance function, see Definition 9.10, or shot-noise fields $M_g(x) = \sum_{i\geq 1} g(x - \xi_i)$ with some response function

$g : \mathbb{R}^d \mapsto \mathbb{R}^1$ and a (stationary Poisson) point process $\{\xi_i, i \geq 1\}$ chosen independently of Ψ (see Sect. 9.2.5 for more details) are suitable examples for mark fields.

3. General Dependently Marked Point Processes

In this case the locations X_i of the marked atoms and their associated marks M_i may depend on each other and, in addition, there are intrinsic interactions within the point field $\{X_i\}$ as well as within the mark field $\{M_i\}$. This means that the k-point Palm mark distribution $Q_{\mathsf{M}}^{x_1,\ldots,x_k}$ must be considered as an conditional distribution, in particular, $Q_{\mathsf{M}}^{o,x}(L \times \mathsf{M})$ does not coincide with $Q_{\mathsf{M}}^{o}(L)$.

Examples

1. Germ-Grain Processes: Germ-Grain Models

A stationary independently MPP $\Psi_{\mathsf{M}} = \{(X_i, \Xi_i), i \geq 1\}$ on $\mathbb{R}^d$ with mark space $\mathsf{M} = \mathcal{K}^d$ (= space of all non-empty compact sets in $\mathbb{R}^d$ equipped with the Hausdorff metric) is called *germ-grain process* or *particle process* driven by the PP $\Psi = \{X_i, i \geq 1\}$ of germs and the *typical grain* $\Xi_0 \sim Q$. The associated random set $\Xi = \bigcup_{i \geq 1}(\Xi_i + X_i)$ is called *germ-grain model*. Note that in general Ξ need not to be closed (**P**-a.s.). The condition

$$\sum_{i \geq 1} \mathbf{P}(\Xi_0 \cap (K - X_i) \neq \emptyset) < \infty \quad \mathbf{P}-\text{a.s. for all } K \in \mathcal{K} \tag{4.11}$$

is sufficient to ensure the **P**-a.s.-closedness of Ξ, see [229]. The most important and best studied germ-grain model is the *Poisson-grain model* (also called *Boolean model*) driven by a Poisson process $\Psi \sim \Pi_\Lambda$ of germs $\{X_i, i \geq 1\}$, see for example [366, 489] for more details.

2. Poisson-Cluster Processes

If the typical grain $\Xi_0 = \{Y_1, \ldots, Y_{N_0}\}$ is a **P**-a.s. finite random point set satisfying (4.11) then the discrete random closed set $\Xi = \bigcup_{i \geq 1}\{Y_1^{(i)} + X_i, \ldots, Y_{N_i}^{(i)} + X_i\}$ coincides with the support of a random locally finite counting measure Ψ_{cl} and is called a *cluster point process* with the PP $\Psi_c = \{X_i, i \geq 1\}$ of cluster centres and the typical cluster $\{Y_1, \ldots, Y_{N_0}\}$. Factorial moment and cumulant measures of any order can be expressed in terms of the corresponding measures of Ψ and the finite PP $\sum_{i=1}^{N_0} \delta_{Y_i}$, see for example [227]. In case of a (stationary) Poisson cluster centre process Ψ_c we get a (stationary) Poisson-cluster process Ψ_{cl}, see Sect. 3.1.4 for more details. In particular, if $\Psi \sim \Pi_{\lambda_c \nu_d}$ and the random number

N_0 with probability generating function $g_0(z)$ is independent of the i.i.d. sequence of random vectors $Y_1, Y_2, \ldots$ in $\mathbb{R}^d$ with common density function f we obtain a so-called *Neyman–Scott process* Ψ_{cl} with intensity $\lambda_{\text{cl}} = \lambda_c \, \mathbf{E} N_0$, second-order product density $\rho^{(2)}(x) = c^{(2)}(x) + \lambda_{\text{cl}}$ and its kth-order cumulant density for $k \geq 2$ takes on the form

$$c^{(k)}(x_1, \ldots, x_{k-1}) = \frac{g_0^{(k)}(1)}{\mathbf{E} N_0} \int_{\mathbb{R}^d} f(y)\, f(y + x_1) \cdots f(y + x_{k-1})\, dy \,. \quad (4.12)$$

Compare Definition 3.11 for its special case.

3. Doubly Stochastic Poisson Processes

Now, let Λ be a (stationary) random measure on $\mathbb{R}^d$, see for example [140] for details. The new unmarked PP $\Psi_\Lambda \sim P_\Lambda$ defined by the finite-dimensional distributions

$$\mathbf{P}(\Psi_\Lambda(B_1) = n_1, \ldots, \Psi_\Lambda(B_k) = n_k) = \mathbf{E}\Big(\prod_{i=1}^{k} \frac{\Lambda^{n_i}(B_i)}{n_i!}\, \mathrm{e}^{-\Lambda(B_i)} \Big)$$

for any disjoint bounded $B_1, \ldots, B_k \in \mathcal{B}_0(\mathbb{R}^d)$ and any $n_1, \ldots, n_k \in \mathbb{N} \cup \{0\}$, is called *doubly stochastic Poisson (or Cox) process* with driving measure $\Lambda(\cdot)$, compare Definition 3.7. In the special case $\Lambda(\cdot) = \lambda\, \nu_d((\cdot) \cap \Xi)$, where Ξ is a (stationary) random closed set, for example a Boolean model, the (stationary) PP Ψ_Λ (called *interrupted Poisson process*) is considered as a Poisson process restricted on the (hidden) realizations of Ξ. The factorial moment and cumulant measures of Ψ_Λ are expressible in terms of the corresponding measures of random driving measure Λ, see for example [289].

4.2 Point Process Statistics in Large Domains

Statistics of stationary point processes is mostly based on a single observation of some point pattern in a sufficiently large domain which is assumed to extend unboundedly in all directions. We demonstrate this concept of asymptotic spatial statistics for several second-order characteristics of point processes including different types of K-functions, product densities and the pair correlation function. Variants of Brillinger-type mixing are considered to obtain consistency and asymptotic normality of the estimators.

The philosophy of large-domain spatial statistics is as follows: Let there be given a single realization of a random point pattern or a more general random set in a sufficiently large sampling window $W_n \subset \mathbb{R}^d$, which is thought to expand in all directions as $n \to \infty$. Further, we assume that there is an adequate model describing the spatial random structure whose distribution is at least shift-invariant (stationary) and sometimes additionally even isotropic. Then only using the information drawn from the available observation in W_n we define empirical counterparts (estimators) of those parameters and non-parametric characteristics which reflect essential properties of our model. To study the asymptotic behaviour of the estimators such as weak or strong consistency and the existence of limit distributions (after suitable centering and scaling) we let W_n increase unboundedly which requires additional weak dependence conditions. Throughout we assume that $\{W_n, n \in \mathbb{N}\}$ is a *convex averaging sequence*, that is,

1. W_n is bounded, compact, convex and $W_n \subseteq W_{n+1}$ for $n \in \mathbb{N}$.
2. $r(W_n) := \sup\{r > 0 : B_r(x) \subseteq W_n \text{ for some } x \in W_n\} \uparrow \infty$.

The second property means that W_n expands unboundedly in all directions and is equivalent to $\nu_{d-1}(\partial W_n)/\nu_d(W_n) \underset{n\to\infty}{\longrightarrow} 0$ as immediate consequence of the geometric inequality

$$\frac{1}{r(W_n)} \le \frac{\nu_{d-1}(\partial W_n)}{\nu_d(W_n)} \le \frac{d}{r(W_n)}, \tag{4.13}$$

see [237].

Exercise 4.1. Show that

$$\nu_d(W_n \setminus (W_n \ominus B_r(o))) = \int_0^r \nu_{d-1}(\partial(W_n \ominus B_s(o)))\, ds \le r\, \nu_{d-1}(\partial W_n)$$

for $0 \le r \le r(W_n)$ from which, together with $\nu_d(W_n \ominus B_{r(W_n)}(o)) = 0$, the l.h.s. of (4.13) immediately follows. The r.h.s. of (4.13) results from an inequality by J.M. Wills, see [519].

From the mathematical view point it is sometimes more convenient to consider rectangles $W_n = \times_{i=1}^d [0, a_i^{(n)}]$ with $a_i^{(n)} \uparrow \infty$ for $i = 1, \ldots, d$ or blown up sets $W_n = n\, W$, where $W \subset \mathbb{R}^d$ is a fixed convex body containing the origin o as inner point.

4.2.1 Empirical K-Functions and Other Summary Statistics of Stationary PP's

Second-order statistical analysis of spatial point patterns is perhaps the most important branch in point process statistics comparable with the spectral density estimation in time series analysis. We assume that the simple unmarked PP $\Psi = \sum_{i\geq 1} \delta_{X_i}$ has finite second moments, i.e. $\mathbf{E}\Psi^2(B) < \infty$ for all bounded $B \in \mathcal{B}(\mathbb{R}^d)$, and is strictly or at least weakly stationary.

Weak (or second-order) stationarity of an unmarked PP $\Psi \sim P$ requires only the shift-invariance of the first- and second-order moment measures, i.e. $\Lambda(B_1 + x) = \Lambda(B_1)$ and $\alpha^{(2)}((B_1 + x) \times (B_2 + x)) = \alpha^{(2)}(B_1 \times B_2)$ for any bounded $B_1, B_2 \in \mathcal{B}(\mathbb{R}^d)$) and all $x \in \mathbb{R}^d$. Obviously, strictly stationary point processes having finite second moments are weakly stationary. Further note that the reduced second factorial moment measure $\alpha^{(2)}_{red}(\cdot)$ is well-defined also under weak stationarity but it can not be expressed as first-order moment measure w.r.t. $P^{o\,!}$ as in (4.9), see [140] for more details. In what follows we assume strict stationarity. By applying the Palm and reduction machinery sketched in Sect. 4.1.2 we can describe the first and second moment properties by the intensity $\lambda = \mathbf{E}\Psi([0,1]^d)$ and the reduced second factorial moment measure $\alpha^{(2)}_{red}(\cdot)$ defined by (4.7) for $k = 2$ resp. (4.9) as first moment measure with respect to the Palm distribution $P^{o!}$ in case of strict stationarity. If Ψ is additionally strictly or at least weakly isotropic, i.e. $R_O \alpha^{(2)}_{red} = \alpha^{(2)}_{red}$ for $O \in SO_d$, then it suffices to know the function $\alpha^{(2)}_{red}(B_r(o))$ for $r \geq 0$. In [424] B. Ripley introduced the *K-function*

$$K(r) := \frac{1}{\lambda}\,\alpha^{(2)}_{red}\big(B_r(o)\big) = \frac{1}{\lambda^2}\,\mathbf{E}\Big(\sum_{i\geq 1} \mathbf{1}(X_i \in [0,1]^d)\,\Psi(B_r(X_i) \setminus \{X_i\})\Big) \tag{4.14}$$

for $r \geq 0$ as basic summary characteristic for the second-order analysis of motion-invariant PP's, see also [22] or [265, Chap. 4.3] for more details and historical background. From (4.9) we see that $\lambda\, K(r)$ coincides with conditional expectation $\mathbf{E}(\Psi\big(B_r(\{o\}) \setminus \{o\}\big) \mid \Psi(\{o\}) = 1)$ giving the mean number of points within the Euclidean distance r from the typical point (which is not counted). If Ψ is a homogeneous Poisson process with intensity λ, then, by Slivnyak's theorem (see Theorem 4.5 in Sect. 4.1.2), $\alpha^{(2)}_{red}(\cdot) = \mathbf{E}\Psi(\cdot) = \lambda\, \nu_d(\cdot)$ and hence we get

$$K(r) = \omega_d\, r^d \quad \text{with } \omega_d := \nu_d(B_1(o)) = \frac{\pi^{d/2}}{\Gamma\big(\frac{d}{2} + 1\big)}\,. \tag{4.15}$$

For better visualization of the Poisson property by a linear function the so-called *L-function* $L(r) := \big(K(r)/\omega_d\big)^{1/d}$ is sometimes preferred instead of the K-function. Both the K- and L-function represent the same information, but they cannot completely characterize the distribution of a (motion-invariant) PP. In other words, there are different point processes having the same K-function. Further note that an explicit description of the family of K-functions does not exist so far. Nevertheless, the K-function and its empirical variants, see below, are used to check point process hypotheses when the K-function of the null hypothesis is known (or generated by simulation on a finite interval $[0, r_0]$), see Sect. 3.3.3 and Figs. 3.47–3.48.

In particular, the simple parabola-shape of the K-function (4.15) facilitates to check the property of *complete spatial randomness* (briefly *CSR*) of a given point pattern. Lemma 3.12 shows the connection between CSR and the Poisson property shows the following

It contains the interpretation of the Poisson point process in statistical mechanics as particle configuration, for example molecules in "ideal gases", modelled as *grand canonical ensemble*, where neither attraction nor repulsion forces between particles occur. Lemma 3.12 also reveals an easy way to simulate homogeneous Poisson processes in bounded domains, see Algorithm 3.6 in Sect. 3.2.2.

Since the K-function is also used to analyze (second-order) stationary, non-isotropic PP's we introduce two generalized versions of Ripley's K-function (4.14). First, the Euclidean d-ball $B_r(o)$ in (4.14) is replaced by $r\,B$, where $B \subset \mathbb{R}^d$ is a compact, convex, centrally symmetric set containing o as inner point. Such set B is called *structuring element* in image analysis and coincides with the unit ball $\{x \in \mathbb{R}^d : \mathrm{N}_B(x) \leq 1\}$ generated by a unique norm $\mathrm{N}_B(\cdot)$ on $\mathbb{R}^d$. Let $K_B(r)$ denote the analogue to (4.14) which equals $\nu_d(B)\,r^d$ if $\Psi \sim \Pi_{\lambda\nu_d}$. In case of a Neyman–Scott process we obtain from (4.12) that

$$K_B(r) = \nu_d(B)\,r^d + \frac{\mathbf{E}N_0(N_0-1)}{\lambda_c\,(\mathbf{E}N_0)^2}\int_{r\,B} f_s(x)\,dx \;\text{ with }\; f_s(x) = \int_{\mathbb{R}^d} f(y)\,f(y+x)\,dy\,.$$

A second generalization of (4.14) is the *multiparameter K-function*, see [231], defined by

$$K(r_1,\ldots,r_d) := \frac{1}{\lambda}\,\alpha^{(2)}_{red}\big(\mathop{\times}_{k=1}^{d}[-r_k,r_k]\big) \;\text{ for }\; r_1,\ldots,r_d \geq 0\,,$$

which contains the same information as the centrally symmetric measure $\alpha^{(2)}_{red}(\cdot)$. For stationary Poisson processes we get

$$K(r_1,\ldots,r_d) = \nu_d\big(\mathop{\times}_{k=1}^{d}[-r_k,r_k]\big) = 2^d\,r_1\cdot\ldots\cdot r_d \;\text{ for }\; r_1,\ldots,r_d \geq 0\,.$$

We next define three slightly different non-parametric estimators of the function $\lambda^2 K_B(r)$ (briefly called *empirical K-functions*):

$$\widehat{(\lambda^2 K_B)}_{n,1}(r) := \frac{1}{\nu_d(W_n)}\sum_{i\geq 1}\mathbf{1}(X_i\in W_n)\,(\Psi-\delta_{X_i})(r\,B+X_i)\,,$$

$$\widehat{(\lambda^2 K_B)}_{n,2}(r) := \frac{1}{\nu_d(W_n)}\sum_{i,j\geq 1}^{\neq}\mathbf{1}(X_i\in W_n)\,\mathbf{1}(X_j\in W_n)\,\mathbf{1}(\mathrm{N}_B(X_j-X_i)\in[0,r])\,,$$

$$\widehat{(\lambda^2 K_B)}_{n,3}(r) := \sum_{i,j\geq 1}^{\neq}\frac{\mathbf{1}(X_i\in W_n)\,\mathbf{1}(X_j\in W_n)\,\mathbf{1}(\mathrm{N}_B(X_j-X_i)\in[0,r])}{\nu_d((W_n-X_i)\cap(W_n-X_j))}\,.$$

Each of these empirical processes is non-decreasing, right-continuous, and piecewise constant with jumps of magnitude $1/\nu_d(W_n)$ (for $i = 1, 2$) at random positions $N_B(X_j - X_i)$ arranged in order of size for $i \neq j$. Quite analogously, by substituting the indicators of the events $\{N_B(X_j - X_i) \in [0, r]\}$ by the indicators of $\{X_j - X_i \in \times_{k=1}^d [-r_k, r_k]\}$ we obtain the multivariate empirical processes $\big(\widehat{\lambda^2 K}\big)_{n,i}(r_1, \ldots, r_d)$ for $i = 1, 2, 3$ as empirical counterparts of $\lambda^2 K(r_1, \ldots, r_d)$. By (4.8) resp. (4.9), $\big(\widehat{\lambda^2 K_B}\big)_{n,1}(r)$ is easily seen to be an unbiased estimator for $\lambda^2 K_B(r)$ but it ignores the edge effect problem, that is, we need information from the dilated sampling window $W_n \oplus r_0 B$ to calculate this estimator for $0 \leq r \leq r_0$. If this information is not available then one has to reduce the original window to the eroded set $W_n \ominus r_0 B$ which is known as *minus sampling*. The second estimator needs only the positions of points within W_n, however, its bias disappears only asymptotically, i.e.

$$\mathbf{E}\big(\widehat{\lambda^2 K_B}\big)_{n,2}(r) = \lambda \int\limits_{rB} \frac{\nu_d(W_n \cap (W_n - x))}{\nu_d(W_n)} \alpha_{red}^{(2)}(dx) \underset{n\to\infty}{\longrightarrow} \lambda^2 K_B(r) . \quad (4.16)$$

Finally, $\big(\widehat{\lambda^2 K_B}\big)_{n,3}(r)$ is a so-called *edge-corrected* or *Horvitz–Thompson-type estimator* which also needs only the points located within W_n. The pairs $(X_i, X_j) \in W_n \times W_n$ are weighted according to the length and direction of the difference vector $X_j - X_i$ providing the unbiasedness of the estimator $\mathbf{E}\big(\widehat{\lambda^2 K_B}\big)_{n,3}(r)$.

Exercise 4.2. Show (4.16) by applying the inequality (4.13) and prove

$$\mathbf{E}\big(\widehat{\lambda^2 K_B}\big)_{n,3}(r) = \lambda^2 K_B(r)$$

by means of (4.8) resp. (4.9).

For further details and more sophisticated edge corrections we refer to [265, 385, 489] and references therein.

Before regarding consistency properties of the empirical K-functions we have a short look at the estimation of the simplest summary characteristic—the intensity λ—and its powers λ^k given by

$$\widehat{\lambda}_n := \frac{\Psi(W_n)}{\nu_d(W_n)} \quad \text{and} \quad (\widehat{\lambda^k})_n := \prod_{j=0}^{k-1} \frac{\Psi(W_n) - j}{\nu_d(W_n)} \quad (4.17)$$

for any fixed integer $k \geq 2$. A simple application of the Campbell formula (4.6) (or (4.8)) and the definition of the kth-order factorial moment measure yields

$$\mathbf{E}\widehat{\lambda}_n = \lambda \quad \text{and} \quad \mathbf{E}(\widehat{\lambda^k})_n = \frac{\alpha^{(k)}(W_n \times \cdots \times W_n)}{\nu_d^k(W_n)}$$

which shows the unbiasedness of $\widehat{\lambda}_n$ for any stationary PP, whereas $(\widehat{\lambda^k})_n$ for $k \geq 2$ is unbiased only for the Poisson process $\Psi \sim \Pi_{\lambda \nu_d}$.

Exercise 4.3. For a stationary Poisson process $\Psi \sim \Pi_{\lambda \nu_d}$ show that

$$\alpha^{(k)}(B_1 \times \cdots \times B_k) = \lambda^k \, \nu_d(B_1) \cdots \nu_d(B_k)$$

for any (not necessarily disjoint) bounded sets $B_1, \ldots, B_k \in \mathcal{B}(\mathbb{R}^d)$.

The decomposition $\alpha^{(2)}(W_n \times W_n) = \gamma^{(2)}(W_n \times W_n) + \lambda^2 \, \nu_d^2(W_n)$ and reduction reveal the asymptotic unbiasedness of $(\widehat{\lambda^2})_n$

$$\mathbf{E}(\widehat{\lambda^2})_n = \lambda^2 + \frac{\lambda}{\nu_d^2(W_n)} \int\limits_{W_n} \gamma_{red}^{(2)}(W_n - x) \, dx \underset{n \to \infty}{\longrightarrow} \lambda^2$$

provided that the total variation $\| \gamma_{red}^{(2)} \|_{\mathrm{TV}}$ is finite. This motivates the assumption of bounded total variation of the reduced factorial cumulant measure $\gamma_{red}^{(k)}(\cdot)$ for some $k \geq 2$ to express short-range correlation of the point process. To be precise, we rewrite the locally finite (in general not finite) signed measure $\gamma_{red}^{(k)}(\cdot)$ on $\big(\mathbb{R}^{d(k-1)}, \mathcal{B}(\mathbb{R}^{d(k-1)})\big)$ as difference of the positive and negative part $\gamma_{red}^{(k)+}(\cdot)$ resp. $\gamma_{red}^{(k)-}(\cdot)$ (Jordan decomposition) and define the corresponding *total variation measure* $\big| \gamma_{red}^{(k)} \big|(\cdot)$ as a sum of the positive and negative part:

$$\gamma_{red}^{(k)}(\cdot) = \gamma_{red}^{(k)+}(\cdot) - \gamma_{red}^{(k)-}(\cdot) \; \text{ and } \; \big| \gamma_{red}^{(k)} \big|(\cdot) := \gamma_{red}^{(k)+}(\cdot) + \gamma_{red}^{(k)-}(\cdot) \, .$$

Note that the locally finite measures $\gamma_{red}^{(k)+}(\cdot)$ and $\gamma_{red}^{(k)-}(\cdot)$ are concentrated on two disjoint Borel sets H^+ resp. H^- with $H^+ \cup H^- = \mathbb{R}^{d(k-1)}$ (Hahn decomposition) which leads to the *total variation* of $\gamma_{red}^{(k)}(\cdot)$:

$$\| \gamma_{red}^{(k)} \|_{\mathrm{TV}} := \big| \gamma_{red}^{(k)} \big|(\mathbb{R}^{d(k-1)}) = \gamma_{red}^{(k)+}(H^+) + \gamma_{red}^{(k)-}(H^-) = \int\limits_{\mathbb{R}^{d(k-1)}} | c^{(k)}(x) | \, d\,x \, ,$$

where $c^{(k)} : \mathbb{R}^{d(k-1)} \mapsto [-\infty, \infty]$ is the kth-order cumulant density, if it exists.

Definition 4.5. A stationary PP $\Psi \sim P$ on $\mathbb{R}^d$ satisfying $\mathbf{E}\Psi^k([0,1]^d) < \infty$ for some integer $k \geq 2$ is said to be $\mathbf{B}_k$*-mixing* if $\| \gamma_{red}^{(j)} \|_{\mathrm{TV}} < \infty$ for $j = 2, \ldots, k$. A $\mathbf{B}_\infty$-mixing stationary PP is called *Brillinger-mixing* or briefly $\mathbf{B}$-mixing.

Example 4.1. From (4.12) it is easily seen that a Neyman–Scott process is $\mathbf{B}_k$-mixing iff $\mathbf{E}N_0^k < \infty$ without restrictions on f. This remains true for any Poisson-cluster process. Moreover, a general cluster process is $\mathbf{B}_k$-mixing if the PP

Ψ_c of cluster centres is $\mathbf{B}_k$-mixing and the typical cluster size N_0 has a finite kth moment, see [227] for details and further examples like Cox processes.

Proposition 4.1. *For any $\mathbf{B}_k$-mixing stationary PP we have $\mathbf{E}(\widehat{\lambda^k})_n \underset{n\to\infty}{\longrightarrow} \lambda^k$ for $k \geq 2$.*

We next state the mean-square consistency of the above-defined empirical K-functions under mild conditions. Furthermore, it can be shown that a possible weak Gaussian limit (after centering with mean and scaling with $\sqrt{\nu_d(W_n)}$) is for each of the estimators of $\lambda^2\, K_B(r)$ the same.

Theorem 4.6. *Let $\Psi \sim P$ be a $\mathbf{B}_4$-mixing stationary PP with intensity λ. Then*

$$\mathbf{E}\Big(\big(\widehat{\lambda^2 K_B}\big)_{n,i}(r) - \lambda^2\, K_B(r)\Big)^2 \underset{n\to\infty}{\longrightarrow} 0 \text{ for } i = 1,2,3$$

$$\nu_d(W_n)\ \mathbf{var}\Big(\big(\widehat{\lambda^2 K_B}\big)_{n,1}(r) - \big(\widehat{\lambda^2 K_B}\big)_{n,i}(r)\Big) \underset{n\to\infty}{\longrightarrow} 0 \text{ for } i = 2,3\,.$$

In other words, the boundary effects are asymptotically neglectable which can be considered as a general rule of thumb in large domain statistics.

Finally, we mention that also higher-order reduced moment measures can be estimated in quite the same way, see for example [274, 289]. Further second-order summary characteristics and their empirical counterparts (called summary statistics) such as the second-order product density $\rho^{(2)}(x)$, the *pair correlation function* $g(r)$ and the asymptotic variance $\sigma^2 := \lim_{n\to\infty} \nu_d(W_n)\,\mathbf{E}(\widehat{\lambda}_n - \lambda)^2$, see (4.17), are briefly discussed in Sect. 4.2.3. Summary statistics are used in all branches of statistics to summarize data sets—in our case data from point patterns or from realizations of random sets—to describe the underlying models by a small number of parametric and non-parametric estimates. Further summary characteristics frequently used in point process statistics are the *empty space function* (or *contact distribution function*) F, the *nearest-neighbour distance function* G and the J*-function* defined for a stationary PP $\Psi = \sum_{i\geq 1} \delta_{X_i} \sim P$ by

$$F(r) = \mathbf{P}(\Psi(B_r(o)) > 0) = P(\{\psi \in \mathcal{N} : \psi(B_r(o)) > 0\})\,,$$

$$G(r) = P^{o!}(\{\psi \in \mathcal{N}^o : \psi(B_r(o)) > 0\}) \quad \text{and} \quad J(r) = (1 - G(r))/(1 - F(r))\,.$$

F is the distribution function of the distance $\operatorname{dist}(x, \Psi)$ from a fixed point $x \in \mathbb{R}^d$ to the nearest atom of Ψ, whereas G is the distribution function of the corresponding distance from a typical atom of Ψ to the nearest other atom of Ψ. Unbiased non-parametric estimators of $F(r)$ and $\lambda\, G(r)$ are

$$\widehat{F}_n(r) = \frac{\nu_d(\bigcup_{i\geq 1} B_r(X_i) \cap W_n)}{\nu_d(W_n)}, \quad (\widehat{\lambda\, G})_n(r) = \sum_{X_i \in W_n} \frac{\mathbf{1}(\operatorname{dist}(X_i, \Psi - \delta_{X_i}) \in [0,r])}{\nu_d(W_n)}.$$

The empirical J-function $\widehat{J}_n(r)$ is defined as ratio $\widehat{\lambda}_n \widehat{F}_n(r)/(\widehat{\lambda G})_n(r)$. To avoid boundary effects we replace W_n by $W_n \ominus B_r(o)$ for $0 \le r \le \mathrm{diam}(W_n)/2$, if the point pattern is observable only inside W_n. In case of $\Psi \sim \Pi_{\lambda \nu_d}$ Slivnyak's theorem yields $F(r) = G(r) = 1 - \exp\{-\lambda\, \omega_d\, r^d\}$ so that $J(r) \equiv 1$. This fact can be used for testing CSR just by regarding the plot of the empirical version $\widehat{J}_n(r)$ in some interval $[0, r_0]$.

4.2.2 *The Role of Ergodicity and Mixing in Point Process Statistics*

The assumption of (strict) stationarity of a point process or random closed set under consideration is frequently accompanied by the requirement of ergodicity. It is beyond the scope of this survey to capture the full depth of this notion. We only say that ergodicity is always connected with a group of measure preserving transformations acting on the probability space. In our situation we take quite naturally the group of translations $\{T_x : x \in \mathbb{R}^d\}$ as defined in Sect. 4.1.1 on the space of (marked) locally-finite counting measures or the corresponding shifts on the space of closed sets in $\mathbb{R}^d$. To be precise, we define besides ergodicity also the somewhat stronger condition of mixing for stationary (unmarked) PP's:

Definition 4.6. A (strictly) stationary PP $\Psi \sim P$ is said to be *ergodic* resp. *mixing* if

$$\frac{1}{\nu_d(W_n)} \int_{W_n} P(T_x Y_1 \cap Y_2)\, dx \underset{n\to\infty}{\longrightarrow} P(Y_1)\, P(Y_2) \;\text{ resp. }\; P(T_x\, Y_1 \cap Y_2) \underset{\|x\|\to\infty}{\longrightarrow} P(Y_1)\, P(Y_2)$$

for any $Y_1, Y_2 \in \mathfrak{N}$.

Loosely speaking, mixing means that two events becomes nearly independent when they occur over parts of $\mathbb{R}^d$ being separated by a great distance and ergodicity weakens this distributional property in the sense of Cesaro limits. In physics and engineering one says that an ergodic stochastic process allows to detect its distribution after very long time of observation which carries over to spatial ergodic processes when the observation window expands unboundedly in all directions. This interpretation is rigorously formulated by ergodic theorems which state the **P**-a.s. convergence of spatial means to expectations with respect to the underlying distribution. The following ergodic theorem by X.X. Nguyen and H. Zessin [382] is of particular importance in the theory as well as in statistics of stationary PP's.

Theorem 4.7. *Let $\Psi \sim P$ be a stationary ergodic PP on $\mathbb{R}^d$ with intensity λ, and let $g : \mathcal{N} \mapsto [0, \infty]$ be $(\mathfrak{N}, \mathcal{B}(\mathbb{R}^d))$-measurable such that $\int_{\mathcal{N}^o} g(\psi) P^{o\,!}(d\psi) < \infty$. Then*

$$\frac{1}{\nu_d(W_n)} \int\limits_{W_n} g(T_x\psi - \delta_o)\,\psi(dx) \underset{n\to\infty}{\longrightarrow} \lambda \int\limits_{\mathcal{N}^o} g(\psi) P^{o!}(d\psi)$$

for P-almost every $\psi \in \mathcal{N}$.

This result can be applied to prove strong consistency for many estimators, in particular, for various empirical Palm characteristics. In the special cases (a) $g(\psi) \equiv 1$ and (b) $g(\psi) = \psi(r\,B)$ we obtain strong consistency of the intensity estimator (4.17) and $\widehat{(\lambda^2 K_B)}_{n,1}(r)$ for any $r \geq 0$, which implies even uniformly strong consistency:

$$\widehat{\lambda}_n \overset{\mathbf{P}-\text{a.s.}}{\underset{n\to\infty}{\longrightarrow}} \lambda \text{ and } \sup_{0\leq r\leq R} \left| \widehat{(\lambda^2 K_B)}_{n,1}(r) - \lambda^2\, K_B(r) \right| \overset{\mathbf{P}-\text{a.s.}}{\underset{n\to\infty}{\longrightarrow}} 0\,.$$

We mention just one asymptotic relationship which requires mixing instead of ergodicity, namely the generalized version of Blackwell's renewal theorem. If the stationary second-order PP $\Psi \sim P$ is mixing, then, for any bounded $B \in \mathcal{B}(\mathbb{R}^d)$ satisfying $\nu_d(\partial B) = 0$, it holds

$$\alpha^{(2)}_{red}(B + x) \underset{\|x\|\to\infty}{\longrightarrow} \lambda\, \nu_d(B),$$

see [140]. Note that a renewal process is just mixing if the length of the typical renewal interval has a non-arithmetic distribution and thus, the latter result (applied to an bounded interval $B = [a, b]$) contains the mentioned classical result from renewal theory. For related results concerning the weak convergence of the shifted Palm distribution $P^{o!}(T_x(\cdot))$ to the stationary distribution $P(\cdot)$ as $\|x\| \to \infty$ we refer the reader to [140, 346].

4.2.3 *Kernel-Type Estimators for Product Densities and the Asymptotic Variance of Stationary Point Processes*

The Lebesgue density $\rho^{(2)}(x)$ of $\alpha^{(2)}_{red}(\cdot)$—introduced in Sect. 4.1.2 as second-order product density—and, if $\Psi \sim P$ is motion-invariant, the *pair correlation function* $g(r)$ defined by

$$g(r) = \frac{\rho^{(2)}(x)}{\lambda} \quad \text{for } \|x\| = r > 0 \quad \text{or equivalently } g(r) = \frac{1}{d\,\omega_d\, r^{d-1}}\, \frac{dK(r)}{dr}$$

are very popular second-order characteristics besides the cumulative K-function. Note that $g(r)$ is understood as derivative (existing for ν_1-almost every $r \geq 0$) of an absolutely continuous K-function (4.14). Since the numerical differentiation of the empirical versions of $K(r)$ as well as of the multiparameter K-function

$K(r_1, \ldots, r_d)$ leads to density estimators of minor quality, the most statisticians prefer the established method of kernel estimation in analogy to probability density functions. The corresponding edge-corrected kernel estimators for $\lambda\, \rho^{(2)}(x)$ and $\lambda^2\, g(r)$ are

$$\widehat{(\lambda\, \rho^{(2)})}_n(x) = \frac{1}{b_n^d} \sum_{i,j\geq 1}^{\neq} \frac{\mathbf{1}(X_i \in W_n)\, \mathbf{1}(X_j \in W_n)}{\nu_d((W_n - X_i) \cap (W_n - X_j))}\, k_d\Big(\frac{X_j - X_i - x}{b_n}\Big)$$

resp.

$$\widehat{(\lambda^2\, g)}_n(r) = \frac{1}{d\, \omega_d\, r^{d-1}\, b_n} \sum_{i,j\geq 1}^{\neq} \frac{\mathbf{1}(X_i \in W_n)\, \mathbf{1}(X_j \in W_n)}{\nu_d((W_n - X_i) \cap (W_n - X_j))}\, k_1\Big(\frac{\|X_j - X_i\| - r}{b_n}\Big),$$

where the *kernel function* $k_d \mid \mathbb{R}^d \mapsto \mathbb{R}$ is integrable (and mostly symmetric, bounded with bounded support) such that $\int_{\mathbb{R}^d} k_d(x)\, dx = 1$ and the sequence of *bandwidths* is chosen such that $b_n \downarrow 0$ and $b_n^d\, \nu_d(W_n) \underset{n\to\infty}{\longrightarrow} \infty$. These conditions imply the pointwise asymptotic unbiasedness of the kernel estimators, namely

$$\mathbf{E}\widehat{(\lambda\, \rho^{(2)})}_n(x) \underset{n\to\infty}{\longrightarrow} \lambda\, \rho^{(2)}(x) \quad \text{and} \quad \mathbf{E}\widehat{(\lambda^2\, g)}_n(r) \underset{n\to\infty}{\longrightarrow} \lambda^2\, g(r)$$

at any continuity point $x \neq o$ of $\rho^{(2)}$ resp. at any continuity point $r > 0$ of g, see e.g. [232–234, 275]. Under some further additional conditions one can show that

$$b_n^d\, \nu_d(W_n)\, \mathbf{var}\, \widehat{(\lambda\, \rho^{(2)})}_n(x) \underset{n\to\infty}{\longrightarrow} \lambda \rho^{(2)}(x) \int_{\mathbb{R}^d} k_d^2(x)\, dx$$

and also central limit theorems (briefly CLT's) and optimal bandwidths can be derived, see for example [232] for an application to testing point process models. Furthermore, various asymptotic results for higher-order kernel-type product density estimators (among them rates of convergence, **P**-a.s. convergence) have been obtained under stronger mixing assumptions, see [234, 275].

Finally, we regard a kernel-type estimator of the limit

$$\sigma^2 = \lim_{n\to\infty} \nu_d(W_n)\, \mathbf{var}\big(\widehat{\lambda}_n\big)$$

which exists for all $\mathbf{B}_2$-mixing stationary PP's. The following estimator has been studied in [238]:

$$\widehat{(\sigma^2)}_n := \widehat{\lambda}_n + \sum_{i,j\geq 1}^{\neq} \frac{\mathbf{1}(X_i \in W_n)\, \mathbf{1}(X_j \in W_n)\, w(\,(X_j - X_i)/c_n\,)}{\nu_d((W_n - X_i) \cap (W_n - X_j))} - c_n^d\, \widehat{(\lambda^2)}_n \int_{\mathbb{R}^d} w(x)\, dx,$$

where $c_n := b_n\,(\nu_d(W_n))^{1/d}$ and $w : \mathbb{R}^d \mapsto \mathbb{R}^1$ is a non-negative, symmetric, bounded function with bounded support satisfying $\lim_{x\to o} w(x) = w(o) = 1$.

Theorem 4.8. *For a* $\mathbf{B}_4$*-mixing stationary PP the estimator* $\widehat{(\sigma^2)}_n$ *is asymptotically unbiased and mean-square consistent if* $b_n \underset{n\to\infty}{\longrightarrow} 0$, $c_n/r(W_n) \underset{n\to\infty}{\longrightarrow} 0$, $c_n \underset{n\to\infty}{\longrightarrow} \infty$, *and* $b_n\,c_n \underset{n\to\infty}{\longrightarrow} 0$. *If the PP is even* **B***-mixing, then* $\sqrt{\nu_d(W_n)}\,\big(\widehat{\lambda}_n - \lambda\big)/\sigma$ *is asymptotically* $N(0,1)$*-distributed, where* σ *can be replaced by the square root of* $\widehat{(\sigma^2)}_n$.

In this way one can construct an asymptotic confidence interval which covers the intensity λ with given probability $1-\alpha$.

4.3 Mixing and m-Dependence in Random Point Processes

Large domain statistics requires weak dependence assumptions of the observed spatial process to derive properties of the estimators and to construct asymptotic tests for checking statistical hypotheses. We formulate and apply a spatial ergodic theorem. The notion of m-dependence plays an important role to prove limits theorems for Poisson-driven models demonstrated in particular for the Boolean model and statistics taken from Poisson procesess. We consider also some examples which exhibit appropriate spatial versions of the α- and β-mixing condition.

4.3.1 Poisson-Based Spatial Processes and m-Dependence

Definition 4.7. A family of random variables $\{\xi(t),\ t \in \mathbb{Z}^d\}$ defined on $(\Omega, \mathcal{F}, \mathbf{P})$ is called *m-dependent (d-dimensional) random field* for some $m \in \mathbb{N}$ if for any finite $U, V \subset \mathbb{Z}^d$ the random vectors $(\xi(u))_{u\in U}$ and $(\xi(v))_{v\in V}$ are independent whenever $\max\limits_{1\le i\le d} |\,u_i - v_i\,| > m$ for all $u = (u_1,\ldots,u_d)^\top \in U$ and $v = (v_1,\ldots,v_d)^\top \in V$, see also Sect. 10.1.2.

For $d = 1$ we use the term "sequence" instead of "field" and in what follows we shall fix the dimension $d \ge 1$. In particular, in the theory of limit theorems for sums of random fields the particular case of m-dependent random variables indexed by a subset of $\mathbb{Z}^d$ plays an important role because most of the classical limit theorems known for sums of independent random variables remain valid with obvious modifications for m-dependent sequences and fields. This includes also a number of refined results such as Berry–Esseen bounds and asymptotic expansions of the remainder term in the CLT, see [226], or Donsker's invariance principle and functional CLT's for empirical m-dependent processes with càdlàg-trajectories, see for example [69].

In stochastic geometry and point process statistics, m-dependent random fields appear in connection with models which are defined by independently marked Poisson processes. We discuss here two examples which exhibit the main idea. This approach has been successfully applied to derive CLT's for functionals of Poisson-cluster processes and Poisson-grain models, see for example [227, 236]. For notational ease, let $W_n = \times_{i=1}^d [0, a_i^{(n)})$ be a rectangle with large enough edges $a_1^{(n)}, \ldots, a_d^{(n)}$.

Example 4.2. Let $\Xi = \bigcup_{i \geq 1} (\Xi_i + X_i)$ be a Boolean model generated by the stationary Poisson process $\Psi \sim \Pi_{\lambda \nu_d}$ and a bounded typical grain satisfying $\Xi_0 \subseteq [-r, r]^d$ **P**-a.s. for some fixed $r > 0$. We are interested in the asymptotic behaviour of the random d-volume $S_n = \nu_d(\Xi \cap W_n)$ which is closely connected with the empirical volume fraction $\widehat{p}_n = S_n / \nu_d(W_n)$.

Example 4.3. We consider the random sum

$$S_n(r) = \sum_{i \geq 1} \mathbf{1}(X_i \in W_n) \left(\Psi - \delta_{X_i} \right)(r\,B + X_i)$$

which coincides up to the scaling factor $1/\nu_d(W_n)$ with the empirical K-function $\left(\widehat{\lambda^2 K_B}\right)_{n,1}(r)$. We are able to derive the Gaussian limit distribution using the CLT for m-dependent field provided that $\Psi \sim \Pi_{\lambda \nu_d}$. For simplicity assume that $B \subseteq [-1, 1]^d$.

In both cases take the smallest number $r_i \geq r$ such that the ratio $v_i^{(n)} = a_i^{(n)}/2r_i$ is an integer for $i = 1, \ldots, d$ and decompose W_n into blocks E_t with $t = (t_1, \ldots, t_d)^\top$ as follows:

$$W_n = \bigcup_{t \in V_n} E_t \ , \quad E_t = \mathop{\times}_{i=1}^{d} \left[2r_i\, t_i, 2r_i\, (t_i + 1) \right) \ , \quad V_n = \mathop{\times}_{i=1}^{d} \{1, \ldots, v_i^{(n)}\} .$$

Then we may write $S_n = \sum_{t \in V_n} \xi(t)$ and $S_n(r) = \sum_{t \in V_n} \xi_r(t)$ with the random variables

$$\xi(t) = \nu_d(\Xi \cap E_t) \ \text{ and } \ \xi_r(t) = \sum_{i \geq 1} \mathbf{1}(X_i \in E_t) \left(\Psi - \delta_{X_i} \right)(r\,B + X_i), \quad t \in V_n \,,$$

forming a stationary 1-dependent random field due to the independence properties of the stationary Poisson process Ψ and the fact that grains $\{\Xi_i,\, i \geq 1\}$ are i.i.d. and independent of Ψ. By the same arguments we get an i.i.d. sequence of random marked counting measures

$$\Psi_t = \sum_{i \geq 1} \mathbf{1}(X_i \in E_t)\, \delta_{(X_i, \Xi_i)} \ \text{ for } t \in \mathbb{Z}_d$$

and, in addition, the $\xi(t)$'s admit a representation $\xi(t) = f(\Psi_y, |t-y| \le 1)$ in terms of a measurable function $f : (\mathcal{N}_{\mathsf{M}}^0)^{3^d} \mapsto \mathbb{R}^1$, where $\mathcal{N}_{\mathsf{M}}^0$ denotes the space of locally-finite marked counting measures on $\times_{i=1}^d [0, 2r_i) \times \mathcal{K}$. In this way $\{\xi(t),\ t \in V_n\}$ becomes a two-dependent random field with *block representation*, see [199, 226] for details. This representation of the field by functions of finite blocks of independent random elements allows to check simple conditions that imply explicit bounds of the remainder terms of asymptotic expansions in the CLT for S_n and $S_n(r)$ as well.

The CLT for (stationary) m-dependent random fields, see for example [69] or references in [226], combined with $|V_n| = \nu_d(W_n)/(2\,r)^d$ and $p = \mathbf{E}\nu_d(\Xi \cap [0,1)^d)$ yields

$$\sqrt{\nu_d(W_n)}\,\big(\widehat{p}_n - p\big) \underset{n\to\infty}{\Longrightarrow} N(0,\sigma_p^2) \ \text{ with } \sigma_p^2 = (1-p)^2 \int_{\mathbb{R}^d} \big(\mathrm{e}^{\lambda\,\mathbf{E}\nu_d(\Xi_0 \cap (\Xi_0 - x))} - 1\big)\,dx.$$

If the compact typical grain Ξ_0 is not strictly bounded, then we first replace Ξ_0 by the truncated grain $\Xi_0 \cap [-r,r]^d$ and apply the above CLT to the corresponding truncated Boolean model $\Xi(r)$. In a second step we show that the ratio

$$\mathbf{var}\big(\nu_d((\Xi \setminus \Xi(r)) \cap W_n\big)/\nu_d(W_n)\big)$$

becomes arbitrarily small uniformly in $n \in \mathbb{N}$ as r grows large provided that $\mathbf{E}\nu_d^2(\Xi_0) < \infty$. Finally, Slutsky's theorem completes the proof of the CLT in the general case.

In Example 4.3 we immediately obtain the normal convergence

$$\sqrt{\nu_d(W_n)}\,\big(\big(\widehat{\lambda^2 K_B}\big)_{n,1}(r) - \lambda^2\,K_B(r)\big) \underset{n\to\infty}{\Longrightarrow} N(0,\sigma_B^2(r)) \tag{4.18}$$

with $\sigma_B^2(r) = 2\,\lambda\,\nu_d(B)\,r^d\,\big(1 + 2\,\lambda\,\nu_d(B)\,r^d\big)$, see also [274] for related CLT's for $\mathbf{B}$-mixing stationary PP's. Using the block representation of the random variables $\xi_r(t),\ t \in V_n$, and the some results in [226], see also references therein, we obtain the optimal Berry–Esseen bound

$$\sup_{x\in\mathbb{R}^1} \Big|\,\mathbf{P}\big(\sqrt{\nu_d(W_n)}\,\big(\big(\widehat{\lambda^2 K_B}\big)_{n,1}(r) - \lambda^2\,K_B(r)\big) \le x\big) - \Phi\Big(\frac{x}{\sigma_B(r)}\Big)\Big| \le \frac{c(\lambda,B,r)}{\sqrt{\nu_d(W_n)}}\,,$$

where $\Phi(x) := \mathbf{P}(N(0,1) \le x)\,,\ x \in \mathbb{R}^1$, denotes the standard normal distribution function.

Moreover, for the random sum

$$\widehat{S}_n(r) = \nu_d(W_n)\,\big(\widehat{\lambda^2 K_B}\big)_{n,2}(r) = \sum_{i\ge 1} \mathbf{1}(X_i \in W_n)\,\big(\Psi - \delta_{X_i}\big)\big((r\,B + X_i) \cap W_n\big)\,,$$

which equals twice the number of pairs of points having N_B-distance less than or equal to r, a local CLT with asymptotic expansion can be proved by methods developed in [199, 226]:

$$(1+|x_n(k,r)|^3)\,\sqrt{\nu_d(W_n)}\,\Big|\frac{1}{2}\sqrt{\mathbf{var}\,\widehat{S}_n(r)}\,\mathbf{P}(\widehat{S}_n(r)=2k)-\varphi_n\big(x_n(k,r)\big)\Big|\underset{n\to\infty}{\longrightarrow}0$$

for any $k=0,1,2,\ldots$, where $x_n(k,r)=\big(2k-\mathbf{E}\widehat{S}_n(r)\big)/\big(\mathbf{var}\,\widehat{S}_n(r)\big)^{1/2}$ and

$$\varphi_n(x)=\frac{1}{\sqrt{2\pi}}\,e^{-x^2/2}\left(1+\frac{(x^3-3x)\,\mathbf{E}\big(\widehat{S}_n(r)-\mathbf{E}\widehat{S}_n(r)\big)^3}{6\sqrt{2\pi}\,\big(\mathbf{var}\,\widehat{S}_n(r)\big)^{3/2}}\right).$$

4.3.2 *Strong Mixing and Absolute Regularity for Spatial Processes*

The quantitative assessment of (weak) dependence between parts of spatial processes (e.g. random fields, point processes, random closed sets) over disjoint subsets of $\mathbb{R}^d$ is based on *mixing coefficients*. These quantities provide uniform bounds of the dependence between σ-algebras generated by the spatial process over these disjoint set which include rates of decay when the distance between these subsets increases. These mixing coefficients permit to derive covariance estimates of the random variables measurable with respect to these σ-algebras. This in turn is essential in proving asymptotic normality for sums of these random fields defined over $(\Omega,\mathcal{F},\mathbf{P})$. Here we shall briefly discuss two of the most relevant mixing coefficients, see also Sect. 10.1.2.

Definition 4.8. For any two sub-σ-algebras $\mathcal{A}$, $\mathcal{B}\subset\mathcal{F}$ the α-*mixing* (or *strong*) coefficient $\alpha(\mathcal{A},\mathcal{B})$ and the β-*mixing* (or *absolute regularity*) coefficient $\beta(\mathcal{A},\mathcal{B})$ are defined by

$$\alpha(\mathcal{A},\mathcal{B}):=\sup_{A\in\mathcal{A},B\in\mathcal{B}}|\mathbf{P}(A\cap B)-\mathbf{P}(A)\,\mathbf{P}(B)|,$$

$$\beta(\mathcal{A},\mathcal{B}):=\mathbf{E}\sup_{B\in\mathcal{B}}|\mathbf{P}(B\mid\mathcal{A})-\mathbf{P}(B)|=\sup_{C\in\mathcal{A}\otimes\mathcal{B}}|\mathbf{P}_{\mathcal{A}\otimes\mathcal{B}}(C)-(\mathbf{P}_{\mathcal{A}}\times\mathbf{P}_{\mathcal{B}})(C)|,$$

where $\mathcal{A}\otimes\mathcal{B}$ is the product σ-algebra generated by $\mathcal{A}$ and $\mathcal{B}$ and $\mathbf{P}_{\mathcal{A}}\times\mathbf{P}_{\mathcal{B}}$ denotes the product measure of the corresponding marginal distributions.

The inequality $2\,\alpha(\mathcal{A},\mathcal{B})\le\beta(\mathcal{A},\mathcal{B})$ is immediately seen from the above definition, see [83] for an all-embracing discussion of mixing coefficients. As already mentioned the covariance $\mathbf{cov}(\xi,\eta)$ can be bounded by means of these mixing coefficient with respect to the σ-algebras $\mathcal{A} = \sigma(\xi)$ and $\mathcal{B} = \sigma(\eta)$ generated by the random variables ξ and η, respectively. Such covariance bounds are known for long time and can be found in many papers and textbooks on limit theorems for sums of weakly dependent random variables. If ξ, η are real-valued and $p\,,\, q \in [1,\infty]$ such that $p^{-1} + q^{-1} \le 1$, then the inequality

$$|\,\mathbf{cov}\{\xi,\eta\}\,| \le C\,\left(\mathbf{E}|\xi|^p\right)^{1/p}\left(\mathbf{E}|\eta|^q\right)^{1/q}\left(2\,\alpha(\sigma(\xi),\sigma(\eta))\right)^{1-1/p-1/q}$$

holds which has been first proved by Yu.A. Davydov [146] with some positive constant $C\ (\ge 10)$. Recently, by improving the approximation technique used in [146], the author and M. Nolde could prove that $C = 2$ is possible, see also [423] for a different approach. A corresponding estimate with $\beta(\sigma(\xi),\sigma(\eta))$ rather than $\alpha(\sigma(\xi),\sigma(\eta))$ on the right-hand side goes back to K. Yoshihara [522], see also [236] for this and further references.

Let us consider a Voronoi-tessellation $V(\Psi) = \bigcup_{i\ge1} \partial C_i(\Psi)$ generated by a simple stationary PP $\Psi = \sum_{i\ge1} \delta_{X_i}$, where $\partial C_i(\Psi)$ denotes the boundary of the cell $C_i(\Psi)$ formed by all point in $\mathbb{R}^d$ which are closest to the atom X_i, i.e. $C_i(\Psi) = \{x \in \mathbb{R}^d : \|x - X_i\| < \|x - X_j\|, j \ne i\}$, and let denote by $\mathcal{A}_\Psi(F)$ resp. $\mathcal{A}_{V(\Psi)}(F)$ the σ-algebra generated by the PP Ψ restricted to $F \subset \mathbb{R}^d$ resp. the σ-algebra generated by the random closed set $V(\Psi) \cap F$. With the notation $F_a = [-a,a]^d$ and $\Delta = b/4$ the estimate

$$\beta\big(\mathcal{A}_{V(\Psi)}(F_a), \mathcal{A}_{V(\Psi)}(F^c_{a+b})\big) \le \beta\big(\mathcal{A}_\Psi(F_{a+\Delta}), \mathcal{A}_\Psi(F^c_{a+3\,\Delta})\big) + R(a,b) \tag{4.19}$$

has been obtained in [230], where $R(a,b)$ is a finite sum of certain void probabilities of the PP Ψ decaying to zero at some rate (depending on a) as $b \to \infty$. A similar estimate of $\beta\big(\mathcal{A}_\Xi(F_a), \mathcal{A}_\Xi(F^c_{a+b})\big)$ could be derived in [236] for stationary grain-germ models $\Xi = \bigcup_{i\ge1}(\Xi_i + X_i)$ in terms of a suitable β-mixing coefficient of the generating stationary PP $\Psi \sim P$ with intensity λ and the distribution function $D(x) = \mathbf{P}(\mathrm{diam}(\Xi_0) \le x)$ of the diameter of the typical grain Ξ_0:

$$\beta\big(\mathcal{A}_\Xi(F_a), \mathcal{A}_\Xi(F^c_{a+b})\big) \le \beta\big(\mathcal{A}_\Psi(F_{a+\Delta}), \mathcal{A}_\Psi(F^c_{a+3\,\Delta})\big) + \lambda\, d\, 2^{d+1}\left[\left(1+\frac{a}{\Delta}\right)^{d-1} + \left(3+\frac{a}{\Delta}\right)^{d-1}\right]\int_\Delta^\infty x^d\, dD(x) \tag{4.20}$$

Note that $\beta\big(\mathcal{A}_\Psi(F_{a+\Delta}), \mathcal{A}_\Psi(F^c_{a+3\,\Delta})\big) = 0$ in (4.19) and (4.20) if $\Psi \sim \Pi_{\lambda\,\nu_d}$, i.e. for the Poisson–Voronoi tessellation and for Boolean models. Furthermore, there exist precise estimates of this β-mixing coefficient for Poisson-cluster and Cox processes and some classes of dependently thinned Poisson processes. We only mention that both of the previous estimates can be reformulated with slight modifications in terms of α-mixing coefficients.

In [236] a CLT for geometric functionals of β-mixing random closed sets has been proved. The conditions of this CLT can be expressed more explicitly for germ-grain models due to (4.20). CLT's for stationary random fields put assumptions on *mixing rates* derived from mixing coefficients between specific σ-algebras, see [267] in the case of PP's. An application of α-mixing to study empirical functionals of geostatistically marked point processes can be found in [392]. Besides the frequently used CLT of E. Bolthausen [74] the following CLT (first proved and applied in [230]) presents a meaningful alternative to verify asymptotic normality of estimators in stochastic-geometric models.

Let $\xi = \{\xi(t),\, t \in V_n\}$ be a stationary random field with index set $V_n = \{t \in \mathbb{Z}^d : ([0,1)^d + t) \subset W_n\}$, where $\{W_n\,,\, n \in \mathbb{N}\}$ is a convex averaging sequence in $\mathbb{R}^d$ implying $|V_n|/\nu_d(W_n) \underset{n\to\infty}{\longrightarrow} 1$. Further, $\mathcal{A}_\xi(F)$ denotes the σ-algebra generated by the random variables $\{\xi(t),\, t \in F \cap \mathbb{Z}^d\}$ and $S_n = \sum_{t\in V_n} \xi(t)$.

Theorem 4.9. *Assume that there are two functions β_ξ^* and β_ξ^{**} on $\mathbb{N}$ such that*

$$\beta\big(\mathcal{A}_\xi(F_p), \mathcal{A}_\xi(F^c_{p+q})\big) \;\le\; \begin{cases} \beta_\xi^*(q) & \text{for } p = 1,\ q \in \mathbb{N} \\[2ex] p^{d-1}\,\beta_\xi^{**}(q) & \text{for } p \in \mathbb{N},\ q = 1,\ldots,p\,. \end{cases}$$

If, for some $\delta > 0$,

$$\mathbf{E}|\,\xi(o)\,|^{2+\delta} < \infty,\ \sum_{r=1}^{\infty} r^{d-1}\,\big(\beta_\xi^*(r)\big)^{\delta/(2+\delta)} < \infty \text{ and } r^{2d-1}\,\beta_\xi^{**}(r) \underset{r\to\infty}{\longrightarrow} 0\,,$$

then the asymptotic variance $\tau^2 = \lim_{n\to\infty} \mathbf{var}\, S_n/\nu_d(W_n) = \sum_{t\in\mathbb{Z}^d} \mathbf{cov}(\xi(o),\xi(t))$ exists and the normal convergence $(\nu_d(W_n))^{-1/2}\big(S_n - |V_n|\,\mathbf{E}\xi(o)\big) \underset{n\to\infty}{\Longrightarrow} N(0,\tau^2)$ holds.

Note that the assertion of Theorem 4.9 remains valid if the slightly weaker α-mixing coefficient is used, see [235] and references therein.

On the other hand, there are situations which require the stronger β-mixing coefficient. For example, $\Psi \sim P$ can be shown to be $\mathbf{B}_k$-mixing for any fixed $k \ge 2$ if $\mathbf{E}\Psi([0,1]^d)^{k+\delta} < \infty$ and

$$\int_1^\infty r^{(k-1)d-1}\,\big(\beta_\Psi(r)\big)^{\delta/(k+\delta)} dr < \infty$$

for some $\delta > 0$, where the β-mixing coefficient $\beta_\Psi : [1,\infty) \to [0,1]$ is defined as a non-increasing function such that $\beta_\Psi(r) \ge \big(\min\{1, \frac{r}{a}\}\big)^{d-1}\,\beta\big(\mathcal{A}_\Psi(F_a), \mathcal{A}_\Psi(F^c_{a+r})\big)$ for all $a, r \ge 1$. This implies that $\Psi \sim P$ is Brillinger-mixing if $\beta_\Psi(r) \le e^{-g(r)}$ with $g : [1,\infty) \to [0,\infty]$ satisfying $g(r)/\log(r) \underset{r\to\infty}{\longrightarrow} \infty$.

4.3.3 Testing CSR Based on Empirical K-Functions

Let $\big(\widehat{\lambda^2 K_B}\big)_n(r)$ be any of the empirical K-functions $\big(\widehat{\lambda^2 K_B}\big)_{n,i}(r)$, $i = 1,2,3$, introduced and discussed in the above Sect. 4.2.1. Below we formulate two functional CLT's for the corresponding centered and scaled empirical process on the interval $[0,R]$ when $\Psi = \{X_i,\, i \geq 1\}$ is a stationary Poisson process. We distinguish between the cases of known intensity λ and estimated intensity $\widehat{\lambda}_n$ which leads to two distinct zero mean Gauss–Markov limit processes (in the sense of weak convergence in the Skorokhod-space D$[0,R]$, see [69]). For both limits the distribution function of the maximal deviation over $[0,R]$ can be calculated. This fact can be used to establish a Kolmogorov–Smirnov-type test for checking the null hypothesis of CSR via testing the suitably scaled maximal deviation of the empirical K-functions from $\lambda^2\, \nu_d(B)\, r^d$ resp. $(\widehat{\lambda^2})_n\, \nu_d(B)\, r^d$, see (4.17). For the details of the proofs (in the particular case $B = B_1(o)$) and some extensions (among them, a Cramér-von Mises-type test for K-functions) the reader is referred to [228].

Theorem 4.10. *Let the stationary Poisson process* $\Psi \sim \Pi_{\lambda\,\nu_d}$ *with intensity* $\lambda > 0$ *be observed in window* $W_n = \times_{i=1}^d [0, a_i^{(n)}]$ *with unboundedly increasing edges. Then*

$$\zeta_n(r) := \sqrt{\nu_d(W_n)/\lambda}\,\Big(\big(\widehat{\lambda^2 K_B}\big)_n(r) - \lambda^2\, \nu_d(B)\, r^d \Big) \underset{n\to\infty}{\Longrightarrow} \zeta(r) \overset{d}{=} \frac{W\big(L(r)\big)}{1 - L(r)}$$

$$\eta_n(r) := \sqrt{\nu_d(W_n)/\widehat{\lambda}_n}\,\Big(\big(\widehat{\lambda^2 K_B}\big)_n(r) - (\widehat{\lambda^2})_n\, \nu_d(B)\, r^d \Big) \underset{n\to\infty}{\Longrightarrow} \eta(r) \overset{d}{=} W(2\,\nu_d(B)\, r^d) \tag{4.21}$$

for $0 \leq r \leq R$, *where* $\underset{n\to\infty}{\Longrightarrow}$ *stands for weak convergence in the Skorokhod-space* D$[0,R]$. *Both weak limits* $\zeta(r)$ *and* $\eta(r)$ *for* $r \in [0,R]$ *are Gaussian diffusion processes with zero means and covariance functions*

$$\mathbf{E}\zeta(s)\zeta(t) = 2\,\lambda\,\nu_d(B)\, s^d\, \big(1 + 2\,\lambda\,\nu_d(B)\, t^d\big) \quad \textit{and} \quad \mathbf{E}\eta(s)\eta(t) = 2\,\nu_d(B)\, s^d$$

for $0 \leq s \leq t \leq R$. *In* (4.21), $\overset{d}{=}$ *means stochastic equivalence,* $W = \{W(t),\, t \geq 0\}$ *denotes the one-dimensional standard Wiener process and* $L(r) = 2\,\lambda\,\nu_d(B)\, r^d / (1 + 2\,\lambda\,\nu_d(B)\, r^d)$.

Corollary 4.1. *The* continuous mapping theorem, *see [69], applied to* (4.21) *implies that*

$$\max_{0\leq r\leq R} |\zeta_n(r)| \underset{n\to\infty}{\Longrightarrow} \max_{0\leq t\leq L} \frac{|W(t)|}{1-t} \sim F_L \quad \textit{and} \quad \frac{\max_{0\leq r\leq R} |\eta_n(r)|}{\sqrt{2\,\nu_d(B)\, R^d}} \underset{n\to\infty}{\Longrightarrow} \max_{0\leq t\leq 1} |W(t)| \sim G\,,$$

where $L = L(R)\,(<1)$ *and*

$$1 - F_L(x) = 2\Big(1 - \Phi\big(x(1-L)/\sqrt{L}\big)\Big) + 2\sum_{n=1}^{\infty}(-1)^{n+1}e^{2nx^2}\Big(\Phi\big(x(2n+1-L)/\sqrt{L}\big) - \Phi\big(x(2n-1-L)/\sqrt{L}\big)\Big)$$

and $$1 - G(x) \;=\; 4\big(1-\Phi(x)\big) + 4\sum_{n=1}^{\infty}(-1)^n\big(1-\Phi(\,(2n+1)x\,)\big).$$

Remark 4.5. The relevant quantiles of F_L and G are known. Obviously, testing the CSR-property via checking the goodness-of-fit of the K-function seems to be easier when λ is unknown. The convergence of the finite-dimensional distributions of $\{\zeta_n(\cdot),\, n \in \mathbb{N}\}$ follows from the CLT for m-dependent fields and the tightness in D$[0, R]$ is seen by an exact bound of the mixed fourth-order moment of two consecutive increments. The convergence of the finite-dimensional distributions of $\{\eta_n(\cdot),\, n \in \mathbb{N}\}$ follows by applying a variant of Stein's method to an asymptotically degenerate U-statistic, see [228, 231]

In [231] an analogous test of CSR based on the multivariate K-function $K(r_1,..,r_d) = 2^d\, r_1 \cdot \ldots \cdot r_d$ and its empirical counterpart in case of a Poisson process has been developed. We only sketch the main result in the case of unknown intensity λ. In [231] the case of known λ is also treated in detail.

Let the assumptions of Theorem 4.10 be satisfied. Setting

$$\widehat{(\lambda^2 K)}_n(\mathbf{r}) := \frac{1}{\nu_d(W_n)}\sum_{j\geq 1}\mathbf{1}(X_j \in W_n)\,(\Psi - \delta_{X_j})\big((\times_{i=1}^{d}[-r_i, r_i] + X_j) \cap W_n\,\big)$$

for $\mathbf{r} = (r_1,\ldots,r_d)^T \in [0,\infty)^d$, and

$$\eta_n(\mathbf{r}) := \sqrt{\frac{\nu_d(W_n)}{\widehat{\lambda}_n}}\,\Big(\,\widehat{(\lambda^2 K)}_n(\mathbf{r}) - \widehat{(\lambda^2)}_n\, 2^d \prod_{i=1}^{d} r_i\,\Big)$$

we obtain a sequence $\{\eta_n(\mathbf{r}),\, \mathbf{r} \in [0, R]^d\}$ of empirical processes belonging to the Skorokhod-space D$([0, R]^d)$ of d-parameter càdlàg-processes that converges weakly to a Gaussian random field $\{\eta(\mathbf{r}),\, \mathbf{r} \in [0, R]^d\} \stackrel{d}{=} \{\sqrt{2^{d+1}}\, W_d(\mathbf{r}),\, \mathbf{r} \in [0, R]^d\,\}$, where $\{\,W_d(\mathbf{r}),\, \mathbf{r} \in [0,\infty)^d\,\}$ denotes the d-dimensional *standard Wiener sheet* with mean value function $\mathbf{E}W_d(\mathbf{r}) = 0$ and covariance function $\mathbf{E}W_d(\mathbf{s})W_d(\mathbf{t}) = \prod_{i=1}^{d}(\,s_i \wedge t_i\,)$ for $\mathbf{s} = (s_1,\ldots,s_d)^\top$, $\mathbf{t} = (t_1,\ldots,t_d)^\top$. Hence, by the continuous mapping theorem it follows that

$$\max_{\mathbf{r}\in[0,R]^d}\,|\,\eta_n(\mathbf{r})\,| \underset{n\to\infty}{\Longrightarrow} \max_{\mathbf{r}\in[0,R]^d}\,|\,\eta(\mathbf{r})\,| \stackrel{d}{=} \sqrt{2^{d+1}\,R^d}\,\max_{\mathbf{r}\in[0,1]^d}\,|\,W_d(\mathbf{r})\,|\,.$$

The α-quantiles of the distribution function

$$G_2(x) = \mathbf{P}(|W_2(r_1, r_2)| \le x \,,\ \forall\, (r_1, r_2) \in [0, 1]^2)$$

can be determined only approximately via large-scale simulations of the planar Wiener sheet. In this way we found $m_{0.95} = 2.1165$, $m_{0.99} = 2.7105$, and $m_{0.995} = 2.9313$, where $G_2(m_\alpha) = \alpha$.

Chapter 5
Random Tessellations and Cox Processes

Florian Voss, Catherine Gloaguen, and Volker Schmidt

Abstract We consider random tessellations $\mathtt{T}$ in $\mathbb{R}^2$ and Cox point processes whose driving measure is concentrated on the edges of $\mathtt{T}$. In particular, we discuss several classes of Poisson-type tessellations which can describe for example the infrastructure of telecommunication networks, whereas the Cox processes on their edges can describe the locations of network components. An important quantity associated with stationary point processes is their typical Voronoi cell Z. Since the distribution of Z is usually unknown, we discuss algorithms for its Monte Carlo simulation. They are used to compute the distribution of the typical Euclidean (i.e. direct) connection length D^o between pairs of network components. We show that D^o converges in distribution to a Weibull distribution if the network is scaled and network components are simultaneously thinned in an appropriate way. We also consider the typical shortest path length C^o to connect network components along the edges of the underlying tessellation. In particular, we explain how scaling limits and analytical approximation formulae can be derived for the distribution of C^o.

5.1 Random Tessellations

In the section we introduce the notion of random tessellations in $\mathbb{R}^2$, where we show that they can be regarded as marked point processes as well as random closed sets, and we discuss some mean-value formulae of stationary random tessellations.

F. Voss
Boehringer-Ingelheim Pharma GmbH & Co., Ingelheim, Germany
e-mail: florian.voss@boehringer-ingelheim.com

C. Gloaguen
Orange Labs, Issy les Moulineaux, France
e-mail: catherine.gloaguen@orange.com

V. Schmidt (✉)
Ulm University, Ulm, Germany
e-mail: volker.schmidt@uni-ulm.de

E. Spodarev (ed.), *Stochastic Geometry, Spatial Statistics and Random Fields*, Lecture Notes in Mathematics 2068, DOI 10.1007/978-3-642-33305-7_5,

Furthermore, we introduce simple tessellation models of Poisson type like Poisson–Voronoi, Poisson–Delaunay and Poisson line tessellations. Confer Chap. 6 for tessellations in arbitrary dimension $d \geq 2$.

5.1.1 Deterministic Tessellations

Intuitively speaking, a tessellation is a subdivision of $\mathbb{R}^2$ into a sequence of convex polygons. However, a tessellation can also be identified with the segment system consisting of the boundaries of these polygons. Because of these different viewpoints, random tessellations introduced later on in Sect. 5.1.2 are flexible models which can be applied in many different fields of science.

We start with the definition of deterministic planar tessellations. A *tessellation* τ in $\mathbb{R}^2$ is a countable family $\{B_n\}_{n\geq 1}$ of convex bodies B_n fulfilling the conditions $\mathring{B}_n \neq \emptyset$ for all n, $\mathring{B}_n \cap \mathring{B}_m = \emptyset$ for all $n \neq m$, $\bigcup_{n\geq 1} B_n = \mathbb{R}^2$ and $\sum_{n\geq 1} \mathbf{1}(B_n \cap C \neq \emptyset) < \infty$ for any $C \in \mathcal{K}^2$, where $\mathring{A}$ denotes the interior of the set $A \subset \mathbb{R}^2$, and $\mathcal{K}^2$ is the family of compact sets in $\mathbb{R}^2$. The sets B_n are called the *cells* of the tessellation τ and are bounded polygons in $\mathbb{R}^2$. In the following, we use the notation $\mathbb{T}$ for the family of all tessellations in $\mathbb{R}^2$. Note that we can identify a tessellation τ with the segment system $\tau^{(1)} = \cup_{n=1}^{\infty} \partial B_n$ constructed from the boundaries of the cells of τ. Thus, a tessellation can be identified with a closed subset of $\mathbb{R}^2$ and hence we can regard $\mathbb{T}$ as a subset of the family G of all closed subsets of $\mathbb{R}^2$. We use this connection in order to define the σ-algebra $\mathcal{T}$ on $\mathbb{T}$ as the trace-σ-algebra of $\mathcal{B}(\mathsf{G})$ in $\mathbb{T}$.

With each cell B_n of τ we can associate some "marker point" in the following way. Consider a mapping $\alpha : \mathcal{K}^2 \setminus \{\emptyset\} \to \mathbb{R}^2$ which satisfies

$$\alpha(K + x) = \alpha(K) + x \quad \text{for all } K \in \mathcal{K}^2,\ K \neq \emptyset \text{ and } x \in \mathbb{R}^2, \tag{5.1}$$

where $\alpha(K)$ is called the *nucleus* of K and can be for example the center of gravity of K.

There are various ways to generate tessellations based on sets of points and lines. Particular models are Voronoi tessellations and Delaunay tessellations as well as line tessellations, namely introduced in the following.

Let $\mathbf{x} = \{x_1, x_2, \ldots\} \subset \mathbb{R}^2$ be a locally finite set with conv($\mathbf{x}$) $= \mathbb{R}^2$, where conv($\mathbf{x}$) denotes the convex hull of the family $\mathbf{x}$. Then the *Voronoi tessellation* τ induced by $\mathbf{x}$ is defined by the nearest-neighbour principle, i.e., the cells B_n of τ are given by

$$B_n = \{x \in \mathbb{R}^2 : |x - x_n| \leq |x - x_m| \text{ for all } m \neq n\}. \tag{5.2}$$

Note that $B_n = \bigcap_{m\neq n} H(x_n, x_m)$, i.e. the cell B_n can be represented as intersection of the half-planes $H(x_n, x_m) = \{x \in \mathbb{R}^2 : |x - x_n| \leq |x - x_m|\}$ for $m \neq n$, where the half-planes $H(x_n, x_m)$ are also called *bisectors*. Since $\mathbf{x}$ is locally finite it is clear that the cells of τ have non-empty interior. Moreover, their union covers $\mathbb{R}^2$ and two different cells can only intersect at their boundaries. Using conv($\mathbf{x}$) $= \mathbb{R}^2$,

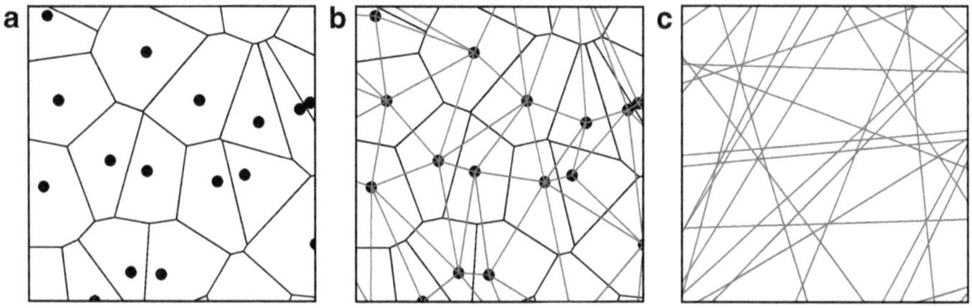

Fig. 5.1 Different types of tessellation models. (**a**) Voronoi tessellation. (**b**) Delaunay tessellation (*red*). (**c**) Line tessellation

it can be shown that the cells are convex polygons which are bounded and locally finite, see Exercise 5.1 below. Thus, the family $\tau = \{\Xi_n\}$ constructed in this way is indeed a tessellation. A Voronoi tessellation together with the generating point set is displayed in Fig. 5.1a.

Exercise 5.1. Let $\mathbf{x} = \{x_1, x_2, \ldots\} \subset \mathbb{R}^2$ be a locally finite set with $\mathrm{conv}(\mathbf{x}) = \mathbb{R}^2$. Show that the Voronoi cells Ξ_n given in (5.2) are convex polygons which are bounded and locally finite.

Now assume that four cocircular points do not exist in $\mathbf{x}$, i.e., we assume that there are no pairwise different points $x_i, x_j, x_k, x_l \in \mathbf{x}$ which are located on a circle. In this case, the *Delaunay tessellation* τ' induced by $\mathbf{x}$ can be generated uniquely as the dual tessellation of the Voronoi tessellation τ which is induced by $\mathbf{x}$. The cells of τ' are triangles which are constructed in the following way. Three points $x_i, x_j, x_k \in \mathbf{x}$ form a triangle of τ' if the corresponding Voronoi cells B_i, B_j and B_k have a common intersection point. This rule is equivalent to the empty circle criterion: three points of $\mathbf{x}$ are the vertices of a triangle of τ' iff the circumcircle of these three points does not contain other points of $\mathbf{x}$. It can be shown that the resulting sequence of triangles forms a tessellation in $\mathbb{R}^2$. In Fig. 5.1b a Delaunay tessellation is displayed together with its generating points and the dual Voronoi tessellation.

Exercise 5.2. Let $\mathbf{x} = \{x_1, x_2, \ldots\} \subset \mathbb{R}^2$ be a locally finite set with $\mathrm{conv}(\mathbf{x}) = \mathbb{R}^2$ and assume that there are no pairwise different points $x_i, x_j, x_k, x_l \in \mathbf{x}$ which are located on a circle. Show that the construction rule described above leads to a sequence of triangles which has the properties of a tessellation in $\mathbb{R}^2$.

Consider a set $\boldsymbol{\ell} = \{\ell_1, \ell_2, \ldots\}$ of lines in $\mathbb{R}^2$ and let $p_i \in \mathbb{R}^2$ denote the orthogonal projection of o onto ℓ_i, where it is assumed that $\mathrm{conv}(\{p_1, p_2, \ldots\}) = \mathbb{R}^2$. Furthermore, we assume that $|\{i : \ell_i \cap B \neq \emptyset\}| < \infty$ for all $B \in \mathcal{K}^2$. Then, in a natural way, we can generate a tessellation with respect to the intersecting lines of $\boldsymbol{\ell}$. Recall that we can identify a tessellation τ with the edge set $\tau^{(1)} = \cup_{n=1}^{\infty} \partial \Xi_n$ given by the union of the cell boundaries. Thus, we define the *line tessellation* τ induced by $\boldsymbol{\ell}$ via the edge set $\tau^{(1)} = \cup_{i=1}^{\infty} \ell_i$ formed by the union of the lines $\ell_1, \ell_2, \ldots$. If the family $\boldsymbol{\ell}$ fulfills the assumptions above, then it is ensured that the resulting cells possess the properties of a tessellation of $\mathbb{R}^2$, see also Fig. 5.1c.

5.1.2 *Random Tessellations*

Usually, a *random tessellation* in $\mathbb{R}^2$ is defined as a measurable mapping $\mathrm{T} : \Omega \to \mathbb{T}$, i.e. as a sequence $\mathrm{T} = \{\Xi_n\}$ of random convex bodies Ξ_n such that $\mathbf{P}(\{\Xi_n\} \in \mathbb{T}) = 1$. Notice that there are various ways to look at tessellations. In particular, they can be viewed as marked point processes and random closed sets. Each of these different points of view leads to new characteristics that can be associated with a tessellation. The tessellation T is said to be stationary and isotropic if $T_x\mathrm{T} = \{T_x\Xi_n\} \stackrel{d}{=} \mathrm{T}$ for all $x \in \mathbb{R}^2$ and $\vartheta_R\mathrm{T} = \{\vartheta_R\Xi_n\} \stackrel{d}{=} \mathrm{T}$ for all rotations ϑ_R around the origin, respectively.

5.1.2.1 Random Tessellations as Marked Point Processes

It is often convenient to represent a random tessellation $\mathrm{T} = \{\Xi_n\}$ as a marked point process with an appropriate mark space. Note that we can associate various point processes with T, for example the point processes of cell nuclei, vertices and edge midpoints. If these point processes are marked with suitable marks, then we can identify T with the corresponding marked point process.

We first consider the *point process of cell nuclei* marked with the cells. Let $\alpha : \mathcal{K}^2\backslash\{\emptyset\} \to \mathbb{R}^2$ be a mapping such that (5.1) holds. Let $\mathcal{P}^o$ denote the family of all convex and compact polygons A with their nucleus $\alpha(A)$ at the origin. Then $\mathcal{P}^o \subset \mathsf{G}$ is an element of $\mathcal{B}(\mathsf{G})$ and we can define the σ-algebra $\mathcal{B}(\mathcal{P}^o) = \mathcal{B}(\mathsf{G}) \cap \mathcal{P}^o$. Furthermore, the random tessellation $\mathrm{T} = \{\Xi_n\}$ can be identified with the marked point process $\{(\alpha(\Xi_n), \Xi_n^o)\}$, where $\Xi_n^o = \Xi_n - \alpha(\Xi_n)$ is the n-th cell shifted to the origin. If T is stationary, then $\{(\alpha(\Xi_n), \Xi_n^o)\}$ is also stationary and we denote its intensity by $\lambda^{(2)}$, where we always assume that $0 < \lambda^{(2)} < \infty$. The typical mark $Z : \Omega \to \mathcal{P}^o$ of $\{(\alpha(\Xi_n), \Xi_n^o)\}$ is a random polygon distributed according to the Palm mark distribution of $\{(\alpha(\Xi_n), \Xi_n^o)\}$ as defined in formula (4.6). We call the random polygon Z the *typical cell* of the tessellation T, see also Sects. 6.1.2 and 7.2.

Another possibility to represent T by a marked point process is the following. Consider the *point process of vertices* $\{V_n\}$ of T. For each vertex V_n we define the *edge star* E_n as the union of all edges of T emanating from V_n. Thus, $E_n^o = E_n - V_n$ is an element of the family $\mathcal{L}^o$ of finite segment systems containing the origin. Since $\mathcal{L}^o \in \mathcal{B}(\mathsf{G})$ we can consider the σ-algebra $\mathcal{B}(\mathcal{L}^o) = \mathcal{B}(\mathsf{G}) \cap \mathcal{L}^o$ on $\mathcal{L}^o$. Hence, we can represent the random tessellation T by the marked point process $\{(V_n, E_n^o)\}$ with mark space $\mathcal{L}^o$. If T is stationary, then $\{(V_n, E_n^o)\}$ is stationary and its intensity is denoted by $\lambda^{(0)}$, where we assume that $0 < \lambda^{(0)} < \infty$. The *typical edge star* $E^o : \Omega \to \mathcal{L}^o$ of T is defined as a random segment system distributed according to the Palm mark distribution of $\{(V_n, E_n^o)\}$.

The random tessellation T can also be represented by the marked *point process of edge midpoints* $\{(Y_n, S_n^o)\}$, where each midpoint Y_n is marked with the centered version $S_n^o = S_n - Y_n \in \mathcal{L}^o$ of the edge S_n corresponding to Y_n. If T is stationary, then it is easy to see that $\{(Y_n, S_n^o)\}$ is stationary. The intensity of edge midpoints is denoted by $\lambda^{(1)}$, where we again assume that $0 < \lambda^{(1)} < \infty$. The *typical edge* $S^o : \Omega \to \mathcal{L}^o$ is defined as the typical mark of the stationary marked point process $\{(Y_n, S_n^o)\}$.

5.1.2.2 Random Tessellations as Random Closed Sets

In the preceding section random tessellations have been represented as marked point processes. Alternatively, random tessellations can be regarded as random closed sets, see Sect. 1.2 for their definition and basic properties. Recall that deterministic tessellations can be identified with their edge sets. Thus, in the random setting, we can identify a random tessellation $\mathrm{T} = \{\Xi_n\}$ with the corresponding *random closed set* of its edges which is defined by $\mathrm{T}^{(1)} = \cup_{n=1}^{\infty} \partial \Xi_n$. If T is stationary and isotropic, then the random closed set $\mathrm{T}^{(1)}$ is stationary and isotropic, respectively. Since, almost surely, $\mathrm{T}^{(1)}$ is a locally finite system of line segments, we can consider the 1-dimensional Hausdorff measure ν_1 on $\mathrm{T}^{(1)}$. Furthermore, if T is stationary, then it is not difficult to see that the mapping $B \mapsto \mathbf{E}\nu_1(B \cap \mathrm{T}^{(1)})$ is a (σ-additive) measure on $\mathcal{B}(\mathbb{R}^2)$, which is invariant with respect to translations. Thus, by Haar's lemma, we get that $\mathbf{E}\nu_1(B \cap \mathrm{T}^{(1)}) = \gamma \nu_2(B)$ for any $B \in \mathcal{B}(\mathbb{R}^2)$ and some constant γ which is called the *length intensity* of $\mathrm{T}^{(1)}$. As for the intensities $\lambda^{(0)}, \lambda^{(1)}$ and $\lambda^{(2)}$ regarded above, we always assume that $0 < \gamma < \infty$.

5.1.2.3 Mean-Value Formulae

We now discuss mean-value formulae for stationary tessellations. These are relationships connecting the intensities of vertices $\lambda^{(0)}$, edge midpoints $\lambda^{(1)}$ and cell nuclei $\lambda^{(2)}$, the length intensity $\gamma = \mathbf{E}\nu_1(\mathrm{T}^{(1)} \cap [0,1)^2)$, the expected area $\mathbf{E}\nu_2(Z)$, perimeter $\mathbf{E}\nu_1(\partial Z)$ and number of vertices $\mathbf{E}\nu_0(Z)$ of the typical cell Z, the expected length of the typical edge $\mathbf{E}\nu_1(S^o)$, and the expected length $\mathbf{E}\nu_1(E^o)$ and number of edges $\mathbf{E}\nu_0(E^o)$ of the typical edge star E^o. It turns out that all these characteristics can be expressed by for example the three parameters $\lambda^{(0)}, \lambda^{(2)}$ and γ.

Theorem 5.1. *It holds that*

$$\lambda^{(1)} = \lambda^{(0)} + \lambda^{(2)}, \qquad \mathbf{E}\nu_0(E^o) = 2 + 2\frac{\lambda^{(2)}}{\lambda^{(0)}}, \qquad \mathbf{E}\nu_1(E^o) = 2\frac{\lambda^{(1)}}{\lambda^{(0)}}\mathbf{E}\nu_1(S^o),$$

$$\mathbf{E}\nu_0(Z) = 2 + 2\frac{\lambda^{(0)}}{\lambda^{(2)}}, \qquad \mathbf{E}\nu_2(Z) = \frac{1}{\lambda^{(2)}}, \qquad \mathbf{E}\nu_1(\partial Z) = 2\frac{\lambda^{(1)}}{\lambda^{(2)}}\mathbf{E}\nu_1(S^o),$$

$$\gamma = \lambda^{(1)}\mathbf{E}\nu_1(S^o) = \frac{\lambda^{(2)}}{2}\mathbf{E}\nu_1(\partial Z), \qquad 3 \le \mathbf{E}\nu_0(Z), \mathbf{E}\nu_0(E^o) \le 6\,.$$

Proof. We show how some of the formulae stated above can be proven using Campbell's theorem for stationary marked point processes; see formula (4.10). For example, consider the marked point process $\{(Y_n, S_n^o)\}$ of edge midpoints Y_n marked with the centered edges S_n^o. Then, formula (4.10) yields

$$\gamma = \mathbf{E}\nu_1(\mathrm{T}^{(1)} \cap [0,1)^2) = \mathbf{E}\sum_{n=1}^{\infty} \nu_1((S_n^o + Y_n) \cap [0,1)^2)$$
$$= \lambda^{(1)} \int_{\mathbb{R}^2} \mathbf{E} \underbrace{\nu_1(S^o \cap [0,1)^2 - x)}_{= \int_{S^o} \mathbf{1}(y \in [0,1)^2 - x)\, \nu_1(dy)} \nu_2(dx)$$
$$= \lambda^{(1)}\mathbf{E} \int_{S^o} \int_{\mathbb{R}^2} \mathbf{1}(x \in [0,1)^2 - y)\, \nu_2(dx)\, \nu_1(dy) = \lambda^{(1)}\mathbf{E}\nu_1(S^o)\,,$$

thus $\gamma = \lambda^{(1)}\mathbf{E}\nu_1(S^o)$. Furthermore,

$$\lambda^{(2)}\mathbf{E}\nu_2(Z) = \lambda^{(2)}\mathbf{E}\int_{\mathbb{R}^2} \mathbf{1}(-x \in Z)\, \nu_2(dx)$$
$$= \mathbf{E} \sum_{(\alpha(\Xi_n), \Xi_n^o) \in \mathrm{T}} \underbrace{\mathbf{1}(-\alpha(\Xi_n) \in \Xi_n^o)}_{= \mathbf{1}(o \in \Xi_n)} = 1\,,$$

which yields $\mathbf{E}\nu_2(Z) = 1/\lambda^{(2)}$. The other statements can be proven similarly. For a complete proof of Theorem 5.1, see for example [133, 350]. □

Exercise 5.3. Show that

$$\lambda^{(1)} = \lambda^{(0)} + \lambda^{(2)}, \qquad \mathbf{E}\nu_0(E^o) = 2 + 2\frac{\lambda^{(2)}}{\lambda^{(0)}}, \qquad \mathbf{E}\nu_1(E^o) = 2\frac{\lambda^{(1)}}{\lambda^{(0)}}\mathbf{E}\nu_1(S^o).$$

5.1.3 Tessellation Models of Poisson Type

In this section we consider several tessellation models of Poisson type, like Poisson–Voronoi, Poisson–Delaunay and Poisson line tessellations. They are based on planar or linear Poisson point processes.

5.1.3.1 Poisson–Voronoi Tessellation

In Sect. 5.1.1 the notion of a deterministic Voronoi tessellation has been introduced for a certain class of locally finite point sets. Since almost every realization of a stationary point process $\Psi = \{X_n\}$ with $\mathbf{P}(\Psi(\mathbb{R}^2) = \infty) = 1$ is a locally finite point set such that $\mathrm{conv}(\Psi) = \mathbb{R}^2$, we can regard the random Voronoi tessellation $\{\Xi_n\}$

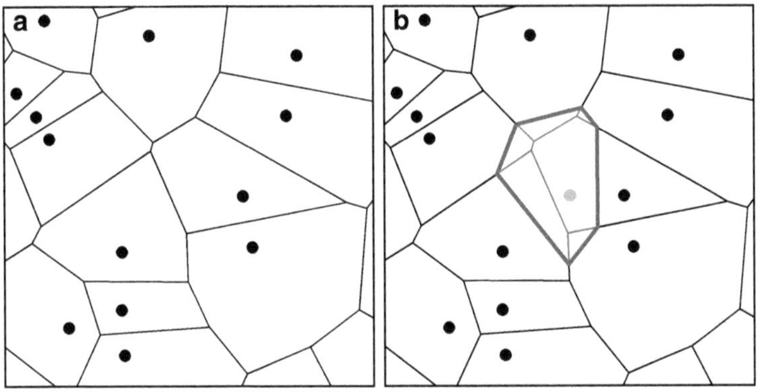

Fig. 5.2 Realization of a PVT and its typical cell. (**a**) Realization of a PVT. (**b**) Typical cell of PVT

with respect to the point process $\{X_n\}$. Thus, in accordance with (5.2), the cells Ξ_n are defined as the random closed sets $\Xi_n = \{x \in \mathbb{R}^2 : |x-X_n| \leq |x-X_m| \ \forall m \neq n\}$. We call $\mathrm{T} = \{\Xi_n\}$ the Voronoi tessellation induced by Ψ. Note that we can consider the point X_n as nucleus of the cell Ξ_n. If the underlying point process Ψ is stationary, then the Voronoi tessellation induced by Ψ is also stationary, see also Exercise 5.4. In particular, if $\Psi = \Pi_\lambda$ is a stationary Poisson process with intensity $\lambda > 0$, then we call the induced Voronoi tessellation a *Poisson–Voronoi tessellation* (PVT). Realizations of PVT are shown in Figs. 5.2 and 5.3a. Note that $\nu_0(E_n) = 3$ for all $n \in \mathbb{N}$, $\lambda^2 = \lambda$, and the intensities $\lambda^{(0)}, \lambda^{(1)}$ and γ can be computed in the following way.

Theorem 5.2. *Let* T *be a PVT induced by a Poisson process with intensity* λ. *Then*

$$\lambda^{(0)} = 2\lambda, \qquad \lambda^{(1)} = 3\lambda, \qquad \lambda^{(2)} = \lambda, \qquad \gamma = 2\sqrt{\lambda}\,.$$

Proof. Applying Theorem 5.1 with $\lambda^{(2)} = \lambda$ and $\mathbf{E}\nu_0(E^o) = 3$ yields $\lambda^{(0)} = 2\lambda$, $\lambda^{(1)} = 3\lambda$, and $\lambda^{(2)} = \lambda$. For the proof of $\gamma = 2\sqrt{\lambda}$ see for example [451, Chap. 10]. □

Consider the random Voronoi tessellation T induced by any stationary (not necessarily Poisson) point process Ψ. Then, the distribution of the typical cell of T is given by the distribution of the Voronoi cell at o with respect to the Palm version Ψ^o of Ψ. In particular, due to Slivnyak's theorem for stationary Poisson processes (see for example Theorem 4.5), we get that the typical cell of a PVT is obtained as the Voronoi cell at o with respect to the point process $\Pi_\lambda^o = \Pi_\lambda \cup \{o\}$, see Fig. 5.2b.

5.1.3.2 Poisson–Delaunay Tessellation

In the same way as in Sect. 5.1.1 for deterministic Voronoi tessellations, we can construct the dual Delaunay tessellation corresponding to a random Voronoi tessellation. If, almost surely, the underlying point process Ψ is locally finite, where

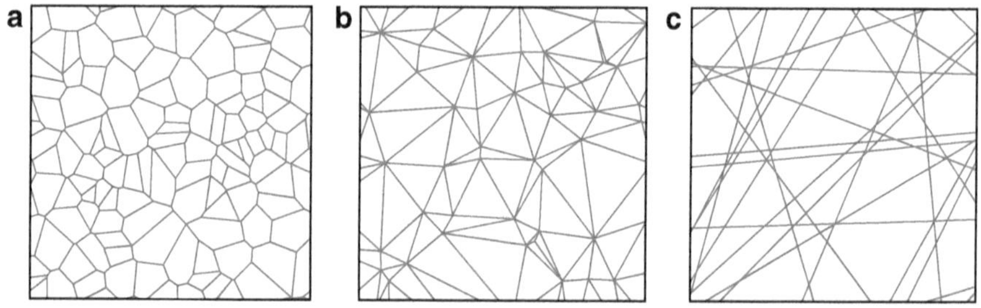

Fig. 5.3 Realizations of tessellation models of Poisson type. (**a**) PVT; (**b**) PDT; (**c**) PLT

$\text{conv}(\Psi) = \mathbb{R}^2$ and no four points of Ψ are cocircular, then the random Delaunay tessellation T induced by Ψ is well-defined. Furthermore, T is stationary if Ψ is stationary, see also Exercise 5.4. In particular, if $\Psi = \Pi_\lambda$ is a stationary Poisson process with intensity $\lambda > 0$, we can generate the Delaunay tessellation $\mathrm{T} = \{\Xi_n\}$ of Ψ as the dual tessellation of the PVT induced by Ψ, where T is called a *Poisson–Delaunay tessellation* (PDT). In Fig. 5.3b a realization of a PDT is shown.

Theorem 5.3. *Let* T *be a PDT induced by a Poisson process with intensity* λ*. Then*

$$\lambda^{(0)} = \lambda, \qquad \lambda^{(1)} = 3\lambda, \qquad \lambda^{(2)} = 2\lambda, \qquad \gamma = \frac{32}{3\pi}\sqrt{\lambda}\,.$$

Proof. Since $\lambda^{(0)} = \lambda$ and $\mathbf{E}\nu_0(Z) = 3$ we get with Theorem 5.1 that $\lambda^{(1)} = 3\lambda, \lambda^{(2)} = 2\lambda$. For the proof of $\gamma = \frac{32}{3\pi}\sqrt{\lambda}$ see for example [451, Chap. 10]. □

If $\mathrm{T} = \{\Xi_n\}$ is a PDT induced by the stationary Poisson process Π_λ, then the vertices of T are given by the points of Π_λ. Moreover, due to Slivnyak's theorem, the random Delaunay tessellation T^o with respect to the Palm version Π_λ^o of Π_λ is given by the dual Delaunay tessellations corresponding to the Voronoi tessellations induced by $\Pi_\lambda^o = \Pi_\lambda \cup \{o\}$. Thus, the union of edges of T^o emanating from o can be regarded as the typical edge star E^o of T.

Exercise 5.4. Assume that T is a random Voronoi tessellation or a random Delaunay tessellation which is induced by some point process $\Psi = \{X_1, X_2, \dots\} \subset \mathbb{R}^2$ such that with probability 1 it holds that $\text{conv}(\Psi) = \mathbb{R}^2$ and there are no pairwise different points $X_i, X_j, X_k, X_l \in X$ which are located on a circle. Show that T is stationary and isotropic if X is stationary and isotropic, respectively.

5.1.3.3 Poisson Line Tessellation

Consider a stationary Poisson process $\{R_n\}$ on the real line $\mathbb{R}$ with (linear) intensity $\widetilde{\gamma} > 0$. Each point R_n is independently marked with a random angle $\Phi_n \sim \text{Unif}[0, \pi)$. Then we can identify each pair (R_i, Φ_i) with the line

$$\ell_{(R_n,\Phi_n)} = \{(x, y) \in \mathbb{R}^2 : x \cos \Phi_n + y \sin \Phi_n = R_n\}.$$

The resulting stationary random closed set $\bigcup_{n\geq 1} \ell_{(R_n,\Phi_n)}$ is called a *Poisson line process* with intensity $\widetilde{\gamma}$. It can be regarded as the edge set $\mathrm{T}^{(1)} = \bigcup_{n\geq 1} \ell_{(R_n,\Phi_n)}$ of a stationary isotropic tessellation T which is called a *Poisson line tessellation* (PLT). A realization of a PLT is displayed in Fig. 5.3c.

Theorem 5.4. *Let* T *be a PLT induced by a Poisson line process with intensity* $\widetilde{\gamma}$. *Then*

$$\gamma = \widetilde{\gamma}, \qquad \lambda^{(0)} = \frac{1}{\pi}\gamma^2, \qquad \lambda^{(1)} = \frac{2}{\pi}\gamma^2, \qquad \lambda^{(2)} = \frac{1}{\pi}\gamma^2.$$

Proof. Note that $\gamma = \mathbf{E}\nu_1(B_1(o) \cap \bigcup_{n\geq 1} \ell_{(R_n,\Phi_n)})/\pi$ does not depend on the distribution of $\Phi_1, \Phi_2, \ldots$, where $B_r(x) = \{y \in \mathbb{R}^2 : |x - y| \leq r\}$ denotes the ball with midpoint $x \in \mathbb{R}^2$ and radius $r > 0$. Thus,

$$\gamma = \mathbf{E}\nu_1(B_1(o) \cap \bigcup_{n\geq 1} \ell_{(R_n,0)})/\pi = \mathbf{E}\nu_1([0,1)^2 \cap \bigcup_{n\geq 1} \ell_{(R_n,0)}) = \widetilde{\gamma}.$$

Theorem 5.1 with $\mathbf{E}\nu_0(E^o) = 4$ yields $\lambda^{(0)} = \lambda^{(2)}$ and $\lambda^{(1)} = 2\lambda^{(0)}$. Furthermore, it holds that $\mathbf{E}\nu_1(S^o) = \gamma/\lambda^{(1)} = \pi/(2\gamma)$, see for example [489]. Thus $\lambda^{(1)} = \frac{2}{\pi}\gamma^2$. □

Exercise 5.5. Show that $\mathbf{E}\nu_1(S^o) = \gamma/\lambda^{(1)} = \pi/(2\gamma)$.

5.2 Cox Processes

The notion of Cox processes, which is closely related with the notion of random measures (called driving measures), has already been mentioned in Sects. 3.1 and 4.1.

In this section we concentrate on stationary Cox processes and on stationary random driving measures associated with this class of point processes, as well as on their Palm distributions. Particular emphasis is put on the case that the driving measure of a Cox process $\{X_n\}$ is concentrated on the edge set $\mathrm{T}^{(1)}$ of a stationary tessellation T, i.e., we assume that $\mathbf{P}(X_n \in \mathrm{T}^{(1)} \text{ for all } n \in \mathbb{N}) = 1$.

5.2.1 Cox Processes and Random Measures

Let $\mathcal{M} = \mathcal{M}(\mathbb{R}^2)$ denote the set of all locally finite measures on $\mathcal{B}(\mathbb{R}^2)$. On $\mathcal{M}$ we define the σ-algebra $\mathfrak{M} = \mathfrak{M}(\mathbb{R}^2)$ as the smallest σ-algebra such that the mappings $\eta \mapsto \eta(B)$ are $(\mathfrak{M}, \mathcal{B}(\mathbb{R}^2))$-measurable for all $B \in \mathcal{B}_0(\mathbb{R}^2)$. Thus, we obtain the measurable space $(\mathcal{M}, \mathfrak{M})$. The shift operator $T_x : \mathcal{M} \to \mathcal{M}$ on $\mathcal{M}$ is defined in the same way as for counting measures, i.e. $T_x\eta(B) = \eta(B + x)$ for all $B \in \mathcal{B}(\mathbb{R}^2)$ and $x \in \mathbb{R}^2$, and we define the rotation operator $\vartheta_R : \mathcal{M} \to \mathcal{M}$ by $\vartheta_R\eta(B) = \eta(\vartheta_R^{-1}B) = \eta(\vartheta_{R^{-1}}B)$ for all rotations $R : \mathbb{R}^2 \to \mathbb{R}^2$ around the origin.

A measurable mapping $\Lambda : \Omega \to \mathcal{M}$ from some probability space $(\Omega, \mathcal{A}, \mathbf{P})$ into the measurable space $(\mathcal{M}, \mathfrak{M})$ is then called *random measure* on $\mathcal{B}(\mathbb{R}^2)$. The random measure Λ is called *stationary* if $T_x\Lambda \stackrel{d}{=} \Lambda$ for all $x \in \mathbb{R}^2$. In this case $\mathbf{E}\Lambda(B) = \lambda\nu_2(B)$ for $B \in \mathcal{B}(\mathbb{R}^2)$, where $\lambda \geq 0$ is some constant which is called the intensity of Λ. Notice that $\lambda = \mathbf{E}\Lambda([0,1]^2)$. If $0 < \lambda < \infty$, we define the *Palm distribution* of Λ as the probability measure $P_\Lambda^0 : \mathfrak{M} \to [0,1]$ given by

$$P_\Lambda^0(A) = \frac{1}{\lambda}\mathbf{E}\left(\int_{[0,1]^2} \mathbf{1}(T_x\Lambda \in A)\,\Lambda(dx)\right), \quad A \in \mathfrak{M}. \tag{5.3}$$

Assume now that a random measure Λ is given. Recall that the Cox process Ψ with random *driving measure* Λ is then defined by

$$\mathbf{P}(\Psi(B_1) = k_1, \ldots, \Psi(B_n) = k_n) = \mathbf{E}\left(\prod_{i=1}^{n} \frac{\Lambda(B_i)^{k_i} \mathrm{e}^{-\Lambda(B_i)}}{k_i!}\right) \tag{5.4}$$

for any $k_1, \ldots, k_n \in \mathbb{N}_0$ and pairwise disjoint $B_1, \ldots, B_n \in \mathcal{B}_0(\mathbb{R}^2)$, see also Sects. 3.1.4 and 4.1.3.

Now we summarize some basic properties of Cox processes. The following result is an immediate consequence of (4.13).

Theorem 5.5. *Let Ψ be a Cox process with random driving measure Λ. Then Ψ is stationary (resp. isotropic) iff Λ is stationary (resp. isotropic). If Ψ is stationary, then its intensity is equal to the intensity λ of Λ.*

Exercise 5.6. Provide a proof of Theorem 5.5.

Classical examples of Cox processes are the Neyman–Scott process and the modulated Poisson process [173].

Recall that the Palm version $\Psi^o = \Psi \cup \{o\}$ of a stationary Poisson process Ψ is obtained by adding the origin o to Ψ, compare Theorem 4.5. This property of Poisson processes can be generalized to get the following result, which is called *Slivnyak's theorem* for Cox processes. Namely, the Palm distribution P_Ψ^o of a stationary Cox process Ψ with random driving measure Λ can be characterized as follows, see for example [489, p. 156].

Theorem 5.6. *Let Ψ be a Cox process with stationary driving measure Λ. Then $P_\Psi^o(A) = \mathbf{P}(\widetilde{\Psi} \cup \{o\} \in A)$ for all $A \in \mathfrak{N}$, where $\widetilde{\Psi}$ is a Cox process with random driving measure Λ^o distributed according to the Palm distribution P_Λ^o of Λ.*

Thus, to simulate the Palm version Ψ^o of a stationary Cox process Ψ, we can use a two-step procedure. First, we generate a realization η^o of Λ^o. Afterwards, adding a point at the origin o, we simulate a Poisson process with intensity measure η^o.

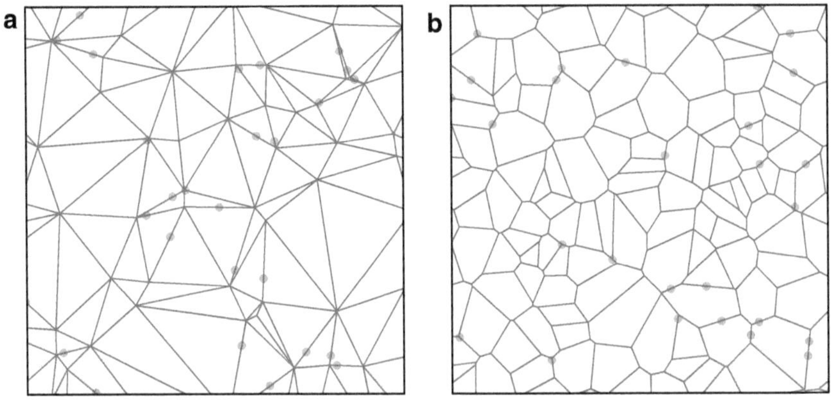

Fig. 5.4 Realizations of Cox processes on PDT and PVT. (**a**) PDT; (**b**) PVT

5.2.2 Cox Processes on the Edges of Random Tessellations

In this section, we introduce a class of Cox processes Ψ whose random driving measures Λ are concentrated on the edge sets of random tessellations. Let T be a stationary random tessellation with length intensity $\gamma = \mathbf{E}\nu_1([0,1]^2 \cap \mathrm{T}^{(1)})$ and, for some $\lambda_\ell > 0$, define the random measure $\Lambda : \mathcal{B}(\mathbb{R}^2) \to [0,\infty]$ by

$$\Lambda(B) = \lambda_\ell \nu_1(B \cap \mathrm{T}^{(1)}), \quad B \in \mathcal{B}(\mathbb{R}^2). \tag{5.5}$$

Notice that Λ is stationary. Its intensity is given by

$$\lambda = \lambda_\ell \mathbf{E}\nu_1([0,1)^2 \cap \mathrm{T}^{(1)}) = \lambda_\ell \gamma. \tag{5.6}$$

Let Ψ be a Cox process whose random driving measure Λ is given by (5.5). Then, a direct application of Theorem 5.5 yields that Ψ is stationary with intensity λ given in (5.6). Furthermore, Ψ is isotropic if T is isotropic. Realizations of Ψ can be generated by simulating first T and then simulating Poisson processes with (linear) intensity λ_ℓ on each segment of $\mathrm{T}^{(1)}$. In Fig. 5.4 realizations of Cox processes on $\mathrm{T}^{(1)}$ are shown for T being a PDT and PVT, respectively.

Recall that in Theorem 5.6 the Palm distribution of stationary Cox processes is characterized, which is uniquely determined by the Palm version Λ^o of the stationary driving measure Λ. For Cox processes on the edge set of stationary tessellations, the result of Theorem 5.6 can be specified in the following way.

Theorem 5.7. *Let Λ be the stationary random measure given in formula* (5.5). *Then $\Lambda^o(B) = \lambda_\ell \nu_1(B \cap \widetilde{\mathrm{T}}^{(1)})$ for each $B \in \mathcal{B}(\mathbb{R}^2)$, where the tessellation $\widetilde{\mathrm{T}}$ is distributed according to the Palm distribution $P^o_{\mathrm{T}^{(1)}}$ with respect to the 1-dimensional Hausdorff measure on $\mathrm{T}^{(1)}$.*

Proof. Let $\tau \in \mathbb{T}$ be an arbitrary tessellation. Then we can identify the measure η given by $\eta(\cdot) = \lambda_\ell \nu_1(\cdot \cap \tau^{(1)})$ with τ, writing η_τ. It is easy to see that $\eta_{T_x\tau} = T_x\eta_\tau$ for all $x \in \mathbb{R}^2$. Furthermore, using the definition of the Palm distribution P^o_Λ of Λ given in (5.3), we get for each $A \in \mathfrak{M}$ that
$P^o_\Lambda(A) = \lambda^{-1} \int_{\mathcal{M}} \int_{[0,1]^2} \mathbf{1}(T_x\eta \in A)\, \eta(dx)\, P_\Lambda(d\eta)$, i.e.

$$P^o_\Lambda(A) = \frac{1}{\gamma} \int_{\mathbb{T}} \int_{[0,1]^2 \cap \tau^{(1)}} \mathbf{1}(T_x\eta_\tau \in A)\, \nu_1(dx)\, P_{\mathrm{T}}(d\tau) = P^o_{\mathrm{T}^{(1)}}(\{\tau \in \mathbb{T} : \eta_\tau \in A\}),$$

where the latter equality immediately follows from the definition of the Palm distribution $P^o_{\mathrm{T}^{(1)}}$ since $\gamma = \mathbf{E}\nu_1(\mathrm{T}^{(1)} \cap [0,1)^2)$. This means that the distributions of Λ^o and $\eta_{\widetilde{\mathrm{T}}}(\cdot) = \lambda_\ell \nu_1(\cdot \cap \widetilde{\mathrm{T}}^{(1)})$ are equal. □

Note that $\widetilde{\mathrm{T}}$ can be viewed as the random tessellation T under the condition that $o \in \mathrm{T}^{(1)}$. Thus, under $P^o_{\mathrm{T}^{(1)}}$, there is an edge $\widetilde{S}$ of $\widetilde{\mathrm{T}}$ through o with probability 1. However, the distributions of $\widetilde{S}$ and the typical edge S^o do not coincide. This can be seen as follows. Assume that $h : \mathcal{L}^o \to [0,\infty)$ is a translation-invariant measurable function and let $S(x)$ denote the segment of $\mathrm{T}^{(1)}$ through x for $x \in \mathrm{T}^{(1)}$. Then,

$$\begin{aligned}
\mathbf{E}h(\widetilde{S}) &= \frac{1}{\gamma}\mathbf{E} \int_{\mathrm{T}^{(1)} \cap [0,1)^2} h(S(x) - x)\, \nu_1(dx) \\
&= \frac{1}{\gamma}\mathbf{E} \sum_{(Y_i, S_i^o) \in \mathrm{T}} h(S_i^o) \int_{S_i} \mathbf{1}(x \in [0,1)^2)\, \nu_1(dx) \\
&= \frac{\lambda^{(1)}}{\gamma}\mathbf{E}h(S^o) \int_{\mathbb{R}^2} \int_{S^o} \mathbf{1}(x \in [0,1)^2 - y)\, \nu_1(dx)\, \nu_2(dy) = \frac{\mathbf{E}\nu_1(S^o)h(S^o)}{\mathbf{E}\nu_1(S^o)},
\end{aligned}$$

where we used the refined Campbell theorem for stationary marked point processes (formula (4.10)) and the mean-value formulae given in Theorem 5.1. Thus, the distribution of $\widetilde{S}$ can be represented as a length-weighted distribution of S^o.

For Cox processes on the edges of random tessellations the following scaling invariance can be observed. Let T be a stationary random tessellation with length intensity 1. Then we define the scaled tessellation T_γ as the random tessellation whose edge set is given by $\mathrm{T}_\gamma^{(1)} = \frac{1}{\gamma}\mathrm{T}^{(1)}$. Thus, the length intensity of T_γ is γ since $\mathbf{E}\nu_1(\mathrm{T}_\gamma^{(1)} \cap [0,1)^2) = \mathbf{E}\nu_1(\mathrm{T}^{(1)} \cap [0,\gamma)^2)/\gamma = \gamma$ due to the homogeneity of the Hausdorff measure ν_1, see also Sect. 2.1.1.

Now let $\Psi = \{X_n\}$ be a Cox process on T_γ with linear intensity λ_ℓ and let $\Psi' = \{X'_n\}$ be a Cox process on $\mathrm{T}_{\gamma'}$ whose linear intensity is given by λ'_ℓ. Moreover, assume that the intensity quotients $\kappa = \gamma/\lambda_\ell$ and $\kappa' = \gamma'/\lambda'_\ell$ are equal, i.e., $\kappa = \kappa'$. Then we get for any $C \in \mathcal{K}^2$ that

$$\begin{aligned}\mathbf{P}(\Psi(C)=0) &= \mathbf{E}\exp\left\{\lambda_\ell \nu_1(C\cap \mathtt{T}_\gamma^{(1)})\right\}\\ &= \mathbf{E}\exp\left\{\frac{\lambda_\ell\gamma'}{\gamma}\nu_1\big(\frac{\gamma}{\gamma'}C\cap \mathtt{T}_{\gamma'}^{(1)}\big)\right\}\\ &= \mathbf{P}\left(\Psi'\big(\frac{\gamma}{\gamma'}C\big)=0\right)=\mathbf{P}\left(\big(\frac{\gamma'}{\gamma}\Psi'\big)(C)=0\right),\end{aligned}$$

where the scaled point process $\frac{\gamma'}{\gamma}\Psi'$ is defined by $\frac{\gamma'}{\gamma}\Psi' = \{\frac{\gamma'}{\gamma}X_n'\}$. Since the distribution of a point process Ψ is uniquely determined by its void probabilities $\mathbf{P}(\Psi(C) = 0)$, $C \in \mathcal{K}^2$, we have that $\Psi \overset{d}{=} \frac{\gamma'}{\gamma}\Psi'$. Thus, for a given tessellation type $\mathtt{T}$, the intensity quotient κ defines the Cox process Ψ on the scaled tessellation $\mathtt{T}_\gamma$ with linear intensity λ_ℓ uniquely up to a certain scaling. We therefore call κ the *scaling factor* of Ψ. For numerical results it is therefore sufficient to focus on single parameter pairs γ and λ_ℓ for each value of κ. For other parameters with the same scaling factor κ the corresponding results can then be obtained by a suitable scaling.

5.3 Cox–Voronoi Tessellations

In this section we consider Voronoi tessellations induced by stationary Cox point processes. The typical cell of these so-called *Cox–Voronoi tessellations* can describe for example the typical serving zone of telecommunication networks. Unfortunately, its distribution is not known analytically. Even for the typical cell of PVT it is hard to obtain closed analytical expressions for the distribution of cell characteristics like the perimeter, the number of vertices, or the area. On the other hand, it is often possible to develop simulation algorithms for the typical Voronoi cell, which can be used to determine the distribution of cell characteristics approximatively. We discuss such simulation algorithms for two examples of Voronoi tessellations. To begin with, in Sect. 5.3.1, we first consider the case of the typical Poisson–Voronoi cell. Then, in Sect. 5.3.2, we show how the typical cell of a Cox–Voronoi tessellation $\mathtt{T}_\Psi$ can be simulated if the random driving measure of the underlying Cox process Ψ is concentrated on the edge set of a certain stationary tessellation $\mathtt{T}$, where we assume that $\mathtt{T}$ is a PLT, see Fig. 5.6a.

In the ergodic case, the distribution of the typical cell can be obtained as the limit of empirical distributions of cells observed in a sequence of unboundedly increasing sampling windows, see Theorem 4.7. Thus, in order to approximate the distribution of the typical cell, we can simulate the random tessellation in a large sampling window W, considering spatial averages of those cells whose associated points belong to W. Alternatively, we can approximate this distribution by simulating independent copies of the typical cell and by taking sample means instead of spatial averages.

Note that there are several advantages of the latter approach. If we simulate the tessellation in a large sampling window, then the cells are correlated and there are edge effects which may be significant if W is not large enough. On the other hand, for large W, runtime and memory problems occur. However, these problems can be avoided if independent copies of the typical cell are simulated locally, but the challenge is then to develop such simulation algorithms. Recall that the typical Voronoi cell Z_Ψ of any stationary point process Ψ can be (locally) represented as $Z_\Psi = \bigcap_{n\in\mathbb{N}} H(o, X_n^o)$, where $\Psi^o = \{X_n^o\}$ is the Palm version of Ψ. Thus, suitable simulation algorithms for the points of Ψ^o have to be developed.

5.3.1 Local Simulation of the Typical Poisson–Voronoi Cell

Recall that due to Slivnyak's theorem, the typical cell of a PVT can be regarded as the Voronoi cell at the origin with respect to $\Psi^o = \Psi \cup \{o\}$, where $\Psi = \{X_n\}$ is the underlying stationary Poisson process. Thus, we can place a point at o, simulate further points X_n of Ψ radially and then construct the typical cell $Z_\Psi = \bigcap_{n\in\mathbb{N}} H(o, X_n)$ as intersection of the bisectors $H(o, X_n)$ for $n \geq 1$.

More precisely, we simulate the points $X_1, X_2, \ldots$ successively, with increasing distance to the origin, until a bounded Voronoi cell at o can be constructed by the simulated points. We call this cell the *initial cell*. Afterwards, we check for each newly simulated point if the initial cell is influenced by points with larger distances from o than the latest generated point. If this is not the case, we stop the algorithm. Otherwise we simulate a further point. This local *simulation algorithm* of the typical Poisson–Voronoi cell is summarized below. The main steps of the algorithm are visualized in Fig. 5.5.

1. Put $\Psi^o = \{o\}$.
2. Simulate random variables $V_1, V_2, \ldots$ and $\Theta_1, \Theta_2, \ldots$ as in Algorithm 3.8, where $V_1, V_2, \ldots$ are the arrival times of a homogeneous Poisson process in $(0, \infty)$ with intensity $\lambda\pi$, and $\Theta_1, \Theta_2, \ldots$ are independent and uniformly distributed on $[0, 2\pi)$.
3. Compute the points $X_1, \ldots, X_n$ by $X_n = (\sqrt{V_n}\cos\Theta_n, \sqrt{V_n}\sin\Theta_n)$ and add them to Ψ^o until a (compact) initial cell Z_Ψ at o can be constructed from Ψ^o.
4. If $\sqrt{V_n} \geq r_{max} = 2\max\{|v_i|\}$, were $\{v_i\}$ is the set of vertices of Z_Ψ, then stop, else add further points to Ψ^o and update Z_Ψ.

When implementing this simulation algorithm we have to take into account some technical details. First, a rule for constructing the initial cell has to be implemented. If for some $n \geq 3$ the points $X_1, \ldots, X_n$ have been generated, then we can use a simple cone criterion in order to check if a bounded Voronoi cell can be constructed around o by these points which says if the intersection of the bisectors $H(o, X_1), \ldots, H(o, X_n)$ is bounded. Once the initial cell Z_Ψ has been generated, points of Ψ^o outside the ball $B_{r_{max}}(o)$ cannot influence the typical cell anymore since the bisector between o and any $x \in B_{r_{max}}(o)^c$ does not intersect Z_Ψ. Thus, the simulation stops if $\sqrt{V_n} \geq r_{max}$.

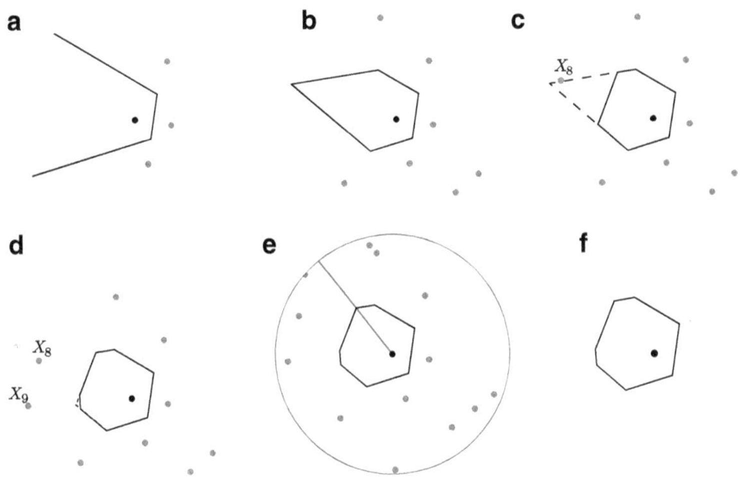

Fig. 5.5 Simulation of the typical cell of PVT. (**a**) Origin (*black*) and radially simulated points X_1, X_2, X_3 (*gray*). Initial cell incomplete. (**b**) Initial cell Z_Ψ around o is constructed using the radially simulated points $X_1, \ldots, X_7$. (**c**) Point X_8 is simulated radially and Z_Ψ is cut by the bisector $H(o, X_8)$. (**d**) Point X_9 is simulated radially and Z_Ψ is cut by $H(o, X_9)$. (**e**) Further points X_n are simulated radially until $|X_n| \geq r_{max}$. (**f**) Realization of the typical cell Z_Ψ of PVT

5.3.2 *Cox Processes on the Edges of PLT*

We now consider Cox processes $\Psi = \{X_n\}$ whose random driving measure Λ is concentrated on the edges of a stationary random tessellation T, where we assume that Λ is given by (5.5). In particular, the typical cell Z_Ψ of the Voronoi tessellation $\mathrm{T}_\Psi = \{\Xi_{\Psi,n}\}$ induced by Ψ will be investigated. Recall that T_Ψ can be identified with the marked point process $\{(X_n, \Xi^o_{\Psi,n})\}$, where $\Xi^o_{\Psi,n} = \Xi_{\Psi,n} - X_n$ denotes the centered version of the Voronoi cell $\Xi_{\Psi,n}$ at X_n with respect to Ψ, see Fig. 5.6.

If the Cox process Ψ models the locations of network components in telecommunication networks, then $\Xi_{\Psi,n}$ can be regarded as the area of influence of the network component at X_n, where $\Xi_{\Psi,n}$ is called the *serving zone* of X_n. Thus, the typical cell Z_Ψ of T_Ψ is an important characteristic in global econometric analysis and planning of telecommunication networks, because various cost functionals of hierarchical network models can be represented as expectations of functionals of Z_Ψ, see also Sect. 5.4.

Suitable simulation algorithms for the points of the Palm version Ψ^o of Ψ have to be developed in order to locally simulate the typical cell Z_Ψ of the Cox–Voronoi tessellation T_Ψ. However, in contrast to the situation discussed in Sect. 5.3.1, we do not simulate the points of Ψ^o radially, at least not at once, when considering Cox processes on PDT, PLT and PVT, respectively. But we simulate the points of the Poisson process radially which induces the Palm version of the underlying

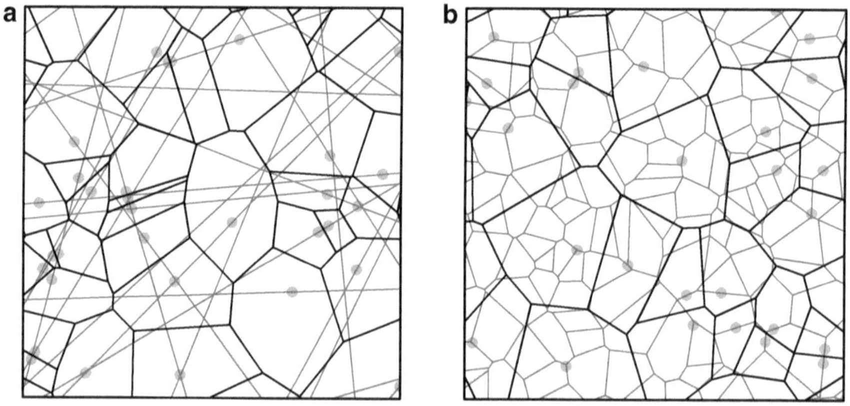

Fig. 5.6 Realizations of T_Ψ for Cox processes on PLT and PVT. (**a**) PLT; (**b**) PVT

tessellation (regarded as random Hausdorff measure), alternatingly with the points of the (linear) Poisson processes on the edges of this tessellation. As an example, we show how this can be done for Cox processes on PLT.

5.3.2.1 Palm Version of PLT

Let Ψ be a stationary Cox process with linear intensity λ_ℓ on a stationary PLT T with length intensity γ. Note that due to Theorems 5.6 and 5.7 it holds that $\Psi^o = \widetilde{\Psi} \cup \{o\}$, where $\widetilde{\Psi}$ is a Cox process on the Palm version $\widetilde{\mathrm{T}}$ of T regarded as the random Hausdorff measure $\nu_1(\cdot \cap \mathrm{T}^{(1)})$ on $\mathrm{T}^{(1)}$. Thus, in a first step, T has to be simulated according to its Palm distribution with respect to $\nu_1(\cdot \cap \mathrm{T}^{(1)})$, i.e., under the condition that $o \in \mathrm{T}^{(1)}$. It turns out that the edge set $\widetilde{\mathrm{T}}^{(1)}$ of this conditional PLT can be constructed just by adding an isotropic line through o to $\mathrm{T}^{(1)}$.

Theorem 5.8. *Let* $\mathrm{T}^{(1)}$ *be the edge set of a stationary PLT with intensity* γ *and let* $\ell(\Phi)$ *be a line through the origin with random direction* $\Phi \sim \mathrm{Unif}[0,\pi)$ *which is independent of* $\mathrm{T}^{(1)}$. *Then* $\widetilde{\mathrm{T}}^{(1)} \stackrel{d}{=} \mathrm{T}^{(1)} \cup \ell(\Phi)$.

Proof. Since the distribution of a random closed set is uniquely determined by its capacity functional (see Definition 1.2 of Sect. 1.2.2), we show that the capacity functionals of $\widetilde{\mathrm{T}}^{(1)}$ and $\mathrm{T}^{(1)} \cup \ell(\Phi)$ coincide. With the notation $\mathrm{T}^{(1)} = \bigcup_{n\ge 1} \ell_{(R_n,\Phi_n)}$ introduced in Sect. 5.1.3.3, the definition of the Palm distribution of stationary random measures (see (5.3)) gives that for each $C \in \mathcal{K}^2$

$$\mathbf{P}(\widetilde{\mathrm{T}}^{(1)} \cap C \neq \emptyset) = \frac{1}{\pi\gamma}\mathbf{E}\int_{\mathrm{T}^{(1)}\cap B_1(o)} \mathbf{1}\Big(\bigcup_{n\ge 1}(\ell_{(R_n,\Phi_n)} - x) \cap C \neq \emptyset\Big)\, \nu_1(dx)\,.$$

Note that the number N of lines of a Poisson line process which intersect a convex compact set $W \subset \mathbb{R}^2$ is $\mathrm{Poi}(\lambda)$-distributed with $\lambda = \gamma\nu_1(\partial W)/\pi$ and, given $N = k$, these k lines $\ell_1,\ldots,\ell_k$ are independent and isotropic uniform random (IUR), see

Sect. 2.1.1. Thus, for $W = B_{R(C)+1}(o)$, where $R(C) = \sup_{x \in C}\{|x|\}$, we get

$$
\begin{aligned}
&\mathbf{P}(\widetilde{\mathrm{T}}^{(1)} \cap C \neq \emptyset) \\
&= \sum_{k=0}^{\infty} \frac{\mathbf{P}(N=k)}{\pi\gamma} \sum_{i=1}^{k} \mathbf{E}\left(\int_{\ell_i \cap B_1(o)} \mathbf{1}(\bigcup_{i=1}^{k}(\ell_i - x) \cap C \neq \emptyset)\nu_1(dx) \mid N = k \right) \\
&= \frac{1}{\pi\gamma} \sum_{k=0}^{\infty} \frac{\mathrm{e}^{-\lambda}\lambda^k}{k!}\, k\, \mathbf{E}\left(\int_{\ell_1 \cap B_1(o)} \mathbf{1}(\bigcup_{i=1}^{k}(\ell_i - x) \cap C \neq \emptyset)\nu_1(dx) \mid N = k \right) \\
&= \frac{\lambda}{\pi\gamma} \sum_{k=0}^{\infty} \frac{\mathrm{e}^{-\lambda}\lambda^k}{k!} \mathbf{E}\left(\int_{\ell_1 \cap B_1(o)} \mathbf{1}(\bigcup_{i=1}^{k+1}(\ell_i - x) \cap C \neq \emptyset)\nu_1(dx) \mid N = k+1 \right).
\end{aligned}
$$

Since the lines $\ell_1, \ell_2, \ldots, \ell_{k+1}$ are independent and IUR, we can consider ℓ_1 separately, where the remaining $\ell_2, \ldots, \ell_{k+1}$ still are independent of each other, IUR, and independent of ℓ_1. This gives

$$
\begin{aligned}
\mathbf{P}(\widetilde{\mathrm{T}}^{(1)} \cap C \neq \emptyset) &= \frac{\lambda}{\pi\gamma} \mathbf{P}\left((\mathrm{T}^{(1)} \cup \ell(\Phi)) \cap C \neq \emptyset \right) \mathbf{E}\nu_1(\ell_1 \cap B_1(o)) \\
&= \mathbf{P}\left((\mathrm{T}^{(1)} \cup \ell(\Phi)) \cap C \neq \emptyset \right),
\end{aligned}
$$

where in the last equality we used that $\mathbf{E}\nu_1(\ell_1 \cap B_1(o)) = \pi^2/\nu_1(\partial W)$, see for example Sect. 2.1.1, formula (2.5). □

5.3.2.2 Local Simulation of the Typical Cox–Voronoi Cell

Using Theorem 5.8, we are able to briefly describe the main idea of an algorithm for local simulation of the typical cell Z_Ψ of the Voronoi tessellation $\mathrm{T}_\Psi = \{\Xi_{\Psi,n}\}$ induced by the Cox process Ψ on PLT.

We first put a line $\ell(\Phi)$ with direction $\Phi \sim \mathrm{Unif}[0,\pi)$ through the origin o and then, on both half-lines of $\ell(\Phi)$ seen from o, we simulate the nearest points to o of a Poisson process with intensity λ_ℓ. Next, we simulate independent random variables Φ_1 and R_1 $(= Y_1)$ with $\Phi_1 \sim \mathrm{Unif}[0,2\pi)$ and $R_1 \sim \mathrm{Exp}(2\gamma)$ and construct the line $\ell_{(R_1,\Phi_1)} = \{(x,y) \in \mathbb{R}^2 : x\cos\Phi_1 + y\sin\Phi_1 = R_1\}$. Note that $\ell_{(R_1,\Phi_1)}$ is the closest line to the origin of a Poisson line process with length intensity γ. Then, on $\ell_{(R_1,\Phi_1)}$, we simulate points of a Poisson process with intensity λ_ℓ. In the next step, we simulate independent random variables Φ_2 and Y_2 with $\Phi_2 \sim \mathrm{Unif}[0,2\pi)$ and $Y_2 \sim \mathrm{Exp}(2\gamma)$ constructing the line $\ell_{(R_2,\Phi_2)} = \{(x,y) \in \mathbb{R}^2 : x\cos\Phi_2 + y\sin\Phi_n = R_2\}$, where $R_2 = R_1 + Y_2$, and so on.

In this way, similar to the algorithm discussed in Sect. 5.3.1, we simulate points of Ψ^o in a neighbourhood of the origin until a bounded Voronoi cell at o can be constructed by the simulated points. Afterwards, we check for each newly simulated point if this initial cell is influenced by points with larger distances from o than the latest generated point. If this is not the case, we stop the algorithm. Otherwise we continue to alternatingly simulate lines and points on them respectively. For further technical details of the simulation algorithm we refer to [189].

Similar algorithms can be constructed for local simulation of the typical Voronoi cell of stationary Cox processes on PVT and PDT, respectively, see [174, 498].

5.4 Typical Connection Lengths in Hierarchical Network Models

We now consider two Cox processes simultaneously. The leading measures of either one or both of these Cox processes are concentrated on the edge set of a stationary tessellation, where we assume that the Cox processes are jointly stationary. We discuss representation formulae which have been derived in [498] for the distribution function and density of the typical Euclidean (i.e. direct) connection length D^o between certain pairs of points, chosen at random, one from each of the Cox processes. Furthermore, the typical shortest path length C^o is considered which is needed to connect such pairs of points along the edges of the underlying tessellation. A useful tool in investigating these characteristics is Neveu's exchange formula (see for example [377]) for jointly stationary marked point processes, which is stated in Sect. 5.4.1. Then, in Sect. 5.4.2, we give a motivation of our investigations, where we explain how the results can be applied for example in econometric analysis and planning of hierarchical telecommunication networks.

5.4.1 *Neveu's Exchange Formula*

Let $\Psi^{(1)} = \{(X_n^{(1)}, M_n^{(1)})\}$ and $\Psi^{(2)} = \{(X_n^{(2)}, M_n^{(2)})\}$ be jointly stationary marked point processes with mark spaces M_1 and M_2, respectively, and let $\mathcal{N}_{\mathsf{M}_1,\mathsf{M}_2} = \mathcal{N}_{\mathsf{M}_1} \times \mathcal{N}_{\mathsf{M}_2}$ denote the product space of the families of simple and locally finite counting measures with marks in M_1 and M_2, respectively, equipped with product-σ-algebra $\mathfrak{N}_{\mathsf{M}_1} \otimes \mathfrak{N}_{\mathsf{M}_2}$. We then put $\zeta = (\Psi^{(1)}, \Psi^{(2)})$ which can be regarded as a random element of $\mathcal{N}_{\mathsf{M}_1,\mathsf{M}_2}$. Let λ_1 and λ_2 denote the intensities of $\Psi^{(1)}$ and $\Psi^{(2)}$, respectively, and assume that the shift operator T_x is defined by $T_x\zeta = (T_x\Psi^{(1)}, T_x\Psi^{(2)})$ for $x \in \mathbb{R}^2$. Thus, T_x shifts the points of both $\Psi^{(1)}$ and $\Psi^{(2)}$ by $-x \in \mathbb{R}^2$. Note that $T_x\zeta \stackrel{d}{=} \zeta$ for each $x \in \mathbb{R}^2$ since $\Psi^{(1)}$ and $\Psi^{(2)}$ are jointly stationary. The *Palm distributions* $P_\zeta^{(i)}, i = 1, 2$ on $\mathfrak{N}_{\mathsf{M}_1} \otimes \mathfrak{N}_{\mathsf{M}_2} \otimes \mathcal{B}(\mathsf{M}_i)$ with respect to the i-th component of ζ are probability measures defined by

$$P_\zeta^{(i)}(A \times G) = \frac{1}{\lambda_i} \mathbf{E}|\{n : X_n^{(i)} \in [0,1)^2, M_n^{(i)} \in G, T_{X_n^{(i)}}\zeta \in A\}| \qquad (5.7)$$

for $A \in \mathfrak{N}_{\mathsf{M}_1} \otimes \mathcal{N}_{\mathsf{M}_2}$ and $G \in \mathcal{B}(\mathsf{M}_i)$. In particular, for $A \in \mathfrak{N}_{\mathsf{M}_i}$, $G \in \mathcal{B}(\mathsf{M}_i)$, we get $P_\zeta^{(1)}(A \times \mathcal{N}_{\mathsf{M}_2} \times G) = P^o_{\Psi^{(1)}}(A \times G)$ if $i = 1$, and $P_\zeta^{(2)}(\mathcal{N}_{\mathsf{M}_1} \times A \times G) = P^o_{\Psi^{(2)}}(A \times G)$ if $i = 2$, where $P^o_{\Psi^{(1)}}$ and $P^o_{\Psi^{(2)}}$ are the ordinary Palm distributions of the marked point processes $\Psi^{(1)}$ and $\Psi^{(2)}$, respectively.

Note that we also use the notation $P^o_{\Psi^{(i)}}$ for the Palm distribution $P_\zeta^{(i)}$ of the vector $(\Psi^{(1)}, \Psi^{(2)})$ in order to emphasize the dependence on $\Psi^{(i)}$ for $i = 1, 2$. With the definitions and notation introduced above, and writing $\psi = (\psi^{(1)}, \psi^{(2)})$ for the elements of $\mathcal{N}_{\mathsf{M}_1, \mathsf{M}_2}$, *Neveu's exchange formula* can be stated as follows, see for example [339].

Theorem 5.9. *For any measurable* $f : \mathbb{R}^2 \times \mathsf{M}_1 \times \mathsf{M}_2 \times \mathcal{N}_{\mathsf{M}_1, \mathsf{M}_2} \to [0, \infty)$, *it holds that*

$$\lambda_1 \int_{\mathcal{N}_{\mathsf{M}_1, \mathsf{M}_2} \times \mathsf{M}_1} \int_{\mathbb{R}^2 \times \mathsf{M}_2} f(x, m_1, m_2, T_x \psi)\ \psi^{(2)}(d(x, m_2))\ P_\zeta^{(1)}(d(\psi, m_1))$$

$$= \lambda_2 \int_{\mathcal{N}_{\mathsf{M}_1, \mathsf{M}_2} \times \mathsf{M}_2} \int_{\mathbb{R}^2 \times \mathsf{M}_1} f(-x, m_1, m_2, \psi)\ \psi^{(1)}(d(x, m_1))\ P_\zeta^{(2)}(d(\psi, m_2))\,.$$

Exercise 5.7. Provide a proof of Theorem 5.9.

Neveu's exchange formula given in Theorem 5.9 allows to express the (conditional) distribution of functionals of a vector $(\Psi^{(1)}, \Psi^{(2)})$ of jointly stationary point processes, seen from the perspective of the Palm distribution $P^o_{\Psi^{(1)}}$, by the distribution of the same functional under $P^o_{\Psi^{(2)}}$. This means that we can switch from the joint distribution of $(\Psi^{(1)}, \Psi^{(2)})$ conditioned on $o \in \{\Psi_n^{(1)}\}$ to the joint distribution of $(\Psi^{(1)}, \Psi^{(2)})$ conditioned on $o \in \{\Psi_n^{(2)}\}$.

5.4.2 Hierarchical Network Models

Models from stochastic geometry have been used since more than 10 years in order to describe and analyze telecommunication networks, see for example [21,221,533]. However, the infrastructure of the network, like road systems or railways, has been included into the model rather seldom.

In this section we introduce spatial stochastic models for telecommunication networks with two hierarchy levels which take the underlying infrastructure of the network into account. In particular, we model the network infrastructure, for example road systems or railways, by the edge set $\mathrm{T}^{(1)}$ of a stationary tessellation T with (length) intensity $\gamma = \mathbf{E}\nu_1(\mathrm{T}^{(1)} \cap [0,1]^2) > 0$. The locations of both high and low level components (HLC, LLC) of the network are modelled by stationary point processes $\Psi^H = \{H_n\}$ and $\Psi^L = \{L_n\}$, respectively, where Ψ^H is assumed to be a Cox process on $\mathrm{T}^{(1)}$ whose random driving measure is given by (5.5), with linear intensity $\lambda_\ell > 0$ and (planar) intensity $\lambda = \lambda_\ell \gamma$. Regarding the point process

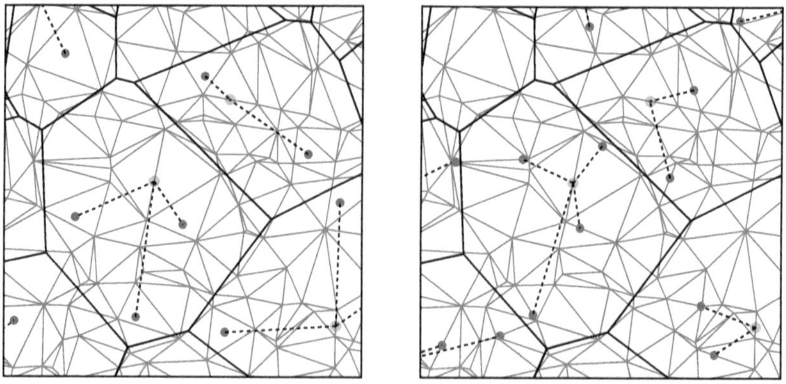

Fig. 5.7 Cox process Ψ^H on PDT with serving zones (*black*) and direct connection lengths (*dashed*) for Ψ^L Poisson (*left*) and Cox (*right*)

Ψ^L we distinguish between two different scenarios. On the one hand, we consider the case that Ψ^L is a stationary (planar) Poisson process with intensity λ' which is independent of T and Ψ^H. On the other hand, we assume that Ψ^L is a Cox process whose random driving measure is concentrated on the same edge set $\mathrm{T}^{(1)}$ as Ψ^H and given by (5.5), but now with linear intensity λ'_ℓ. Furthermore, we assume that Ψ^L is conditionally independent of Ψ^H given T. Thus, in the latter case, the planar intensity λ' of Ψ^L is given by $\lambda' = \lambda'_\ell \gamma$.

5.4.2.1 Typical Serving Zone

Each LLC of the network is connected with one of the HLC, i.e., each point L_n of Ψ^L is linked to some point H_n of Ψ^H. In order to specify this connection rule, so-called serving zones are considered, which are domains associated to each HLC such that the serving zones of distinct HLC do not overlap, but their union covers the whole region considered. Then a LLC is linked to that HLC in whose serving zone it is located. In the following, we assume that the *serving zones* of HLC are given by the cells of the stationary Voronoi tessellation $\mathrm{T}_H = \{\Xi_{H,n}\}$ induced by Ψ^H. Thus, the point L_n is linked to the point H_j iff $L_n \in \Xi_{\Psi^H,j}$, i.e., all LLC inside $\Xi_{\Psi^H,j}$ are linked to H_j, see Fig. 5.7. The typical cell Z_H of T_H is called the *typical serving zone*.

However, note that more complex models for (not necessarily convex) serving zones can be considered as well, like Laguerre tessellations [324] or aggregated Voronoi tessellations [491].

Furthermore, we define the stationary marked point process $\Psi_S^H = \{(H_n, S_{H,n}^o)\}$, where the marks are given by $S_{H,n}^o = (\mathrm{T}^{(1)} \cap \Xi_{H,n}) - H_n$. Thus, each point H_n of Ψ^H is marked with the segment system contained inside its serving zone. If Ψ^L is a Cox process on T, then the point L_n of Ψ^L is connected to H_j iff $L_n \in S_{H,j}^o + H_j$.

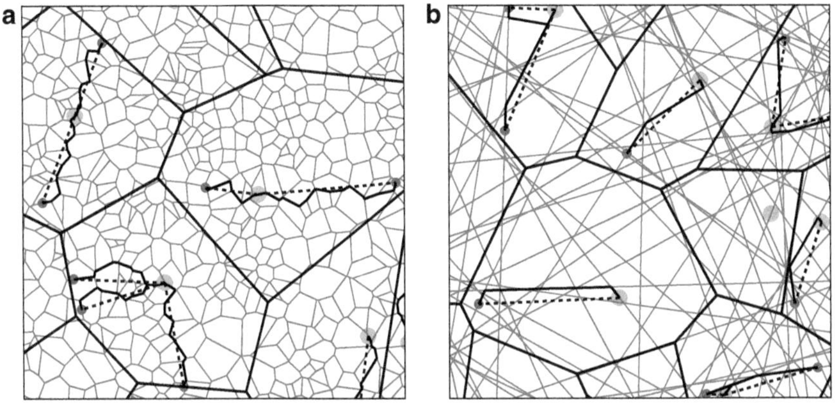

Fig. 5.8 Euclidean distances and shortest paths along the edge system. (**a**) PVT (**b**) PLT

It is easy to see that Ψ_S^H is a stationary marked point process with intensity λ whose mark space is given by the family of finite segment systems $\mathcal{L}^o$ which contain the origin. In particular, the typical mark $S_H^o : \Omega \to \mathcal{L}^o$ of Ψ_S^H is a random segment system which contains the origin, where S_H^o is called the *typical segment system* within the typical serving zone Z_H^o, see also Sect. 5.3.2.

5.4.2.2 Typical Connection Lengths

So far, we introduced the four modelling components T, Ψ^L, Ψ_S^H and T_H. They can be used in order to define the stationary marked point process $\Psi_D^L = \{(L_n, D_n)\}$, where $D_n = |L_n - H_j|$ is the Euclidean distance between L_n and H_j provided that $L_n \in \Xi_{H,j}$. We are then interested in the distribution of the typical mark D^o of Ψ_D^L which we call the *typical direct connection length* or, briefly, the *typical Euclidean distance*.

Realizations of the distances D_n for two different models of Ψ^L are displayed in Fig. 5.7, where the underlying tessellation T is a PDT; see also Fig. 5.8. Note that realizations of the marked point process Ψ_D^L can be constructed from realizations of Ψ^L and Ψ_S^H if Ψ^L is a Cox process and from realizations of Ψ^L and T_H if Ψ^L is a Poisson process. Hence, instead of Ψ_D^L, we can consider the vectors $\zeta = (\Psi^L, \Psi_S^H)$ and $\zeta = (\Psi^L, \mathrm{T}_{\Psi^H})$, respectively, together with the Palm distribution $P^o_{\Psi^L}$ of ζ with respect to the first component Ψ^L introduced in (5.7).

Suppose now that $(\Psi^{L^o}, \widetilde{\Psi_S^H})$ and $(\Psi^{L^o}, \mathrm{T}_{\widetilde{\Psi^H}})$, respectively, are distributed according to the Palm distribution $P^o_{\Psi^L}$, where we use the notation $\widetilde{\Psi^H} = \{\widetilde{H}_n\}$, $\widetilde{\Psi_S^H} = \{(\widetilde{H}_n, \widetilde{S}^o_{H,n})\}$ and $\widetilde{\mathrm{T}}^{(1)} = \bigcup_{n\geq 1}\big(\widetilde{S}^o_{H,n} + \widetilde{H}_n\big)$. Then D^o can be regarded as the distance from o to the point $\widetilde{H}_n$ of $\widetilde{\Psi^H}$ in whose serving zone o is located. Note that $\Psi^{L^o} \setminus \{o\}$ is a stationary Poisson process resp. a Cox process on $\widetilde{\mathrm{T}}$ if Ψ^L is a Poisson process resp. a Cox process on T. In the same way we regard the vectors $(\widetilde{\Psi^L}, \Psi_S^{H^o})$ and $(\widetilde{\Psi^L}, \mathrm{T}_H^o)$ which are distributed according to the Palm distributions $P^o_{\Psi_S^H}$ and $P^o_{\mathrm{T}_H}$, respectively. Here we denote with $\mathrm{T}^{o(1)}$ the edge set of $\Psi_S^{H^o}$. On the one hand, if Ψ^L is a Cox process on T, then $\widetilde{\Psi^L}$ is a (non-stationary) Cox process on $\mathrm{T}^{o(1)}$ with linear intensity λ'_ℓ, which is conditionally independent of Ψ^{H^o} given $\mathrm{T}^{0(1)}$. On the other hand, if Ψ^L is a stationary Poisson process which is independent of T and Ψ^H, then $\widetilde{\Psi^L} \stackrel{d}{=} \Psi^L$.

If Ψ^L is a Cox process on T, then besides $\Psi_D^L = \{(L_n, D_n)\}$, we consider the point process $\Psi_C^L = \{(L_n, C_n)\}$, where C_n is the shortest path length from L_n to H_j along the edges of T, provided that $L_n \in \Xi_{\Psi^H, j}$, see Fig. 5.8. We are interested in the distribution of the *typical shortest path length*, i.e., the typical mark C^o of Ψ_C^L.

5.4.3 Distributional Properties of D^o and C^o

We show that the distribution function and density of the typical (direct) connection length D^o can be expressed as expectations of functionals of the typical serving zone and its typical segment system. Furthermore, the density of the typical shortest path length C^o is considered.

Applying Neveu's exchange formula stated in Theorem 5.9 we can represent the distribution function of D^o in terms of the typical Voronoi cell Z_H^o of Ψ^H if Ψ^L is a planar Poisson process, and in terms of the typical segment system $S^o_{\Psi^H}$ if Ψ^L is a Cox process on $\mathrm{T}^{(1)}$. This shows that the distribution of D^o is uniquely determined by T_H and Ψ_S^H, respectively.

5.4.3.1 Distribution Function of D^o

Note that the representation formulae stated in Theorem 5.10 below do not depend on Ψ^L at all. The random closed sets $Z_H^o \cap B_x(o)$ and $S_H^o \cap B_x(o)$ occurring on the right-hand sides of (5.8) and (5.9) are illustrated in Fig. 5.9.

Theorem 5.10. (i) *If Ψ^L is a planar Poisson process that is independent of* T *and Ψ^H, then the distribution function $F_{D^o} : [0, \infty) \to [0, 1]$ of D^o is given by*

$$F_{D^o}(x) = \lambda_\ell \, \gamma \, \mathbf{E} \, \nu_2(Z_H^o \cap B_x(o)) \,, \qquad x \geq 0 \,, \tag{5.8}$$

where $\nu_2(Z_H^o \cap B_x(o))$ denotes the area of Z_H^o intersected with the ball $B_x(o) \subset \mathbb{R}^2$.
(ii) *If Ψ^L is a Cox processes on $\mathrm{T}^{(1)}$ which is conditionally independent of Ψ^H given* T, *then the distribution function of D^o is given by*

$$F_{D^o}(x) = \lambda_\ell \, \mathbf{E} \, \nu_1(S_H^o \cap B_x(o)) \,, \qquad x \geq 0 \,. \tag{5.9}$$

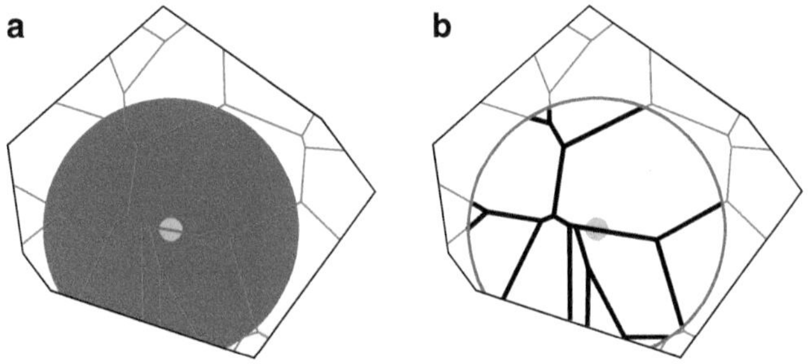

Fig. 5.9 Typical serving zone and its typical segment system intersected by $B_x(o)$. (**a**) $Z_H^o \cap B_x(o)$ (*blue*). (**b**) $S_H^o \cap B_x(o)$ (*black*)

Proof. Let us first assume that Ψ^L is a planar Poisson process with intensity λ' and regard the vector $\zeta = (\Psi_D^L, \mathrm{T}_H)$ as a random element of $\mathcal{N}_{[0,\infty),\mathcal{P}^o}$, where the definition of $\mathcal{P}^o$ and $\mathcal{N}_{[0,\infty),\mathcal{P}^o}$ has been introduced in Sects. 5.1.2.1 and 5.4.1, respectively. Furthermore, we use the notation $(\Psi_D^{L^o}, \mathrm{T}_{\widetilde{\Psi H}})$ and $(\widetilde{\Psi_D^L}, \mathrm{T}^o_{\Psi H})$ introduced in Sect. 5.4.2.2 for the Palm versions of ζ distributed according to $P^o_{\Psi_D^L}$ and $P^0_{\mathrm{T}_H}$, respectively. For some measurable function $h : [0,\infty) \to [0,\infty)$ we consider $f : \mathbb{R}^2 \times [0,\infty) \times \mathcal{P}^o \times \mathcal{N}_{[0,\infty),\mathcal{P}^o} \to [0,\infty)$ defined by

$$f(x, m, \Xi, \psi) = \begin{cases} h(m), & \text{if } o \in \Xi + x, \\ 0, & \text{otherwise.} \end{cases}$$

Then, applying Theorem 5.9, we get

$$\begin{aligned} \mathbf{E}\, h(D^o) &= \int_{\mathcal{N}_{[0,\infty),\mathcal{P}^o}} \int_{\mathbb{R}^2 \times \mathcal{P}^o} f(x, \Xi, m, \psi)\, \psi^{(2)}(d(x, \Xi))\, P^0_{\Psi_D^L}(d(\psi, m)) \\ &= \frac{\lambda}{\lambda'} \int_{\mathcal{N}_{[0,\infty),\mathcal{P}^o}} \int_{\mathbb{R}^2 \times [0,\infty)} f(-x, \Xi, m, T_x\psi)\, \psi^{(1)}(d(x, m))\, P^0_{\mathrm{T}_H}(d(\psi, \Xi)) \\ &= \frac{\lambda}{\lambda'} \int_{\mathcal{N}_{[0,\infty),\mathcal{P}^o}} \int_{\mathbb{R}^2 \times [0,\infty)} h(|x|)\mathbf{1}(x \in \Xi)\, \psi^{(1)}(d(x, m))\, P^0_{\mathrm{T}_H}(d(\psi, \Xi)) \\ &= \frac{\lambda}{\lambda'} \mathbf{E}\left(\mathbf{E}\left(\sum_{\widetilde{\Psi^L}_n \in Z_H^o} h(|\widetilde{\Psi^L}_n|) \mid Z_H^o \right)\right). \end{aligned}$$

Since Ψ^L and Ψ^H are independent, we get that T^o_H and $\widetilde{\Psi^L}$ are also independent and in addition that $\widetilde{\Psi^L} \stackrel{d}{=} \Psi^L$. Thus, given Z_H, we get that $\widetilde{\Psi^L}$ is a stationary Poisson process of intensity λ'. Using Campbell's formula (see formula (4.10)), we obtain

$$\mathbf{E}\left(\sum_{\widetilde{L}_n \in Z_H^o} h(|\widetilde{L}_n|) \mid Z_H^o\right) = \lambda' \int_{Z_H} h(|u|)\, \nu_2(du)$$

which yields for $h(|u|) = \mathbf{1}(|u| \in [0, x])$ that

$$F_{D^o}(x) = \mathbf{E}\,\mathbf{1}(D^o \in [0, x]) = \lambda\,\mathbf{E}\,\nu_2(Z_H^o \cap B_x(o)).$$

On the other hand, if Ψ^L is a Cox process on $\mathrm{T}^{(1)}$, then we regard the vector $\zeta = (\Psi_D^L, \Psi_S^H)$ as a random element of $\mathcal{N}_{[0,\infty),\mathcal{L}^o}$. Recall that we use the notation $(\Psi_D^{L^o}, \widetilde{\Psi_S^H})$ and $(\widetilde{\Psi^L}_D, \Psi_S^{H^o})$ for the Palm versions of ζ with respect to the Palm distributions $P^o_{\Psi_D^L}$ and $P^o_{\Psi_S^H}$, respectively. Similarly as above, an appropriate application of Neveu's exchange formula stated in Theorem 5.9 yields

$$\mathbf{E}\,h(D^o) = \frac{\lambda}{\lambda'}\,\mathbf{E}\left(\mathbf{E}\left(\sum_{\widetilde{L}_n \in S_H^o} h(|\widetilde{L}_n|) \mid S_H^o\right)\right).$$

Note that $\widetilde{\Psi^L}$ is independent of $\Psi_S^{H^o}$ under $P^o_{\Psi_S^H}$ given S_H^o. Furthermore, $\lambda' = \lambda'_\ell \gamma$ and $\widetilde{\Psi^L} \cap S_H^o$ is a Cox process whose random intensity measure is given by $\lambda'_\ell \nu_1(B \cap S_H^o)$ for $B \in \mathcal{B}(\mathbb{R}^2)$. Thus, Campbell's formula (see formula (4.10)) yields

$$\mathbf{E}\left(\sum_{\widetilde{L}_n \in S_H^o} h(|\widetilde{L}_n|) \mid S_H^o\right) = \lambda'_\ell \int_{S_H^o} h(|u|)\, \nu_1(du)$$

and, for $h(|u|) = \mathbf{1}(|u| \in [0, x])$, formula (5.9) follows. □

5.4.3.2 Probability Density of D^o

Using Theorem 5.10 we can derive analogous representation formulae for the probability density of D^o.

Theorem 5.11. (i) *If Ψ^L is a planar Poisson process, which is independent of* T *and Ψ^H, then the probability density $f_{D^o} : [0,\infty) \to [0,\infty)$ of D^o is given by*

$$f_{D^o}(x) = \lambda_\ell\, \gamma\, \mathbf{E}\,\nu_1(Z_H \cap \partial B_x(o)), \qquad x \geq 0, \tag{5.10}$$

where $\nu_1(Z_H \cap \partial B_x(o))$ denotes the curve length of the circle $\partial B_x(o)$ inside Z_H.
(ii) *If Ψ^L is a Cox processes on $\mathrm{T}^{(1)}$ which is conditionally independent of Ψ^H given* T, *then the probability density of D^o is given by*

$$f_{D^o}(x) = \lambda_\ell \, \mathbf{E}\left(\sum_{i=1}^{N_x^o} \frac{1}{\sin \alpha_i^o}\right), \qquad x \geq 0, \tag{5.11}$$

where $N_x^o = |S_H^o \cap \partial B_x(o)|$ *is the number of intersection points of the segment system* S_H^o *with* $\partial B_x(o)$ *and* $\alpha_1^o, \ldots, \alpha_{N_x^o}^o$ *are the angles at the corresponding intersection points between their tangents to* $\partial B_x(o)$ *and the intersecting segments.*

Proof. Assuming that Ψ^L is a Poisson process and using the polar decomposition of the two-dimensional Lebesgue measure, we get from (5.8) that

$$\begin{aligned} F_{D^o}(x) &= \lambda_\ell \gamma \, \mathbf{E} \int_{\mathbb{R}^2} \mathbf{1}(y \in Z_H \cap B_x(o)) \, \nu_2(dy) \\ &= \lambda_\ell \gamma \, \mathbf{E} \int_0^x \int_0^{2\pi} r \mathbf{1}((r \cos t, r \sin t) \in Z_H) \, dt \, dr \\ &= \int_0^x \lambda_\ell \gamma \mathbf{E} \nu_1(Z_H \cap \partial B_r(o)) \, dr, \end{aligned}$$

i.e., (5.10) is shown. If Ψ^L is a Cox process on $\mathrm{T}^{(1)}$, then we get from (5.9) that

$$\begin{aligned} F_{D^o}(x) &= \lambda_\ell \, \mathbf{E} \, \nu_1(S_H^o \cap B_x(o)) \\ &= \lambda_\ell \, \mathbf{E} \int_0^\infty \sum_{i=1}^{N_y^o} \frac{1}{\sin \alpha_i^o} \mathbf{1}(y \in [0,x]) \, dy \\ &= \int_0^x \lambda_\ell \, \mathbf{E}\left(\sum_{i=1}^{N_y^o} \frac{1}{\sin \alpha_i^o}\right) dy, \end{aligned}$$

decomposing the Hausdorff measure ν_1 similarly as in the proof of (5.10). □

5.4.3.3 Representation Formulae for C^o

Theorem 5.12. *Let* Ψ^L *be a Cox processes on* $\mathrm{T}^{(1)}$ *which is conditionally independent of* Ψ^H *given* T. *Then, for any measurable function* $h : \mathbb{R} \to [0, \infty)$ *it holds that*

$$\mathbf{E} h(C^o) = \lambda_\ell \mathbf{E} \int_{S_H^o} h(c(y)) \, \nu_1(dy), \tag{5.12}$$

where $c(y)$ *is the shortest path length from* y *to* o *along the edges of the Palm version* $\Psi_S^{H^o}$ *of* Ψ_S^H *and* S_H^o *is the (typical) segment system of* $\Psi_S^{H^o}$ *centered at* o.

Proof. It is similar to the proof of Theorem 5.10. Note that formula (5.12) can be written as

$$\mathbf{E} h(C^o) = \lambda_\ell \mathbf{E} \sum_{i=1}^{N} \int_{c(A_i)}^{c(B_i)} h(u) \, du,$$

where the segment system S_H^o is decomposed into line segments $S_1, \dots, S_N$ with endpoints $A_1, B_1, \dots, A_N, B_N$ such that $S_H^o = \bigcup_{i=1}^N S_i$, $\nu_1(S_i \cap S_j) = 0$ for $i \neq j$, and $c(A_i) < c(B_i) = C(A_i) + \nu_1(S_i)$. Furthermore, putting $h(x) = \mathbf{1}(x \in B)$ for any Borel set $B \subset \mathbb{R}$, we get that $\mathbf{P}(C^o \in B) = \int_B \lambda_\ell \, \mathbf{E} \sum_{i=1}^N \mathbf{1}(u \in [c(A_i), c(B_i)) \, du$. Thus, the following formulae for the probability density $f_{C^o} : \mathbb{R} \to [0, \infty)$ of C^o are obtained:

$$f_{C^o}(x) = \begin{cases} 2\lambda_\ell & \text{if } x = 0, \\ \lambda_\ell \, \mathbf{E} \sum_{i=1}^N \mathbf{1}(x \in [c(A_i), c(B_i)) & \text{if } x > 0. \end{cases} \tag{5.13}$$

The proof is complete. □

5.5 Scaling Limits

In this section we assume that Ψ^L is a Cox process on $\mathrm{T}^{(1)}$ with random driving measure given by (5.5). We investigate the asymptotic behaviour of the distributions of the typical connection lengths D^o and C^o as the parameters of the stochastic network model introduced in Sect. 5.4.2 tend to some extremal values. The resulting limit theorems for the distributions of D^o and C^o can be used in order to derive parametric approximation formulae for the distribution of C^o, see Sect. 5.6.

5.5.1 Asymptotic Behaviour of D^o

We consider the asymptotic behaviour of the distribution of $D^o = D^o(\gamma, \lambda_\ell)$ if the scaling factor $\kappa = \gamma/\lambda_\ell$ introduced in Sect. 5.2.2 tends to ∞, where we assume that $\gamma \to \infty$ and $\lambda_\ell \to 0$ such that $\lambda_\ell \gamma = \lambda$ is fixed. This means that the planar intensity λ of Ψ^H is constant, but the edge set of T_γ gets unboundedly dense as $\kappa \to \infty$; see Fig. 5.10 for realizations of the network model for small and large values of κ. In particular, we show that D^o converges in distribution to the (random) Euclidean distance ξ from the origin to the nearest point of a stationary Poisson process in $\mathbb{R}^2$ with intensity λ.

Theorem 5.13. *Let* T *be ergodic and* $\xi \sim \mathrm{Wei}(\lambda\pi, 2)$ *for some* $\lambda > 0$. *If* $\kappa \to \infty$, *where* $\gamma \to \infty$ *and* $\lambda_\ell \to 0$ *such that* $\lambda = \gamma\lambda_\ell$, *then*

$$D^o(\gamma, \lambda_\ell) \xrightarrow{d} \xi \, . \tag{5.14}$$

In the proof of Theorem 5.13 given below, we use two classical results regarding weak convergence of point processes, which are stated separately in Sect. 5.5.1.1. For further details on weak convergence of point processes, see for example [140, 282, 346].

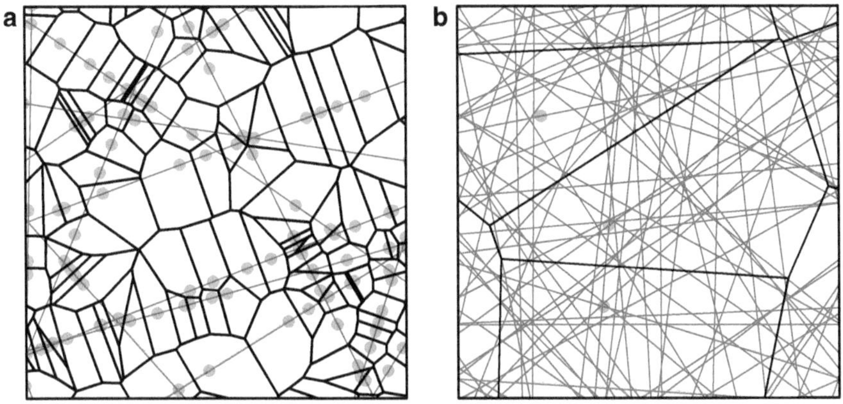

Fig. 5.10 Realizations of the network model for extreme values of κ. (**a**) $\kappa = 1$; (**b**) $\kappa = 1,000$

5.5.1.1 Weak Convergence of Point Processes

A sequence of point processes $\Psi^{(1)}, \Psi^{(2)}, \ldots$ in $\mathbb{R}^2$ is said to *converge weakly* to a point process Ψ in $\mathbb{R}^2$ iff

$$\lim_{m\to\infty} \mathbf{P}(\Psi^{(m)}(B_1) = i_1, \ldots, \Psi^{(m)}(B_k) = i_k) = \mathbf{P}(\Psi(B_1) = i_1, \ldots, \Psi(B_k) = i_k)$$

for any $k \geq 1$, $i_1, \ldots, i_k \geq 0$ and for any continuity sets $B_1, \ldots, B_k \in \mathcal{B}_0(\mathbb{R}^2)$ of Ψ, where $B \in \mathcal{B}(\mathbb{R}^2)$ is called a *continuity set* of Ψ if $\mathbf{P}(\Psi(\partial B) > 0) = 0$. If the sequence $\Psi^{(1)}, \Psi^{(2)}, \ldots$ converges weakly to Ψ, we briefly write $\Psi^{(m)} \Longrightarrow \Psi$.

Now let $\Psi = \{X_n\}$ be an arbitrary ergodic point process in $\mathbb{R}^2$ with intensity $\lambda \in (0, \infty)$. Then the following limit theorem for independently thinned and appropriately re-scaled versions of Ψ can be shown. For each $p \in (0, 1)$, let $\Psi^{(p)}$ denote the point process which is obtained from Ψ by an independent thinning, where each point X_n of Ψ survives with probability p and is removed with probability $1 - p$ independently of the other points of Ψ. Furthermore, assume that $\overline{\Psi}^{(p)}$ is a re-scaled version of the thinned process $\Psi^{(p)}$, which is defined by $\overline{\Psi}^{(p)}(B) = \overline{\Psi}^{(p)}(B/\sqrt{p})$ for each $B \in \mathcal{B}(\mathbb{R}^2)$. Thus, for each $p \in (0, 1)$, the point processes $\overline{\Psi}^{(p)}$ and Ψ are both stationary with the same intensity λ since $\mathbf{E}\overline{\Psi}^{(p)}([0, 1)^2) = \mathbf{E}\Psi^{(p)}([0, 1/\sqrt{p})^2) = \lambda$.

Theorem 5.14. *Let Π_λ be a stationary Poisson process in $\mathbb{R}^2$ with intensity λ. Then,*

$$\overline{\Psi}^{(p)} \Longrightarrow \Pi_\lambda \qquad \text{as } p \to 0. \tag{5.15}$$

Proof. See for example [140, Sect. 11.3] or [346, Theorem 7.3.1]. □

Intuitively, the statement of Theorem 5.14 can be explained as follows. The dependence between the points of Ψ in two sets $A, B \in \mathcal{B}_0(\mathbb{R}^2)$ decreases with increasing distance between A and B. Thus, if the point process is thinned independently only points far away of each other survive with high probability which in the limit yields a point process with complete spatial randomness, that is a Poisson process.

The following continuity property with respect to weak convergence of Palm distributions of stationary point processes holds.

Theorem 5.15. *Let Ψ, $\Psi^{(1)}, \Psi^{(2)}, \ldots$ be stationary point processes in $\mathbb{R}^2$ with intensity λ. If $\Psi^{(m)} \Longrightarrow \Psi$ as $m \to \infty$, then the Palm versions $\Psi^{(1)o}, \Psi^{(2)o}, \ldots$ of $\Psi^{(1)}, \Psi^{(2)}, \ldots$ converge weakly to the Palm version Ψ^o of Ψ, i.e.,*

$$\Psi^{(m)o} \Longrightarrow \Psi^o \qquad \text{as } m \to \infty. \tag{5.16}$$

Proof. See for example [346, Proposition 10.3.6]. □

5.5.1.2 Proof of Theorem 5.13

We now are able to prove Theorem 5.13 using the auxiliary results stated above, where we first show that the Cox process Ψ^H on $\mathrm{T}^{(1)}$ converges weakly to a stationary Poisson process with intensity λ if $\kappa \to \infty$ provided that $\lambda_\ell \gamma = \lambda$ is constant. This result is then used in order to investigate the asymptotic behaviour of the typical Euclidean distance $D^o = |\widetilde{H}_0|$, where $\widetilde{H}_0$ denotes that point of $\widetilde{\Psi}^H = \{\widetilde{H}_n\}$ which is closest to the origin (see Sect. 5.4.2.2).

Lemma 5.1. *If $\kappa = \gamma/\lambda_\ell \to \infty$, where $\lambda_\ell \gamma = \lambda$ for some constant $\lambda \in (0, \infty)$, then $\Psi^H \Longrightarrow \Pi_\lambda$, where Π_λ is a stationary Poisson process in $\mathbb{R}^2$ with intensity λ.*

Proof. For each $\gamma > 1$, let $\Psi^H = \Psi^H(\gamma)$ be the Cox process on the scaled version T_γ of T with linear intensity λ_ℓ, where $\lambda_\ell = \lambda/\gamma$ for some constant $\lambda \in (0, \infty)$. Then the Cox process $\Psi^H(\gamma)$ can be obtained from $\Psi^H(1)$ by an independent thinning with survival probability $p = 1/\gamma$ followed by a re-scaling with the scaling factor $\sqrt{1/\gamma}$, i.e., $\Psi^H(\gamma) \stackrel{d}{=} \Psi^H(1)^{(p)}$. Furthermore, the Cox process $\Psi^H(1)$ is ergodic since the underlying tessellation T and hence the random intensity measure of $\Psi^H(1)$ is ergodic. Thus we can apply Theorem 5.14 which yields $\Psi^H(\gamma) \Longrightarrow \Pi_\lambda$ as $\gamma \to \infty$. □

Lemma 5.2. *Let $\xi \sim \mathrm{Wei}(\lambda\pi, 2)$ for some $\lambda > 0$. Then $D^o \stackrel{d}{\to} \xi$ as $\kappa \to \infty$ provided that $\gamma \to \infty$ and $\lambda_\ell \to 0$ such that $\lambda_\ell \gamma = \lambda$.*

Proof. Assume that $\Psi^{H^o} = \Psi^{H^o}(\gamma)$ is the Palm version of the stationary point process $\Psi^H = \Psi^H(\gamma)$. Then the distribution of $\Pi_\lambda \cup \{o\}$ is equal to the Palm distribution of Π_λ due to Slivnyak's theorem. Thus, Lemma 5.1 and Theorem 5.15

yield that

$$\Psi^{H^o}(\gamma) \Longrightarrow \Pi_\lambda \cup \{o\} \tag{5.17}$$

if $\gamma \to \infty$ and $\lambda_\ell \to 0$ with $\lambda_\ell \gamma = \lambda$. Since both Ψ^L and Ψ^H are Cox processes on $\mathrm{T}_\gamma^{(1)}$ conditionally independent given T_γ, we get that $\widetilde{\Psi^H} \cup \{o\}$ and the Palm version Ψ^{H^o} of Ψ^H have the same distributions. This is a consequence of Slivnyak' theorem for stationary Cox processes, see Theorem 5.6. Thus, using (5.17), for each $r > 0$ we get

$$\begin{aligned}
\lim_{\gamma\to\infty} \mathbf{P}(|\widetilde{\Psi^H}_0| > r) &= \lim_{\gamma\to\infty} \mathbf{P}(\widetilde{\Psi^H}(B_r(o)) = 0) \\
&= \lim_{\gamma\to\infty} \mathbf{P}((\widetilde{\Psi^H} \cup \{o\})(B_r(o)) = 1) \\
&= \lim_{\gamma\to\infty} \mathbf{P}(\Psi^{H^o}(B_r(o)) = 1) \\
&= \mathbf{P}((\Pi_\lambda \cup \{o\})(B_r(o)) = 1) \\
&= \mathbf{P}(\Pi_\lambda(B_r(o)) = 0)\,.
\end{aligned}$$

Hence, $\lim_{\gamma\to\infty} \mathbf{P}(|\widetilde{\Psi^H}_0| > r) = \mathbf{P}(\Pi_\lambda(B_r(o)) = 0) = \exp\{-\lambda\pi r^2\}$ for each $r > 0$, which shows that $D^o = |\widetilde{\Psi^H}_0| \xrightarrow{d} \xi \sim \mathrm{Wei}(\lambda\pi, 2)$. □

5.5.2 Asymptotic Behaviour of C^o

The results presented in the preceding section can be extended to further cost functionals of the stochastic network model introduced in Sect. 5.4.2. For instance, if T is isotropic, mixing and $\mathbf{E}\nu_1^2(\partial Z) < \infty$, where $\nu_1^2(\partial Z)$ denotes the circumference of the typical cell Z of T, then it can be shown that

$$C^o \xrightarrow{d} a\xi \tag{5.18}$$

as $\kappa = \gamma/\lambda_\ell \to \infty$ provided that $\lambda = \gamma\lambda_\ell$ is fixed. Here, $\xi \sim \mathrm{Wei}(\lambda\pi, 2)$ and $a \in [1, \infty)$ is some constant which depends on type of the underlying tessellation T. In the proof of (5.18), the result of Theorem 5.13 is used. This is then combined with fact that under the additional conditions on T mentioned above, one can show that $C^o - a D^o$ converges in probability to 0. Moreover, it can be shown that

$$C^o \xrightarrow{d} \xi' \tag{5.19}$$

as $\kappa = \gamma/\lambda_\ell \to 0$, where λ_ℓ is fixed and $\xi' \sim \mathrm{Exp}(2\lambda_\ell)$. For further details, see [500].

5.6 Monte Carlo Methods and Parametric Approximations

The representation formulae (5.8)–(5.10) can easily be used to obtain simulation-based approximations for the distribution function and probability density of D^o, see Sect. 5.6.1. These estimates can be computed based on samples of Z_H and S_H^o which are generated by Monte Carlo simulation, using algorithms like those discussed in Sect. 5.3.2.2. Note that we do not have to simulate any points of Ψ^L.

Similarly, we can use formula (5.13) to get a Monte Carlo estimator for the density of C^o. However, note that the density formula (5.11) for D^o is not suitable in this context, because it would lead to an estimator which is numerically instable.

Moreover, the scaling limits for C^o stated in (5.18) and (5.19) can be used in order to determine parametric approximation formulae for the density of C^o, which are surprisingly accurate for a wide range of (non-extremal) model parameters, see Sect. 5.6.2.

5.6.1 *Simulation-Based Estimators*

Assume that $Z_{H,1}, \ldots, Z_{H,n}$ and $S_{H,1}^o, \ldots, S_{H,n}^o$ are n independent copies of Z_H and S_H^o, respectively. If Ψ^L is a stationary Poisson process in $\mathbb{R}^2$, then we can use (5.8) and (5.10) to define the estimators for $F_{D^o}(x)$ and $f_{D^o}(x)$ by

$$\widehat{F}_{D^o}(x;n) = \frac{\lambda_\ell \gamma}{n} \sum_{i=1}^{n} \nu_2(Z_{H,i} \cap B_x(o)) \tag{5.20}$$

and

$$\widehat{f}_{D^o}(x;n) = \frac{\lambda_\ell \gamma}{n} \sum_{i=1}^{n} \nu_1(Z_{H,i} \cap \partial B_x(o)), \tag{5.21}$$

respectively. If Ψ^L is a Cox process on $\mathrm{T}^{(1)}$, then we can use (5.9) to define an estimator for $F_{D^o}(x)$ by

$$\widehat{F}_{D^o}(x;n) = \frac{\lambda_\ell}{n} \sum_{i=1}^{n} \nu_1(S_{H,i}^o \cap B_x(o)). \tag{5.22}$$

Similarly, using formula (5.13), we can define an estimator for $f_{C^o}(x)$ by

$$\widehat{f}_{C^o}(x;n) = \frac{\lambda_\ell}{n} \sum_{j=1}^{n} \sum_{i=1}^{N_j} \mathbf{1}(x \in [c(A_i^{(j)}), c(B_i^{(j)})), \tag{5.23}$$

where the independent copies $S_{H,1}^o, \ldots, S_{H,n}^o$ of S_H^o are decomposed into the line segments $S_1^{(j)}, \ldots, S_{N_j}^{(j)}$ with endpoints $A_1^{(j)}, B_1^{(j)}, \ldots, A_{N_j}^{(j)}, B_{N_j}^{(j)}$. It is not difficult to see that the estimators given in (5.20)–(5.23) are unbiased and in addition strongly consistent for fixed $x \geq 0$. However, if Ψ^L is a Cox process, then it

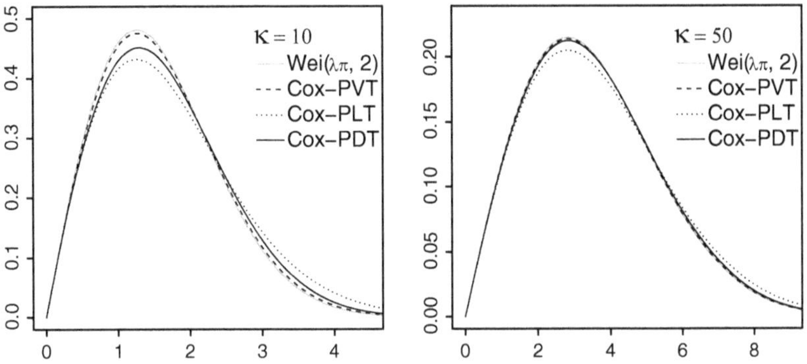

Fig. 5.11 Estimated density of D^o if Ψ^L is a stationary Poisson process in $\mathbb{R}^2$

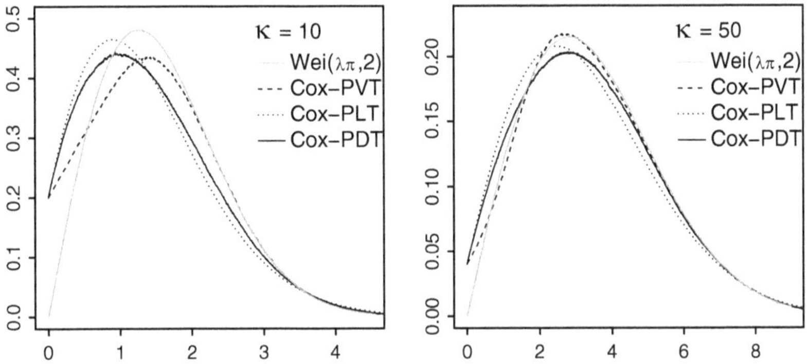

Fig. 5.12 Estimated density of D^o if Ψ^L is a Cox process on $\mathrm{T}^{(1)}$

is not recommended to construct an estimator $\widehat{f}_{D^o}(x;n)$ for $f_{D^o}(x)$ based on (5.11) by just omitting the expectation in (5.11). This estimator is numerically unstable since infinitely small angles can occur. In this case, it is better to first compute the distribution function $\widehat{F}_{D^o}(x;n)$ using formula (5.22) and afterwards considering difference quotients obtained from this estimated distribution function as estimator $\widehat{f}_{D^o}(x;n)$ for $f_{D^o}(x)$, see [499]. Some examples of estimated densities are shown in Figs. 5.11 and 5.12, together with the corresponding (scaling) limit as $\kappa = \gamma/\lambda_\ell \to \infty$ with $\lambda_\ell\gamma$ $(=\lambda)$ fixed., i.e. the density of the Wei$(\lambda\pi, 2)$-distribution.

5.6.2 Parametric Approximation Formulae

For practical applications it is useful to have parametric approximation formulae for the distribution of C^o, where the parameters depend on the model type of T and the

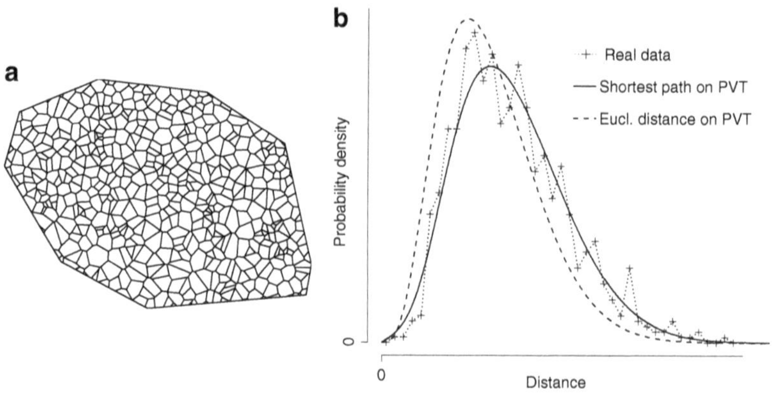

Fig. 5.13 (**a**) Typical serving zone for $\kappa = 1,000$ and (**b**) parametric density of C^o for PVT fitted to infrastructure data (*solid line*), compared with histogram of connection lengths estimated from real network data, showing that the assumption of direct physical connections (*dashed line*) is incorrect

scaling factor κ. Therefore, the problem arises to fit suitable classes of parametric densities to the densities of C^o which have been computed by the simulation-based algorithm discussed in Sect. 5.6.1. In [190] truncated Weibull distributions were used for this purpose since the scaling limits for the distribution of C^o, i.e. the exponential and Weibull distributions mentioned in Sect. 5.5.2, belong to this parametric family. It turned out that the fitted densities approximate the estimated densities surprisingly well for different types of $\mathtt{T}$ and for a wide range of κ. These parametric densities can be used in order to efficiently analyze and plan telecommunication networks. In a first step, a suitable tessellation model has to be fitted to real infrastructure data. Afterwards, the scaling factor κ must be estimated computing length intensity of the infrastructure and the number of HLC in the network per unit area. Then the distribution of the typical shortest path length C^o is directly available via the parametric densities in order to analyze connection lengths of existing or planned telecommunication networks. In Fig. 5.13 the parametric density chosen in this way is compared to a histogram of connection lengths of real network data of Paris. One can see that there is a quite good fit, see [190] for details and further results.

Chapter 6
Asymptotic Methods for Random Tessellations

Pierre Calka

Abstract In this chapter, we are interested in two classical examples of random tessellations which are the Poisson hyperplane tessellation and Poisson–Voronoi tessellation. The first section introduces the main definitions, the application of an ergodic theorem and the construction of the so-called typical cell as the natural object for a statistical study of the tessellation. We investigate a few asymptotic properties of the typical cell by estimating the distribution tails of some of its geometric characteristics (inradius, volume, fundamental frequency). In the second section, we focus on the particular situation where the inradius of the typical cell is large. We start with precise distributional properties of the circumscribed radius that we use afterwards to provide quantitative information about the closeness of the cell to a ball. We conclude with limit theorems for the number of hyperfaces when the inradius goes to infinity.

6.1 Random Tessellations: Distribution Estimates

This section is devoted to the introduction of the main notions related to random tessellations and to some examples of distribution tail estimates. In the first subsection, we define the two main examples of random tessellations, namely the *Poisson hyperplane tessellation* and the *Poisson–Voronoi tessellation*. The next subsection is restricted to the stationary tessellations for which it is possible to construct a statistical object called *typical cell* Z via several techniques (ergodicity, Palm measures, explicit realizations). Having isolated the cell Z, i.e. a random polyhedron which represents a cell "picked at random" in the whole tessellation, we can investigate its geometric characteristics. In the last subsection, we present techniques for estimating their distribution tails.

P. Calka (✉)
Université de Rouen, Rouen, France
e-mail: pierre.calka@univ-rouen.fr

E. Spodarev (ed.), *Stochastic Geometry, Spatial Statistics and Random Fields*, Lecture Notes in Mathematics 2068, DOI 10.1007/978-3-642-33305-7_6,

This section is not intended to provide the most general definitions and results. Rather, it is aimed at emphasizing some basic examples. Quite often, we shall consider the particular case of the plane. A more exhaustive study of random tessellations can be found in the books [368, 386, 451, 489] as well as the surveys [108, 252].

6.1.1 Definitions

Definition 6.1 (Convex tessellation). A *convex tessellation* is a locally finite collection $\{\Xi_n\}_{n\in\mathbb{N}}$ of convex polyhedra of $\mathbb{R}^d$ such that $\cup_{n\in\mathbb{N}}\Xi_n = \mathbb{R}^d$ and Ξ_n and Ξ_m have disjoint interiors if $n \neq m$. Each Ξ_n is called a *cell* of the tessellation.

The set $\mathbb{T}$ of convex tessellations is endowed with the σ-algebra generated by the sets

$$\{\{\Xi_n\}_{n\in\mathbb{N}} : [\cup_{n\in\mathbb{N}}\partial\Xi_n] \cap K = \emptyset\}$$

where K is any compact set of $\mathbb{R}^d$.

Definition 6.2 (Random convex tessellation). A *random convex tessellation* is a random variable with values in $\mathbb{T}$.

Remark 6.1. We can equivalently identify a tessellation $\{\Xi_n\}_{n\in\mathbb{N}}$ with its skeleton $\cup_{n\in\mathbb{N}}\partial\Xi_n$ which is a random closed set of $\mathbb{R}^d$.

Definition 6.3 (Stationarity, isotropy). A random convex tessellation is *stationary* (resp. *isotropic*) if its skeleton is a translation-invariant (resp. rotation-invariant) random closed set.

We describe below the two classical constructions of random convex tessellations, namely the hyperplane tessellation and the Voronoi tessellation. In the rest of the section, we shall only consider these two particular examples even though many more can be found in the literature (Laguerre tessellations [324], iterated tessellations [340], Johnson–Mehl tessellations [367], crack STIT tessellations [376], etc.).

Definition 6.4 (Hyperplane tessellation). Let Ψ be a point process which does not contain the origin almost surely. For every $x \in \Psi$, we define its *polar hyperplane* as $H_x = \{y \in \mathbb{R}^d : \langle y - x, x\rangle = 0\}$. The associated *hyperplane tessellation* is the set of the closure of all connected components of $\mathbb{R}^d \setminus \cup_{x\in\Psi} H_x$.

We focus on the particular case where Ψ is a Poisson point process. The next proposition provides criteria for stationarity and isotropy.

Proposition 6.1 (Stationarity of Poisson hyperplane tessellations). *Let $\Psi = \Pi_\Lambda$ be a Poisson point process of intensity measure Λ.*

The associated hyperplane tessellation is stationary iff Λ can be written in function of spherical coordinates $(u, t) \in \mathbb{S}^{d-1} \times \mathbb{R}_+$ as

$$\Lambda(du, dt) = \lambda\, dt\, \varphi(du) \tag{6.1}$$

where φ is a probability measure on $\mathbb{S}^{d-1}$.

It is additionally isotropic iff φ is the uniform measure σ_{d-1} on $\mathbb{S}^{d-1}$.

A so-called *Poisson hyperplane tessellation* (*Poisson line tessellation* in dimension two) is a hyperplane tessellation generated by a Poisson point process but it is quite often implied in the literature that it is also stationary and isotropic. Up to rescaling, we will assume in the rest of the chapter that its intensity λ is equal to one (Fig. 6.1).

Exercise 6.1. Verify that a stationary and isotropic Poisson hyperplane tessellation satisfies the following property with probability one: for $0 \le k \le d$, each k-dimensional face of a cell is the intersection of exactly $(d-k)$ hyperplanes H_x, $x \in \Pi_\Lambda$, and is included in exactly 2^{d-k} cells.

This tessellation has been introduced for studying trajectories in bubble chambers by S.A. Goudsmit in [200] in 1945. It has been used in numerous applied works since then. For instance, R.E. Miles describes it as a possible model for the fibrous structure of sheets of paper [356, 357].

Definition 6.5 (Voronoi tessellation). Let Ψ be a point process. For every $x \in \Psi$, we define the *cell associated with* x as

$$Z(x|\Psi) = \{y \in \mathbb{R}^d : \|y - x\| \le \|y - x'\| \ \forall x' \in \Psi, x' \neq x\}.$$

The associated *Voronoi tessellation* is the set $\{Z(x|\Psi)\}_{x\in\Psi}$.

Proposition 6.2 (Stationarity of Voronoi tessellations). *The Voronoi tessellation associated with a point process Ψ is stationary iff Ψ is stationary.*

A so-called *Poisson–Voronoi tessellation* is a Voronoi tessellation generated by a homogeneous Poisson point process. Up to rescaling, we will assume in the rest of the chapter that its intensity is equal to one.

Exercise 6.2. Show that a Poisson–Voronoi tessellation is *normal* with probability one, i.e. every k-dimensional face of a cell, $0 \le k \le d$, is included in exactly $d - k + 1$ cells.

This tessellation has been introduced in a deterministic context by R. Descartes in 1644 as a description of the structure of the universe (see also the more recent work [514]). It has been developed since then for many applications, for example in telecommunications [21, 173], image analysis [162] and molecular biology [185].

We face the whole population of cells in a random tessellation. How to study them? One can provide two possible answers:

1. Either you isolate one particular cell.
2. Or you try conversely to do a statistical study over all the cells by taking means.

An easy way to fix a cell consists in considering the one containing the origin.

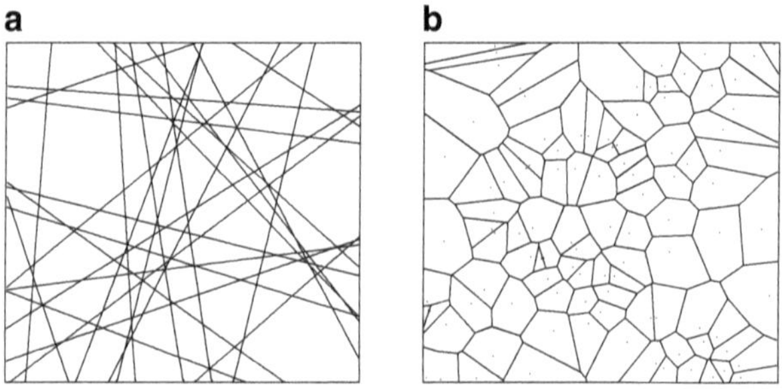

Fig. 6.1 Realizations of the isotropic and stationary Poisson line tessellation (**a**) and the planar stationary Poisson–Voronoi tessellation (**b**) in the unit square

Definition 6.6 (Zero-cell). If $o \notin \cup_{n\in\mathbb{N}} \partial \Xi_n$ a.s., then the *zero-cell* (denoted by Z_0) is the cell containing the origin. In the case of an isotropic and stationary Poisson hyperplane tessellation, it is called the *Crofton cell.*

The second point above will be developed in the next section. It is intuitively clear that it will be possible to show the convergence of means over all cells only if the tessellation is translation-invariant.

6.1.2 Empirical Means and Typical Cell

This section is restricted to the stationary Poisson–Voronoi and Poisson hyperplane tessellations. We aim at taking means of certain characteristics over all the cells of the tessellation. But of course, we have to restrict the mean to a finite number of these cells due to technical reasons. A natural idea is to consider those contained in or intersecting a fixed window, for example the ball $B_R(o)$, then take the limit when the size of the window goes to infinity. Such an argument requires the use of an ergodic theorem and the first part of the section will be devoted to prepare and show an ergodic result specialized to our set-up. In the second part of the section, we use it to define the notion of the typical cell and we investigate several equivalent ways of defining it.

6.1.2.1 Ergodic Theorem for Tessellations

The first step is to realize the measurable space $(\Omega, \mathcal{F})$ as $(\mathcal{N}, \mathfrak{N})$ where $\mathcal{N}$ is the set of locally finite sets of $\mathbb{R}^d$ and $\mathfrak{N}$ is the σ-algebra generated by the functions $\#(\cdot \cap A)$, where A is any bounded Borel set. We define the *shift* $T_a : \mathcal{N} \longrightarrow \mathcal{N}$ as the operation over the points needed to translate the tessellation by a vector $a \in \mathbb{R}^d$. In other words, for every locally finite set $\{x_n\}_{n\in\mathbb{N}}$, the underlying tessellation generated by $T_a(\{x_n\}_{n\in\mathbb{N}})$ is the translate by a of the initial tessellation generated by $\{x_n\}_{n\in\mathbb{N}}$.

Proposition 6.3 (Explicit shifts). *For a Voronoi tessellation, T_a is the function which associates to every locally finite set $\{x_n\}_{n\in\mathbb{N}} \in \mathcal{N}$ the set $\{x_n + a\}_{n\in\mathbb{N}}$.*

For a hyperplane tessellation, T_a is the function which associates to every locally finite set $\{x_n\}_{n\in\mathbb{N}} \in \mathcal{N}$ (which does not contain the origin) the set $\{x_n + \langle x_n/\|x_n\|, a\rangle x_n/\|x_n\|\}_{n\in\mathbb{N}}$.

Proof. In the Voronoi case, the translation of the skeleton is equivalent with the translation of the nuclei which generate the tessellation.

In the case of a hyperplane tessellation, the translation of a fixed hyperplane preserves its orientation but modifies the distance from the origin. To prove the proposition, it suffices to notice that a polar hyperplane H_x is sent by a translation of vector a to H_y with $y = x + \langle u, a\rangle u$ where $u = \frac{x}{\|x\|}$. □

Proposition 6.4 (Ergodicity of the shifts). *In both cases, T_a preserves the measure* **P** *(i.e. the distribution of the Poisson point process) and is ergodic.*

Sketch of proof. Saying that T_a preserves the measure is another way of expressing the stationarity of the tessellation.

To show ergodicity, it is sufficient to prove that $\{T_a : a \in \mathbb{R}^d\}$ is mixing (cf. Definition 4.6), i.e. that for any bounded Borel sets A, B and $k, l \in \mathbb{N}$, we have as $|a| \to \infty$

$$\mathbf{P}(\#(\Psi\cap A) = k;\ \#(T_a(\Psi)\cap B) = l) \to \mathbf{P}(\#(\Psi\cap A) = k)\mathbf{P}(\#(\Psi\cap B) = l). \quad (6.2)$$

In the Voronoi case, for $|a|$ large enough, the two events $\{\#(\Phi \cap A) = k\}$ and $\{\#(T_a(\Phi) \cap B) = l\}$ are independent since $A \cap (B - a) = \emptyset$. Consequently, the two sides of (6.2) are equal.

In the hyperplane case, the same occurs as soon as B is included in a set

$$\{x \in \mathbb{R}^d : |\langle x, a\rangle| \geq \varepsilon\|x\|\|a\|\}$$

for some $\varepsilon > 0$. Otherwise, we approximate B with a sequence of Borel subsets which satisfy this condition. □

In the next theorem, the main application of ergodicity for tessellations is derived.

Theorem 6.1 (Ergodic theorem for tessellations). *Let N_R be the number of cells which are included in the ball $B_R(o)$. Let $h : \mathcal{K}^d_{\text{conv}} \to \mathbb{R}$ be a measurable, bounded and translation-invariant function over the set $\mathcal{K}^d_{\text{conv}}$ of convex and compact sets of $\mathbb{R}^d$. Then almost surely,*

$$\lim_{R\to\infty} \frac{1}{N_R} \sum_{\Xi \subset B_R(o)} h(\Xi) = \frac{1}{\mathbf{E}(\nu_d(Z_0)^{-1})}\mathbf{E}\left(\frac{h(Z_0)}{\nu_d(Z_0)}\right). \quad (6.3)$$

Proof. The proof is done in three steps: use of Wiener's continuous ergodic theorem, then rewriting the mean of h over cells included in $B_R(o)$ as the sum of an integral and the rest, finally proving that the rest is negligible.

Step 1. The main ingredient is Wiener's ergodic theorem applied to the ergodic shifts $\{S_x : x \in \mathbb{R}^d\}$. We have almost surely

$$\lim_{R\to\infty} \frac{1}{\nu_d(B_R(o))} \int_{B_R(o)} \frac{h(Z_0(T_{-x}\omega))}{\nu_d(Z_0(T_{-x}\omega))} dx = \mathbf{E}\left(\frac{h(Z_0)}{\nu_d(Z_0)}\right).$$

This can be roughly interpreted by saying that the mean in space (in the left-hand side) for a fixed sample ω is asymptotically close to the mean with respect to the probability law $\mathbf{P}$.

Step 2. We have for almost every $\omega \in \Omega$ that

$$\frac{1}{\nu_d(B_R(o))} \int_{B_R(o)} \frac{h(Z_0(T_{-x}\omega))}{\nu_d(Z_0(T_{-x}\omega))} dx = \frac{1}{\nu_d(B_R(o))} \sum_{\Xi\subset B_R(o)} h(\Xi) + \text{Rest}(R) \tag{6.4}$$

where

$$\text{Rest}(R) = \frac{1}{\nu_d(B_R(o))} \sum_{\Xi:\Xi\cap\partial B_R(o)\neq\emptyset} \frac{\nu_d(\Xi\cap B_R(o))}{\nu_d(\Xi)} h(\Xi).$$

In particular, if we define N'_R as the number of cells which intersect the boundary of the ball $B_R(o)$, then there is a positive constant K depending only on h such that

$$|\text{Rest}(R)| \le K \frac{N'_R}{\nu_d(B_R(o))}.$$

We observe that in order to get (6.3), it is enough to prove that the rest goes to 0. Indeed, when $h \equiv 1$, the equality (6.4) will provide that

$$\frac{N_R}{\nu_d(B_R(o))} = \frac{1}{\nu_d(B_R(o))} \int_{B_R(o)} \frac{1}{\nu_d(Z_0(T_{-x}\omega))} dx - \text{Rest}(R) \to \mathbf{E}(\nu_d(Z_0)^{-1}).$$

Step 3. We have to show that Rest(R) goes to 0, that is what R. Cowan calls the insignificance of edge effects [132, 133]. In the sequel, we use his argument to show it and for sake of simplicity, we only consider the particular case of the two-dimensional Voronoi tessellation. Nevertheless, the method can be extended to any dimension by showing by induction that the number of k-faces hitting the boundary of the ball is negligible for every $0 \le k \le d$.

Here, in the two-dimensional case, let us fix $\varepsilon > 0$ and consider $R > \varepsilon$. Let us denote by V_R the number of vertices of the tessellation in $B_R(o)$ and by L_R the sum of the edge lengths inside $B_R(o)$. A direct use of Wiener's theorem applied to the functionals V_ε and L_ε on the sets $B_{R-\varepsilon}(o)$ and $B_{R+\varepsilon}(o)$ shows that the quantities $V_R/\nu_d(B_R(o))$ and $L_R/\nu_d(B_R(o))$ tend almost surely to constants.

We recall that a Voronoi tessellation is *normal* which means in particular that a fixed edge (resp. vertex) is contained in exactly two (resp. three) cells.

If a cell intersects the boundary of $B_R(o)$, then we are in one of the three following cases:

1. No edge of the cell intersects $B_{R+\varepsilon}(o) \setminus B_R(o)$.
2. Some edges but no vertex of the cell intersect $B_{R+\varepsilon}(o) \setminus B_R(o)$.
3. At least one vertex of the cell is in $B_{R+\varepsilon}(o) \setminus B_R(o)$.

The first case is satisfied by at most one cell, the second case by at most $\frac{2}{\varepsilon}(L_{R+\varepsilon} - L_R)$ cells and the last one by at most $3(S_{R+\varepsilon} - S_R)$. Consequently, we get when $R \to \infty$

$$\frac{N'_R}{V_2(B_R(o))} \leq \frac{1}{V_2(B_R(o))} + 2\,\frac{L_{R+\varepsilon} - L_R}{V_2(B_R(o))} + 3\,\frac{V_{R+\varepsilon} - V_R}{V_2(B_R(o))} \longrightarrow 0,$$

which completes the proof. □

Exercise 6.3. Show a similar result for a Johnson–Mehl tessellation (defined in [367]).

Remark 6.2. The statement of Theorem 6.1 still holds if condition "h bounded" is replaced with $\mathbf{E}(|h(Z_0)|^p) < \infty$ for a fixed $p > 1$ (see for example [196, Lemma 4]).

Remark 6.3. When using this ergodic theorem for tessellations in practice, it is needed to have also an associated central limit theorem. Such second-order results have been proved for some particular functionals in the Voronoi case [20] and for the Poisson line tessellation [389] in dimension two. Recently, a more general central limit result for hyperplane tessellations has been derived from the use of U-statistics in [239].

The limit in the convergence (6.3) suggests the next definition for the typical cell, i.e. a cell which represents an "average individual" from the whole population.

Definition 6.7 (Typical cell 1). The *typical cell* Z is defined as a random variable with values in $\mathcal{K}^d_{\mathrm{conv}}$ and such that for every translation-invariant measurable and bounded function $h : \mathcal{K}^d_{\mathrm{conv}} \to \mathbb{R}$, we have

$$\mathbf{E}(h(Z)) = \frac{1}{\mathbf{E}(\nu_d(Z_0)^{-1})}\mathbf{E}\left(\frac{h(Z_0)}{\nu_d(Z_0)}\right).$$

Remark 6.4. Taking for h any indicator function of geometric events (for example {the cell is a triangle}, {the area of the cell is greater than 2}, etc.), we can define via the equality above the distribution of any geometric characteristic of the typical cell.

Remark 6.5. One should keep in mind that the typical cell Z is not distributed as the zero-cell Z_0. Indeed, the distribution of Z has a density proportional to ν_2^{-1} with respect to the distribution of Z_0. In particular, since it has to contain the origin, Z_0 is larger than Z. This is a d-dimensional generalization of the famous bus paradox in renewal theory which states that at your arrival at a bus stop, the time interval between the last bus you missed and the first bus you'll get is actually bigger than the typical waiting time between two buses. Moreover, it has been proved in the case of a Poisson hyperplane tessellation that Z and Z_0 can be coupled in such a way that $Z \subset Z_0$ almost surely (see [352] and Proposition 6.6 below).

Looking at Definition 6.7, we observe that it requires to know either the distribution of Z_0 or the limit of the ergodic means in order to get the typical cell. The next definition is an alternative way of seeing the typical cell without the use of any convergence result. It is based on the theory of Palm measures [323, 348]. For sake of simplicity, it is only written in the case of the Poisson–Voronoi tessellation but it can be extended easily to any stationary Poisson hyperplane tessellation.

Definition 6.8 (Typical cell 2 (Poisson–Voronoi tessellation)). The *typical cell* Z is defined as a random variable with values in $\mathcal{K}^d_{\text{conv}}$ such that for every bounded and measurable function $h : \mathcal{K}^d_{\text{conv}} \to \mathbb{R}$ and every Borel set B with $0 < \nu_d(B) < \infty$, we have

$$\mathbf{E}(h(Z)) = \frac{1}{\nu_d(B)} \mathbf{E}\left(\sum_{x \in B \cap \Pi} h(Z(x|\Pi) - x) \right). \tag{6.5}$$

This second definition is still an intermediary and rather unsatisfying one but via the use of Slivnyak–Mecke formula for Poisson point processes (see Theorem 4.5), it provides a way of realizing the typical cell Z.

Exercise 6.4. Verify that the relation (6.5) does not depend on B.

Proposition 6.5 (Typical cell 3 (Poisson–Voronoi tessellation)). *The typical cell Z is equal in distribution to the set $Z(o) = Z(o|\Pi \cup \{o\})$, i.e. the Voronoi cell associated with a nucleus at the origin when this nucleus is added to the original Poisson point process.*

Remark 6.6. The cell $Z(o)$ defined above is not a particular cell isolated from the original tessellation. It is a cell extracted from a different Voronoi tessellation but which has the right properties of a cell "picked at random" in the original tessellation. For any $x \in \Pi$, we define the bisecting hyperplane of $[o, x]$ as the hyperplane containing the midpoint $x/2$ and orthogonal to x. Since $Z(o)$ is bounded by portions of bisecting hyperplanes of segments $[o, x]$, $x \in \Pi$, we remark that $Z(o)$ can be alternatively seen as the zero-cell of a (non-stationary) Poisson hyperplane tessellation associated with the homogeneous Poisson point process up to a multiplicative constant.

The Poisson–Voronoi tessellation is not the only tessellation such that the associated typical cell can be realized in an elementary way. There exist indeed several ways of realizing the typical cell of a stationary and isotropic Poisson hyperplane tessellation. We present below one of the possible constructions of Z, which offers the advantage of satisfying $Z \subset Z_0$ almost surely. It is based on a work [106] which is an extension in any dimension of an original idea in dimension two due to R.E. Miles [359] (Fig. 6.2).

Proposition 6.6 (Typical cell 3 (Poisson hyperplane tessellation)). *The radius R_m of the largest ball included in the typical cell Z is an exponential variable of parameter the area of the unit sphere. Moreover, conditionally on R_m, the typical cell Z is equal in distribution to the intersection of the two independent following random sets:*

(i) a random simplex with inscribed ball $B_{R_m}(o)$ such that the vector $(u_0, \ldots, u_d)$ of the $d + 1$ normal unit-vectors is independent of R_m and has a density proportional to the volume of the simplex spanned by $u_0, \ldots, u_d$.
(ii) the zero-cell of an isotropic Poisson hyperplane tessellation outside $B_{R_m}(o)$ of intensity measure $\Lambda(du, dt) = \mathbf{1}(B_{R_m}(o)^c)\, dt\, d\sigma_{d-1}(u)$ (in spherical coordinates).

Exercise 6.5. When $d = 2$, let us denote by α, β, γ the angles between u_0 and u_1, u_1 and u_2, u_2 and u_0 respectively. Write the explicit density in (i) in function of α, β and γ.

Exercise 6.6. We replace each hyperplane H_x from a Poisson hyperplane tessellation by a ε-thickened hyperplane $H_x^{(\varepsilon)} = \{y \in \mathbb{R}^d : d(y, H_x) \leq \varepsilon\}$ where $\varepsilon > 0$ is fixed. Show that the distribution of the typical cell remains unchanged, i.e. is the same as for $\varepsilon = 0$.

We conclude this subsection with a very basic example of calculation of a mean value: it is well-known that in dimension two, the mean number of vertices of the typical cell is 4 for an isotropic Poisson line tessellation and 6 for a Poisson–Voronoi tessellation. We give below a small heuristic justification of this fact: for a Poisson line tessellation, each vertex is in four cells exactly and there are as many cells as vertices (each vertex is the highest point of exactly one cell) whereas in the Voronoi case, each vertex is in three cells exactly and there are twice more vertices than cells (each vertex is either the highest point or the lowest point of exactly one cell).

In the next subsection, we estimate the distribution tails of some geometric characteristics of the typical cell.

6.1.3 Examples of Distribution Tail Estimates

Example 6.1 (Poisson hyperplane tessellation, Crofton cell, inradius). We consider a stationary and isotropic Poisson hyperplane tessellation, i.e. with an intensity measure equal to $\Lambda(du, dt) = dt\, d\sigma_{d-1}(u)$ in spherical coordinates (note that the constant λ appearing in (6.1) is chosen equal to one).

Let us denote by R_m the radius of the largest ball included in the Crofton cell and centered at the origin. Since it has to be centered at the origin, the ball $B_{R_m}(o)$ is not

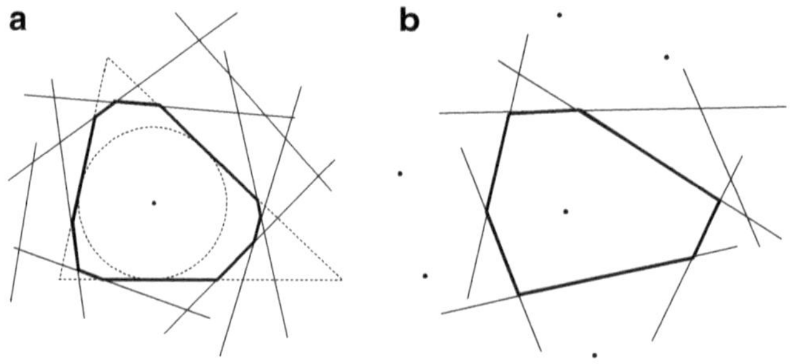

Fig. 6.2 Realizations of the typical cells of the stationary and isotropic Poisson line tessellation (**a**) and the homogeneous planar Poisson–Voronoi tessellation (**b**)

the real inball of the Crofton cell. Nevertheless, we shall omit that fact and call R_m the *inradius* in the rest of the chapter.

For every $r > 0$, we have

$$\mathbf{P}(R_m \geq r) = \mathbf{P}(\Pi \cap B_r(o) = \emptyset)$$
$$= \exp\left\{-\int_0^r \int_{\mathbb{S}^{d-1}} dt\, d\sigma_{d-1}(u)\right\} = \mathrm{e}^{-d\kappa_d r}$$

where κ_d is the Lebesgue measure of the d-dimensional unit-ball. We can remark that it is the same distribution as the real inradius of the typical cell, i.e. the radius of the largest ball included in the typical cell with unfixed center (see [106, 356] and Proposition 6.6 above).

This result can be extended by showing that for every deterministic convex set K containing the origin, the probability $\mathbf{P}(K \subset Z_0)$ is equal to $\exp\{-\frac{d}{2}\kappa_d V_1(K)\}$ where $V_1(K)$ is the mean width of K. In dimension two, the probability reduces to $\exp\{-P(K)\}$ where $P(K)$ is the perimeter of K.

Example 6.2 (Poisson–Voronoi tessellation, typical cell, inradius). We consider a homogeneous Poisson–Voronoi tessellation of intensity one in the rest of the section.

We realize its typical cell as $Z(o) = Z(o|\Pi \cup \{o\})$ (see Proposition 6.5). We consider the radius R_m of the largest ball included in $Z(o)$ and centered at the origin. We call it *inradius* with the same abuse of language as in the previous example. The radius R_m is larger than r iff for every x, $\|x\| = r$, x is in $Z(o)$, i.e. $B_r(x)$ does not intersect the Poisson point process Π. In other words, for every $r > 0$, we have

$$\mathbf{P}(R_m \geq r) = \mathbf{P}\left(\Pi \cap \bigcup_{x:\|x\|=r} B_r(x) = \emptyset\right)$$
$$= \mathbf{P}(\Pi \cap B_{2r}(o) = \emptyset) = \mathrm{e}^{-2^d \kappa_d r^d}. \tag{6.6}$$

In general, for a deterministic convex set K containing the origin, we define the *Voronoi flower* $F(K) = \bigcup_{x \in K} B_{\|x\|}(x)$ (Fig. 6.3). We can show the following equality:

$$\mathbf{P}(K \subset Z(o)) = \exp\{-\nu_d(F(K))\}.$$

Exercise 6.7. Verify that for any compact subset A of $\mathbb{R}^d$, the Voronoi flowers of A and of its convex hull coincide.

Example 6.3 (Poisson–Voronoi tessellation, typical cell, volume). The next proposition comes from a work due to E.N. Gilbert [187].

Proposition 6.7 (E.N. Gilbert [187]). *For every $t > 0$, we have*

$$\mathrm{e}^{-2^d t} \le \mathbf{P}(\nu_d(Z) \ge t) \le \frac{t-1}{\mathrm{e}^{t-1} - 1}.$$

Proof. Lower bound: It suffices to notice that $\nu_d(Z) \ge \nu_d(B_{R_m}(o))$ and apply (6.6).

Upper bound: Using Markov's inequality, we get for every $\alpha, t \ge 0$

$$\mathbf{P}(\nu_d(Z) \ge t) \le (\mathrm{e}^{\alpha t} - 1)^{-1}(\mathbf{E}(\mathrm{e}^{\alpha \nu_d(Z)}) - 1). \tag{6.7}$$

Let us consider now the quantity $f(\alpha) = \mathbf{E}\left(\int_{Z(o)} \mathrm{e}^{\alpha \kappa_d \|x\|^d}\, dx\right)$. On one hand, we can show by Fubini's theorem that for every $\alpha < 1$,

$$f(\alpha) = \int_{\mathbb{R}^d} \mathrm{e}^{\alpha \kappa_d \|x\|^d} \mathbf{P}(x \in Z(o))\, dx = \int_{\mathbb{R}^d} \mathrm{e}^{(\alpha-1)\kappa_d \|x\|^d}\, dx = \frac{1}{1-\alpha}. \tag{6.8}$$

On the other hand, when comparing $Z(o)$ with the ball centered at the origin and of same volume, we use an isoperimetric inequality to get a lower bound for the same quantity:

$$f(\alpha) \ge \mathbf{E}\left(\int_{B_{(\nu_d(Z(o))/\kappa_d)^{1/d}}(o)} \mathrm{e}^{\alpha \kappa_d \|x\|^d}\, dx\right) = \frac{1}{\alpha}\mathbf{E}(\mathrm{e}^{\alpha \nu_d(Z)}). \tag{6.9}$$

Combining (6.7), (6.8) with (6.9), we obtain that for every $t > 0$,

$$\mathbf{P}(\nu_d(Z) \ge t) \le (\mathrm{e}^{\alpha t} - 1)^{-1} \frac{\alpha}{1-\alpha}$$

and it remains to optimize the inequality in α by taking $\alpha = \frac{t-1}{t}$. □

Exercise 6.8. Show the isoperimetric inequality used above.

Remark 6.7. It has been proved since then (see Theorem 7.10 and [259]) that the lower bound provides the right logarithmic equivalent, i.e.

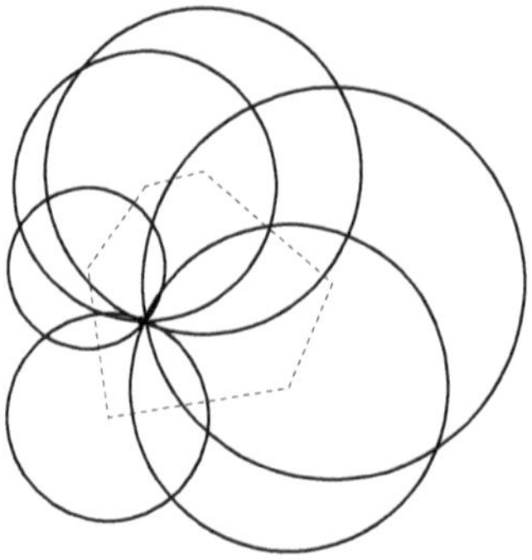

Fig. 6.3 Example of the Voronoi flower of a convex polygon

$$\lim_{t\to\infty} \frac{1}{t} \log \mathbf{P}(\nu_d(Z) \geq t) = -2^d.$$

In other words, distribution tails of the volumes of both the typical cell Z and its inball have an analogous asymptotic behaviour. This is due to D.G. Kendall's conjecture (see the foreword of the book [489]) which was historically written for the two-dimensional Crofton cell. Indeed, it roughly states that cells with a large volume must be approximately spherical. After a first proof by I.N. Kovalenko [312], this conjecture has been rigorously reformulated and extended in many directions by D. Hug, M. Reitzner and R. Schneider (see Theorems 7.9 and 7.11 as well as [255, 256]).

Example 6.4 (Poisson–Voronoi tessellation, typical cell, fundamental frequency in dimension two). This last more exotic example is motivated by the famous question due to Kac [279] back in 1966: "Can one hear the shape of a drum?". In other words, let us consider the Laplacian equation on $Z(o)$ with a Dirichlet condition on the boundary, that is

$$\begin{cases} \Delta f(x) = -\lambda f(x), & \text{if } x \in Z(o), \\ f(x) = 0, & \text{if } x \in \partial Z(o). \end{cases}$$

It has been proved that the eigenvalues satisfy

$$0 < \lambda_1 \leq \lambda_2 \leq \cdots \leq \lambda_n \leq \cdots < \infty.$$

Is it possible to recover the shape of $Z(o)$ by knowing only its spectrum? In particular, λ_1 is called the *fundamental frequency* of $Z(o)$. It is a decreasing function of the convex set considered. When the volume of the domain is fixed, Faber–Krahn's inequality [40] says that it is minimal iff the domain is a ball. In such a case, we have $\lambda_1 = j_0^2/r^2$ where r is the radius of the ball and j_0 is the first positive zero of the Bessel function J_0 [80].

The next theorem which comes from a collaboration with A. Goldman [198] provides an estimate for the distribution function of λ_1 in the two-dimensional case.

Theorem 6.2 (Fundamental frequency of the typical Poisson–Voronoi cell). *Let μ_1 denote the fundamental frequency of the inball of $Z(o)$. Then when $d = 2$, we have*

$$\lim_{t \to 0} t \cdot \log \mathbf{P}(\lambda_1 \leq t) = \lim_{t \to 0} t \cdot \log \mathbf{P}(\mu_1 \leq t) = -4\pi j_0^2.$$

Remark 6.8. The larger $Z(o)$ is, the smaller is λ_1. When evaluating the probability of the event $\{\lambda_1 \leq t\}$ for small t, the contribution comes from the largest cells $Z(o)$. Consequently, the fact that the distribution functions for small λ_1 and small μ_1 have roughly the same behaviour is a new contribution for justifying D.G. Kendall's conjecture.

Remark 6.9. An analogous result holds for the Crofton cell of a Poisson line tessellation in the plane [196].

Sketch of proof.

Step 1. By a Tauberian argument (see [172], Vol. 2, Chap. 13, pages 442–448), we only have to investigate the behaviour of the Laplace transform $\mathbf{E}(\mathrm{e}^{-t\lambda_1})$ when t goes to infinity.

Step 2. We get a lower bound by using the monotonicity of the fundamental frequency ($\lambda_1 \leq \mu_1$) and the explicit distribution of $\mu_1 = j_0^2/R_m^2$.

Step 3. In order to get an upper bound, we observe that almost surely

$$\mathrm{e}^{-t\lambda_1} \leq \varphi(t) = \sum_{n \geq 1} \mathrm{e}^{-t\lambda_n}$$

where φ is called the *spectral function* of $Z(o)$. It is known that the spectral function of a domain is connected to the probability that a two-dimensional Brownian bridge stays in that domain (see for example [279]). More precisely, we denote by W the trajectory of a standard two-dimensional Brownian bridge between 0 and 1 (i.e. a planar Brownian motion starting at 0 and conditioned on being at 0 at time 1) and independent from the point process. We have

$$\varphi(t) = \frac{1}{4\pi t} \int_{Z(o)} \mathbf{P}^W(x + \sqrt{2t}W \subset Z(o))\, dx$$

where $\mathbf{P}^W$ denotes the probability with respect to the Brownian bridge W. We then take the expectation of the equality above with respect to the point process and we get the Laplace transform of the area of the Voronoi flower of the convex hull of W. We conclude by using results related to the geometry of the two-dimensional Brownian bridge [197]. □

Exercise 6.9. In the case of the Crofton cell, express $\varphi(t)$ in function of the Laplace transform of the mean width of W.

6.2 Asymptotic Results for Zero-Cells with Large Inradius

In this section, we shall focus on the example of the Poisson–Voronoi typical cell, but the reader should keep in mind that the results discussed below can be extended to the Crofton cell and more generally to zero-cells of certain isotropic hyperplane tessellations (see [108, Sect. 3]). This section is devoted to the asymptotic behaviour of the typical cell, under the condition that it has large inradius. Though it may seem at first sight a very artificial and restrictive choice, we shall see that it falls in the more general context of D.G. Kendall's conjecture and that this particular conditioning allows us to obtain very precise estimations for the geometry of the cell.

In the first subsection, we are interested in the distribution tail of a particular geometric characteristic that we did not consider before, the so-called *circumscribed radius*. We deduce from the general techniques involved an asymptotic result for the joint distribution of the two radii. In the second subsection, we make the convergence of the cell to the spherical shape more precise by showing limit theorems for some of its characteristics when the inradius goes to infinity. In this section, two fundamental models from stochastic geometry will be introduced as tools for understanding the geometry of the typical cell: random coverings of the circle/sphere and random polytopes generated as convex hulls of Poisson point processes in the ball.

6.2.1 *Circumscribed Radius*

We consider a homogeneous Poisson–Voronoi tessellation of intensity one and we realize its typical cell as $Z(o)$ according to Proposition 6.5. With the same misuse of language as for the inradius, we define the *circumscribed radius* R_M of the typical cell $Z(o)$ as the radius of the smallest ball containing $Z(o)$ and centered at the origin. We first propose a basic way of estimating its distribution and we proceed with a more precise calculation through a technique based on coverings of the sphere which provides satisfying results essentially in dimension two.

6.2.1.1 Estimation of the Distribution Tail

For the sake of simplicity, the following argument is written only in dimension two and comes from an intermediary result of a work due to S. Foss and S. Zuyev [176]. We observe that R_M is larger than $r > 0$ iff there exists x, $\|x\| = r$, which is in $Z(o)$, i.e. such that $B_{\|x\|}(x)$ does not intersect the Poisson point process Π. Compared to the event $\{R_m \geq r\}$, the only difference is that "there exists x" is replacing "for every x" (see Example 2 of Sect. 6.1.3).

In order to evaluate this probability, the idea is to discretize the boundary of the circle and consider a deterministic sequence of balls $B_{\|z_k\|}(z_k)$, $0 \le k \le (n-1)$, $n \in \mathbb{N} \setminus \{0\}$ with $z_k = r(\cos(2\pi k/n), \sin(2\pi k/n))$. We call the intersection of two consecutive such disks a petal. If $R_M \ge r$, then one of these n petals has to be empty. We can calculate the area of a petal and conclude that for every $r > 0$, we have

$$\mathbf{P}(R_M \ge r) \le n \exp\{-r^2(\pi - \sin(2\pi/n) - (2\pi/n))\}. \tag{6.10}$$

In particular, when we look at the chord length in one fixed direction, i.e. the length l_u of the largest segment emanating from the origin in the direction u and contained in $Z(o)$, we have directly for every $r > 0$,

$$\mathbf{P}(l_u \ge r) = \exp\{-\pi r^2\},$$

which seems to provide the same logarithmic equivalent as the estimation (6.10) when n goes to infinity. This statement will be reinforced in the next section.

6.2.1.2 Calculation and New Estimation

This section and the next one present ideas and results contained in [107, 108]. The distribution of R_M can be calculated explicitly: let us recall that $Z(o)$ can be seen as the intersection of half-spaces delimited by random bisecting hyperplanes and containing the origin. We then have $R_M \ge r$ $(r > 0)$ iff the half-spaces do not cover the sphere $\partial B_r(o)$. Of course, only the hyperplanes which are at a distance less than r are necessary and their number is finite and Poisson distributed. The trace of a half-space on the sphere is a spherical cap with a (normalized) angular diameter α which is obviously less than $1/2$ and which has an explicit distribution. Indeed, α can be written in function of the distance L from the origin to the hyperplane via the formula $\alpha = \arccos(L/r)/\pi$. Moreover, the obtained spherical caps are independent. For any probability measure ν on $[0, 1/2]$ and $n \in \mathbb{N}$, we denote by $P(\nu, n)$ the probability to cover the unit-sphere with n i.i.d. isotropic spherical caps such that their normalized angular diameters are ν distributed. Following this reasoning, the next proposition connects the distribution tail of R_M with some covering probabilities $P(\nu, n)$ (Fig. 6.4).

Proposition 6.8 (Rewriting of the distribution tail of R_M). *For every $r > 0$, we have*

$$\mathbf{P}(R_M \ge r) = e^{-2^d \kappa_d r^d} \sum_{n=0}^{\infty} \frac{(2^d \kappa_d r^d)^n}{n!} (1 - P(\nu, n)) \tag{6.11}$$

where ν is a probability measure on $[0, 1/2]$ with the density

$$f_\nu(\theta) = d\pi \sin(\pi\theta) \cos^{d-1}(\pi\theta), \quad \theta \in [0, 1/2].$$

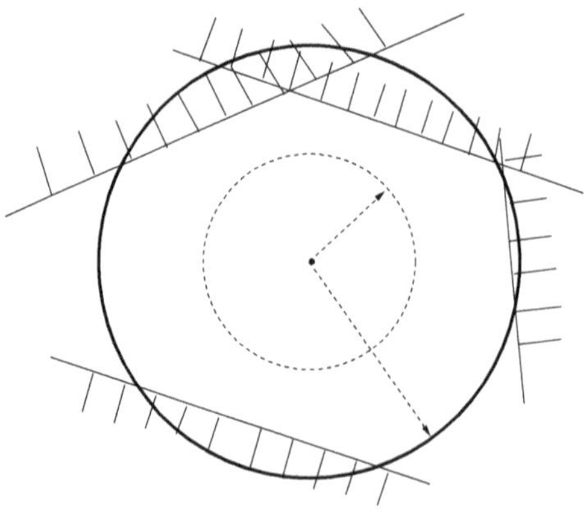

Fig. 6.4 Covering of the circle of radius r when $R_M \geq r$

Exercise 6.10. Verify the calculation of ν and do the same when the Poisson–Voronoi typical cell is replaced by the Crofton cell of a Poisson hyperplane tessellation.

The main question is now to evaluate the covering probability $P(\nu, n)$. In the two-dimensional case, it is known explicitly [475] so the preceding proposition provides in fact the exact calculation of the distribution tail of R_M. Unfortunately, the formula for $P(\nu, n)$ when ν is a continuous measure is not easy to handle but in the particular case where ν is simply a Dirac measure at $a \in [0, 1]$ (i.e. all circular arcs have a fixed length equal to a), then it has been proved by W.L. Stevens [485] with very elementary arguments that for every $n \in \mathbb{N}^*$, we have

$$P(\delta_a, n) = \sum_{k=0}^{n} (-1)^k \binom{n}{k} (1 - ka)_+^{n-1} \tag{6.12}$$

where $x_+ = \max(x, 0)$ for every $x \in \mathbb{R}$. In particular, it implies the following relation for every $a \in [0, 1]$

$$\lim_{n \to \infty} \frac{1 - P(\delta_a, n)}{n(1 - a)^{n-1}} = 1.$$

Exercise 6.11. Calculate $P((1 - p)\delta_0 + p\delta_a, n)$ for $a, p \in [0, 1], n \in \mathbb{N}^*$

In higher dimensions, no closed formula is currently available for $P(\nu, n)$. The case where $\nu = \delta_a$ with $a > 1/2$ has been solved recently [103], otherwise bounds do exist in the particular case of a deterministic radius of the spherical caps [188, 222].

In dimension two, we can use Proposition 6.8 in order to derive an estimation of the distribution tail which is better than (6.10).

Theorem 6.3 (Distribution tail estimate of R_M in dimension two). *For a sufficiently large r, we have*

$$2\pi r^2 e^{-\pi r^2} \leq \mathbf{P}(R_M \geq r) \leq 4\pi r^2 e^{-\pi r^2}. \tag{6.13}$$

Sketch of proof. When using (6.11), we have to estimate $P(\nu, n)$, with ν chosen as in Proposition 6.8, but possibly without considering a too complicated explicit formula. In particular, since the asymptotic equivalent (6.12) for $P(\delta_a, n)$ seems to be quite simple, we aim at replacing the covering probability $P(\nu, n)$ with a covering probability $P(\delta_a, n)$ where a is equal to $1/4$, i.e. the mean of ν.

The problem is reduced to the investigation under which conditions we can compare two different covering probabilities $P(\mu_1, n)$ and $P(\mu_2, n)$ where μ_1, μ_2 are two probability measures on [0, 1]. We recall that μ_1 and μ_2 are said to be *ordered according to the convex order*, i.e. $\mu_1 \prec_{\text{conv}} \mu_2$, if $\langle f, \mu_1\rangle \leq \langle f, \mu_2\rangle$ for every convex function $f : [0,1] \to \mathbb{R}$ [374] (where $\langle f, \mu_1\rangle = \int f d\mu_1$). In particular, Jensen's inequality says that $\delta_a \prec_{\text{conv}} \nu$ and we can easily prove that $\nu \prec_{\text{conv}} \frac{1}{2}(\delta_0 + \delta_{2a})$. The next proposition shows how the convex ordering of distributions implies the ordering of the underlying covering probabilities.

Proposition 6.9 (Ordering of covering probabilities). *If $\nu_1 \prec_{\text{conv}} \nu_2$, then for every $n \in \mathbb{N}$, $P(\nu_1, n) \leq P(\nu_2, n)$.*

Exercise 6.12. Find a heuristic proof of Proposition 6.9.

Thanks to this proposition and the remark above, we can write

$$P(\delta_{1/4}, n) \leq P(\nu, n) \leq P((\delta_0 + \delta_{1/2})/2, n),$$

then insert the two bounds in the equality (6.11) and evaluate them with Stevens' formula (6.12). □

Remark 6.10. Numerical estimates of $\mathbf{P}(R_M \geq r)$ with the formula (6.11) indicate that $\mathbf{P}(R_M \geq r)$ should be asymptotically equivalent to the upper bound of (6.13).

6.2.1.3 Distribution Conditionally on the Inradius

Why should we be interested in the behaviour of the typical cell when conditioned on the value of its inradius?

First, it is one of the rare examples of conditioning of the typical cell which can be made completely explicit. Indeed, conditionally on $\{R_m \geq r\}$, any point x of the Poisson point process is at distance larger than $2r$ from o so the typical cell $Z(o)$ is equal in distribution to the zero-cell associated with the bisecting hyperplanes of the segments $[o, x]$ where x is any point of a homogeneous Poisson point process in $B_{2r}(o)^c$.

Conditionally on $\{R_m = r\}$, the distribution of $Z(o)$ is obtained as previously, but with an extra-bisecting hyperplane generated by a deterministic point x_0 at distance $2r$.

The second reason for investigating this particular conditioning is that a large inradius implies a large typical cell. In other words, having R_m large is a particular case of the general setting of D.G. Kendall's conjecture (see Remark 6.7). But we can be more precise about how the typical cell is converging to the spherical shape. Indeed, the boundary of the polyhedron is included in an annulus between the two radii R_m and R_M and so the order of decreasing of the difference $R_M - R_m$ will provide a satisfying way of measuring the closeness of $Z(o)$ to a sphere. The next theorem provides a result in this direction.

Theorem 6.4 (Asymptotic joint distribution of (R_m, R_M)). *There exists a constant $c > 0$ such that for every $\frac{d-1}{d+1} < \delta < 1$, we have*

$$\mathbf{P}(R_M \geq r + r^{\delta} \mid R_m = r) = \mathcal{O}(\exp\{-cr^{\beta}\}), \quad r \to \infty, \tag{6.14}$$

where $\beta = \frac{1}{2}[(d-1) + \delta(d+1)]$.

Sketch of proof in dimension two. The joint distribution of the couple (R_m, R_M) can be obtained explicitly via the same method as in Proposition 6.8. Indeed, the quantity $\mathbf{P}(R_M \geq r + s \mid R_m = r)$ can be rewritten as the probability of not covering the unit-sphere with random i.i.d. and uniform spherical caps. The only difference lies in the common distribution of the angular diameters of the caps which will now depend upon r since bisecting hyperplanes have to be at least at distance r from the origin. In dimension two, the covering probability can be estimated with an upper-bound due to L. Shepp [470], which implies the estimation (6.14).

Unfortunately, the method does not hold in higher dimensions because of the lack of information about random coverings of the sphere. Nevertheless, a different approach will be explained in the next section in order to extend (6.14) to $d \geq 3$.

□

Exercise 6.13. For $d = 2$, estimate the minimal number of sides of the Poisson–Voronoi typical cell conditioned on $\{R_m = r\}$.

Remark 6.11. This roughly means that the boundary of the cell is included in an annulus centered at the origin and of thickness of order $R_m^{-(d-1)/(d+1)}$. The next problem would be to describe the shape of the polyhedron inside this annulus. For instance, in dimension two, a regular polygon which would be exactly "inscribed" in an annulus of thickness $R_m^{-1/3}$ would have about $R_m^{2/3}$ sides. Is it the same growth rate as the number of sides of the typical cell? The next section will be devoted to this problem. In particular, we shall see that indeed, this quantity behaves roughly as if the typical cell would be a deterministic regular polygon.

6.2.2 *Limit Theorems for the Number of Hyperfaces*

This section is based on arguments and results which come from a collaboration with T. Schreiber and are developed in [108–110]. When the inradius goes to infinity, the shape of the typical Poisson–Voronoi cell becomes spherical. In particular its boundary is contained in an annulus with a thickness going to zero and thus we aim at being more specific about the evolution of the geometry of the cell when R_m is large. For the sake of simplicity, we focus essentially on a particular quantity, which is the number of hyperfaces, but our methods can be generalized to investigate other characteristics, as emphasized in the final remarks.

6.2.2.1 Connection with Random Convex Hulls in the Ball

We start with the following observation: in the literature, there are more limit theorems available for random polytopes constructed as convex hulls of a Poisson point process than for typical cells of stationary tessellations (cf. Sects. 7.1 and 8.4.2). Models of random convex hulls have been probably considered as more natural objects to be constructed and studied. Our aim is first to connect our model of typical Poisson–Voronoi cell with a classical model of a random convex hull in the ball and then work on this possible link between the two in order to extend what is known about random polytopes and solve our current problem.

Conditionally on $\{R_m \geq r\}$, the rescaled typical cell $\frac{1}{r}Z(o)$ is equal in distribution to the zero-cell of a hyperplane tessellation generated by a Poisson point process of intensity measure $(2r)^d \mathbf{1}(x \in B_1(o)^c)\, \nu_d(dx)$ [109]. In other words, via this scaling we fix the inradius of the polyhedron whereas the intensity of the underlying hyperplane process outside of the inball is now the quantity which goes to infinity.

The key idea is then to apply a geometric transformation to $\frac{1}{r}Z(o)$ in order to get a random convex hull inside the unit-ball $B_1(o)$. Let us indeed consider the inversion I defined by

$$I(x) = \frac{x}{\|x\|^2}, \quad x \in \mathbb{R}^d \setminus \{o\}.$$

In the following lines, we investigate the action of I on points, hyperplanes and the cell itself. The Poisson point process of intensity measure $(2r)^d \mathbf{1}(x \notin B_1(o))\, \nu_d(dx)$ is sent by I to a new Poisson point process Y_r of intensity measure $(2r)^d \mathbf{1}(x \in B_1(o)) \frac{1}{\|x\|^{2d}}\, \nu_d(dx)$. The hyperplanes are sent to spheres containing the origin and included in the unit ball, i.e. spheres $\partial B_{\|x\|/2}(x/2)$ where x belongs to the new Poisson point process Y_r in $B_1(o)$. The boundary of the rescaled typical cell $\frac{1}{r}Z(o)$ is sent to the boundary of a certain Voronoi flower, i.e. the union of balls $B_{\|x\|/2}(x/2)$ where x belongs to Y_r. In particular, the number of hyperfaces of the typical cell $Z(o)$ remains unchanged after rescaling and can also be seen through the action of the inversion I as the number of portions of spheres on the boundary of the Voronoi flower in $B_1(o)$, that is the number of extreme points of the convex hull of the process Y_r. Indeed, it can be verified that the ball $B_{\|x\|/2}(x/2)$ intersects the boundary of the Voronoi flower of Y_r iff there exists a support hyperplane of the convex hull of Y_r which contains x.

6.2.2.2 Results

Let N_r be a random variable distributed as the number of hyperfaces of the typical cell $Z(o)$ conditioned on $\{R_m = r\}$.

We are now ready to derive limit theorems for the behaviour of N_r when r goes to infinity:

Theorem 6.5 (Limit theorems for the number of hyperfaces). *There exists a constant $a > 0$ (known explicitly) depending only on d such that*

$$ar^{-\frac{d(d-1)}{d+1}} N_r \to 1 \quad \text{in } L^1 \text{ and a.s. as } r \to \infty.$$

Moreover, the number N_r satisfies a central limit theorem when $r \to \infty$ as well as a moderate-deviation result: for every $\varepsilon > 0$,

$$\liminf_{r\to\infty} \frac{1}{\log(r)} \log\left(-\log\left(\mathbf{P}\left\{\left|\frac{N_r}{\mathbf{E}N_r} - 1\right| \geq \varepsilon\right\}\right)\right) \geq \frac{d-1}{3d+5}.$$

Sketch of proof. The first two steps are devoted to proving the results for the number $\widetilde{N}_r$ of hyperfaces of $Z(o)$ conditioned on $\{R_m \geq r\}$. In the last step, we explain how to adapt the arguments for the number N_r.

Step 1. We use the action of the inversion I to rewrite N_r as the number of vertices of the convex hull of the Poisson point process Y_r of intensity measure $(2r)^d \mathbf{1}(x \in B_1(o)) \frac{1}{\|x\|^{2d}} \nu_d(dx)$ in the unit ball. Limit theorems for the number of extreme points of a homogeneous set of points in the ball are classically known in the literature: indeed, a first law of large numbers has been established in [421] and generalized in [418]. A central limit theorem has been proved in [419] and extended by a precise variance estimate in [459]. Finally, a moderate deviations-type result has been provided in [110,502] (see also Sects. 7.1 and 8.4.2 for more details).

Step 2. The only problem here is that we are not in the classical setting of all these previous works since the process Y_r is not homogeneous. Nevertheless, it can be overcome by emphasizing two points: first, when $\|x\|$ is close to one, the intensity measure of Y_r is close to $(2r)^d \nu_d(dx)$ and secondly, with high probability, only the points near the boundary of the unit sphere will be useful to construct the convex hull. Indeed, for any Poisson point process of intensity measure $\lambda f(\|x\|) \nu_d(dx)$ with $f : (0,1) \to \mathbb{R}_+$ a function such that

$\lim_{t\to 1} f(t) = 1$, it can be stated that the associated convex hull K_λ satisfies the following: there exist constants $c, c' > 0$ such that for every $\alpha \in (0, \frac{2}{d+1})$, we have

$$\mathbf{P}(B_{1-c\lambda^{-\alpha}}(o) \not\subseteq K_\lambda) = \mathcal{O}(\exp\{-c'\lambda^{1-\alpha(d+1)/2}\}). \tag{6.15}$$

The asymptotic result (6.15) roughly means that all extreme points are near $\mathbb{S}^{d-1}$ and included in an annulus of thickness $\lambda^{-2/(d+1)}$. It can be shown in the following way: we consider a deterministic covering of an annulus $B_1(o) \setminus B_{1-\lambda^{-\alpha}}(o)$ with a polynomial number of full spherical caps. When the Poisson point process intersects each of these caps, its convex hull contains $B_{1-c\lambda^{-\alpha}}(o)$ where $c > 0$ is a constant. Moreover, an estimation of the probability that at least one of the caps fails to meet the point process provides the right-hand side of (6.15).

To conclude, the estimation (6.15) allows us to apply the classical limit theory of random convex hulls in the ball even if the point process Y_r is not homogeneous.

Step 3. We recall that the difference between the constructions of $Z(o)$ conditioned either on $\{R_m \geq r\}$ or on $\{R_m = r\}$ is only an extra deterministic hyperplane at distance r from the origin. After the use of a rescaling and of the inversion I, we obtain that $\widetilde{N}_r$ (obtained with conditioning on $\{R_m \geq r\}$) is the number of extreme points of Y_r whereas N_r (obtained with conditioning on $\{R_m = r\}$) is the number of extreme points of $Y_r \cup \{x_0\}$ where x_0 is a deterministic point on $\mathbb{S}^{d-1}$. A supplementary extreme point on $\mathbb{S}^{d-1}$ can "erase" some of the extreme points of Y_r but it can be verified that it will not subtract more than the number of extreme points contained in a d-dimensional polyhedron. Now the growth of extreme points of random convex hulls in a polytope has been shown to be logarithmic so we can consider that the effect of the extra point x_0 is negligible (see in particular [48,49,375] about limit theorems for random convex hulls in a fixed polytope). Consequently, results proved for $\widetilde{N}_r$ in Steps 1–2 hold for N_r as well. □

Remark 6.12. Up to now, the bounds on the conditional distribution of the circumscribed radius (6.14) was only proved in dimension two through techniques involving covering probabilities of the circle. Now applying the action of the inversion I once again, we deduce from (6.15) the generalization of the asymptotic result (6.14) to higher dimensions.

Remark 6.13. The same type of limit theorems occurs for the Lebesgue measure of the region between the typical cell and its inball. Indeed, after application of I, this volume is equal to the μ-measure of the complementary of the Voronoi flower of the Poisson point process in the unit ball, where μ is the image of the Lebesgue measure under I. Limit theorems for this quantity have been obtained in [455,456].

In a recent paper [111], this work is extended in several directions, including variance estimates and a functional central limit result for the volume of the typical cell. Moreover [111] contains an extreme value-type convergence for R_M which adds to (6.14) by providing a three-terms expansion of $(R_M - r)$ conditionally on $\{R_m \geq r\}$, when r goes to infinity. More precisely, it is proved that there exist explicit constants $c_1, c_2, c_3 > 0$ (depending only on the dimension) such that conditionally on $\{R_m \geq r\}$, the quantity

$$\frac{2^{\frac{3d+1}{2}}\kappa_{d-1}}{d+1} r^{\frac{d-1}{2}} (R_M - r)^{\frac{d+1}{2}} - c_1 \log(r) - c_2 \log(\log(r)) - c_3$$

converges in distribution to the Gumbel law when $r \to \infty$.

Chapter 7
Random Polytopes

Daniel Hug

Abstract Random polytopes arise naturally as convex hulls of random points selected according to a given distribution. In a dual way, they can be derived as intersections of random halfspaces. Still another route to random polytopes is via the consideration of special cells and faces associated with random mosaics. The study of random polytopes is based on the fruitful interplay between geometric and probabilistic methods. This survey describes some of the geometric concepts and arguments that have been developed and applied in the context of random polytopes. Among these are duality arguments, geometric inequalities and stability results for various geometric functionals, associated bodies and zonoids as well as methods of integral geometry. Particular emphasis is given to results on the shape of large cells in random tessellations, as suggested in Kendall's problem.

7.1 Random Polytopes

In this chapter, we consider a particular class of set-valued random variables which are denoted as random polytopes. Such random sets usually arise by the application of fundamental geometric operations to basic (random) geometric objects such as (random) points or hyperplanes. The best known example of a random polytope is obtained by taking the convex hull of n random points in Euclidean space. A brief outline of this most common model and of typical problems considered in this context is provided in the introductory Sect. 7.1.1. In Sect. 7.1.2 we describe some new geometric techniques for determining asymptotic mean values of geometric functionals of random polytopes. Then variance estimates and consequences for limit results are briefly discussed in Sect. 7.1.3. Related questions and methods for

D. Hug (✉)
Karlsruhe Institute of Technology, Karlsruhe, Germany
e-mail: daniel.hug@kit.edu

E. Spodarev (ed.), *Stochastic Geometry, Spatial Statistics and Random Fields*, Lecture Notes in Mathematics 2068, DOI 10.1007/978-3-642-33305-7_7,

a dual model involving random hyperplanes are described in Sect. 7.1.4. A different view on random polytopes is provided in Sect. 7.2. Here random polytopes are derived in various ways from random tessellations such as hyperplane tessellations or Voronoi tessellations. For such random polytopes we explore the effect of large size on the shape of typical cells and faces.

7.1.1 Introduction

Random polytopes are basic geometric objects that arise as the result of some random experiment. This experiment may consist in choosing randomly n points in the whole space or in a fixed subdomain according to a given distribution. The convex hull of these points then yields a random polytope. There exist various other definitions of random polytopes, but this definition provides the best known and most extensively studied class of models. The study of random polytopes naturally connects geometry and probability theory and, therefore, recent investigations make use of tools from both disciplines. Moreover, random polytopes are also related and have applications to other fields such as the average case analysis of algorithms, computational complexity theory, optimization theory [158, 178, 179, 529], error correction in signal processing [157], extreme value theory [322, 347], random matrices [335] and asymptotic geometric analysis [10, 139, 303–305]. Various aspects of random polytopes are also discussed in the surveys [420, 447] and in [43, 44].

In order to be more specific, let $K \subset \mathbb{R}^d$ be a convex body (a compact, convex set with nonempty interior) of unit volume. Let $X_1, \dots, X_n$ be independent and uniformly distributed random points in K. The convex hull of $X_1, \dots, X_n$ is a *random polytope* which is denoted by K_n. In dimension $d = 2$ and for special bodies K such as a square, a circle or a triangle, it is indeed possible to determine explicitly the mean number of vertices of K_4, $\mathbf{E} f_0(K_4)$, or the average area of K_3, $\mathbf{E}A(K_3)$. But already in dimension $d = 3$ and for a simplex K, the calculation of the mean volume of K_n is a formidable task which was finally accomplished by Buchta and Reitzner [90]. Since no explicit results can be expected for mean values or even distributions of functionals of random polytopes for an arbitrary convex body K in general dimensions, one is interested in sharp estimates, i.e. the solution of extremal problems with respect to K (cf. [451, Sect. 8.6]), or in asymptotic results as the number n of points increases. As an example of the former problem, one may ask for the minimum of $\mathbf{E}V_d(K_n)$ among all convex bodies K of unit volume. It is known that the minimum is attained precisely if K is an ellipsoid; see [204, 224]. It has been conjectured that $\mathbf{E}V_d(K_n)$ is maximal if K is a simplex, but despite substantial progress due to Bárány and Buchta [45], this is still an unresolved problem. In fact, it is known that a solution to this problem will also resolve the slicing problem (hyperplane conjecture) [362]. Asymptotic results for geometric functionals of random polytopes as the number of random points goes to infinity will be discussed in some detail subsequently.

In the second part of this contribution, we consider random tessellations and certain random polytopes which can be associated with random tessellations. These random polytopes are obtained by selecting specific cells of the tessellation. The cell containing the origin or some kind of average cell are two common choices that will be considered. Lower-dimensional random polytopes that are derived from a given random tessellation by selecting a typical face provide another important choice. We describe the asymptotic or limit shape of certain random polytopes associated with random tessellations given the size of these polytopes is large. For the determination of limit shapes, geometric results and constructions for these convex bodies turn out to be crucial.

7.1.2 Asymptotic Mean Values

Our basic setting is d-dimensional Euclidean space $\mathbb{R}^d$ with scalar product $\langle \cdot, \cdot \rangle$ and induced norm $\| \cdot \|$. A convex body is a compact convex set with nonempty interior. The set of all convex bodies in $\mathbb{R}^d$ is denoted by $\mathcal{K}^d_{\text{conv}}$. For simplicity, in the following we often assume that $K \in \mathcal{K}^d_{\text{conv}}$ has unit volume. The extension to the general case is straightforward. Let $X_1, \ldots, X_n$, $n \in \mathbb{N}$, be independent random points which are uniformly distributed in a convex body $K \in \mathcal{K}^d_{\text{conv}}$ with unit volume, that is $\mathbf{P}(X_i \in A) = \mathcal{H}^d(A \cap K)$ for all $A \in \mathcal{B}(\mathbb{R}^d)$ and $i \in \{1, \ldots, n\}$. Here we write $\mathcal{H}^s$ for the s-dimensional Hausdorff measure in $\mathbb{R}^d$. From such a sample of n random points, we obtain a *random polytope* K_n (see Fig. 7.1) which is defined as the convex hull $K_n := \text{conv}(X_1, \ldots, X_n)$ of $X_1, \ldots, X_n$. The convexity of K ensures that $K_n \subset K$. It is also clear that as n increases, the random polytope K_n should approximate K with increasing precision. The degree of approximation can be quantified by evaluating suitable functionals of convex bodies at K and K_n, respectively. An important functional in this respect is the volume V_d (or d-dimensional Hausdorff measure), since it is continuous, isometry invariant (i.e. invariant with respect to rigid motions) and increasing with respect to set inclusion. Whereas the volume is defined for arbitrary measurable sets, there are other functionals of convex bodies that share all these properties. These are the *intrinsic volumes* V_i, $i = 0, \ldots, d$, which are distinguished among all additive functionals on $\mathcal{K}^d_{\text{conv}}$ by also being continuous (respectively, monotone) and isometry invariant. The intrinsic volumes naturally arise as coefficients of the *Steiner formula*, cf. (2.2). Other functionals of convex bodies that are natural to consider are the diameter or the (Hausdorff, Banach–Mazur, symmetric difference) distance of a convex body to some fixed convex body. Since our focus is on random polytopes, the number f_i of i-dimensional faces of a polytope or the total edge length are other functionals on which random polytopes can be evaluated.

The prototype of a result comparing the volumes of K (which is 1) and K_n is stated in the following theorem on the asymptotic behaviour of the expected value $\mathbf{E}V_d(K_n)$ as $n \to \infty$. It shows that the speed of convergence to $V_d(K)$ is at least of order $n^{-2/(d+1)}$. The result also explains how geometric properties of K determine

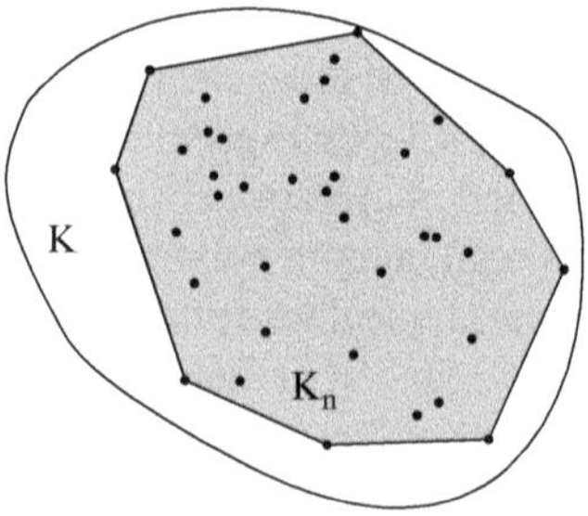

Fig. 7.1 Convex hull of random points

the speed of convergence. In this case, the relevant geometric quantity is based on the generalized Gauss curvature function $x \mapsto H_{d-1}(x)$ with $x \in \partial K$ (the boundary of K). We do not indicate in our notation the dependence on K, if K is clear from the context. For a convex body K with a twice continuously differentiable boundary, the Gauss curvature of K at $x \in \partial K$ is the product of the principal curvatures of K at x (see [445, Sect. 2.5]). For a general convex body, the Gauss curvature is still defined for $\mathcal{H}^{d-1}$-almost all boundary points in Aleksandrov's sense. We explain this notion more carefully. It is a basic fact in convex geometry that $\mathcal{H}^{d-1}$-almost all boundary points y of K are smooth in the sense that there exists a unique support plane H of K through y (see [445, p. 73]). (The nonsmooth boundary points may still form a dense subset of the boundary.) If y is a *smooth boundary point* of K, then there exists a neighbourhood U of y and a nonnegative convex function f defined on H such that $\partial K \cap U$ is the graph of $f|_{U \cap H}$. In particular, we have $f(y) = 0$ and $df(y) = 0$. A much deeper fact, which is a version of Aleksandrov's theorem on the second order differentiability of a convex set (or function, resp.) states that for $\mathcal{H}^{d-1}$-almost all smooth $y \in \partial K$ the corresponding function f even has a second order Taylor expansion at y in the sense that

$$f(y+h) = \frac{1}{2} q(h,h) + o(\|h\|^2),$$

where $h \in u^{\perp}$ (the orthogonal complement of u), u is the unique exterior unit normal of K at y and q is a symmetric bilinear form on $u^{\perp}$ which may be denoted by $d^2 f(y)$. The principal curvatures of K at y are then defined as the eigenvalues of q (or rather of the associated symmetric linear map from $u^{\perp}$ to itself) and the Gauss curvature $H_{d-1}(y)$ is defined as the product of these $d-1$ principal curvatures. Further details and proofs can be found in [445, Notes for Sect. 1.5] and [206, Theorem 2.9 and Lemma 5.2]. A boundary point y as described above is usually called a point of second order differentiability or simply a *normal boundary point* of K.

Theorem 7.1. *For an arbitrary convex body $K \in \mathcal{K}^d_{\mathrm{conv}}$ with $V_d(K) = 1$, it holds*

$$\lim_{n\to\infty} (V_d(K) - \mathbf{E}V_d(K_n))\, n^{\frac{2}{d+1}} = c_d \cdot \Omega_d(K),$$

where

$$\Omega_d(K) := \int_{\partial K} H_{d-1}(x)^{\frac{1}{d+1}} \mathcal{H}^{d-1}(dx)$$

and c_d is an explicitly known constant.

Since the constant is independent of K, it can be determined for a ball. The required calculations can be carried out explicitly by methods of integral geometry (such as the affine Blaschke–Petkantschin formula; cf. Sect. 2.1.2 and [451, Theorem 7.2.7]), and thus one obtains

$$c_d := \frac{(d^2+d+2)(d^2+1)}{2(d+3)\cdot(d+1)!} \cdot \Gamma\left(\frac{d^2+1}{d+1}\right) \cdot \left(\frac{d+1}{\kappa_{d-1}}\right)^{2/(d+1)},$$

where κ_n denotes the volume of the n-dimensional unit ball. The functional Ω_d on convex bodies is known as the *affine surface area*. It has been extensively studied in the last two decades as an important affine invariant in convex geometry; see, e.g. [250,251,337] and the literature cited therein. One of the features that distinguishes the affine surface area from the intrinsic volumes is that it is affine invariant and zero for polytopes and therefore not a continuous functional (though, it is still upper-semicontinuous).

Exercise 7.1. 1. Restate Theorem 7.1 without the assumption $V_d(K) = 1$ and deduce the general result from the given special case by a scaling argument.
2. Give a proof of Theorem 7.1 in the case where K is a ball and $d = 2$, and determine the constant c_2 explicitly.

Theorem 7.1 was proved by Rényi and Sulanke [421] in the plane ($d = 2$) for convex bodies with sufficiently smooth boundaries (three times differentiable) and everywhere positive Gauss curvature. Later it was generalized to higher dimensions (Wieacker considered the unit ball [517]) and established under relaxed smoothness assumptions by Bárány [42]. Affentranger [2] explored the convex hull of random points in the unit ball, but for more general rotation invariant distributions and a larger class of functionals. In the form given here, the result was first stated by Schütt [461]. However, one lemma in the paper by Schütt seems to require that K has a unique tangent plane at each boundary point, i.e. K is smooth (of class C^1). Below we outline a modification and extension of the approach by Schütt which was developed to prove a more general result and which works for arbitrary convex bodies (see [75] for further comments).

Writing $f_0(K_n)$, the number of vertices of K_n, as a sum of indicator functions,

$$f_0(K_n) = \sum_{i=1}^{n} \mathbf{1}(X_i \notin \text{conv}(X_1, \ldots, X_{i-1}, X_{i+1}, \ldots, X_n)),$$

we obtain *Efron's identity*

$$\mathbf{E} f_0(K_n) = n \cdot \mathbf{E} V_d(K \setminus K_{n-1}). \tag{7.1}$$

Exercise 7.2. Carry out the details of the derivation of Efron's identity.

From (7.1) and Theorem 7.1 we deduce an interesting corollary.

Corollary 7.1. *For an arbitrary convex body $K \in \mathcal{K}^d_{\text{conv}}$ with $V_d(K) = 1$, we have*

$$\lim_{n\to\infty} \mathbf{E} f_0(K_n) n^{-\frac{d-1}{d+1}} = c_d \cdot \Omega_d(K).$$

Equation (7.1) explains why the constants in Theorem 7.1 and Corollary 7.1 are the same.

The following results concerning the asymptotic behaviour of various functionals of random polytopes complement the picture. Recall that for a polytope P in $\mathbb{R}^d$ we denote by $f_i(P)$, $i \in \{0, \dots, d-1\}$, the number of i-dimensional faces of P.

1. For a convex body K of class C^2_+ (twice differentiable boundary and positive Gauss curvature), we have

$$\lim_{n\to\infty} \mathbf{E} f_i(K_n) n^{-\frac{d-1}{d+1}} = c_i \cdot \Omega_d(K)$$

   with a constant c_i which is independent of K and $i \in \{0, \dots, d-1\}$ (here and below we assume again that $V_d(K) = 1$).
2. For a polytope $P \subset \mathbb{R}^d$, the convergence in limit results is of a different order compared to the case of a smooth convex body. This is shown by

$$\lim_{n\to\infty} (V_d(P) - \mathbf{E} V_d(P_n)) \frac{n}{\log^{d-1} n} = \tilde{c} \cdot T(P),$$

$$\lim_{n\to\infty} \mathbf{E} f_i(P_n) \frac{1}{\log^{d-1} n} = \hat{c}_i \cdot T(P),$$

   where T is a geometric-combinatorial functional on polytopes, $\tilde{c}$, $\hat{c}_i$ are constants independent of P, and $i \in \{0, \dots, d-1\}$. More explicitly, for a polytope P a tower (or flag) is a sequence $F_0 \subset F_1 \subset \dots \subset F_{d-1}$ of faces F_i of P of dimension i and $T(P)$ is the number of all such towers (flags) of P. These results have been proved in full generality in [45]. It is again an easy exercise to deduce the result for the volume functional from the one for the number of vertices, and vice versa, by applying Efron's identity.
3. For an arbitrary convex body $K \in \mathcal{K}^d_{\text{conv}}$, we have

$$c_1 \cdot \frac{\log^{d-1} n}{n} \le V_d(K) - \mathbf{E} V_d(K_n) \le c_2 \cdot n^{-\frac{2}{d+1}}$$

with constants $0 < c_1 < c_2 < \infty$ which may depend on K but not on n. The order of these estimates is optimal in general. It is known that there exist convex bodies for which the sequence $V(K) - \mathbf{E}V(K_n)$, $n \in \mathbb{N}$, oscillates between these two bounds infinitely often. The upper bound follows from the statement of Theorem 7.1, but it was first deduced by the powerful "cap covering technique" which was explored in depth by Bárány and Larman; see for instance [47] and the contribution by Bárány in [43].

We now explain an extension of Theorem 7.1 which was obtained in [75]. Let ϱ be a probability density with respect to Lebesgue measure on $K \in \mathcal{K}^d_{\mathrm{conv}}$ (if not stated otherwise, speaking of densities we consider densities with respect to Lebesgue measure). The random points $X_1, \dots, X_n$ are assumed to be independent and distributed according to

$$\mathbf{P}(X_i \in A) = \int_A \varrho(x)\,\mathcal{H}^d(dx),$$

where $A \in \mathcal{B}(\mathbb{R}^d)$ with $A \subset K$. We assume that ϱ is positive and continuous on ∂K. Moreover, let $\lambda : K \to \mathbb{R}$ be an integrable function which is continuous on ∂K. In order to indicate the dependence of the expected value and of the probability on the choice of ϱ, we use a corresponding index.

Theorem 7.2. *With the preceding notation, we have*

$$n^{\frac{2}{d+1}}\,\mathbf{E}_\varrho \int_{K\setminus K_n} \lambda(x)\,\mathcal{H}^d(dx) \xrightarrow[n\to\infty]{} c_d \int_{\partial K} \varrho(x)^{\frac{-2}{d+1}} \lambda(x) H_{d-1}(x)^{\frac{1}{d+1}}\,\mathcal{H}^{d-1}(dx). \tag{7.2}$$

The special case where ϱ and λ are constant functions yields the statement of Theorem 7.1. For this reason, the constant here is the same as the one in that previous theorem. It is remarkable that the right-hand side of (7.2) only depends on the values of ϱ and λ on ∂K. The fact that Theorem 7.2 allows us to choose λ appropriately is crucial for the derivation of the subsequent corollary and for the investigation of a dual model of random polytopes determined by random half-spaces, which will be considered below.

A straightforward generalization of Efron's identity, in the present setting, is

$$\mathbf{E}_\varrho f_0(K_n) = n \cdot \mathbf{E}_\varrho \int_{K\setminus K_{n-1}} \varrho(x)\,\mathcal{H}^d(dx), \tag{7.3}$$

which yields the following consequence of Theorem 7.2.

Corollary 7.2. *For $K \in \mathcal{K}^d_{\mathrm{conv}}$ and for a probability density function ϱ on K which is continuous and positive at each point of ∂K, we have*

$$\lim_{n\to\infty} n^{-\frac{d-1}{d+1}}\,\mathbf{E}_\varrho f_0(K_n) = c_d \int_{\partial K} \varrho(x)^{\frac{d-1}{d+1}} H_{d-1}(x)^{\frac{1}{d+1}}\,\mathcal{H}^{d-1}(dx).$$

Exercise 7.3. Verify relation (7.3) and then deduce Corollary 7.2.

Let us give the idea of the proof of Theorem 7.2. Details can be found in [75]. The starting point is the relation

$$\mathbf{E}_{\varrho} \int_{K \setminus K_n} \lambda(x)\, \mathcal{H}^d(dx) = \int_K \mathbf{P}_{\varrho}(x \notin K_n)\lambda(x)\, \mathcal{H}^d(dx), \tag{7.4}$$

which is an immediate consequence of Fubini's theorem. For the proof of Theorem 7.2, it can be assumed that $o \in \mathrm{Int}(K)$. (It is an instructive exercise to deduce the general case by using the translation invariance of Hausdorff measures.) The asymptotic behaviour, as $n \to \infty$, of the right-hand side of (7.4) is determined by points $x \in K$ which are sufficiently close to the boundary of K. In order to make this statement precise, we introduce scaled copies $K(t) := (1-t)K$, $t \in (0,1)$, of K and define $y_t := (1-t)y$ for $y \in \partial K$, see Fig. 7.2. The cap $G(y,t)$ is defined below.

Then it can be shown that

$$\lim_{n\to\infty} n^{\frac{2}{d+1}} \int_{K(n^{-\frac{1}{d+1}})} \mathbf{P}_{\varrho}(x \notin K_n)\lambda(x)\, \mathcal{H}^d(dx) = 0. \tag{7.5}$$

This auxiliary result is based on a geometric estimate of $\mathbf{P}_{\varrho}(x \notin K_n)$, which states that if $t > 0$ is sufficiently small and $y \in \partial K$, then

$$\mathbf{P}_{\varrho}(y_t \notin K_n) \leq \gamma_0 \left(1 - \gamma_1 r(y)^{\frac{d-1}{2}} t^{\frac{d+1}{2}}\right)^n,$$

where $r(y)$ is the radius of the largest ball which contains y and is contained in K, and γ_0, γ_1 are constants independent of y, t and n.

Exercise 7.4. 1. Describe the boundary points y of a polytope for which $r(y) > 0$.
2. Prove that $r(y) > 0$ for $\mathcal{H}^{d-1}$ almost all $y \in \partial K$. A short proof can be based on the fact (mentioned before) that almost all boundary points are normal boundary points.

Moreover, the proof of (7.5) makes essential use of the following disintegration result. In order to state it, let $u(y)$, for $y \in \partial K$, denote an exterior unit normal of K at y. Such an exterior unit normal is uniquely determined for $\mathcal{H}^{d-1}$-almost all boundary points of K.

Lemma 7.1. *If $\delta > 0$ is sufficiently small, $0 \leq t_0 \leq t_1 < \delta$ and $h : K \to [0,\infty]$ is a measurable function, then*

$$\int_{K(t_0)\setminus K(t_1)} h(x)\, \mathcal{H}^d(dx) = \int_{\partial K} \int_{t_0}^{t_1} (1-t)^{d-1} \langle y, u(y)\rangle h(y_t)\, dt\, \mathcal{H}^{d-1}(dy).$$

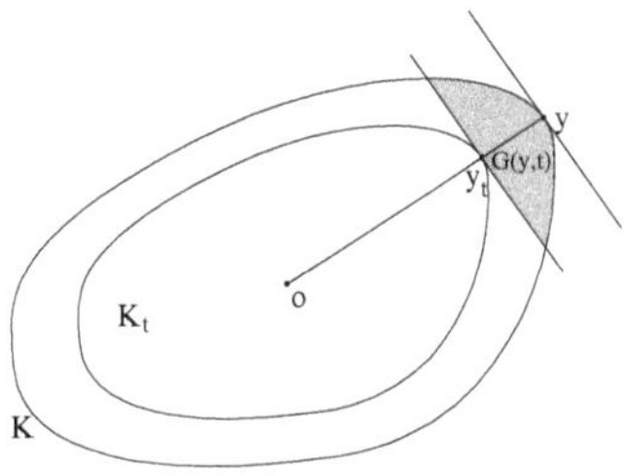

Fig. 7.2 Convex body K with scaled copy $K(t)$ and cap $G(y,t)$

This follows by a straightforward application of the area formula, see [75]. Applying the disintegration result again and using Lebesgue's dominated convergence result, we finally get

$$\lim_{n\to\infty} n^{\frac{2}{d+1}} \cdot \mathbf{E}_\varrho \int_{K\setminus K_n} \lambda(x)\,\mathcal{H}^d(dx) = \int_{\partial K} \lambda(y) J_\varrho(y)\,\mathcal{H}^{d-1}(dy),$$

where

$$J_\varrho(y) = \lim_{n\to\infty} \int_0^{n^{-1/(d+1)}} n^{\frac{2}{d+1}} \langle y, u(y)\rangle\, \mathbf{P}_\varrho(y_t \notin K_n)\,dt$$

for $\mathcal{H}^{d-1}$-almost all $y \in \partial K$. It is now clear that it is sufficient to determine $J_\varrho(y)$ for a normal boundary point y. For this we distinguish two cases. In the case $H_{d-1}(y) = 0$, it can be shown by a direct estimate that $J_\varrho(y) = 0$. The main case is $H_{d-1}(y) > 0$. The assumption of positive Gauss curvature yields some control over the boundary of K in a neighbourhood of y, in particular it admits a local second order approximation of ∂K be osculating paraboloids. For the corresponding analysis, the basic strategy is to show that it is sufficient to consider a small cap $G(y,t)$ (see Fig. 7.2) of K at $y \in \partial K$ whose bounding hyperplane passes through y_t. More precisely, we proceed as follows. First, we reparametrize y_t as $\tilde{y}_s$, in terms of the probability content $s = \int_{G(y,t)} \varrho(x)\,\mathcal{H}^d(dx)$ of the cap $G(y,t)$. This transformation leads to

$$\begin{aligned} J_\varrho(y) &= (d+1)^{-\frac{d-1}{d+1}} (\kappa_{d-1}\varrho(y))^{-\frac{2}{d+1}} H_{d-1}(y)^{\frac{1}{d+1}} \\ &\quad \times \lim_{n\to\infty} \int_0^{n^{-1/2}} n^{\frac{2}{d+1}} \mathbf{P}_\varrho(\tilde{y}_s \notin K_n)\, s^{-\frac{d-1}{d+1}}\,ds. \end{aligned}$$

It is then another crucial step in the proof to show that the remaining integral asymptotically is independent of the particular convex body K, and thus the limit of the integral is the same as for a Euclidean ball. To achieve this, the integral is first approximated, up to a prescribed error of order $\varepsilon > 0$, by replacing $\mathbf{P}_\varrho(\tilde{y}_s \notin K_n)$ by the probability of an event that depends only on a small cap of K at y and on a small number of random points. For this final step in the proof, it is essential that the boundary of K near the normal boundary point y can be suitably approximated by the osculating paraboloid of K at y. In a very rough sense, the proof boils down to reducing the assertion for a general convex body to the case of the Euclidean ball for which it is well known.

It is natural to explore the asymptotic behaviour of not just the volume functional of a random polytope but of more general intrinsic volumes $V_i(K_n)$ as $n \to \infty$. A first result in general dimensions, which concerns the case of the mean width, was obtained by Schneider and Wieacker [452] and later generalized by Böröczky et al. [76]. The mean width of a convex body $K \in \mathcal{K}^d_{\text{conv}}$ is proportional to the first intrinsic volume of K, i.e. $W(K) = \frac{2\kappa_{d-1}}{d\kappa_d} V_1(K)$. More explicitly, it can be described in terms of an average of the support function h_K of K (cf. Chap. 1)

$$W(K) = \frac{2}{d\kappa_d} \int_{\mathbb{S}^{d-1}} h_K(u)\, \mathcal{H}^{d-1}(du) = \frac{1}{d\kappa_d} \int_{\mathbb{S}^{d-1}} (h_K(u) + h_K(-u))\, \mathcal{H}^{d-1}(du).$$

Now we consider again independent and uniformly distributed random points in a convex body K with $V_d(K) = 1$. A major difference in the statement of the next result as compared to the case of the volume functional is the occurrence of an additional assumption on K. We say that a ball rolls freely inside K if for each boundary point $y \in \partial K$ there is a ball of fixed radius $r > 0$ which contains y and is contained in K. In other words, we have $r(y) \geq r$ for all $y \in \partial K$. This condition is equivalent to requiring that K is the Minkowski sum of a ball and another convex body. In particular, all boundary points are smooth points (in fact, the normal map is even Lipschitz).

Theorem 7.3. *Let $K \in \mathcal{K}^d_{\text{conv}}$ be a convex body in which a ball rolls freely. Then*

$$\lim_{n\to\infty} n^{\frac{2}{d+1}} \left(V_1(K) - \mathbf{E}V_1(K_n)\right) = c_d \int_{\partial K} H_{d-1}(x)^{\frac{d+2}{d+1}}\, \mathcal{H}^{d-1}(dx),$$

where c_d is a constant which is explicitly known.

It was shown by the example of a convex body K that is smooth and of class C^∞_+ except for one point that the assumption of a rolling ball cannot be completely removed. In fact, K is constructed in such a way that in a neighbourhood of the origin the graph of the function $f(x) := \|x\|^{1+1/(3d)}$, $x \in \mathbb{R}^{d-1}$, is part of the boundary, so that o is the critical boundary point of K. See Example 2.1 in [76] for further details.

However, the general bounds

$$c_1 \cdot n^{-\frac{2}{d+1}} \leq V_1(K) - \mathbf{E}V_1(K_n) \leq c_2 \cdot n^{-\frac{1}{d}},$$

where c_1, c_2 are positive constants possibly depending on K, are provided by Schneider [443].

A result for all intrinsic volumes was finally established in [77]. In addition to the Gauss curvature, the limit involves the i-th (normalized) elementary symmetric function $H_i(x)$ of the (generalized) principal curvatures of the convex body K at its boundary points $x \in \partial K$. See [445, Sect. 2.5, (2.5.5)] for an introduction to these curvature functions in the framework of convex sets.

Theorem 7.4. *Let $K \subset \mathbb{R}^d$ be a convex body with $V_d(K) = 1$ in which a ball rolls freely, and let $j \in \{2, \dots, d-1\}$. Then*

$$\lim_{n\to\infty} n^{\frac{2}{d+1}} \left(V_j(K) - \mathbf{E}V_j(K_n)\right) = c_{d,j} \int_{\partial K} H_{d-1}(x)^{\frac{1}{d+1}} H_{d-j}(x)\, \mathcal{H}^{d-1}(dx)$$

with a constant $c_{d,j} > 0$ which is independent of K.

As before it is an easy exercise to remove the assumption $V_d(K) = 1$ in the statement of the theorem (cf. [77]). The proof is based on an integral geometric representation of the intrinsic volumes as average projection volumes, that is

$$V_j(K) = \frac{\binom{d}{j}\kappa_d}{\kappa_j\, \kappa_{d-j}} \int_{\alpha_j^d} V_j(K|L)\, \nu_j(dL),$$

where α_j^d is the Grassmannian of all j-dimensional linear subspaces of $\mathbb{R}^d$, ν_j is the (unique) rotation invariant (Haar) probability measure on α_j^d, and, for $L \in \alpha_j^d$, $K|L$ denotes the orthogonal projection of K to L. Here, $V_j(K|L)$ is just the j-dimensional volume (Lebesgue measure) of $K|L$.

7.1.3 Variance Estimates and Limit Results

The results described so far concern mean values and therefore first order properties of random polytopes. Methods of integral geometry are a natural and appropriate tool in this context. Good estimates or even exact results for higher moments have been out of reach until very recently. Here we mention just two surprising results and methods that have been discovered within the last decade.

The first is a far reaching generalization of Efron's identity. It relates the kth moment of the volume $V_d(K_n)$ to moments of the number of vertices $f_0(K_{n+k})$. In particular, it thus follows that $\mathbf{E}(V_d(K_n))^k$ is determined by the distribution of $f_0(K_{n+k})$. Since the result follows from an unexpectedly simple double counting argument, we indicate the approach.

Let $X_1, \dots, X_{n+k}$ be independent and uniformly distributed random points in K. It is sufficient to consider a realization of mutually different points, since this situation is available almost surely. Let $P_{n,k}$ denote the number of k-element subsets of $\{X_1, \dots, X_{n+k}\}$ contained in the convex hull of the other n points. Clearly, $P_{n,k}$ is equal to the number of possibilities of choosing k elements from $\{X_1, \dots, X_{n+k}\} \setminus f_0(\mathrm{conv}(X_1, \dots, X_{n+k}))$, that is from the set of those points which are not vertices, and therefore we get

$$\mathbf{E}P_{n,k} = \mathbf{E}\binom{n+k-f_0(\mathrm{conv}(X_1,,\dots,X_{n+k}))}{k}.$$

On the other hand, by symmetry we easily deduce that

$$\begin{aligned}\mathbf{E}P_{n,k} &= \binom{n+k}{k}\mathbf{P}(X_1,\dots,X_k \in \mathrm{conv}(X_{k+1},\dots,X_{n+k}))\\ &= \binom{n+k}{k}\mathbf{E}(V_d(K_n))^k/(V_d(K))^k.\end{aligned}$$

A comparison of the right-hand sides of these two equations leads to the following result due to Buchta [89].

Theorem 7.5. *Let $K \in \mathcal{K}^d_{\mathrm{conv}}$ be a convex body, and let $n, k \in \mathbb{N}$. Then*

$$\frac{\mathbf{E}(V_d(K_n))^k}{(V_d(K))^k} = \mathbf{E}\left(\prod_{i=1}^{k}\left(1-\frac{f_0(K_{n+k})}{n+i}\right)\right). \tag{7.6}$$

Equation (7.6) can be inverted so that $\mathbf{P}(f_0(K_n) = k)$ is expressed in terms of the moments of the form $\mathbf{E}(V_d(K_j)^{n-j})$, $j = d+1,\dots,k$, whenever $n \geq d+1$ and $k \in \{1,\dots,n\}$. Buchta discusses several consequences for the determination of variances that can be deduced from his relation.

The second example is a method for estimating variances of geometric functionals of random polytopes. This new method which was first discovered by Reitzner [418] is based on the classical Efron–Stein jackknife inequality from statistics (see [165]). It can be described as follows. Let $Y_1, Y_2, \dots$ be a sequence of independent and identically distributed random vectors. Let $S = S(Y_1,\dots,Y_n)$ be a real symmetric function of the first n of these vectors, put $S_i = S(Y_1,\dots,Y_{i-1},Y_{i+1},\dots,Y_{n+1})$, $i \in \{1,\dots,n+1\}$, as well as $S_{(\cdot)} = \frac{1}{n+1}\sum_{i=1}^{n+1} S_i$. The Efron–Stein jackknife inequality then states that

$$\mathbf{var}\, S \leq \mathbf{E}\sum_{i=1}^{n+1}(S_i - S_{(\cdot)})^2 = (n+1)\mathbf{E}(S_{n+1} - S_{(\cdot)})^2. \tag{7.7}$$

Moreover, it is clear that the right-hand side is not decreased if $S_{(\cdot)}$ is replaced by any other function of $Y_1,\dots,Y_{n+1}$.

In a geometric framework, this can be used to show that if f is a functional on convex polytopes (such as volume or number of vertices) and $X_1, X_2, \dots$ are independent and identically distributed random points in K, then

$$\mathbf{var}\, f(K_n) \leq (n+1)\mathbf{E}\left(f(K_{n+1}) - f(K_n)\right)^2.$$

To obtain this, we use $S_{n+1} = f(\text{conv}(X_1, \ldots, X_n)) = f(K_n)$ and replace $S_{(\cdot)}$ by $f(K_{n+1})$ in (7.7). Thus an estimate of the variance follows from an estimate of the cost for adding one further point. Combining integral geometric arguments and geometric estimates, Reitzner thus managed to establish the upper bounds

$$\mathbf{var}\, V_d(K_n) \le c_1(K) \cdot n^{-\frac{d+3}{d+1}}, \qquad \mathbf{var}\, f_0(K_n) \le c_2(K) \cdot n^{\frac{d-1}{d+1}},$$

which later were complemented by lower bounds of the same order with respect to n (see [419] and, for $K = B_1(o)$, formulae (8.68)–(8.70) in Sect. 8.4.2). The upper bound of the variance can now be used to obtain an almost sure convergence result. In fact, let $X_1, X_2, \ldots$ be independent and identically distributed random points in K with $V_d(K) = 1$. Chebyshev's inequality and the variance bound then yield that

$$\begin{aligned} p(n, \epsilon) :=& \mathbf{P}\left(|(V_d(K) - V_d(K_n)) - \mathbf{E}(V_d(K) - V_d(K_n))| \cdot n^{\frac{2}{d+1}} \ge \epsilon\right) \\ \le& c_1(K) \cdot \epsilon^{-2} \cdot n^{-\frac{d-1}{d+1}}. \end{aligned}$$

For $n_k := k^4, k \in \mathbb{N}$, the sum $\sum_{k \ge 1} n_k^{-\frac{d-1}{d+1}}$ is finite and therefore also $\sum_{k \ge 1} p(n_k, \epsilon)$. Hence an application of the Borel–Cantelli theorem together with Theorem 7.1 show that

$$\left(V_d(K) - V_d(K_{n_k})\right) n_k^{\frac{2}{d+1}} \to c_d \cdot \Omega(K)$$

with probability 1 as $k \to \infty$. Since the volume functional is monotone, we have

$$(V_d(K) - V_d(K_{n_k}))\, n_{k-1}^{\frac{2}{d+1}} \le (V_d(K) - V_d(K_n))\, n^{\frac{2}{d+1}} \le (V_d(K) - V_d(K_{n_{k-1}}))\, n_k^{\frac{2}{d+1}}$$

for $n_{k-1} \le n \le n_k$. Using that $n_{k+1}/n_k \to 1$ as $k \to \infty$, we finally obtain that

$$(V_d(K) - V_d(K_n))\, n^{\frac{2}{d+1}} \to c_d \cdot \Omega(K)$$

with probability 1 as $n \to \infty$. Clearly, the constant c_d here must be the same as in Theorem 7.1. Since the number of vertices of the convex hull does not necessarily increase if one point is added, the corresponding argument for f_0 is more delicate. A general convergence result along with various estimates and refinements was finally obtained by Vu [503] who thus completed the investigation in [418].

The Efron–Stein method can also be used for studying the asymptotic behaviour in different models of random polytopes. An important example is provided by Gaussian polytopes P_n which arise as convex hulls of a Gaussian sample, that is of independent random points $X_1, \ldots, X_n$ that follow a d-dimensional standard normal distribution with mean zero and covariance matrix $\frac{1}{2} I_d$. The asymptotics of the mean values $\mathbf{E} f_i(P_n)$ and $\mathbf{E} V_i(P_n)$ have been investigated by Raynaud [414], and by Baryshnikov and Vitale [58], where the latter is based on previous work by Affentranger and Schneider [4]. A direct approach exhibiting the asymptotics of mean values of these and other functionals of Gaussian polytopes is developed in

[253, 254]. There it is shown that, for $i \in \{0, \ldots, d-1\}$,

$$f_i(P_n)(\log n)^{-(d-1)/2} \to \frac{2^d}{\sqrt{d}} \binom{d}{i+1} \beta_{i,d-1} \pi^{(d-1)/2}$$

in probability as $n \to \infty$. Here the constant $\beta_{i,d-1}$ is the internal angle of a regular $(d-1)$-simplex at one of its i-dimensional faces.

An important tool for the proof is the variance estimate

$$\mathbf{var}\, f_i(P_n) \leq c'_d \cdot (\log n)^{(d-1)/2}$$

which again is based on the Efron–Stein method, applied to a functional which counts the number of facets of P_n that can be seen from a random point X not contained in P_n. Another ingredient in the proof are new integral formulae of Blaschke–Petkantschin type. These are also the right tools for showing that

$$\mathbf{var}\, V_i(P_n) \leq c''_d \cdot (\log n)^{(i-3)/2}$$

for $i \in \{1, \ldots, d\}$. From this, one finally deduces that

$$V_i(P_n)(\log n)^{-i/2} \to \binom{d}{i} \frac{\kappa_d}{\kappa_{d-i}}$$

with probability 1 as $n \to \infty$.

More recently, Bárány and Vu [50] obtained central limit theorems for various functionals of Gaussian polytopes. Previously, Hueter [249] had already established a central limit theorem for the number of vertices of P_n. The recent results are based on new techniques for proving central limit theorems involving geometric functionals of random polytopes that were initiated by Vu [502, 503] and developed further, e.g. in [48, 49].

7.1.4 Random Polyhedral Sets

In this subsection, we consider a model of a random polytope which is dual to the classical model of the convex hull of a sample of random points. Instead of points, now the basic ingredients are random hyperplanes, and convex hulls are replaced by intersections of halfspaces bounded by hyperplanes and containing the origin (cf. Fig. 7.3). For a more specific and formal description, let again $K \in \mathcal{K}^d_{\text{conv}}$ be given. We fix a point in the interior of K, for the sake of simplicity we take the origin, hence $o \in \text{Int}(K)$. The parallel body of K of radius 1 is $K_1 := K \oplus B_1(o)$. Let $\mathcal{H}$ denote the space of hyperplanes (with its usual topology) in $\mathbb{R}^d$, and let $\mathcal{H}_K$ be the subset of hyperplanes meeting K_1 but not the interior of K. For $H \in \mathcal{H}_K$, the closed halfspace bounded by H that contains K is denoted by H^-. Let μ denote the motion invariant

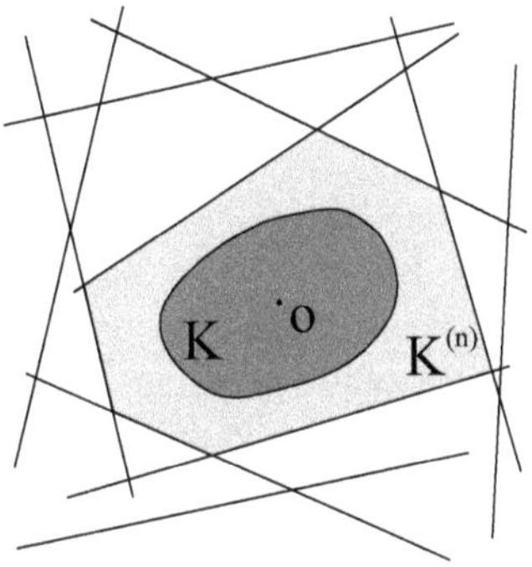

Fig. 7.3 Construction of a random polyhedral set

Borel measure on $\mathcal{H}$, normalized so that $\mu(\{H \in \mathcal{H} : H \cap M \neq \emptyset\})$ is the mean width $W(M)$ of M, for $M \in \mathcal{K}^d_{\text{conv}}$. Let $2\mu_K$ be the restriction of μ to $\mathcal{H}_K$. Since

$$\mu(\mathcal{H}_K) = W(K \oplus B_1(o)) - W(K) = W(B_1(o)) = 2,$$

the measure μ_K is a probability measure. For $n \in \mathbb{N}$, let $H_1, \ldots, H_n$ be independent random hyperplanes in $\mathbb{R}^d$, i.e. independent $\mathcal{H}$-valued random variables on some probability space $(\Omega, \mathcal{A}, \mathbf{P})$, each with distribution μ_K. The possibly unbounded intersection

$$K^{(n)} := \bigcap_{i=1}^{n} H_i^-$$

of the halfspaces H_i^- is a *random polyhedral set*. Subsequently, we shall describe the asymptotic behaviour of the expected value $\mathbf{E}W(K^{(n)} \cap K_1)$. The intersection with K_1 is taken, since $K^{(n)}$ is unbounded with positive probability. Instead of $\mathbf{E}W(K^{(n)} \cap K_1)$, we could consider $\mathbf{E}_1 W(K^{(n)})$, the conditional expectation of $W(K^{(n)})$ under the condition that $K^{(n)} \subset K_1$. Since $\mathbf{E}W(K^{(n)} \cap K_1) = \mathbf{E}_1 W(K^{(n)}) + O(\gamma^n)$ with $\gamma \in (0,1)$, there is no difference in the asymptotic behaviour of both quantities, as $n \to \infty$. We also remark that, for the asymptotic results, the parallel body K_1 could be replaced by any other convex body containing K in its interior; this would only affect some normalization constants. At first thought, one is inclined to believe that a random hyperplane with uniform distribution μ corresponds to a random point with uniform distribution in K. However, the precise connection, which is described below, is more subtle. Subsequently, hyperplanes are written in the form $H(u,t) := \{x \in \mathbb{R}^d : \langle x, u \rangle = t\}$, $u \in \mathbb{R}^d \setminus \{o\}$ and $t \in \mathbb{R}$. More generally, we now consider random hyperplanes with distribution

$$\mu_q := \int_{\mathbb{S}^{d-1}} \int_0^\infty \mathbf{1}\{H(u,t) \in \cdot\} q(t,u)\, dt\, \sigma(du), \tag{7.8}$$

where σ is the rotation invariant probability distribution on $\mathbb{S}^{d-1}$ and $q : [0,\infty) \times \mathbb{S}^{d-1} \to [0,\infty)$ is a measurable function which

1. is concentrated on $D_K := \{(t, u) \in [0, \infty) \times \mathbb{S}^{d-1} : h(K, u) \le t \le h(K_1, u)\}$,
2. is positive and continuous at each point $(h(K, u), u)$ with $u \in \mathbb{S}^{d-1}$,
3. and satisfies $\mu_q(\mathcal{H}_K) = 1$.

Probabilities and expectations with respect to μ_q are denoted by $\mathbf{P}_{\mu_q}$ and $\mathbf{E}_{\mu_q}$, respectively.

In order to obtain results concerning random polyhedral sets, it is appropriate to use the duality between points and hyperplanes, and between convex hulls and intersections of halfspaces. It turns out that in this way also volume and mean width can be related to each other. This rough idea of duality is specified by introducing the polar set A^* of a given nonempty set $A \subset \mathbb{R}^d$. It is defined by

$$A^* := \{x \in \mathbb{R}^d : \langle x, y \rangle \le 1 \text{ for all } y \in A\}.$$

In the particular case when $K \in \mathcal{K}^d_{\text{conv}}$ with $o \in \text{Int}(K)$, we get $K^* \in \mathcal{K}^d_{\text{conv}}$ and also $o \in \text{Int}(K^*)$. Since the realizations of our random polyhedral set are unbounded with positive probability, a definition of the polar of an arbitrary set is needed.

Exercise 7.5. 1. Show that the formation of the polar set is inclusion reversing and always yields a closed set.
2. Determine the polar set of a ball of radius r with centre at the origin.

With a given function q related to the convex body K as described above, we now associate a density ϱ on K^* by

$$\varrho(x) := \begin{cases} \omega_d^{-1} q\left(\|x\|^{-1}, \|x\|^{-1}x\right) \|x\|^{-(d+1)}, & x \in K^* \setminus (K_1)^*, \\ 0, & x \in (K_1)^*, \end{cases} \tag{7.9}$$

where $\omega_d := \mathcal{H}^{d-1}(\mathbb{S}^{d-1})$. The density ϱ is defined in such a way that the distribution of the random polyhedral set $K^{(n)}$, based on K, μ_q, is equal to the distribution of the random polyhedral set $((K^*)_n)^*$ which is obtained as the polar set of the random polytope $(K^*)_n$, based on K^*, ϱ, where ϱ is defined by (7.9); see [75, Proposition 5.1]. For points $x_1, \ldots, x_n \in K^* \setminus (K_1)^*$, we have

$$\begin{aligned} &\mathbf{1}(\text{conv}(x_1, \ldots, x_n)^* \subset K_1)(W(\text{conv}(x_1, \ldots, x_n)^*) - W(K)) \\ &\quad = 2 \cdot \mathbf{1}(\text{conv}(x_1, \ldots, x_n)^* \subset K_1) \int_{K^* \setminus \text{conv}(x_1, \ldots, x_n)} \lambda(x)\, \mathcal{H}^d(dx), \end{aligned}$$

where

$$\lambda(x) := \begin{cases} \omega_d^{-1} \|x\|^{-(d+1)}, & x \in K^* \setminus (K_1)^*, \\ 0, & x \in (K_1)^*. \end{cases}$$

To justify this relation, we need some preparation. For a convex body $L \in \mathcal{K}^d_{\text{conv}}$ with $o \in \text{Int}(L)$, we define the *radial function* of L by

$$\rho(L, u) = \max\{t \ge 0 : tu \in L\}, \qquad u \in \mathbb{S}^{d-1}.$$

Instead of $h_L(u)$ we also write $h(L,u)$ for the support function of L.

Exercise 7.6. Show that for a convex body $L \in \mathcal{K}^d_{\text{conv}}$ with $o \in \text{Int}(L)$ we have $h(L,u) = \rho(L^*,u)$ for $u \in \mathbb{S}^{d-1}$. (See [451, p. 43, Remark 1.7.7])

Now assume that $x_1,\dots,x_n \in K^*\backslash(K_1)^*$ are such that $\text{conv}(x_1,\dots,x_n)^* \subset K_1$. Then $o \in \text{Int}(\text{conv}(x_1,\dots,x_n))$ and we can conclude that

$$\begin{aligned}
&W(\text{conv}(x_1,\dots,x_n)^*) - W(K)\\
&= \frac{2}{d\kappa_d}\int_{\mathbb{S}^{d-1}} h(\text{conv}(x_1,\dots,x_n)^*,u) - h(K,u)\,\mathcal{H}^{d-1}(du)\\
&= \frac{2}{d\kappa_d}\int_{\mathbb{S}^{d-1}} \rho(\text{conv}(x_1,\dots,x_n),u)^{-1} - \rho(K^*,u)^{-1}\,\mathcal{H}^{d-1}(du)\\
&= 2\int_{\mathbb{S}^{d-1}}\int_{\rho(\text{conv}(x_1,\dots,x_n),u)}^{\rho(K^*,u)} \lambda(tu)t^{d-1}\,dt\,\mathcal{H}^{d-1}(du)\\
&= 2\int_{K^*\backslash\text{conv}(x_1,\dots,x_n)} \lambda(x)\,\mathcal{H}^d(dx).
\end{aligned}$$

It is now apparent how Theorem 7.2 can be applied to get

$$\begin{aligned}
&\lim_{n\to\infty} n^{\frac{2}{d+1}}\left(\mathbf{E}_{\mu_q} W(K^{(n)}\cap K_1) - W(K)\right)\\
&= 2\cdot\lim_{n\to\infty} n^{\frac{2}{d+1}}\cdot\mathbf{E}_{\varrho,K^*}\int_{K^*\backslash(K^*)_n}\lambda(x)\,\mathcal{H}^d(dx)\\
&= 2\,c_d\int_{\partial K^*}\varrho(x)^{-\frac{2}{d+1}}\lambda(x)H_{d-1}(K^*,x)^{\frac{1}{d+1}}\,\mathcal{H}^{d-1}(dx)\\
&= 2\,c_d\,\omega_d^{-\frac{d-1}{d+1}}\int_{\partial K^*}\tilde{q}(x)^{-\frac{2}{d+1}}\|x\|^{-d+1}H_{d-1}(K^*,x)^{\frac{1}{d+1}}\,\mathcal{H}^{d-1}(dx),
\end{aligned}$$

where $\tilde{q}(x) := q(\|x\|^{-1},\|x\|^{-1}x)$. The integral thus obtained extends over ∂K^* and involves the Gauss curvature $H_{d-1}(K^*,\cdot)$ of the polar body K^* of K. By the subsequent lemma, the last integral can be transformed into an integral over ∂K which only involves the Gauss curvature $H_{d-1}(K,\cdot)$ of the original body K. Here $h(K,\cdot)$ denotes the support function of K and $\sigma_K(x)$ is the exterior unit normal of K at a smooth boundary point x (recall that almost all boundary points are smooth).

Lemma 7.2. *Let $K \in \mathcal{K}^d_{\text{conv}}$ with $o \in$ Int(K). If $f : [0,\infty)\times\mathbb{S}^{d-1} \to [0,\infty)$ is a measurable function and $\tilde{f}(x) := f\left(\|x\|^{-1},\|x\|^{-1}x\right)$, $x \in \partial K^*$, then*

$$\int_{\partial K^*} \tilde{f}(x)\|x\|^{-d+1} H_{d-1}(K^*, x)^{\frac{1}{d+1}} \mathcal{H}^{d-1}(dx)$$
$$= \int_{\partial K} f(h(K, \sigma_K(x)), \sigma_K(x)) H_{d-1}(K, x)^{\frac{d}{d+1}} \mathcal{H}^{d-1}(dx).$$

Thus we finally arrive at the following theorem which is established in [75].

Theorem 7.6. *Let $K \in \mathcal{K}^d_{\text{conv}}$ with $o \in \text{Int}(K)$, and let $q : [0, \infty) \times \mathbb{S}^{d-1} \to [0, \infty)$ be as described above. Then*

$$\lim_{n\to\infty} n^{\frac{2}{d+1}} \left(\mathbf{E}_{\mu_q} W(K^{(n)} \cap K_1) - W(K)\right)$$
$$= 2\, c_d\, \omega_d{}^{-\frac{d-1}{d+1}} \int_{\partial K} q(h(K, \sigma_K(x)), \sigma_K(x))^{-\frac{2}{d+1}} H_{d-1}(K, x)^{\frac{d}{d+1}} \mathcal{H}^{d-1}(dx).$$

Observe that in the special case $q \equiv 1$ the integral on the right-hand side simplifies considerably. A similar reasoning also implies our next result for the average number of facets of a random polyhedral set.

Theorem 7.7. *Let $K \in \mathcal{K}^d_{\text{conv}}$ with $o \in \text{Int}(K)$, and let $q : [0, \infty) \times \mathbb{S}^{d-1} \to [0, \infty)$ be as described above. Then*

$$\lim_{n\to\infty} n^{-\frac{d-1}{d+1}} \cdot \mathbf{E}_{\mu_q} f_{d-1}(K^{(n)})$$
$$= c_d\, \omega_d{}^{-\frac{d-1}{d+1}} \int_{\partial K} q(h(K, \sigma_K(x)), \sigma_K(x))^{\frac{d-1}{d+1}} H_{d-1}(K, x)^{\frac{d}{d+1}} \mathcal{H}^{d-1}(dx).$$

As for the volume of a random polytope, there are general estimates for the mean width of a random polyhedral set containing a given convex body K,

$$c_1 \cdot \frac{\log^{d-1} n}{n} \leq \mathbf{E} W(K^{(n)}) - W(K) \leq c_2 \cdot n^{-\frac{2}{d+1}}, \tag{7.10}$$

where c_1, c_2 are positive constants, possibly depending on K. These estimates were recently obtained in [79].

Exercise 7.7. Show that the upper bound in (7.10) is implied by Theorem 7.6.

Böröczky and Schneider provide further precise results on the asymptotic behaviour of random polyhedral sets containing a simple polytope, in the case of the mean width functional and the number of i-dimensional faces. The method described here can also be used to improve results for the volume of random polyhedral sets previously obtained by Kaltenbach [284].

7.2 From Random Polytopes to Random Mosaics

In the previous section, random polytopes were obtained by a rather direct construction. Now we consider random polytopes that arise as special cells in random tessellations. Each realization of such a random tessellation is a countable collection of polytopes.

Let G denote the class of nonempty closed subsets of $\mathbb{R}^d$. A *tessellation* of $\mathbb{R}^d$ is a locally finite collection of convex polytopes which cover $\mathbb{R}^d$ and have mutually disjoint interiors. It will be convenient to consider a random tessellation as a particle process X such that $X(\omega)$ is a tessellation of $\mathbb{R}^d$ for $\mathbf{P}$-almost all $\omega \in \Omega$, see Sect. 5.1.2. In addition to the cells, we are also interested in the k-dimensional faces of the cells, for $k = 0, \dots, d$. The collection of all these cells leads to the process $X^{(k)}$ of k-faces of the given tessellation. In particular, we have $X^{(d)} = X$. As an overall assumption, we require the intensity measures $\Lambda^{(k)} := \mathbf{E}X^{(k)}$ to be locally finite. If X is stationary, then $\Lambda^{(k)}$ is a translation invariant measure. In this case, we introduce a centre function (such as the centre of mass), that is a map $c : \mathcal{K}^d_{\text{conv}} \to \mathbb{R}^d$ which is translation covariant in the sense that $c(K + x) = c(K) + x$ for all $K \in \mathcal{K}^d_{\text{conv}}$ and $x \in \mathbb{R}^d$. Then we define the class of centred bodies $\mathcal{K}^0_{\text{conv}} := \{K \in \mathcal{K}^d_{\text{conv}} : c(K) = o\}$ and obtain a representation

$$\Lambda^{(k)} = \lambda^{(k)} \int_{\mathcal{K}^0_{\text{conv}}} \int_{\mathbb{R}^d} \mathbf{1}(x + K \in \cdot)\, \mathcal{H}^d(dx)\, \mathbf{P}_0^{(k)}(dK), \tag{7.11}$$

where $\lambda^{(k)}$ is called the *intensity* of $X^{(k)}$ and $\mathbf{P}_0^{(k)}$ is a probability measure which is called the distribution of the *typical k-face* of X. Instead of $\lambda^{(d)}$ we also write λ and similarly for $\mathbf{P}_0^{(d)}$ and $\mathbf{P}_0$.

Exercise 7.8. Derive the decomposition (7.11) from the fact that the intensity measure $\Lambda^{(k)}$ is translation invariant and locally finite (see Sect. 4.1 in [451]).

A description of these quantities in terms of a spatial average is provided by

$$\lambda^{(k)} \mathbf{P}_0^{(k)} = \lim_{r \to \infty} \frac{1}{V_d(B_r(o))} \mathbf{E} \sum_{K \in X^{(k)}} \mathbf{1}(K - c(K) \in \cdot) \mathbf{1}(K \subset B_r(o)), \tag{7.12}$$

where $B_r(o)$ denotes a ball of radius r with centre at the origin. For an introduction to these notions and for further details, we refer to the monograph by Schneider and Weil [451]. In particular, the representation (7.12) can be deduced from [451, Theorem 4.1.3].

The *typical cell* Z of a stationary random tessellation X is a random polytope with distribution $\mathbf{P}_0$, the *typical k-face* $Z^{(k)}$ is a random k-dimensional polytope with distribution $\mathbf{P}_0^{(k)}$. In addition to these random polytopes obtained by a spatial average, the *zero cell* Z_0 is the cell containing the origin (by stationarity, it is unique almost surely). A basic relation between zero cell and typical cell is

$$\mathbf{E}f(Z_0) = \frac{\mathbf{E}(f(Z)V_d(Z))}{\mathbf{E}V_d(Z)}, \qquad \lambda = (\mathbf{E}V_d(Z))^{-1}, \tag{7.13}$$

which holds for any translation invariant, measurable function $f : \mathcal{K}^d_{\text{conv}} \to [0, \infty]$ (see also Sect. 6.1.2). This relation describes the fact that the distribution of the zero cell is (up to translations) equal to the volume weighted distribution of the typical cell.

Exercise 7.9. Apply Campbell's theorem to

$$\sum_{K \in X} f(K)\mathbf{1}\{o \in \text{Int}(K)\}$$

in order to deduce (7.13); see also [451, Theorem 10.4.3, p. 493].

As a nontrivial consequence, we obtain that

$$\mathbf{E}V_d^k(Z_0) \geq \mathbf{E}V_d^k(Z),$$

which describes in a quantitative way that the zero cell is stochastically larger than the typical cell. This and a more general statement are provided in [451, Theorems 10.4.2 and 10.4.3]. Relation (7.12) obviously implies that

$$\mathbf{P}(Z^{(k)} \in \cdot) = \lim_{r\to\infty} \frac{\mathbf{E}\sum_{K\in X^{(k)}} \mathbf{1}(K - c(K) \in \cdot)\mathbf{1}(K \subset B_r(o))}{\mathbf{E}\sum_{K\in X^{(k)}} \mathbf{1}(K \subset B_r(o))}.$$

In view of (7.13), we get

$$\mathbf{P}(Z_0 - c(Z_0) \in \cdot) = \frac{\mathbf{E}\left[\mathbf{1}(Z - c(Z) \in \cdot)V_d(Z)\right]}{\mathbf{E}V_d(Z)}.$$

If we apply [451, Theorem 4.1.3] to the numerator and to the denominator with the translation invariant functionals $\varphi_1(K) = \mathbf{1}(K - c(K) \in \cdot)V_d(K)$ and resp. $\varphi_2(K) = V_d(K)$, then we obtain

$$\mathbf{P}(Z_0 - c(Z_0) \in \cdot) = \lim_{r\to\infty} \frac{\mathbf{E}\sum_{K\in X} \mathbf{1}(K - c(K) \in \cdot)\mathbf{1}(K \subset B_r(o))V_d(K)}{\mathbf{E}\sum_{K\in X} \mathbf{1}(K \subset B_r(o))V_d(K)}.$$

This discussion also suggests to introduce a *volume-weighted typical k-face* $Z_0^{(k)}$ in such a way that

$$\mathbf{P}(Z_0^{(k)} - c(Z_0^{(k)}) \in \cdot) = \lim_{r\to\infty} \frac{\mathbf{E}\sum_{K\in X^{(k)}} \mathbf{1}(K - c(K) \in \cdot)\mathbf{1}(K \subset B_r(o))V_k(K)}{\mathbf{E}\sum_{K\in X^{(k)}} \mathbf{1}(K \subset B_r(o))V_k(K)}.$$

This can indeed be justified, and an extension of (7.13) for k-faces can be obtained.

There exist very few general relations between the intensities of the k-faces of a stationary random tessellation. A quite general result is the Euler-type relation

$$\sum_{i=0}^{d}(-1)^i \lambda^{(i)} = 0.$$

A more detailed discussion of these random polytopes by means of Palm theory is provided in the recent contributions by Baumstark and Last [59, 60], Schneider [442, 448] and in [261–263].

7.2.1 Hyperplane Tessellations

A prominent example of a tessellation is generated by a random system of hyperplanes. Let X be a *hyperplane process*, that is a point process in $\mathsf{G}(\mathbb{R}^d)$ which is concentrated on the set $\mathcal{H}$ of hyperplanes in $\mathbb{R}^d$. It can be written in the form

$$X = \sum_{H \in X} \delta_H = \sum_{i \geq 1} \delta_{H_i}.$$

The intensity measure $\Lambda = \mathbf{E}X$ is again assumed to be a locally finite measure on $\mathcal{H}$. If X is stationary, then Λ is translation invariant, and therefore can be decomposed in the form

$$\Lambda = \lambda \int_{\mathbb{S}^{d-1}} \int_{\mathbb{R}} \mathbf{1}(H(u,t) \in \cdot)\, dt\, \varphi(du).$$

The measure φ can be chosen as an even probability measure on $\mathcal{B}(\mathbb{S}^{d-1})$. We call λ the intensity and φ the *direction distribution* of X. If X is also isotropic, then Λ is rotation invariant and therefore φ is normalized spherical Lebesgue measure.

For a stationary random hyperplane tessellation, a remarkable result by J. Mecke [351] states that, for $k = 0, \dots, d$,

$$\lambda^{(k)} = \binom{d}{k} \lambda^{(0)}, \qquad \mathbf{E} f_k(Z) = 2^{d-k}\binom{d}{k}.$$

In particular, there are no assumptions on the underlying direction distribution. Despite the similarities between typical and zero cell, a corresponding result for the zero cell only holds under a Poisson assumption. Moreover, there is an essential dependence on the direction distribution φ of the given stationary Poisson hyperplane tessellation X with intensity λ. In order to express this dependence, we introduce the *associated zonoid* Z_X of X via its support function

$$h_{Z_X}(u) = \frac{\lambda}{2} \int_{\mathbb{S}^{d-1}} |\langle u, v\rangle| \, \varphi(dv), \qquad u \in \mathbb{R}^d.$$

Clearly, X is isotropic iff φ is rotation invariant, which is tantamount to Z_X being a Euclidean ball. Recall that the radial function of a convex body K containing the origin in its interior is $\rho(K, u) = \max\{\lambda \geq 0 : \lambda u \in K\}$, $u \in \mathbb{R}^d$. Hence we have $\rho(K, \cdot) = h_{K^*}(\cdot)^{-1}$. Let $\mathcal{H}_{[o,ru]}$ denote the set of all hyperplanes hitting the segment $[o, ru]$. Using the definition of the zero cell, the Poisson assumption, the decomposition of the intensity measure of X and the definition of the associated zonoid, we get

$$\mathbf{P}(\rho(Z_0, u) \leq r) = 1 - \exp\{-\mathbf{E}X(\mathcal{H}_{[o,ru]})\} = 1 - \exp\{-2r h_{Z_X}(u)\}.$$

This finally leads to

$$\begin{aligned}
\mathbf{E}V_d(Z_0) &= \frac{1}{d} \int_{\mathbb{S}^{d-1}} \mathbf{E}\rho(Z_0, u)^d \, \mathcal{H}^{d-1}(du) \\
&= \frac{1}{d} \int_{\mathbb{S}^{d-1}} \frac{d!}{2^d} h_{Z_X}(u)^{-d} \, \mathcal{H}^{d-1}(du) \\
&= 2^{-d} d! V_d(Z_X^*).
\end{aligned}$$

On the other hand, by the Slivnyak–Mecke formula [451, Corollary 3.2.3], we have

$$\begin{aligned}
\mathbf{E} f_0(Z_0) &= \frac{1}{d!} \mathbf{E} \sum_{(H_1,\dots,H_d) \in X^d_{\neq}} \operatorname{card}(Z_0 \cap H_1 \cap \ldots \cap H_d) \\
&= \frac{1}{d!} \int_{\mathcal{H}} \cdots \int_{\mathcal{H}} \mathbf{E} \operatorname{card}(Z_0 \cap H_1 \cap \ldots \cap H_d) \, \Lambda(dH_1) \cdots \Lambda(dH_d) \\
&= \frac{\lambda^d}{d!} \int_{\mathbb{S}^{d-1}} \cdots \int_{\mathbb{S}^{d-1}} \int_{-\infty}^{\infty} \cdots \int_{-\infty}^{\infty} \mathbf{E} \operatorname{card}(Z_0 \cap H(u_1, t_1) \cap \ldots \cap H(u_d, t_d)) \\
&\quad \times dt_1 \ldots dt_d \varphi(du_1) \cdots \varphi(du_d),
\end{aligned}$$

where $X^d_{\neq}$ denotes the set of ordered d-tuples of pairwise distinct hyperplanes from the hyperplane process X. Note that the integral does not change its value if we assume in addition that these hyperplanes have linearly independent normal vectors. Next we carry out a transformation in the d-fold inner integral. For this we can assume that $u_1, \dots, u_d$ are linearly independent. Then the transformation $T : (t_1, \dots, t_d) \mapsto x$ with $\{x\} = H(u_1, t_1) \cap \ldots \cap H(u_d, t_d)$ is injective with inverse $T^{-1}(x) = (\langle x, u_1\rangle, \dots, \langle x, u_d\rangle)$ and Jacobian $JT^{-1}(x) = |\det(u_1, \dots, u_d)|$. Hence we get

$$\mathbf{E} f_0(Z_0) = \frac{\lambda^d}{d!} \mathbf{E} V_d(Z_0) \int_{\mathbb{S}^{d-1}} \dots \int_{\mathbb{S}^{d-1}} |\det(u_1, \dots, u_d)| \, \varphi(du_1) \dots \varphi(u_d).$$

By Schneider and Weil [451, (14.35)], we also have a formula for the volume of a zonoid in terms of its generating measure, i.e.

$$V_d(Z_X) = \frac{\lambda^d}{d!} \int_{\mathbb{S}^{d-1}} \dots \int_{\mathbb{S}^{d-1}} |\det(u_1, \dots, u_d)| \, \varphi(du_1) \dots \varphi(u_d).$$

Altogether, we thus arrive at

$$\mathbf{E} f_0(Z_0) = \frac{d!}{2^d} V_d(Z_X) V_d(Z_X^*).$$

Now we are in the position to apply to fundamental geometric inequalities. If Z is an arbitrary centred zonoid, then the Blaschke–Santaló inequality yields that

$$V_d(Z) V_d(Z^*) \leq \kappa_d^2 \tag{7.14}$$

with equality iff Z is an ellipsoid (see [445, p. 421]). The Mahler inequality for zonoids states that

$$\frac{4^d}{d!} \leq V_d(Z) V_d(Z^*) \tag{7.15}$$

with equality iff Z is an affine cube (parallelepiped) (see [445, p. 427]). It should be observed that the volume product $V_d(K) V_d(K^*)$ for centred convex bodies is a fundamental affine invariant that has been studied intensively. As an immediate consequence of the preceding purely geometric inequalities, we obtain the next theorem. Here a hyperplane mosaic X is called a parallel mosaic if there are d linearly independent vectors such that for almost all realizations of X each hyperplane H of X is orthogonal to one of these vectors.

Theorem 7.8. *Let X be a stationary Poisson hyperplane tessellation with intensity λ and direction distribution φ. Then*

$$2^d \leq \mathbf{E} f_0(Z_0) \leq 2^{-d} d! \kappa_d^2,$$

where equality holds on the left iff X is a parallel mosaic, and equality holds on the right iff up to a linear transformation X is isotropic.

This uniqueness result has recently been strengthened in the form of two stability results in [78]. There it was shown in a precise quantitative form that X must be close to a parallel mosaic if $\mathbf{E} f_0(Z_0)$ is close to 2^d. A similar stability statement has been proved for the upper bound in Theorem 7.8. The crucial ingredients for these improvements are stability versions of the geometric inequalities (7.14) and (7.15) and a stability result for the cosine transform. Extensions of the uniqueness assertions to lower-dimensional faces are explored by Schneider [449].

Next we consider the shape of large cells in Poisson hyperplane tessellations. One motivation is provided by a conjecture due to D.K. Kendall from the 1940s asking whether the conditional law for the shape of the zero cell of a stationary isotropic Poisson line process in the plane, given its area, converges weakly, as the area converges to infinity, to the degenerate law concentrated at the circular shape. While the statement of this conjecture is rather vague, we are interested in

1. Providing a rigorous framework for treating the problem.
2. Extensions to higher dimensions, with line processes replaced by hyperplane processes.
3. Understanding the situation of non-isotropic hyperplane tessellations.
4. Explicit, quantitative estimates—not just limit results.
5. Asymptotic distributions of basic functionals such as $V_d(Z_0)$.

Before we describe in more detail a solution of Kendall's problem, we ask for the shape of a convex body K of given volume such that the inclusion probability $\mathbf{P}(K \subset Z_0)$ is maximal. First, we observe that

$$\mathbf{P}(K \subset Z_0) = \exp\left\{-2\lambda \int_{\mathbb{S}^{d-1}} h_K(u)\,\varphi(du)\right\}.$$

In order to rewrite the right-hand side in geometric terms, we use Minkowski's theorem, which yields the existence of a centrally symmetric convex body $B_X \in \mathcal{K}^d_{\text{conv}}$ such that the surface area measure of B_X equals $\lambda\varphi$; cf. [445, p. 392]. The central symmetry of B_X follows, since φ is an even measure. The convex body B_X is called the *Blaschke body* of X. Then up to the sign the expression in the exponential function is just $2d\,V(B_X[d-1], K)$, that is a multiple of a certain mixed volume (cf. [445, (5.1.18) and (5.3.7)]). By Minkowski's inequality (see [445, p. 317]), we know that the latter can be estimated from below such that we obtain

$$\mathbf{P}(K \subset Z_0) \leq \exp\left\{-2d\,V(B_X)^{\frac{d-1}{d}}\,V(K)^{\frac{1}{d}}\right\}.$$

Equality holds iff K and B_X are homothetic. Hence the inclusion probability is maximized by bodies having the same shape as the Blaschke body of X.

A similar conclusion is available for the typical cell. This was established in [263] by using an idea of R. Miles. The crucial step in the proof consists in showing that, for a convex body $C \in \mathcal{K}^d_{\text{conv}}$,

$$\mathbf{P}(Z \text{ contains a translate of } C) = \exp\{-2d\,V(B_X[d-1], C)\}.$$

The need to compare the shapes of two convex bodies in a quantitative way has led to different deviation measures. One version of such a measure is

$$\vartheta(K, B) := \min\left\{\frac{t}{s} - 1 : sB \subset K + z \subset tB \text{ for some } z \in \mathbb{R}^d,\, s, t > 0\right\},$$

for $K, B \in \mathcal{K}^d_{\text{conv}}$ with nonempty interiors and $o \in \text{Int}(B)$, another one is the Banach–Mazur distance for centrally symmetric convex bodies or norms (cf. [206, p. 207]. Clearly, we have $\vartheta(K, B) = 0$ iff K, B are homothetic.

The following theorem provides a generalized resolution of Kendall's conjecture, which was obtained in [256].

Theorem 7.9. *Let X be a stationary Poisson hyperplane tessellation with intensity λ and Blaschke body B_X. Then there are positive constants $c_0 = c_0(B_X)$ and c_ε such that the following is true: If $\varepsilon \in (0, 1)$ and $a \geq 1$, then*

$$\mathbf{P}(\vartheta(Z_0, B_X) \geq \varepsilon \mid V_d(Z_0) \geq a) \leq c_\varepsilon \cdot \exp\{-c_0 \varepsilon^{d+1} a^{\frac{1}{d}} \lambda\}.$$

The preceding result does not only yield a limit result but provides explicit estimates for fixed parameters ϵ and a. The conditional probability is defined in an elementary way. But it should be observed that the probability of the event $V_d(Z_0) > a$ is decreasing exponentially fast as $a \to \infty$. As a simple consequence we deduce that

$$\lim_{a\to\infty} \mathbf{P}(\vartheta(Z_0, B_X) \geq \varepsilon \mid V_d(Z_0) \geq a) = 0.$$

To state a weak convergence result, we introduce the factor space $\mathcal{S} := \mathcal{K}^d_{\text{conv}}/\sim$, where $K \sim L$ means that K and L are homothetic convex bodies. Note that $\sim$ is an equivalence relation. Hence, if we define the classes $[K] := \{L \in \mathcal{K}^d_{\text{conv}} : K \sim L\}$, $K \in \mathcal{K}^d_{\text{conv}}$, the set $\mathcal{S}$ of all such classes then yields a decomposition of $\mathcal{K}^d_{\text{conv}}$. Let $s_H : \mathcal{K}^d_{\text{conv}} \to \mathcal{S}$, $K \mapsto [K]$, denote the canonical projection. Then, as $a \to \infty$, we have

$$\mathbf{P}(s_H(Z_0) \in \cdot \mid V_d(Z_0) \geq a) \to \delta_{s_H(B_X)},$$

in the sense of the weak convergence of measures. Further details are provided in [256, p. 1144] and a more general framework is depicted in [259, Sect. 4].

For the proof of Theorem 7.9, one writes

$$\mathbf{P}(\vartheta(Z_0, B_X) \geq \varepsilon \mid V_d(Z_0) \geq a) = \frac{\mathbf{P}(\vartheta(Z_0, B_X) \geq \varepsilon, V_d(Z_0) \geq a)}{\mathbf{P}(V_d(Z_0) \geq a)}.$$

The basic aim is to estimate the numerator from above and the denominator from below. This is easy for the denominator, but it turns out that the estimation of the numerator requires first a more general estimate for

$$\mathbf{P}(\vartheta(Z_0, B_X) \geq \varepsilon, V_d(Z_0) \in [a, a(1+h)])$$

for a (sufficiently small) $h > 0$. Hence, a corresponding more general expression has to be treated in the denominator as well. On the geometric side, this analysis uses stability results for Minkowski's inequality and results on the approximation of a convex body by polytopes having few vertices.

A modification of the proof for Theorem 7.9 also leads to the following result, which generalizes a result by Goldman [197] who treated the two-dimensional and isotropic case by completely different methods.

Theorem 7.10. *Under the preceding assumptions, we have*

$$\lim_{a\to\infty} a^{-\frac{1}{d}} \log \mathbf{P}(V_d(Z_0) \geq a) = -2d\, V_d(B_X)^{\frac{d-1}{d}} \lambda .$$

The above results have been generalized subsequently in an axiomatic framework. Here "axiomatic" means that the functionals and distances involved in the statement of results can be chosen quite generally and are subject only to certain natural requirements (axioms) as indicated below. In particular, the following directions have been explored in [259]:

1. Instead of the volume functional, quite general *size functionals* have been considered. These include the intrinsic volumes, the inradius, the thickness, and the minimal width, as particular examples. The class of admissible size functionals is only restricted by a couple of natural conditions such as continuity, homogeneity and monotonicity.
2. Along with more general size functionals various deviation measures turn out to be useful which measure the deviation of shapes. Again an axiomatic treatment is possible which admits a general class of deviation measures to be considered. In particular, these deviation measures should be continuous, nonnegative, homogeneous of degree zero and they should allow to identify certain extremal shapes.
3. Abstract isoperimetric inequalities and corresponding stability results are considered in this general context.
4. The analysis can be extended to not necessarily stationary Poisson hyperplane processes. Thus Poisson–Voronoi tessellations can be studied as well; see [255, 259]. Poisson–Delaunay tessellations can be treated more directly, but also in a very general framework (see [257, 258]).
5. In the same spirit, results for the typical cell are obtained in [256, 260].

Very recently, results for lower-dimensional typical faces have been established. We describe the framework and selected results in Sect. 7.2.3.

7.2.2 *Poisson–Voronoi Mosaics*

In this subsection, we briefly introduce *Voronoi tessellations* which are induced by a Poisson point process and discuss Kendall's problem in this framework. For this purpose, we adjust our notation and denote by $\widetilde{X}$ a stationary Poisson point process in $\mathbb{R}^d$ with intensity λ. The induced random Voronoi tessellation $X := V(\widetilde{X}) := \{C(x \mid \widetilde{X}) : x \in \widetilde{X}\}$ is called Poisson–Voronoi tessellation, compare Sect. 5.1.3.1

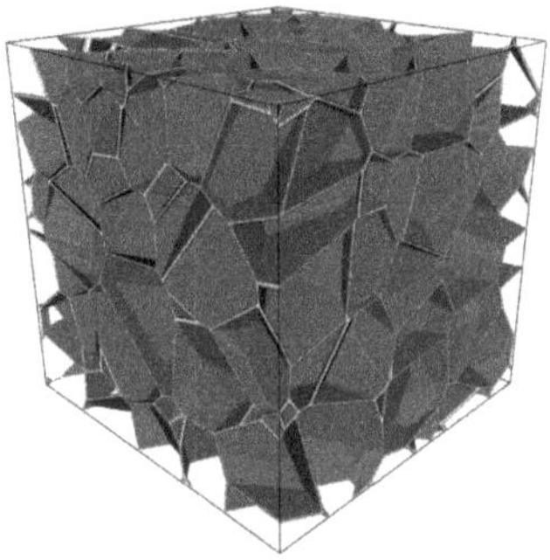

Fig. 7.4 Realization of a three-dimensional Poisson–Voronoi tessellation (Courtesy of Claudia Redenbach)

and Definition 6.5. See Fig. 7.4 for a realization in $\mathbb{R}^3$. Here $C(x \mid \widetilde{X})$ is the cell with nucleus x. As before, the process of k-faces of $V(\widetilde{X})$ is denoted by $X^{(k)}$. Since $\widetilde{X}$ is stationary, so are $X^{(k)}$, $k = 0, \dots, d$. The intensity of $X^{(k)}$ is denoted by $\lambda^{(k)}$.

Let $B \subset \mathbb{R}^d$ be a Borel set with volume 1. Then the distribution of the typical cell Z of X is given by

$$\mathbf{P}(Z \in \cdot) = \frac{1}{\lambda} \cdot \mathbf{E} \sum_{x \in \widetilde{X}} \mathbf{1}(C(x \mid \widetilde{X}) - x \in \cdot)\mathbf{1}(x \in B).$$

Slivnyak's theorem shows that $Z \overset{d}{=} C(o \mid \widetilde{X} \cup \{o\})$, and therefore $Z \overset{d}{=} Z_0(Y)$, where Y is the hyperplane process defined by

$$Y := \left\{ H\left(\|x\|^{-1} x, 2^{-1} \|x\| \right) : x \in \widetilde{X} \setminus \{o\} \right\}.$$

The intensity measure of Y is then given by

$$\mathbf{E}Y(\cdot) = 2^d \lambda \int_{\mathbb{S}^{d-1}} \int_0^\infty \mathbf{1}(H(u,t) \in \cdot) t^{d-1} \, dt \, \mathcal{H}^{d-1}(du); \qquad (7.16)$$

see, for instance, [255].

Exercise 7.10. Derive the representation (7.16) of the intensity measure of the non-stationary Poisson hyperplane process Y.

To estimate the size of the typical cell Z of X, we use the intrinsic volumes $V_1, \dots, V_d$ or the centred inradius R_m. For $K \in \mathcal{K}^d_{\text{conv}}$ with $o \in K$, the latter is defined by $R_m(K) := \max\{r \geq 0 : B_r(o) \subset K\}$. The centred circumradius R_M is defined similarly by $R_M(K) := \min\{r \geq 0 : K \subset B_r(o)\}$. Now the deviation from spherical shape can be measured in terms of

$$\vartheta := \frac{R_M - R_m}{R_M + R_m}.$$

The following result for Poisson–Voronoi tessellations (see [255]) is in the spirit of Kendall's problem.

Theorem 7.11. *Let X be the Poisson–Voronoi tessellation derived from a stationary Poisson point process $\widetilde{X}$ with intensity λ in $\mathbb{R}^d$. Let $k \in \{1, \dots, d\}$. Then there is a constant $c_0 = c_0(d)$ such that the following is true: If $\varepsilon \in (0, 1)$ and $a \geq 1$, then*

$$\mathbf{P}(\vartheta(Z) \geq \varepsilon \mid V_k(Z) \geq a) \leq c \cdot \exp\left\{-c_0 \varepsilon^{(d+3)/2} a^{d/k} \lambda\right\}$$

and

$$\mathbf{P}(\vartheta(Z) \geq \varepsilon \mid R_m(Z) \geq a) \leq c \cdot \exp\left\{-c_0 \varepsilon^{(d+1)/2} a^d \lambda\right\},$$

where $c = c(d, \varepsilon)$.

Further results for Poisson hyperplane tessellations and the functionals considered in Theorem 7.11 are contained in [255]. Related work is due to Kovalenko [312], who considered the two-dimensional case and the area functional, Calka [107], Calka and Schreiber [109, 110] and Baumstark and Last [59]. Since a generalization of the result by Calka will be given subsequently, we provide the result from [107] for comparison.

Theorem 7.12. *For a planar Poisson–Voronoi tessellation derived from a Poisson point process in the plane of intensity 1, there are constants $c_0, c > 0$ such that, for $0 < \alpha < 1/3$,*

$$\mathbf{P}\left(R_M(Z) \geq r + r^{-\alpha} \mid R_m(Z) = r\right) \leq c \cdot \exp\{-c_0 r^{(1-3\alpha)/2}\},$$

as $r \to \infty$.

It is easy to see that $\mathbf{P}(R_m(Z) \geq r) = \exp\{-4\pi r^2\}$. For the conditional probability $\mathbf{P}(R_M(Z) \geq r + s \mid R_m(Z) = r)$ Calka obtains a series representation. The derivation is based on probabilities for the coverage of a circle by random independent and identically distributed arcs. However, the method seems to be restricted to the planar case.

7.2.3 The Shape of Typical Faces

In this section, we describe some of the recent results which were obtained for the distribution of typical faces of Poisson hyperplane and Poisson–Voronoi tessellations. Since various distributional properties of the k-faces of a tessellation depend on the direction of the faces, we define the direction $D(F)$ of a k-dimensional convex set F as the linear subspace parallel to it. Moreover, we write α_k^d for the Grassmann space of k-dimensional linear subspace of $\mathbb{R}^d$. Speaking of distributional properties it should be kept in mind that faces of a given direction may appear with probability zero. Therefore we introduce the condition that the direction is in a small neighbourhood of a fixed direction, or we consider the

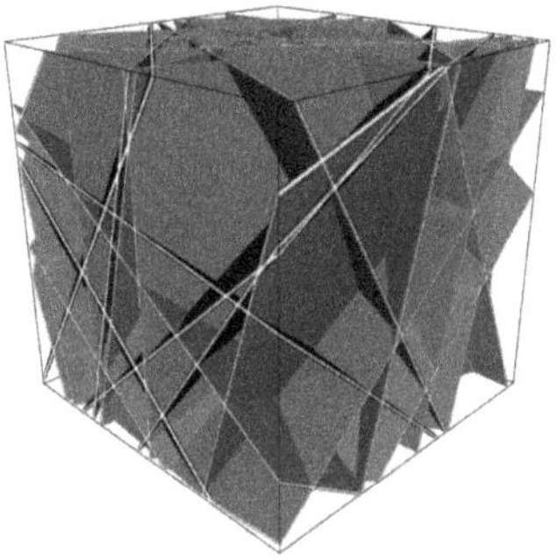

Fig. 7.5 Realization of a three-dimensional Poisson hyperplane tessellation (Courtesy of Claudia Redenbach)

regular conditional probability distribution of the (weighted) typical k-face under the hypothesis that its direction is a given subspace.

We start with a stationary Poisson hyperplane process X with intensity λ and direction distribution φ. A realization in $\mathbb{R}^3$ showing the edges and faces of the induced cells of the hyperplane process is provided in Fig. 7.5.

The *intersection process* of X of order $d-k$ is a stationary process of k-flats obtained by intersecting any $d-k$ of the hyperplanes of X for which the intersection is k-dimensional. Let $\mathbb{Q}_{d-k}$ be the directional distribution of this intersection process. It is known (cf. [451, Sect. 4.4]) that

$$\lambda_{d-k}\mathbb{Q}_{d-k}(A) = \frac{\lambda^{d-k}}{(d-k)!}\int_{(\mathbb{S}^{d-1})^{d-k}} \mathbf{1}(u_1^\perp \cap \ldots \cap u_{d-k}^\perp \in A)$$
$$\times \nabla_{d-k}(u_1,\ldots,u_{d-k})\,\varphi^{d-k}(d(u_1,\ldots,u_{d-k})),$$

where $A \in \mathcal{B}(\alpha_k^d)$, $u^\perp$ is the orthogonal complement of u and $\nabla_{d-k}(u_1,\ldots,u_{d-k})$ is the volume of the parallelepiped spanned by $u_1,\ldots,u_{d-k}$. Then the distribution of the volume-weighted typical k-face of X satisfies

$$\mathbf{P}(Z_0^{(k)} \in A) = \int_{\alpha_k^d} \mathbf{P}(Z_0 \cap L \in A)\,\mathbb{Q}_{d-k}(dL), \quad A \in \mathcal{B}(\alpha_k^d), \tag{7.17}$$

which was established by Schneider [448]. Thus, if $\mathcal{L}$ is a random k-subspace, independent of X, with distribution $\mathbb{Q}_{d-k}$, then $Z_0^{(k)}$ and $Z_0 \cap \mathcal{L}$ are equal in distribution.

For a fixed k-dimensional linear subspace L, we also consider the *section process* $X \cap L$, which is obtained by intersecting all hyperplanes of X with the fixed subspace L. Then $X \cap L$ is a stationary Poisson process of $(k-1)$-flats in L (hence hyperplanes in L); see [451, Sect. 4.4]. It should also be observed that in the preceding formula (7.17), $Z_0 \cap L$ can be replaced by $Z_0(X \cap L)$, where the latter denotes the zero cell of the hyperplane process $X \cap L$ with respect to L.

The important relation (7.17) was recently complemented in [263] by a corresponding relation for the distribution of the typical k-face of X,

$$\mathbf{P}(Z^{(k)} \in A) = \int_{\alpha_k^d} \mathbf{P}(Z(X \cap L) \in A)\, \mathbb{R}_{d-k}(dL), \tag{7.18}$$

where the directional distribution $\mathbb{R}_{d-k}$ of $Z^{(k)}$ is given by

$$\mathbb{R}_{d-k}(A) = \frac{V_{d-k}(Z_X)}{\binom{d}{k} V_d(Z_X)} \int_A V_k(Z_X|L)\, \mathbb{Q}_{d-k}(dL)$$

and Z_X is the zonoid associated with the hyperplane process X and defined by

$$h_{Z_X}(u) = \frac{\lambda}{2} \int_{\mathbb{S}^{d-1}} |\langle u, v\rangle|\, \varphi(dv), \qquad u \in \mathbb{S}^{d-1}.$$

The directional distributions of $Z^{(k)}$ and $Z_0^{(k)}$ are mutually absolutely continuous measures. From these results, one can deduce that the regular conditional distributions of the volume-weighted typical face $Z_0^{(k)}$ and resp. of the typical face $Z^{(k)}$, given the direction of that typical face is equal to L, can be expressed in terms of the section process $X \cap L$, that is

$$\mathbf{P}(Z_0^{(k)} \in A \mid D(Z_0^{(k)}) = L) = \mathbf{P}(Z_0(X \cap L) \in A),$$
$$\mathbf{P}(Z^{(k)} \in A \mid D(Z^{(k)}) = L) = \mathbf{P}(Z(X \cap L) \in A),$$

for $\mathbb{Q}_{d-k}$-almost all $L \in \alpha_k^d$. Here $Z(X \cap L)$ denotes the typical cell of the Poisson hyperplane process $X \cap L$ in L.

Exercise 7.11. Deduce the preceding two relations from (7.17) and (7.18).

In order to extend some of the preceding results to typical k-faces of X, we consider the Blaschke body $B_{X\cap L}$ of the section process $X \cap L$, where $L \in \alpha_k^d$ is chosen from the support of $\mathbb{Q}_{d-k}$, i.e. $L \in \operatorname{supp}(\mathbb{Q}_{d-k})$. The Blaschke body can also be defined as the origin symmetric convex body in L whose area measure with respect to L is given by

$$S^L(B_{X\cap L}, \omega) = \lambda \int_{\mathbb{S}^{d-1}\setminus L^\perp} \mathbf{1}\left(\frac{u|L}{\|u|L\|} \in \omega\right) \|u|L\|\, \varphi(du),$$

for Borel subsets $\omega \subset \mathbb{S}^{d-1} \cap L$.

In the following results taken from [261, 263], the Blaschke body $B_{X\cap L}$ controls the shape of large (weighted) typical k-faces, under the condition that L is the direction of the face.

Theorem 7.13. *Let X be a stationary Poisson hyperplane process with intensity λ and direction distribution φ. Let $k \in \{1, \dots, d-1\}$. For each $L \in \alpha_k^d$, let $C_L \subset L$ be a convex body with $V_k(C_L) = 1$. Then, for $\mathbb{Q}_{d-k}$-almost all $L \in \alpha_k^d$, the conditional probability that the typical k-face $Z^{(k)}$ contains a translate of C_L, under the hypothesis that $D(Z^{(k)}) = L$, is at most*

$$\exp\{-2k V_k(B_{X\cap L})^{1-\frac{1}{k}}\}, \tag{7.19}$$

and it is equal to this value iff C_L is homothetic to $B_{X\cap L}$.

For $a > 0$ and $\mathbb{Q}_{d-k}$-almost all $L \in \alpha_k^d$, let

$$\mathbf{P}(Z_0^{(k)} \in \cdot \mid V_k(Z_0^{(k)}) \geq a,\ D(Z_0^{(k)}) = L)$$

denote the regular conditional probability distribution of $Z_0^{(k)}$ under the hypothesis that $V_k(Z_0^{(k)}) \geq a$ and $D(Z_0^{(k)}) = L$.

Theorem 7.14. *Let X be a stationary Poisson hyperplane process with intensity λ and direction distribution φ. Let $\varepsilon \in (0,1)$ and $a \geq 1$. There exist positive constants $c_0 = c_0(\varphi)$ and $c = c(\varphi, \varepsilon)$ such that*

$$\mathbf{P}(\vartheta(Z_0^{(k)}, B_{X\cap L}) \geq \varepsilon \mid V_k(Z_0^{(k)}) \geq a,\ D(Z_0^{(k)}) = L) \leq c \cdot \exp\{-c_0 \varepsilon^{k+1} a^{1/k}\},$$

for $\mathbb{Q}_{d-k}$-almost all $L \in \alpha_k^d$.

The conditional probability in the next theorem, which is taken from [261], is defined in an elementary way.

Theorem 7.15. *Let X be a stationary Poisson hyperplane process with intensity λ and direction distribution φ. Let $\varepsilon \in (0,1)$ and $a \geq 1$. Let $L \in \alpha_k^d$ be in the support of $\mathbb{Q}_{d-k}$. There exist a constant $c > 0$ and a neighbourhood $N(L)$ of L, both depending only on φ and ε, and a constant $c_0 = c_0(\varphi) > 0$ such that*

$$\mathbf{P}(\vartheta(Z_0^{(k)}, B_{X\cap L}) \geq \varepsilon \mid V_k(Z_0^{(k)}) \geq a,\ D(Z_0^{(k)}) \in N(L)) \leq c \cdot \exp\{-c_0 \varepsilon^{k+1} a^{1/k}\}.$$

The proof requires, in particular, to estimate the deviation of the Blaschke bodies $B_{X\cap L}$ and $B_{X\cap U}$ for different k-dimensional subspaces L and U of $\mathbb{R}^d$. One ingredient of the proof is a geometric stability estimate for Minkowski's uniqueness theorem which had been provided previously. Roughly speaking, for a given even and non-degenerate Borel measure on the unit sphere there exists a unique symmetric convex body having this measure as its area measure. In a stability version of this result it is shown in a quantitative form that the Hausdorff distance of two symmetric convex bodies must be small if the Prohorov distance of the associated area measures is small. We refer to [261] for a detailed argument and further references.

Finally, we turn again to Poisson–Voronoi tessellations and state an extension of a result by Calka [107] mentioned before, to general dimensions and typical k-faces. Let $\widetilde{X}$ be again a stationary Poisson point process in $\mathbb{R}^d$ with intensity λ. The associated Poisson–Voronoi tessellation is $X = V(\widetilde{X})$. In order to explore the asymptotic behaviour of the typical k-face $Z^{(k)}$ of X, some preparations are needed. As a first step, we introduce and describe the k-faces F of an admissible point set $\eta \subset \mathbb{R}^d$, and then the generalized nucleus and the k-co-radius of F with respect to η. In a second step, we start with a Poisson point process (equivalently, the resulting Poisson–Voronoi tessellation) and consider the joint distribution of the typical k-face $Z^{(k)}$ and the typical k-co-radius $R^{(k)}$ of $\widetilde{X}$ with respect to a suitable centre function. For all this we need some more notation.

Let $\eta \subset \mathbb{R}^d$ be a locally finite set in general position (admissible set) whose convex hull equals the whole space. By 'general position' we mean that any $p+1$ points of η are not contained in a $(p-1)$-dimensional plane, for $p = 1, \ldots, d$, and that no $d+2$ points of η lie on some sphere.

Exercise 7.12. Show that the realizations of a stationary Poisson point process almost surely are admissible sets.

Let $k \in \{0, \ldots, d\}$, and choose $d-k+1$ points $x_0, \ldots, x_{d-k} \in \eta$. Then,

1. let $B^{d-k}(x_0, \ldots, x_{d-k})$ be the unique $(d-k)$-ball which contains $x_0, \ldots, x_{d-k}$ in its boundary;
2. let $z(x_0, \ldots, x_{d-k})$ denote the centre of this ball;
3. let $E(x_0, \ldots, x_{d-k})$ be the k-flat through $z(x_0, \ldots, x_{d-k})$ which is orthogonal to the linear subspace $D(B^{d-k}(x_0, \ldots, x_{d-k}))$ parallel to $B^{d-k}(x_0, \ldots, x_{d-k})$.

The set

$$S(x_0, \ldots, x_{d-k}; \eta) := \{y \in E(x_0, \ldots, x_{d-k}) : \mathrm{Int}(B_{\|y-x_0\|}(y)) \cap \eta = \emptyset\}$$

is nonempty iff $F = S(x_0, \ldots, x_{d-k}; \eta)$ is a k-face of the Voronoi mosaic derived from η. Moreover, each k-face of the Voronoi mosaic is obtained in this way. We call $z(x_0, \ldots, x_{d-k}) =: \mathsf{z}(F, \eta)$ the *generalized nucleus* of F, the radius of $B^{d-k}(x_0, \ldots, x_{d-k})$ is called the *co-radius* $R(F, \eta)$ of F. The co-radius of the face F is equal to the distance of the affine hull of F from the nuclei of the neighbouring cells of F. The latter are just the cells whose intersection is equal to F. It should be observed that the generalized nucleus of a k-face need not be contained in that face if $k < d$; see Fig. 7.6. This fact is the reason why the investigation of lower-dimensional faces is much more involved than the case $k = d$ of cells and requires new ideas.

Now the typical k-face and the typical k-co-radius of the Voronoi tessellation $V(\widetilde{X})$, associated with a stationary Poisson point process $\widetilde{X}$ in $\mathbb{R}^d$ with intensity λ, can be introduced by means of Palm distributions. We do not give the details of the construction here, but just describe the result. For the joint distribution of the typical k-face $Z^{(k)}$ and of the typical k-co-radius $R^{(k)}$ of the given Poisson–Voronoi

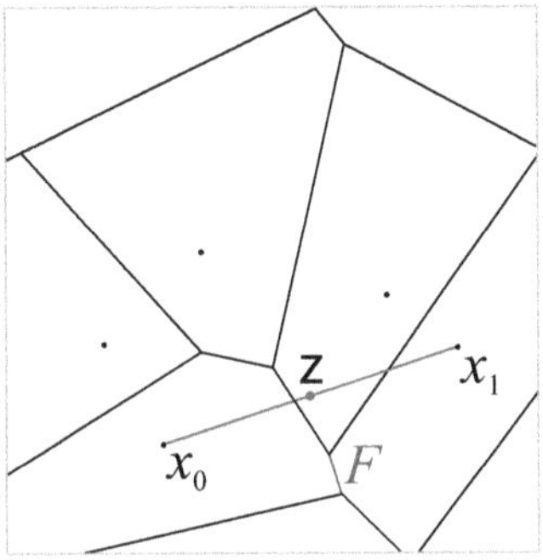

Fig. 7.6 Example for $d = 2$ and $k = 1$: the generalized nucleus $\mathsf{z} = \mathsf{z}(x_0, x_1)$ of a face does not necessarily lie in that face. The midpoint z of the segment $[x_0, x_1]$ does not lie on the associated face F, since the size of F is limited by neighbouring edges

tessellation X, both with respect to the generalized nucleus as centre function, one thus obtains that

$$\begin{aligned}&\mathbf{P}(Z^{(k)} \in A,\ R^{(k)} \in I)\\&= \frac{1}{\lambda^{(k)}} \mathbf{E} \sum_{F \in X^{(k)}} \mathbf{1}_A(F - \mathsf{z}(F, \widetilde{X}))\, \mathbf{1}_I(R(F, \widetilde{X}))\, \mathbf{1}_B(\mathsf{z}(F, \widetilde{X})),\end{aligned}$$

where $A \in \mathcal{B}(\mathcal{K}^d_{\mathrm{conv}})$, $I \in \mathcal{B}([0, \infty))$ and $B \in \mathcal{B}(\mathbb{R}^d)$ with $\lambda^d(B) = 1$. This was rigorously derived in [262, Sect. 2], but the relationship thus obtained also describes the intuitive meaning of both, typical k-face and typical k-co-radius, with respect to the generalized nucleus as centre function. Applying integral geometric transformations of Blaschke–Petkantschin type (in fact a combination of the affine Blaschke–Petkantschin formula and an integral formula involving spheres) and the Slivnyak–Mecke theorem, we can describe the joint distribution by

$$\begin{aligned}&\mathbf{P}(Z^{(k)} \in A,\ R^{(k)} \in I)\\&= C(d,k) \frac{\lambda^{d-k+1}}{\lambda^{(k)}} \int_I \int_{\alpha^d_{d-k}} \mathbf{P}(Z(L, r; \widetilde{X}) \in A) \nu_{d-k}(dL)\, r^{d(d-k)-1}\, dr,\end{aligned}$$

where

$$C(d,k) := \frac{1}{(d-k+1)!} \omega^{d-k}_{d+1} \omega_{k+1} \frac{\kappa_{(d-k)d-2}}{\kappa_{(d-k+1)(d-1)}}$$

and $Z(L, r; \eta)$ is defined by

$$Z(L, r; \eta) = \left\{ y \in L^{\perp} : B^o_{\sqrt{\|y\|^2 + r^2}}(y) \cap \eta = \emptyset \right\}$$

for any admissible set η; here $B_t^o(y)$ denotes an open ball centred at y with radius t and ν_{d-k} is the Haar probability measure on α_{d-k}^d. It can be shown that $Z(L, r; \widetilde{X})$ is almost surely a random polytope or the empty set. This explicit description of the joint distribution is a crucial ingredient in the proof of the following result from [262].

Theorem 7.16. *Let $\widetilde{X}$ be a stationary Poisson point process in $\mathbb{R}^d$ with intensity λ. Let $r \geq 1$, let $k \in \{1, \ldots, d\}$, and choose α with*

$$0 < \alpha < \tfrac{2d-k-1}{k+1}, \qquad \text{so that} \quad \beta := d - (1+\alpha)\tfrac{k+1}{2} > 0.$$

Then there exist constants $c_1 = c_1(d, \lambda)$ and $c_2 = c_2(d)$ such that

$$\mathbf{P}(R_M(Z^{(k)}) > r + r^{-\alpha} \mid R_m(Z^{(k)}) \geq r) \leq c_1 \cdot \exp\{-c_2 \lambda r^{\beta}\}.$$

Observe that the theorem implies that if $rB^{(k)} \subset Z^{(k)}$, then

$$B^{(k)} \subset r^{-1} Z^{(k)} \subset (1 + r^{-1-\alpha}) B^{(k)}$$

with overwhelming probability, where $B^{(k)}$ is a k-dimensional unit ball centred at the origin o.

In the special case $d = k = 2$, we obtain a variant of the result by Calka, since then $0 < \alpha < 1/3$ and $\beta = (1 - 3\alpha)/2$.

Chapter 8
Limit Theorems in Discrete Stochastic Geometry

Joseph Yukich

Abstract We survey two general methods for establishing limit theorems for functionals in discrete stochastic geometry. The functionals are linear statistics with the general representation $\sum_{x \in \mathcal{X}} \xi(x, \mathcal{X})$, where $\mathcal{X}$ is finite and where the interactions of x with respect to $\mathcal{X}$, given by $\xi(x, \mathcal{X})$, are spatially correlated. We focus on subadditive methods and stabilization methods as a way to obtain weak laws of large numbers, variance asymptotics, and central limit theorems for normalized and re-scaled versions of $\sum_{i=1}^{n} \xi(\eta_i, \{\eta_j\}_{j=1}^{n})$, where η_j, $j \geq 1$, are i.i.d. random variables. The general theory is applied to deduce the limit theory for functionals arising in Euclidean combinatorial optimization, convex hulls of i.i.d. samples, random sequential packing, and dimension estimation.

8.1 Introduction

This overview surveys two general methods for establishing limit theorems, including weak laws of large numbers, variance asymptotics, and central limit theorems, for functionals of large random geometric structures. By geometric structures, we mean for example networks arising in computational geometry, graphs arising in Euclidean optimization problems, models for random sequential packing, germ-grain models, and the convex hull of high density point sets. Such diverse structures share only the common feature that they are defined in terms of random points belonging to Euclidean space $\mathbb{R}^d$. The points are often the realization of i.i.d. random variables, but they could also be the realization of Poisson point processes or even Gibbs point processes. There is scope here for generalization to functionals of point processes in more general spaces, including manifolds and general metric

J. Yukich (✉)
Lehigh University, Bethlehem, PA 18015, USA
e-mail: joseph.yukich@lehigh.edu

E. Spodarev (ed.), *Stochastic Geometry, Spatial Statistics and Random Fields*, Lecture Notes in Mathematics 2068, DOI 10.1007/978-3-642-33305-7_8,

spaces, but for ease of exposition we shall usually restrict attention to point processes in $\mathbb{R}^d$. As such, this introductory overview makes few demands involving prior familiarity with the literature.

Our goals are to provide an accessible survey of asymptotic methods involving (a) subadditivity and (b) stabilization and to illustrate the applicability of these methods to problems in discrete stochastic geometry. The treatment of subadditivity parallels that in [524].

8.1.1 Functionals of Interest

Functionals of geometric structures are often formulated as *linear statistics* on finite point sets $\mathcal{X}$ of $\mathbb{R}^d$, that is to say consist of sums represented as

$$H(\mathcal{X}) := H^{\xi}(\mathcal{X}) := \sum_{x \in \mathcal{X}} \xi(x, \mathcal{X}), \tag{8.1}$$

where the function ξ, defined on all pairs $(x, \mathcal{X})$, $x \in \mathcal{X}$, represents the *interaction* of x with respect to input $\mathcal{X}$.

The focus of this chapter is to develop the large n limit theory for the normalized sums

$$n^{-1} H^{\xi}(\{\eta_i\}_{i=1}^n), \tag{8.2}$$

where $\eta_i, i \geq 1$, are i.i.d. with values in $[0,1]^d$. We seek mean and variance asymptotics for (8.2) as well as central limit theorems for $n^{-1/2}(H^{\xi}(\{\eta_i\}_{i=1}^n) - \mathbf{E}H^{\xi}(\{\eta_i\}_{i=1}^n))$, as $n \to \infty$. In nearly all problems of interest, the values of $\xi(x, \mathcal{X})$ and $\xi(y, \mathcal{X})$, $x \neq y$, are not unrelated but, loosely speaking, become more related as the Euclidean distance $\|x - y\|$ becomes smaller. This "spatial dependency" is the chief source of difficulty when developing the limit theory for H^{ξ} on random point sets.

Typical questions motivating this survey, which may be framed in terms of the linear statistics (8.1), include the following:

1. Given i.i.d. points $\eta_1, \ldots, \eta_n$ in the unit cube $[0,1]^d$, what is the asymptotic length of the shortest tour through $\eta_1, \ldots, \eta_n$? To see that this question fits into the framework of (8.1), it suffices to let $\xi(x, \mathcal{X})$ be one half the sum of the lengths of edges incident to x in the shortest tour on $\mathcal{X}$. $H^{\xi}(\mathcal{X})$ is the length of the shortest tour through $\mathcal{X}$.
2. Given i.i.d. points $\eta_1, \ldots \eta_n$ in the unit volume d-dimensional ball, what is the asymptotic distribution of the number of k-dimensional faces, $k \in \{0, 1, \ldots, d-1\}$, in the random polytope given by the convex hull of $\eta_1, \ldots, \eta_n$? To fit this question into the framework of (8.1), we let $\xi_k(x, \mathcal{X})$ be zero if x is not a vertex in the convex hull of $\mathcal{X}$ and otherwise we let it be the product of $(k+1)^{-1}$

and the number of k-dimensional faces containing x. $H^{\xi_k}(\mathcal{X})$ is the number of k-faces in the convex hull of $\mathcal{X}$.

3. Open balls $B_1, \ldots, B_n$ of volume n^{-1} arrive sequentially and uniformly at random in $[0,1]^d$. The first ball B_1 is *packed*, and recursively for $i = 2, 3, \ldots$, the i-th ball B_i is packed iff B_i does not overlap any ball in $B_1, \ldots, B_{i-1}$ which has already been packed. If not packed, the i-th ball is discarded. The process continues until no more balls can be packed. As $n \to \infty$, what is the asymptotic distribution of the number of balls which are packed in $[0,1]^d$? To fit this into the set-up of (8.1), we let $\xi(x, \mathcal{X})$ be equal to one or zero depending on whether the ball with center at $x \in \mathcal{X}$ is accepted or not. $H^{\xi}(\mathcal{X})$ is the total number of accepted balls.

When $\mathcal{X}$ is the realization of a growing point set of random variables, the large scale asymptotic analysis of the sums (8.1) is sometimes handled by M-dependent methods, ergodic theory, or mixing methods; see for example Chap. 10. However, these classical methods, when applicable, may not give explicit asymptotics in terms of the underlying interaction and point densities, they may not yield second order results, or they may not easily yield rates of convergence. Our goal is to provide an abridged treatment of two alternate methods suited to the asymptotic theory of the sums (a) subadditivity and stabilization.

Subadditive methods lean heavily on the self-similarity of the unit cube, but to obtain distributional results, variance asymptotics, and explicit limiting constants in laws of large numbers, one needs tools going beyond subadditivity. When the spatial dependency may be localized, in a sense to be made precise, then this localization yields distributional and second order results, and it also shows that the *large scale macroscopic behaviour of* H^{ξ} on random point sets, for example laws of large numbers and central limit theorems, *is governed by the local interactions involving* ξ.

The subadditive approach, described in detail in the monographs [482, 524], yields a.s. laws of large numbers for problems in Euclidean combinatorial optimization, including the length of minimal spanning trees, minimal matchings, and shortest tours on random point sets. Formal definitions of these archetypical problems are given below. Subadditive methods also yield the a.s. limit theory of problems in computational geometry, including the total edge length of nearest neighbour graphs, the Voronoi and Delaunay graphs, the sphere of influence graph, as well as graphs arising in minimal triangulations and the k-means problem. The approach based on stabilization, originating in Penrose and Yukich [398] and further developed in [57, 395, 396, 400, 402], is useful in proving laws of large numbers, central limit theorems, and variance asymptotics for many of these functionals; as such it provides closed form expressions for the limiting constants arising in the mean and variance asymptotics. This approach has been used to study linear statistics arising in random packing [400], convex hulls [459], ballistic deposition models [57, 400], quantization [460, 525], loss networks [460], high-dimensional spacings [56], distributed inference in random networks [12], and geometric graphs in Euclidean combinatorial optimization [398, 399].

8.1.2 Examples

Letting input $\mathcal{X} := \{x_1, \ldots, x_n\}$ be a finite point set in $\mathbb{R}^d$, functionals and graphs of interest include:

1. *Traveling salesman functional (TSP).* A closed tour on $\mathcal{X}$ or closed Hamiltonian tour is a closed path traversing each vertex in $\mathcal{X}$ exactly once. Let TSP($\mathcal{X}$) be the length of the shortest closed tour T on $\mathcal{X}$. Thus

$$\text{TSP}(\mathcal{X}) := \min_T \sum_{e \in T} |e|, \tag{8.3}$$

where the minimum is over all tours T on $\mathcal{X}$ and where $|e|$ denotes the Euclidean edge length of the edge e. Thus,

$$\text{TSP}(\mathcal{X}) := \min_\sigma \left\{ \|x_{\sigma(n)} - x_{\sigma(1)}\| + \sum_{i=1}^{n-1} \|x_{\sigma(i)} - x_{\sigma(i+1)}\| \right\},$$

where the minimum is taken over all permutations σ of the integers $1, 2, \ldots, n$ and where $\| \cdot \|$ denotes the Euclidean norm.

2. *Minimum spanning tree (MST).* Let MST($\mathcal{X}$) be the length of the shortest spanning tree on $\mathcal{X}$, namely

$$\text{MST}(\mathcal{X}) := \min_T \sum_{e \in T} |e|, \tag{8.4}$$

where the minimum is over all spanning trees T on $\mathcal{X}$.

3. *Minimal matching (MM).* The *minimal matching* on $\mathcal{X}$ has length given by

$$\text{MM}(\mathcal{X}) := \min_\sigma \sum_{i=1}^{n/2} \|x_{\sigma(2i-1)} - x_{\sigma(2i)}\|, \tag{8.5}$$

where the minimum is over all permutations of the integers $1, 2, \ldots, n$. If n has odd parity, then the minimal matching on $\mathcal{X}$ is the minimum of the minimal matchings on the n distinct subsets of $\mathcal{X}$ of size $n - 1$.

4. *k-nearest neighbours graph.* Let $k \in \mathbb{N}$. The k-nearest neighbours (undirected) graph on $\mathcal{X}$, here denoted $G^N(k, \mathcal{X})$, is the graph with vertex set $\mathcal{X}$ obtained by including $\{x, y\}$ as an edge whenever y is one of the k nearest neighbours of x and/or x is one of the k nearest neighbours of y. The k-nearest neighbours (directed) graph on $\mathcal{X}$, denoted $\overrightarrow{G}^N(k, \mathcal{X})$, is the graph with vertex set $\mathcal{X}$

obtained by placing an edge between each point and its k nearest neighbours. Let $\mathrm{NN}(k, \mathcal{X})$ denote the total edge length of $G^N(k, \mathcal{X})$, i.e.,

$$\mathrm{NN}(k, \mathcal{X}) := \sum_{e \in G^N(k, \mathcal{X})} |e|, \tag{8.6}$$

with a similar definition for the total edge length of $\overrightarrow{G}^N(k, \mathcal{X})$.

5. *Steiner minimal spanning tree.* Consider the problem of finding the graph of shortest length which connects the vertices of $\mathcal{X}$. Such a graph is a tree, known as the Steiner minimal spanning tree, and it may include vertices other than those in $\mathcal{X}$. If not, the graph coincides with the minimal spanning tree graph. The total edge length of the Steiner minimal spanning tree on $\mathcal{X}$ is

$$\mathrm{ST}(\mathcal{X}) := \min_S \sum_{e \in S} |e|, \tag{8.7}$$

where the minimum ranges over all connected graphs S on $\mathcal{X}$.

6. *Minimal semi-matching.* A semi-matching on $\mathcal{X}$ is a graph in which all vertices have degree 2, with the understanding that an isolated edge between two vertices represents two copies of that edge. The graph thus contains tours with an odd number of edges as well as isolated edges. The minimal semi-matching functional on $\mathcal{X}$ is

$$\mathrm{SM}(\mathcal{X}) := \min_{\mathrm{SM}} \sum_{e \in \mathrm{SM}} |e|, \tag{8.8}$$

where the minimum ranges over all semi-matchings SM on $\mathcal{X}$.

7. *k-TSP functional.* Fix $k \in \mathbb{N}$. Let $\mathcal{C}$ be a collection of k sub-tours on points of $\mathcal{X}$, each sub-tour containing a distinguished shared vertex x_0 and such that each $x \in \mathcal{X}$ belongs to exactly one sub-tour. $T(k; \mathcal{C}, \mathcal{X})$ is the sum of the combined lengths of the k sub-tours in $\mathcal{C}$. The k-TSP functional is the infimum

$$T(k; \mathcal{X}) := \inf_{\mathcal{C}} T(k; \mathcal{C}, \mathcal{X}). \tag{8.9}$$

Power-weighted edge versions of these functionals are found in [524]. For example, $MST^{(p)}(\mathcal{X})$ is the length of the shortest spanning tree on $\mathcal{X}$ with pth power weighted edges, namely

$$\mathrm{MST}^{(p)}(\mathcal{X}) := \min_T \sum_{e \in T} |e|^p, \tag{8.10}$$

where the minimum is over all spanning trees T on $\mathcal{X}$.

To allow for power weighted edges, we henceforth let the interaction ξ depend on a parameter $p \in (0, \infty)$ and we will write $\xi(\cdot, \cdot) := \xi_p(\cdot, \cdot)$. We henceforth work in this context, but to lighten the notation we shall suppress mention of p.

8.2 Subadditivity

This section gives an introductory account of asymptotic methods based on the subadditivity of the functionals H^ξ defined at (8.1). It culminates with a general umbrella theorem providing an a.s. law of large numbers for H^ξ.

8.2.1 Subadditive Functionals

Let $x_n \in \mathbb{R}$, $n \geq 1$, satisfy the "subadditive inequality"

$$x_{m+n} \leq x_m + x_n \quad \text{for all } m, n \in \mathbb{N}. \tag{8.11}$$

Subadditive sequences are nearly additive in the sense that they satisfy the *subadditive limit theorem*, namely $\lim_{n\to\infty} x_n/n = \alpha$ where $\alpha := \inf\{x_m/m : m \geq 1\} \in [-\infty, \infty)$. This classic result, proved in Hille [245], may be viewed as a limit result about subadditive functions indexed by intervals.

For certain choices of the interaction ξ, the functionals H^ξ defined at (8.1) satisfy geometric subadditivity over *rectangles* and, as we will see, consequently satisfy a subadditive limit theorem analogous to the classic one just mentioned.

Let $\mathcal{R} := \mathcal{R}(d)$ denote the collection of d-dimensional rectangles in $\mathbb{R}^d$. Recall that $\xi(\cdot,\cdot) := \xi_p(\cdot,\cdot)$ depends on the parameter p. Write $H^\xi(\mathcal{X}, R)$ for $H^\xi(\mathcal{X} \cap R)$, $R \in \mathcal{R}$. Say that H^ξ is *geometrically subadditive*, or simply *subadditive*, if there is a constant $c_1 := c_1(p) < \infty$ such that for all $R \in \mathcal{R}$, all partitions of R into rectangles R_1 and R_2, and all finite point sets $\mathcal{X}$ we have

$$H^\xi(\mathcal{X}, R) \leq H^\xi(\mathcal{X}, R_1) + H^\xi(\mathcal{X}, R_2) + c_1(\operatorname{diam}(R))^p. \tag{8.12}$$

Unlike scalar subadditivity (8.11), the relation (8.12) carries an error term.

Classic optimization problems as well as certain functionals of Euclidean graphs, satisfy geometric subadditivity (8.12). For example, the length of the minimal spanning tree defined at (8.4) satisfies (8.12) when p is set to 1, which may be seen as follows. Put $\mathrm{MST}(\mathcal{X}, R)$ to be the length of the minimal spanning tree on $\mathcal{X} \cap R$. Given a finite set $\mathcal{X}$ and a rectangle $R := R_1 \cup R_2$, let $\mathcal{T}_i$ denote the minimal spanning tree on $\mathcal{X} \cap R_i$, $1 \leq i \leq 2$. Tie together the two spanning trees $\mathcal{T}_1$ and $\mathcal{T}_2$ with an edge having a length bounded by the sum of the diameters of the rectangles R_1 and R_2. Performing this operation generates a feasible spanning tree on $\mathcal{X}$ at a total cost bounded by $\mathrm{MST}(\mathcal{X}, R_1)+ \mathrm{MST}(\mathcal{X}, R_2) + \operatorname{diam}(R)$. Putting $p = 1$, (8.12) follows by minimality.

Exercise 8.1. Using edge deletion and insertion techniques, show that the TSP functional (8.3), minimal matching functional (8.5), and nearest neighbour functionals (8.6) satisfy geometric subadditivity (8.12) with $p = 1$.

8.2.2 Superadditive Functionals

If geometric functionals H^ξ were to simultaneously satisfy a superadditive relation analogous to (8.12), then the resulting "near additivity" of H^ξ would lead directly to laws of large numbers. This is too much to expect. On the other hand, many geometric functionals $H^\xi(\cdot, R)$ admit a "dual" version—one which essentially treats the boundary of the rectangle R as a single point, that is to say edges on the boundary ∂R have zero length or "zero cost". This boundary version, introduced in [415] and used in [416, 417] and here denoted $H^\xi_B(\cdot, R)$, closely approximates $H^\xi(\cdot, R)$ in a sense to be made precise (see (8.18) below) and is *superadditive without any error term*. More exactly, the boundary version $H^\xi_B(\cdot, R)$ satisfies

$$H^\xi_B(\mathcal{X}, R) \geq H^\xi_B(\mathcal{X} \cap R_1, R_1) + H^\xi_B(\mathcal{X} \cap R_2, R_2). \tag{8.13}$$

Boundary functionals are defined on a case-by-case basis. For example, the boundary minimal spanning tree functional is defined as follows. For all rectangles $R \in \mathcal{R}$ and finite sets $\mathcal{X} \subset R$ put

$$\mathrm{MST}_B(\mathcal{X}, R) := \min \left(\mathrm{MST}(\mathcal{X}, R), \ \inf \sum_i \mathrm{MST}(\mathcal{X}_i \cup \{a_i\}) \right),$$

where the infimum ranges over all partitions $(\mathcal{X}_i)_{i \geq 1}$ of $\mathcal{X}$ and all sequences of points $(a_i)_{i \geq 1}$ belonging to ∂R. When $\mathrm{MST}_B(\mathcal{X}, R) \neq \mathrm{MST}(\mathcal{X}, R)$ the graph realizing the boundary functional $\mathrm{MST}_B(\mathcal{X}, R)$ may be thought of as a collection of small trees connected via the boundary ∂R into a single large tree, where the connections on ∂R incur no cost. See Fig. 8.1. It is a simple matter to see that the boundary MST functional satisfies subadditivity (8.12) with $p = 1$ and is also superadditive (8.13). Later we will see that the boundary MST functional closely approximates the standard MST functional.

Exercise 8.2. Show that the TSP (8.3), minimal matching (8.5), and nearest neighbour functionals (8.6) have boundary versions which are superadditive (8.13).

8.2.3 Subadditive and Superadditive Euclidean Functionals

Recall that $\xi(\cdot, \cdot) := \xi_p(\cdot, \cdot)$. The following conditions endow the functional $H^\xi(\cdot, \cdot)$ with a *Euclidean structure*:

$$H^\xi(\mathcal{X}, R) = H^\xi(\mathcal{X} + y, R + y) \tag{8.14}$$

for all $y \in \mathbb{R}^d$, $R \in \mathcal{R}$, $\mathcal{X} \subset R$ and

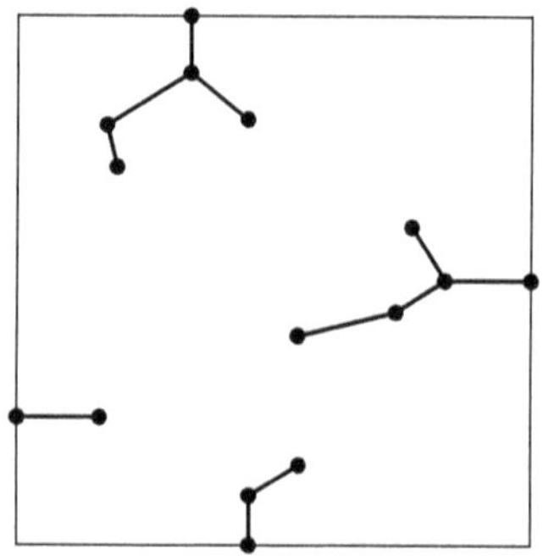

Fig. 8.1 The boundary MST graph; edges on boundary have zero cost

$$H^{\xi}(\alpha\mathcal{X},\alpha R) = \alpha^{p} H^{\xi}(\mathcal{X}, R) \tag{8.15}$$

for all $\alpha > 0$, $R \in \mathcal{R}$ and $\mathcal{X} \subset R$. By αB we understand the set $\{\alpha x,\ x \in B\}$ and by $y + \mathcal{X}$ we mean $\{y + x \ :\ \ x \in \mathcal{X}\}$. Conditions (8.14) and (8.15) express the *translation invariance* and *homogeneity of order* p of H^{ξ}, respectively. Homogeneity (8.15) is satisfied whenever the interaction ξ is itself homogeneous of order p, that is to say whenever

$$\xi(\alpha x,\alpha\mathcal{X}) = \alpha^{p}\xi(x,\mathcal{X}),\ \ \alpha > 0. \tag{8.16}$$

Functionals satisfying translation invariance and homogeneity of order 1 include the total edge length of graphs, including those defined at (8.3)–(8.9).

Exercise 8.3. Show that the TSP functional (8.3), MST functional (8.4), and minimal matching functional (8.5) are homogeneous of order 1 and are thus subadditive Euclidean functionals.

Definition 8.1. Let $H^{\xi}(\emptyset, R) = 0$ for all $R \in \mathcal{R}$ and suppose H^{ξ} satisfies geometric subadditivity (8.12), translation invariance (8.14), and homogeneity of order p (8.15). Then H^{ξ} is a *subadditive Euclidean functional.*

If a functional $H^{\xi}(\mathcal{X}, R)$, $(\mathcal{X}, R) \in \mathcal{N} \times \mathcal{R}$, is superadditive over rectangles and has a Euclidean structure over $\mathcal{N} \times \mathcal{R}$, where $\mathcal{N}$ is the collection of locally finite point sets in $\mathbb{R}^d$, then we say that H^{ξ} is a *superadditive Euclidean functional*, formally defined as follows:

Definition 8.2. Let $H^{\xi}(\emptyset, R) = 0$ for all $R \in \mathcal{R}$ and suppose H^{ξ} satisfies (8.14) and (8.15). If H^{ξ} satisfies

$$H^{\xi}(\mathcal{X}, R) \geq H^{\xi}(\mathcal{X} \cap R_1, R_1) + H^{\xi}(\mathcal{X} \cap R_2, R_2), \tag{8.17}$$

whenever $R \in \mathcal{R}$ is partitioned into rectangles R_1 and R_2 then H^{ξ} is a *superadditive Euclidean functional.*

It may be shown that the functionals TSP, MST and MM are subadditive Euclidean functionals and that they admit dual boundary versions which are superadditive Euclidean functionals; see Chap. 2 of [524].

Pointwise Close Property

To be useful in establishing asymptotics, dual boundary functionals must closely approximate the corresponding functional. The following closeness condition is sufficient for these purposes. Recall that we suppress the dependence of ξ on p, writing $\xi(\cdot,\cdot) := \xi_p(\cdot,\cdot)$.

Definition 8.3. Say that $H^\xi := H^{\xi_p}$ and the boundary version $H_B^\xi := H_B^{\xi_p}$, $p \in (0,\infty)$, are *pointwise close* if for all finite subsets $\mathcal{X} \subset [0,1]^d$ we have

$$|H^\xi(\mathcal{X},[0,1]^d) - H_B^\xi(\mathcal{X},[0,1]^d)| = o\left((\mathrm{card}(\mathcal{X}))^{(d-p)/d}\right). \tag{8.18}$$

The TSP, MST, MM and nearest neighbour functionals all admit respective boundary versions which are pointwise close in the sense of (8.18); see Lemma 3.7 of [524]. See [524] for description of other functionals having boundary versions which are pointwise close in the sense of (8.18).

Growth Bounds

Iteration of geometric subadditivity (8.12) leads to growth bounds on subadditive Euclidean functionals H^ξ, namely for all $p \in (0,d)$ there is a constant $c_2 := c_2(\xi_p, d)$ such that for all rectangles $R \in \mathcal{R}$ and all $\mathcal{X} \subset R$, $\mathcal{X} \in \mathcal{N}$, we have

$$H^\xi(\mathcal{X},R) \le c_2(\mathrm{diam}(R))^p(\mathrm{card}\,\mathcal{X})^{(d-p)/d}. \tag{8.19}$$

Smooth of Order p

Subadditivity (8.12) and growth bounds (8.19) by themselves do not provide enough structure to yield the limit theory for Euclidean functionals; one also needs to control the oscillations of these functionals as points are added or deleted. Some functionals, such as TSP, necessarily increase with increasing argument size, whereas others, such as MST, do not have this property. A useful continuity condition goes as follows.

Definition 8.4. A Euclidean functional $H^\xi := H^{\xi_p}$, $p \in (0,\infty)$, is *smooth of order* p if there is a finite constant $c_3 := c_3(\xi_p,d)$ such that for all finite sets $\mathcal{X}_1, \mathcal{X}_2 \subset [0,1]^d$ we have

$$|H^\xi(\mathcal{X}_1 \cup \mathcal{X}_2) - H^\xi(\mathcal{X}_1)| \le c_3(\mathrm{card}(\mathcal{X}_2))^{(d-p)/d}. \tag{8.20}$$

8.2.4 Examples of Functionals Satisfying Smoothness (8.20)

1. Let TSP be as in (8.3). For all finite sets $\mathcal{X}_1$ and $\mathcal{X}_2 \subset [0,1]^d$ we have

$$\mathrm{TSP}(\mathcal{X}_1) \le \mathrm{TSP}(\mathcal{X}_1 \cup \mathcal{X}_2) \le \mathrm{TSP}(\mathcal{X}_1) + \mathrm{TSP}(\mathcal{X}_2) + c\mathrm{diam}([0,1]^d),$$

   where the first inequality follows by monotonicity and the second by subadditivity (8.12). By (8.19) we have $\mathrm{TSP}(\mathcal{X}_2) \le c_2\sqrt{d}(\mathrm{card}\,\mathcal{X}_2)^{(d-1)/d}$ and since clearly $c\mathrm{diam}([0,1]^d) \le cd^{1/2}(\mathrm{card}(\mathcal{X}_2)^{(d-1)/d}$, it follows that the TSP is smooth of order 1.
2. Let MST be as in (8.4). Subadditivity (8.12) and the growth bounds (8.19) imply that for all finite sets $\mathcal{X}_1, \mathcal{X}_2 \subset [0,1]^d$ we have $\mathrm{MST}(\mathcal{X}_1 \cup \mathcal{X}_2) \le \mathrm{MST}(\mathcal{X}_1) + (c_1\sqrt{d} + c_2\sqrt{d}(\mathrm{card}\,\mathcal{X}_2)^{(d-1)/d} \le \mathrm{MST}(\mathcal{X}_1) + c(\mathrm{card}\,\mathcal{X}_2)^{(d-1)/d}$. It follows that the MST is smooth of order 1 once we show the reverse inequality

$$\mathrm{MST}(\mathcal{X}_1 \cup \mathcal{X}_2) \ge \mathrm{MST}(\mathcal{X}_1) - c(\mathrm{card}\,\mathcal{X}_2)^{(d-1)/d}. \tag{8.21}$$

   To show (8.21) let $\mathcal{T}$ denote the graph of the minimal spanning tree on $\mathcal{X}_1 \cup \mathcal{X}_2$. Remove the edges in $\mathcal{T}$ which contain a vertex in $\mathcal{X}_2$. Since each vertex has bounded degree, say D, this generates a subgraph $\mathcal{T}_1 \subset \mathcal{T}$ which has at most $D \cdot \mathrm{card}\,\mathcal{X}_2$ components. Choose one vertex from each component and form the minimal spanning tree $\mathcal{T}_2$ on these vertices. By the growth bounds (8.19), the edge length of $\mathcal{T}_2$ is bounded by $c(D \cdot \mathrm{card}\,\mathcal{X}_2)^{(d-1)/d}$. Since the union of the trees $\mathcal{T}_1$ and $\mathcal{T}_2$ is a feasible spanning tree on $\mathcal{X}_1$, it follows that

$$\mathrm{MST}(\mathcal{X}_1) \le \sum_{e \in \mathcal{T}_1 \cup \mathcal{T}_2} |e| \le \mathrm{MST}(\mathcal{X}_1 \cup \mathcal{X}_2) + c(D \cdot \mathrm{card}\,\mathcal{X}_2)^{(d-1)/d}.$$

   Thus smoothness (8.20) holds for the MST functional.

It may be shown that a modification of the Steiner functional (8.7) is smooth of order 1 (see Chap. 10 of [524]). Smoothness is a common property of geometric functionals, as indicated in the next exercise.

Exercise 8.4. Show that the minimal matching functional MM defined at (8.5) is smooth of order 1. Likewise, show that the semi-matching, nearest neighbour, and k-TSP functionals are smooth of order 1. Hints; see Chap. 3.3 of [524]), Sects. 8.2, 8.3 and 8.4 of [524], respectively.

The functionals TSP, MST and MM defined at (8.3)–(8.5) are thus smooth subadditive Euclidean functionals which are pointwise close to a canonical boundary functional. The functionals (8.6)–(8.9) satisfy the same properties. Now we give some limit theorems for such functionals.

8.2.5 *Laws of Large Numbers for Superadditive Euclidean Functionals*

We state a basic law of large numbers for Euclidean functionals on i.i.d. uniform random variables $U_1, \dots, U_n$ in $[0,1]^d$. Recall that a sequence of random variables ζ_n converges completely, here denoted c.c., to a limit random variable ζ, if for all $\varepsilon > 0$, we have $\sum_{n=1}^{\infty} \mathbf{P}(|\zeta_n - \zeta| > \varepsilon) < \infty$.

Theorem 8.1. *Let $p \in [1,d)$. If $H_B^{\xi} := H_B^{\xi_p}$ is a smooth superadditive Euclidean functional of order p on $\mathbb{R}^d$, then*

$$\lim_{n\to\infty} n^{(p-d)/d} H_B^{\xi}(U_1,\dots,U_n) = \alpha(H_B^{\xi}, d) \ \ c.c., \tag{8.22}$$

where $\alpha(H_B^{\xi}, d)$ is a positive constant. If $H^{\xi} := H^{\xi_p}$ is a subadditive Euclidean functional which is pointwise close to $H_B^{\xi} := H_B^{\xi_p}$ as in (8.18), then

$$\lim_{n\to\infty} n^{(p-d)/d} H^{\xi}(U_1,\dots,U_n) = \alpha(H_B^{\xi}, d) \ \ c.c. \tag{8.23}$$

Remarks.

1. In practice, Theorem 8.1 involves taking $H_B^{\xi} := H_B^{\xi_p}$ to be a boundary version of $H^{\xi} := H^{\xi_p}$, but it is conceivable that there are functionals $H_B^{\xi_p}$ which satisfy the conditions of Theorem 8.1 and which are not boundary versions. By considering boundary functionals, Theorem 8.1 gives laws of large numbers for the functionals (8.3)–(8.9); see [524] for details.
2. Smooth subadditive Euclidean functionals which are point-wise close to smooth superadditive Euclidean functionals are "nearly additive" and consequently satisfy Donsker–Varadhan-style *large deviation principles*, as shown in [463].
3. The papers [242, 295] provide further accounts of the limit theory for subadditive Euclidean functionals.

Proof of Theorem 8.1. We only prove a mean version of (8.22), namely

$$\lim_{n\to\infty} n^{(p-d)/d} \mathbf{E} L_B^p(U_1,\dots,U_n) = \alpha(L_B^p, d), \tag{8.24}$$

referring the reader to [524] for a complete proof. To prove (8.24), we will follow the proof of Theorem 4.1 of [524]. Fix $1 \le p < d$ and set $\varphi(n) := \mathbf{E} L_B^p(U_1,\dots,U_n)$. The number of points from the sample $(U_1,\dots,U_n)$ belonging to a given subcube of $[0,1]^d$ of volume m^{-d} is a binomial random variable $\mathrm{Binom}(n, m^{-d})$ with parameters n and m^{-d}. Superadditivity of L_B^p, homogeneity (8.15), smoothness (8.20), and Jensen's inequality in this order yield

$$\varphi(n) \geq m^{-p} \sum_{i \leq m^d} \varphi(\text{Binom}(n, m^{-d}))$$

$$\geq m^{-p} \sum_{i \leq m^d} \left(\varphi(nm^{-d}) - c_3\mathbf{E}(|\text{Binom}(n, m^{-d}) - nm^{-d}|^{(d-p)/d})\right)$$

$$\geq m^{-p} \sum_{i \leq m^d} \left(\varphi(nm^{-d}) - c_3(nm^{-d})^{(d-p)/2d}\right).$$

Simplifying, we get

$$\varphi(n) \geq m^{d-p}\varphi(nm^{-d}) - c_3 m^{(d-p)/2} n^{(d-p)/2d}.$$

Dividing by $n^{(d-p)/d}$ and replacing n by nm^d yields the homogenized relation

$$\frac{\varphi(nm^d)}{(nm^d)^{(d-p)/d}} \geq \frac{\varphi(n)}{n^{(d-p)/d}} - \frac{c_3}{n^{(d-p)/2d}}. \tag{8.25}$$

Set $\alpha := \alpha(L_B^p, d) := \limsup_{n\to\infty} \varphi(n)/n^{(d-p)/d}$ and note that $\alpha \leq c_3$ by the assumed smoothness. For all $\epsilon > 0$, choose n_o such that for all $n \geq n_o$ we have $c_3/n^{(d-p)/2d} \leq \epsilon$ and $\varphi(n_o)/n_o^{(d-p)/d} \geq \alpha - \epsilon$. Thus, for all $m = 1, 2, \ldots$ it follows that

$$\frac{\varphi(n_o m^d)}{(n_o m^d)^{(d-p)/d}} \geq \alpha - 2\epsilon.$$

To now obtain (8.24) we use the smoothness of L and an interpolation argument. For an arbitrary integer $k \geq 1$ find the unique integer m such that

$$n_o m^d < k \leq n_o(m+1)^d.$$

Then $|n_o m^d - k| \leq C n_o m^{d-1}$ and by smoothness (8.20) we therefore obtain

$$\frac{\varphi(k)}{k^{(d-p)/d}} \geq \frac{\varphi(n_o m^d)}{(n_o(m+1)^d)^{(d-p)/d}} - \frac{(Cn_o m^{d-1})^{(d-p)/d}}{(m+1)^{d-p}\, n_o^{(d-p)/d}}$$

$$\geq (\alpha - 2\epsilon)\left(\frac{m}{m+1}\right)^{d-p} - \frac{(Cn_o m^{d-1})^{(d-p)/d}}{(m+1)^{d-p}\, n_o^{(d-p)/d}}.$$

Since the last term in the above goes to zero as m goes to infinity, it follows that

$$\liminf_{k\to\infty} k^{(p-d)/d}\varphi(k) \geq \alpha - 2\epsilon.$$

Now let ϵ tend to zero to see that the liminf and the limsup of the sequence $\varphi(k)/k^{(d-p)/d}$, $k \geq 1$, coincide, that is

$$\lim_{k \to \infty} k^{(p-d)/d} \varphi(k) = \alpha.$$

We have thus shown $\lim_{n \to \infty} n^{(p-d)/d} \mathbf{E} L_B^p(U_1, \ldots, U_n) = \alpha$ as desired. This completes the proof of (8.24). □

8.2.6 Rates of Convergence of Euclidean Functionals

Recall that we write $\xi(\cdot, \cdot) := \xi_p(\cdot, \cdot)$. If a subadditive Euclidean functional H^ξ is *close in mean* (cf. Definition 3.9 in [524]) to the associated superadditive Euclidean functional H_B^ξ, namely if

$$|\mathbf{E} H^\xi(U_1, \ldots, U_n) - \mathbf{E} H_B^\xi(U_1, \ldots, U_n)| = o(n^{(d-p)/d}), \tag{8.26}$$

where we recall that U_i are i.i.d. uniform on $[0, 1]^d$, then we may upper bound $|\mathbf{E} H^\xi(U_1, \ldots, U_n) - \alpha(H_B^\xi, d) n^{(d-p)/d}|$, thus yielding rates of convergence of

$$n^{(p-d)/d} \mathbf{E} H^\xi(U_1, \ldots, U_n)$$

to its mean. Since the TSP, MST, and MM functionals satisfy closeness in mean $(p \neq d-1,\ d \geq 3)$ the following theorem immediately provides rates of convergence for our prototypical examples.

Theorem 8.2 (Rates of convergence of means). *Let H^ξ and H_B^ξ be subadditive and superadditive Euclidean functionals, respectively, satisfying the close in mean approximation (8.26). If H^ξ is smooth of order $p \in [1, d)$ as defined at (8.20), then for $d \geq 2$ and for $\alpha(H_B^\xi, d)$ as at (8.22), we have*

$$|\mathbf{E} H^\xi(U_1, \ldots, U_n) - \alpha(H_B^\xi, d) n^{(d-p)/d}| \leq c \left(n^{(d-p)/2d} \vee n^{(d-p-1)/d}\right). \tag{8.27}$$

For a complete proof of Theorem 8.2, we refer to [524]. Koo and Lee [309] give conditions under which Theorem 8.2 can be improved.

8.2.7 General Umbrella Theorem for Euclidean Functionals

Here is the main result of this section. Let $\eta_1, \ldots, \eta_n$ be i.i.d. random variables with values in $[0, 1]^d$, $d \geq 2$, and put $\mathcal{X}_n := \{\eta_i\}_{i=1}^n$.

Theorem 8.3 (Umbrella theorem for Euclidean functionals). *Let H^ξ and H_B^ξ be subadditive and superadditive Euclidean functionals, respectively, both smooth of order $p \in [1, d)$. Assume that H^ξ and H_B^ξ are close in mean (8.26). Then*

$$\lim_{n\to\infty} n^{(p-d)/d} H^{\xi}(\mathcal{X}_n) = \alpha(H_B^{\xi}, d) \int_{[0,1]^d} \kappa(x)^{(d-p)/d}\, dx \quad c.c., \tag{8.28}$$

where κ *is the density of the absolutely continuous part of the law of* η_1.

Remarks.

1. The above theorem captures the limit behavior of the total edge length of the functionals described in Sect. 8.1.1, hence the term "umbrella". Indeed, the TSP functional satisfies the conditions of Theorem 8.3 and we thus recover as a corollary the Beardwood–Halton–Hammersley theorem [61]. See [524] for details.
2. Umbrella theorems for Euclidean functionals satisfying monotonicity and other assumptions not involving boundary functionals appear in Theorem 2 of [481]. Theorem 8.3 has its origins in [415,416].
3. Theorem 8.3 is used by Baltz et al. [39] to analyze asymptotics for the multiple vehicle routing problem; Costa and Hero [130] show asymptotics similar to Theorem 8.3 for the MST on suitably regular Riemannian manifolds and they apply their results to estimation of Rényi entropy and manifold dimension. Costa and Hero [131], using the theory of subadditive and superadditive Euclidean functionals, obtain asymptotics for the total edge length of k-nearest neighbour graphs on manifolds. The paper [242] provides further applications to imaging and clustering.
4. If the η_i fail to have a density then the right-hand side of (8.28) vanishes. On the other hand, Hölder's inequality shows that the right-hand side of (8.28) is largest when κ is uniform on $[0,1]^d$.
5. See Chap. 7 of [524] for extensions of Theorem 8.3 to functionals of random variables on unbounded domains.

Proof (Sketch of proof of Theorem 8.3). The Azuma–Hoeffding concentration inequality shows that it is enough to prove convergence of means in (8.28). Smoothness then shows that it is enough to prove convergence of $n^{(p-d)/d}\mathbf{E}H^{\xi}(\mathcal{X}_n)$ for the so-called blocked distributions, i.e. those whose absolutely continuous part is a linear combination of indicators over congruent sub-cubes forming a partition of $[0,1]^d$. To establish convergence for the blocked distributions, one combines Theorem 8.1 with the subadditive and superadditive relations. We refer to [524] for complete details of these standard methods. □

The limit (8.28) exhibits the asymptotic dependency of the total edge length of graphs on the underlying point density κ. Still, (8.28) is unsatisfying in that we don't have a closed form expression for the constant $\alpha(H_B^{\xi}, d)$. Stabilization methods, described below, are used to explicitly identify $\alpha(H_B^{\xi}, d)$.

8.3 Stabilization

Subadditive methods yield a.s. limit theory for the functionals H^ξ defined at (8.1) but they do not express the macroscopic behaviour of H^ξ in terms of the local interactions described by ξ. Stabilization methods overcome this limitation, they yield second order and distributional results, and they also provide limit results for the empirical measures

$$\sum_{x \in \mathcal{X}} \xi(x, \mathcal{X})\delta_x, \tag{8.29}$$

where δ_x is the point mass at x. The empirical measure (8.29) has total mass given by H^ξ.

We will often assume that the interaction or "score" function ξ, defined on pairs $(x, \mathcal{X})$, with $\mathcal{X}$ locally finite in $\mathbb{R}^d$, is translation invariant, i.e., for all $y \in \mathbb{R}^d$ we have $\xi(x + y, \mathcal{X} + y) = \xi(x, \mathcal{X})$. When $x \in \mathbb{R}^d \setminus \mathcal{X}$, we abbreviate notation and write $\xi(x, \mathcal{X})$ instead of $\xi(x, \mathcal{X} \cup \{x\})$.

When $\mathcal{X}$ is random the range of spatial dependence of ξ at $x \in \mathcal{X}$ is random and the purpose of stabilization is to quantify this range in a way useful for asymptotic analysis. There are several notions of stabilization, with the simplest being that of stabilization of ξ with respect to a rate τ homogeneous Poisson point process Π_τ on $\mathbb{R}^d$, defined as follows. Let $B_r(x)$ denote the Euclidean ball centered at x with radius r and let o denote a point at the origin of $\mathbb{R}^d$.

8.3.1 *Homogeneous Stabilization*

We say that a translation invariant ξ is *homogeneously stabilizing* if for all τ and almost all realizations Π_τ there exists $R := R(\Pi_\tau) < \infty$ such that

$$\xi(o, (\Pi_\tau \cap B_R(o)) \cup \mathcal{A}) = \xi(o, \Pi_\tau \cap B_R(o)) \tag{8.30}$$

for all locally finite $\mathcal{A} \subset \mathbb{R}^d \setminus B_R(o)$. Thus the value of ξ at o is unaffected by changes in the configuration outside $B_R(o)$. The random range of dependency given by R depends on the realization of Π_τ. When ξ is homogeneously stabilizing we may write

$$\xi(o, \Pi_\tau) = \lim_{r \to \infty} \xi(o, \Pi_\tau \cap B_r(o)).$$

Examples of homogeneously stabilizing functionals.

1. *Nearest neighbour distances.* Recalling (8.6), consider the nearest neighbour graph $G^N(1, \mathcal{X})$ on the point set $\mathcal{X}$ and let $\xi(x, \mathcal{X})$ denote one half the sum of the lengths of edges in $G^N(1, \mathcal{X})$ which are incident to x. Thus $H^\xi(\mathcal{X})$ is the sum of edge lengths in $G^N(1, \mathcal{X})$. Partition $\mathbb{R}^2$ into six congruent cones $K_i, 1 \leq i \leq 6$, having apex at the origin of $\mathbb{R}^2$ and for all $1 \leq i \leq 6$, put

R_i to be the distance between the origin and the nearest point in $\Pi_\tau \cap K_i$. We assert that $R := 2 \max_{1 \le i \le 6} R_i$ is a radius of stabilization, i.e., points in $B_{2R}^c(o)$ do not change the value of $\xi(o, \Pi_\tau)$. Indeed, edges in $G^N(1, \Pi_\tau)$ incident to the origin are not changed by the addition of points in $B_{2R}^c(o)$. Such points will be closer to at least one point in $\Pi_\tau \cap B_R(o)$ than to the origin and so will not connect to the origin. Also, edges between points in $\Pi_\tau \cap B_R(o)$ and the origin will not be affected by the insertion of points in $B_{2R}^c(o)$.

2. *Voronoi graphs*. Consider the graph of the Voronoi tessellation of $\mathcal{X}$ and let $\xi(x, \mathcal{X})$ be one half the sum of the lengths of the edges in the Voronoi cell $C(x)$ around x. The Voronoi flower around x, or fundamental region, is the union of those balls having as center a vertex of $C(x)$ and exactly two points of $\mathcal{X}$ on their boundary and no points of $\mathcal{X}$ inside. Then it may be shown (see Zuyev [532]) that the geometry of $C(x)$ is completely determined by the Voronoi flower and thus the radius of a ball centered at x containing the Voronoi flower qualifies as a stabilization radius.
3. *Minimal spanning trees*. Let $\mathcal{X} \subset \mathbb{R}^d, d \ge 2$, be locally finite. Given $a > 0$, let $\mathcal{G}_a(\mathcal{X})$ be the graph with vertex set $\mathcal{X}$ and with edge set $\{\{x, y\} : |x - y| < a\}$. Let $\mathcal{G}_{\mathrm{MST}}(\mathcal{X})$ be the graph with vertex set $\mathcal{X}$ obtained by including each edge $\{x, y\}$ such that x and y lie in different components of $\mathcal{G}_{|x-y|}(\mathcal{X})$ and at least one of the components is finite. When $\mathcal{X}$ is finite, then $\mathcal{G}_{\mathrm{MST}}(\mathcal{X})$ is the minimal spanning tree graph on $\mathcal{X}$, with total edge length $\mathrm{MST}(\mathcal{X})$, as in (8.4). Let $\xi(x, \mathcal{X})$ be one half the sum of the lengths of the edges in $\mathcal{G}_{\mathrm{MST}}(\mathcal{X})$ which are incident to x. Then ξ is homogeneously stabilizing, which follows from arguments involving the uniqueness of the infinite component in continuum percolation [401].

Given $\mathcal{X} \subset \mathbb{R}^d$ and $a > 0$, recall that $a\mathcal{X} := \{ax : x \in \mathcal{X}\}$. For all $\lambda > 0$ define the λ *re-scaled version* of ξ by

$$\xi_\lambda(x, \mathcal{X}) := \xi(\lambda^{1/d} x, \lambda^{1/d} \mathcal{X}). \tag{8.31}$$

Re-scaling is natural when considering point sets in compact sets K having cardinality roughly λ; dilation by $\lambda^{1/d}$ means that unit volume subsets of $\lambda^{1/d} K$ host on the average one point.

It is useful to consider point processes on $\mathbb{R}^d$ more general than the homogeneous Poisson point processes. In what follows, let $\eta_1, \ldots, \eta_n$ be i.i.d., with a distribution which is absolutely continuous with respect to Lebesgue measure on $\mathbb{R}^d$, with density κ having support K. For all $\lambda > 0$, let $\Pi_{\lambda\kappa}$ denote a Poisson point process in $\mathbb{R}^d$ with intensity measure $\lambda\kappa(x)\, dx$. We shall assume throughout that κ is bounded with supremum denoted $\|\kappa\|_\infty$.

Homogeneous stabilization is an example of "point stabilization" [457] in that ξ is required to stabilize around a given point $x \in \mathbb{R}^d$ with respect to homogeneously distributed Poisson points Π_τ. A related "point stabilization" requires that the re-scaled $\xi_\lambda, \lambda \in [1, \infty)$, stabilize around x, but now with respect to $\Pi_{\lambda\kappa}$ uniformly in $\lambda \in [1, \infty)$. This goes as follows.

8.3.2 Stabilization with Respect to the Probability Density κ

ξ is *stabilizing with respect to the probability density* κ *and the subset* K *of* $\mathbb{R}^d$ if for all $\lambda \in [1,\infty)$ and all $x \in K$, there exists almost surely a $R := R(x,\lambda) < \infty$ (*a radius of stabilization* for ξ_λ at x) such that for all locally finite $\mathcal{A} \subset (\mathbb{R}^d \setminus B_{\lambda^{-1/d}R}(x))$, we have

$$\xi_\lambda\left(x, [\Pi_{\lambda\kappa} \cap B_{\lambda^{-1/d}R}(x)] \cup \mathcal{A}\right) = \xi_\lambda\left(x, \Pi_{\lambda\kappa} \cap B_{\lambda^{-1/d}R}(x)\right). \tag{8.32}$$

If the tail probability $\tau(t)$ defined for $t > 0$ by $\tau(t) := \sup_{\lambda\geq 1,\, x\in K} \mathbf{P}(R(x,\lambda) > t)$ satisfies $\limsup_{t\to\infty} t^{-1}\log\tau(t) < 0$ then we say that ξ is *exponentially stabilizing with respect to* κ *and* K.

Roughly speaking, $R := R(x,\lambda)$ is a radius of stabilization if for all $\lambda \in [1,\infty)$, the value of $\xi_\lambda(x, \Pi_{\lambda\kappa})$ is unaffected by changes in point configurations outside $B_{\lambda^{-1/d}R}(x)$. In most examples of interest, methods showing that functionals ξ homogeneously stabilize are easily modified to show stabilization of ξ with respect to densities κ. While it is straightforward to determine conditions under which the interaction function ξ from examples 1 and 2 stabilizes exponentially fast, it is not known whether the interaction ξ from example 3 stabilizes exponentially fast.

Exercise 8.5. Show that the interaction function ξ from examples 1 and 2 stabilizes exponentially fast when κ is bounded away from zero on its support K, assumed compact and convex.

We may weaken homogeneous stabilization by requiring that the point sets $\mathcal{A}$ in (8.30) belong to the homogeneous Poisson point process Π_τ. This weaker version of stabilization, called *localization*, is used in [111,459] to establish variance asymptotics and central limit theorems for functionals of convex hulls of random samples in the unit ball. Given $r > 0$, let $\xi^r(x,\mathcal{X}) := \xi(x, \mathcal{X} \cap B_r(x))$.

Say that $\hat{R} := \hat{R}(x,\Pi_\tau)$ is a *radius of localization* for ξ at x with respect to Π_τ if almost surely $\xi(x,\Pi_\tau) = \xi^{\hat{R}}(x,\Pi_\tau)$ and for all $s > \hat{R}$ we have $\xi^s(x,\Pi_\tau) = \xi^{\hat{R}}(x,\Pi_\tau)$.

8.3.3 A Weak Law of Large Numbers for Stabilizing Functionals

Recall that $\Pi_{\lambda\kappa}$ is the Poisson point process on $\mathbb{R}^d$ with intensity measure $\lambda\kappa(x)dx$. It is easy to show that $\lambda^{1/d}(\Pi_{\lambda\kappa} - x_0)$ converges to $\Pi_{\kappa(x_0)}$ as $\lambda \to \infty$, where convergence is in the sense of weak convergence of point processes. If $\xi(\cdot,\cdot)$ is a functional defined on $\mathbb{R}^d \times \mathcal{N}$, where we recall that $\mathcal{N}$ is the space of locally finite point sets in $\mathbb{R}^d$, one might hope that ξ is *continuous* on the pairs $(o, \lambda^{1/d}(\Pi_{\lambda\kappa} - x_0))$ in the sense that $\xi(o, \lambda^{1/d}(\Pi_{\lambda\kappa} - x_0))$ converges in distribution to $\xi(o, \Pi_{\kappa(x_0)})$ as $\lambda \to \infty$. This turns out to be the case whenever ξ is homogeneously stabilizing as in (8.30). This is the content of the next lemma; for a complete proof see Sect. 3 of [395]. Recall that almost every $x \in \mathbb{R}^d$ is a *Lebesgue point* of κ, that is to say for almost all $x \in \mathbb{R}^d$ we have that $\varepsilon^{-d} \int_{B_\varepsilon(x)} |\kappa(y) - \kappa(x)|\, dy$ tends to zero as ε tends to zero.

Lemma 8.1. *Let x_0 be a Lebesgue point for κ. If ξ is homogeneously stabilizing as in (8.30), then as $\lambda \to \infty$*

$$\xi_\lambda(x_0, \Pi_{\lambda\kappa}) \xrightarrow{d} \xi(o, \Pi_{\kappa(x_0)}). \tag{8.33}$$

Proof (Sketch). We have $\xi_\lambda(x_0, \Pi_{\lambda\kappa}) = \xi(o, \lambda^{1/d}(\Pi_{\lambda\kappa} - x_0))$ by translation invariance of ξ. By the stabilization of ξ, it may be shown [394] that $(o, \Pi_{\kappa(x_0)})$ is a continuity point for ξ with respect to the product topology on $\mathbb{R}^d \times \mathcal{N}$, when the space of locally finite point sets $\mathcal{N}$ in $\mathbb{R}^d$ is equipped with the metric

$$d(\mathcal{X}_1, \mathcal{X}_2) := (\max\{k \in \mathbb{N} : \mathcal{X}_1 \cap B_k(o) = \mathcal{X}_2 \cap B_k(o)\})^{-1}.$$

The result follows by the weak convergence $\lambda^{1/d}(\Pi_{\lambda\kappa} - x_0) \xrightarrow{d} \Pi_{\kappa(x_0)}$ and the continuous mapping theorem (Theorem 2.7 of [69]). □

Recall that $\mathcal{X}_n := \{\eta_i\}_{i=1}^n$, where $\eta_1, \ldots, \eta_n$ are i.i.d. with density κ. Limit theorems for the sums $\sum_{x \in \Pi_{\lambda\kappa}} \xi_\lambda(x, \Pi_{\lambda\kappa})$ as well as for the weighted empirical measures

$$\mu_\lambda := \mu_\lambda^\xi := \sum_{x \in \Pi_{\lambda\kappa}} \xi_\lambda(x, \Pi_{\lambda\kappa})\delta_x \text{ and } \rho_n := \rho_n^\xi := \sum_{i=1}^n \xi_n(\eta_i, \mathcal{X}_n)\delta_{\eta_i} \tag{8.34}$$

naturally require moment conditions on the summands, thus motivating the next definition.

Definition 8.5. ξ has a moment of order $p > 0$ (with respect to κ and K) if

$$\sup_{\lambda \geq 1,\ x \in K, \mathcal{A} \in K} \mathbf{E}[|\xi_\lambda(x, \Pi_{\lambda\kappa} \cup \mathcal{A})|^p] < \infty, \tag{8.35}$$

where $\mathcal{A}$ ranges over all finite subsets of K.

Exercise 8.6. Show that the interaction function ξ from Examples 1 and 2 has moments of all orders when κ is bounded away from zero on its support.

Let $\mathbb{B}(K)$ denote the class of all bounded $f : K \to \mathbb{R}$ and for all measures μ on $\mathbb{R}^d$ let $\langle f, \mu \rangle := \int f d\mu$. Put $\bar{\mu} := \mu - \mathbf{E}\mu$. For all $f \in \mathbb{B}(K)$ we have by Palm theory for the Poisson process (see e.g Theorem 1.6 in [394]) that

$$\mathbf{E}[\langle f, \mu_\lambda \rangle] = \lambda \int_K f(x)\mathbf{E}[\xi_\lambda(x, \Pi_{\lambda\kappa})]\kappa(x)\, dx. \tag{8.36}$$

If (8.35) holds for some $p > 1$, then uniform integrability and Lemma 8.1 show that for all Lebesgue points x of κ one has $\mathbf{E}\xi_\lambda(x, \Pi_{\lambda\kappa}) \to \mathbf{E}\xi(o, \Pi_{\kappa(x)})$ as $\lambda \to \infty$. The set of points failing to be Lebesgue points has measure zero and so when the moment condition (8.35) holds for some $p > 1$, the bounded convergence theorem gives

$$\lim_{\lambda\to\infty} \lambda^{-1}\mathbf{E}[\langle f, \mu_\lambda\rangle] = \int_K f(x)\mathbf{E}[\xi(o, \Pi_{\kappa(x)})]\kappa(x)\, dx.$$

This simple convergence of means $\mathbf{E}[\langle f, \mu_\lambda\rangle]$ is now upgraded to convergence in L^q, $q = 1$ or 2.

Theorem 8.4. *Put $q = 1$ or 2. Let ξ be a homogeneously stabilizing (8.30) translation invariant functional satisfying the moment condition (8.35) for some $p > q$. Then for all $f \in \mathbb{B}(K)$ we have*

$$\lim_{n\to\infty} n^{-1}\langle f, \rho_n\rangle = \lim_{\lambda\to\infty} \lambda^{-1}\langle f, \mu_\lambda\rangle = \int_K f(x)\mathbf{E}[\xi(o, \Pi_{\kappa(x)})]\kappa(x)\, dx \ \ in\ L^q. \tag{8.37}$$

If ξ is homogeneous of order p as defined at (8.16), then for all $\alpha \in (0, \infty)$ and $\tau \in (0, \infty)$ we have $\Pi_{\alpha\tau} \stackrel{d}{=} \alpha^{-1/d}\Pi_\tau$; see for example the mapping theorem on p. 18 of [298]. Consequently, if ξ is homogeneous of order p, it follows that $\mathbf{E}\xi(o, \Pi_{\kappa(x)}) = \kappa(x)^{-p/d}\mathbf{E}\xi(o, \Pi_1)$, whence the following weak law of large numbers.

Corollary 8.1. *Put $q = 1$ or 2. Let ξ be a homogeneously stabilizing (8.30) translation invariant functional satisfying the moment condition (8.35) for some $p > q$. If ξ is homogeneous of order p as at (8.16), then for all $f \in \mathbb{B}(K)$ we have*

$$\lim_{n\to\infty} n^{-1}\langle f, \rho_n\rangle = \lim_{\lambda\to\infty} \lambda^{-1}\langle f, \mu_\lambda\rangle = \mathbf{E}[\xi(o, \Pi_1)] \int_K f(x)\kappa(x)^{(d-p)/d}\, dx \tag{8.38}$$

where the convergence is in the L^q sense.

Remarks.

1. The proofs of the above laws of large numbers are given in [394, 401].
2. The closed form limit (8.38) links the macroscopic limit behaviour of the point measures ρ_n and μ_λ with (i) the local interaction of ξ at a point at the origin inserted into the point process Π_1 and (ii) the underlying point density κ.
3. Going back to the minimal spanning tree at (8.4), the limiting constant $\alpha(\mathrm{MST}_B, d)$ can be found by putting ξ in (8.38) to be ξ_{MST}, letting $f \equiv 1$ in (8.38), and consequently deducing that $\alpha(\mathrm{MST}_B, d) = \mathbf{E}[\xi_{\mathrm{MST}}(o, \Pi_1)]$, where $\xi_{\mathrm{MST}}(x, \mathcal{X})$ is one half the sum of the lengths of the edges in the graph $\mathcal{G}_{\mathrm{MST}}(\mathcal{X} \cup \{x\})$ incident to x.
4. Donsker–Varadhan-style large deviation principles for stabilizing functionals are proved in [460] whereas moderate deviations for bounded stabilizing functionals are proved in [55].

8.3.4 Variance Asymptotics and Central Limit Theorems for Stabilizing Functionals

Asymptotic distribution results for $\langle f, \mu_\lambda \rangle$ and $\langle f, \rho_n \rangle$, $f \in \mathbb{B}(K)$, as λ and n tend to infinity respectively, require additional notation. For all $\tau > 0$, put

$$V^\xi(\tau) := \mathbf{E}[\xi(o, \Pi_\tau)^2] + \tau \int_{\mathbb{R}^d} \{\mathbf{E}[\xi(o, \Pi_\tau \cup \{z\})\xi(z, \Pi_\tau \cup o)] - (\mathbf{E}[\xi(o, \Pi_\tau)])^2\}\, dz \tag{8.39}$$

and

$$\Delta^\xi(\tau) := \mathbf{E}[\xi(o, \Pi_\tau)] + \tau \int_{\mathbb{R}^d} \{\mathbf{E}[\xi(o, \Pi_\tau \cup \{z\}) - \mathbf{E}[\xi(o, \Pi_\tau)]\}\, dz. \tag{8.40}$$

The scalars $V^\xi(\tau)$, $\tau > 0$, should be interpreted as mean pair correlation functions for the functional ξ on homogenous Poisson points Π_τ. By the translation invariance of ξ, the scalars $\Delta^\xi(\tau)$, $\tau > 0$, satisfy

$$\Delta^\xi(\tau) = \mathbf{E}[\xi(o, \Pi_\tau)] + \mathbf{E}\left[\sum_{x \in \Pi_\tau \cup \{z\}} \xi(x, \Pi_\tau \cup \{z\}) - \sum_{x \in \Pi_\tau} \xi(x, \Pi_\tau) \right],$$

which suggests that $\Delta^\xi(\tau)$ may be viewed as the expected "add-one cost" for $\sum_{x \in \Pi_\tau} \xi(x, \Pi_\tau)$ when the point set Π_τ is augmented to $\Pi_\tau \cup \{z\}$.

By extending Lemma 8.1 to an analogous result giving the weak convergence of the joint distribution of $\xi_\lambda(x, \Pi_{\lambda\kappa})$ and $\xi_\lambda(x + \lambda^{-1/d} z, \Pi_{\lambda\kappa})$ for all pairs of points x and z in $\mathbb{R}^d$, we may show for exponentially stabilizing ξ and for bounded K that $\lambda^{-1}\, \mathbf{var}[\langle f, \mu_\lambda \rangle]$ converges as $\lambda \to \infty$ to a weighted average of the mean pair correlation functions.

Furthermore, recalling that $\overline{\mu}_\lambda := \mu_\lambda - \mathbf{E}[\mu_\lambda]$, and by using either Stein's method [395, 402] or the cumulant method [57], we may establish variance asymptotics and asymptotic normality of $\langle f, \lambda^{-1/2} \overline{\mu}_\lambda \rangle$, $f \in \mathbb{B}(K)$, as shown by the next result, proved in [57, 395, 402].

Theorem 8.5 (Variance asymptotics and CLT for Poisson input). *Assume that κ is Lebesgue-almost everywhere continuous. Let ξ be a homogeneously stabilizing (8.30) translation invariant functional satisfying the moment condition (8.35) for some $p > 2$. Suppose further that K is bounded and that ξ is exponentially stabilizing with respect to κ and K as in (8.32). Then for all $f \in \mathbb{B}(K)$ we have*

$$\lim_{\lambda \to \infty} \lambda^{-1}\, \mathbf{var}[\langle f, \mu_\lambda \rangle] = \sigma^2(f) := \int_K f(x)^2 V^\xi(\kappa(x))\kappa(x)\, dx < \infty \tag{8.41}$$

as well as convergence of the finite-dimensional distributions

$$(\langle f_1, \lambda^{-1/2}\overline{\mu}_\lambda\rangle, \dots, \langle f_k, \lambda^{-1/2}\overline{\mu}_\lambda\rangle),$$

$f_1, \dots, f_k \in \mathbb{B}(K)$, *to those of a mean zero Gaussian field with covariance kernel*

$$(f, g) \mapsto \int_K f(x)g(x)V^\xi(\kappa(x))\kappa(x)\,dx. \tag{8.42}$$

Extensions of Theorem 8.5

1. For an extension of Theorem 8.5 to manifolds, see [403]; for extensions to functionals of Gibbs point processes, see [460]. Theorems 8.4 and 8.5 also extend to treat functionals of point sets having i.i.d. marks [57, 395].
2. Rates of convergence. Suppose $\|\kappa\|_\infty < \infty$. Suppose that ξ is exponentially stabilizing and satisfies the moments condition (8.35) for some $p > 3$. If $\sigma^2(f) > 0$ for $f \in \mathbb{B}(K)$, then there exists a finite constant c depending on d, ξ, κ, p and f, such that for all $\lambda \geq 2$,

$$\sup_{t\in\mathbb{R}} \left| \mathbf{P}\left[\frac{\langle f, \mu_\lambda\rangle - \mathbf{E}[\langle f, \mu_\lambda\rangle]}{\sqrt{\mathbf{var}[\langle f, \mu_\lambda\rangle]}} \leq t \right] - \mathbf{P}(N(0,1) \leq t) \right| \leq c(\log\lambda)^{3d}\lambda^{-1/2}. \tag{8.43}$$

   For details, see Corollary 2.1 in [402]. For rates of convergence in the multivariate central limit theorem, see [397].
3. Translation invariance. For ease of exposition, Theorems and 8.4 and 8.5 assume translation invariance of ξ. This assumption may be removed (see [57, 394, 395]), provided that we put $\xi_\lambda(x, \mathcal{X}) := \xi(x, x + \lambda^{1/d}(-x + \mathcal{X}))$ and provided that we replace $V^\xi(\tau)$ and $\Delta^\xi(\tau)$ defined at (8.39) and (8.40) respectively, by

$$V^\xi(x,\tau) := \mathbf{E}[\xi(x, \Pi_\tau)^2] + \tau \int_{\mathbb{R}^d} \{\mathbf{E}[\xi(x, \Pi_\tau \cup \{z\})\xi(x, -z + (\Pi_\tau \cup o))] - (\mathbf{E}[\xi(x, \Pi_\tau)])^2\}\,dz \tag{8.44}$$

   and

$$\Delta^\xi(x,\tau) := \mathbf{E}[\xi(x, \Pi_\tau)] + \tau \int_{\mathbb{R}^d} \{\mathbf{E}[\xi(x, \Pi_\tau \cup \{z\}) - \mathbf{E}[\xi(x, \Pi_\tau)]\}\,dz. \tag{8.45}$$

4. The moment condition (8.35) may be weakened to one requiring only that $\mathcal{A}$ range over subsets of K having at most one element; see [395].

Proof of Variance Asymptotics (8.41)

The proof of (8.41) depends in part on the following generalization of Lemma 8.1, a proof of which appears in [395].

Lemma 8.2. *Let x be a Lebesgue point for κ. If ξ is homogeneously stabilizing as in (8.30), then for all $z \in \mathbb{R}^d$, we have as $\lambda \to \infty$*

$$(\xi_\lambda(x, \Pi_{\lambda\kappa}), \xi_\lambda(x + \lambda^{-1/d} z, \Pi_{\lambda\kappa})) \xrightarrow{d} (\xi(o, \Pi_{\kappa(x)}), \xi(z, \Pi_{\kappa(x)})). \tag{8.46}$$

Given Lemma 8.2 we sketch a proof of the variance convergence (8.41). For simplicity we assume that f is a.e. continuous. By Palm theory for the Poisson process $\Pi_{\lambda\kappa}$ we have

$$\begin{aligned}
&\lambda^{-1} \mathbf{var}[\langle f, \mu_\lambda \rangle] \\
&= \lambda \int_K \int_K f(x) f(y) \{\mathbf{E}[\xi_\lambda(x, \Pi_{\lambda\kappa} \cup \{y\}) \xi_\lambda(y, \Pi_{\lambda\kappa} \cup \{x\})] \\
&\quad - \mathbf{E}[\xi_\lambda(x, \Pi_{\lambda\kappa})] \mathbf{E}[\xi_\lambda(y, \Pi_{\lambda\kappa})]\} \kappa(x) \kappa(y)\, dx\, dy \\
&\quad + \int_K f(x)^2 \mathbf{E}[\xi_\lambda^2(x, \Pi_{\lambda\kappa})] \kappa(x)\, dx.
\end{aligned} \tag{8.47}$$

Putting $y = x + \lambda^{-1/d} z$ in the right-hand side in (8.47) reduces the double integral to

$$\int_K \int_{-\lambda^{1/d} x + \lambda^{1/d} K} f(x) f(x + \lambda^{-1/d} z) \{\ldots\} \kappa(x) \kappa(x + \lambda^{-1/d} z) dz\, dx \tag{8.48}$$

where

$$\begin{aligned}
\{\ldots\} := \big\{&\mathbf{E}[\xi_\lambda(x, \Pi_{\lambda\kappa} \cup \{x + \lambda^{-1/d} z\}) \xi_\lambda(x + \lambda^{-1/d} z, \Pi_{\lambda\kappa} \cup \{x\})] \\
&- \mathbf{E}[\xi_\lambda(x, \Pi_{\lambda\kappa})] \mathbf{E}[\xi_\lambda(x + \lambda^{-1/d} z, \Pi_{\lambda\kappa})]\big\}
\end{aligned}$$

is the two point correlation function for ξ_λ.

The moment condition and Lemma 8.2 imply for all Lebesgue points $x \in K$ that the two point correlation function for ξ_λ converges to the two point correlation function for ξ as $\lambda \to \infty$. Moreover, by exponential stabilization, the integrand in (8.48) is dominated by an integrable function of z over $\mathbb{R}^d$ (see Lemma 4.2 of [395]). The double integral in (8.47) thus converges to

$$\begin{aligned}
&\int_K \int_{\mathbb{R}^d} f(x)^2 \cdot \mathbf{E}[\xi(o, \Pi_{\kappa(x)} \cup \{z\}) \xi(z, \Pi_{\kappa(x)} \cup o)] \\
&\quad - (\mathbf{E}\xi(o, \Pi_{\kappa(x)}))^2 \kappa(x)^2\, dz\, dx
\end{aligned} \tag{8.49}$$

by dominated convergence, the continuity of f, and the assumed moment bounds.

By Theorem 8.4, the assumed moment bounds, and dominated convergence, the single integral in (8.47) converges to

$$\int_K f(x)^2 \mathbf{E}[\xi^2(o, \Pi_{\kappa(x)})]\kappa(x)\,dx. \tag{8.50}$$

Combining (8.49) and (8.50) and using the definition of V^ξ, we obtain the variance asymptotics (8.41) for continuous test functions f. To show convergence for general $f \in \mathbb{B}(K)$ we refer to [395].

8.3.5 *Proof of Asymptotic Normality in Theorem 8.5; Method of Cumulants*

Now we sketch a proof of the central limit theorem part of Theorem 8.5. There are three distinct approaches to proving the central limit theorem:

1. Stein's method, in particular consequences of Stein's method for dependency graphs of random variables, as given by [120]. This approach, spelled out in [402], gives the rates of convergence to the normal law in (8.43).
2. Methods based on martingale differences are applicable when κ is the uniform density and when the functional H^ξ satisfies a stabilization criteria involving the insertion of single point into the sample; see [295, 398] for details.
3. The method of cumulants may be used [57] to show that the k-th order cumulants c_λ^k of $\lambda^{-1/2}\langle f, \overline{\mu}_\lambda\rangle$, $k \geq 3$, vanish in the limit as $\lambda \to \infty$. This method makes use of the standard fact that if the cumulants c^k of a random variable ζ vanish for all $k \geq 3$, then ζ has a normal distribution. This approach assumes additionally that ξ has moments of all orders, i.e. (8.35) holds for all $p \geq 1$.

Here we describe the third method, which, when suitably modified yields moderate deviation principles [55] as well as limit theory for functionals over Gibbs point processes [460].

To show vanishing of cumulants of order three and higher, we follow the proof of Theorem 2.4 in section five of [57] and take the opportunity to correct a mistake in the exposition, which also carried over to [55], and which was first noticed by Mathew Penrose. We assume the test functions f belong to the class $C(K)$ of continuous functions on K and we will show for all continuous test functions f on K, that

$$\langle f, \lambda^{-1/2}\overline{\mu}_\lambda\rangle \xrightarrow{d} N(0, \sigma^2(f)), \tag{8.51}$$

where $\sigma^2(f)$ is at (8.41). The convergence of the finite-dimensional distributions (8.42) follows by standard methods involving the Cramér–Wold device.

We first recall the formal definition of cumulants. Put $K := [0, 1]^d$ for simplicity. Write

$$
\begin{aligned}
&\mathbf{E}\exp\left(\lambda^{-1/2}\langle -f,\overline{\mu}_\lambda\rangle\right) \\
&\quad = \exp\left(\lambda^{-1/2}\langle f,\mathbf{E}\mu_\lambda\rangle\right)\mathbf{E}\exp\left(\lambda^{-1/2}\langle -f,\mu_\lambda\rangle\right) \qquad (8.52)\\
&\quad = \exp\left(\lambda^{-1/2}\langle f,\mathbf{E}\mu_\lambda\rangle\right)\left[1+\sum_{k=1}^{\infty}\frac{\lambda^{-k/2}}{k!}\langle(-f)^k, M_\lambda^k\rangle\right],
\end{aligned}
$$

where $f^k : \mathbb{R}^{dk} \to \mathbb{R}$, $k = 1,2,\ldots$ is given by $f^k(v_1,\ldots,v_k) = f(v_1)\cdots f(v_k)$, and $v_i \in K$, $1 \le i \le k$. $M_\lambda^k := M_{\lambda\kappa}^k$ is a measure on $\mathbb{R}^{dk}$, the k-th moment measure (Chap. 9.5 of [140]), and has the property that

$$
\langle f^k, M_\lambda^k\rangle = \int_{K^k}\mathbf{E}\left[\prod_{i=1}^k \xi_\lambda(x_i,\Pi_{\lambda\kappa})\right]\prod_{i=1}^k f(x_i)\kappa(x_i)\,d(\lambda^{1/d}x_i).
$$

In general M_λ^k is not continuous with respect to Lebesgue measure on K^k, but rather it is continuous with respect to sums of Lebesgue measures on the diagonal subspaces of K^k, where two or more coordinates coincide.

In Sect. 5 of [57], the moment and cumulant measures considered there are with respect to the centered functional $\overline{\xi}$, whereas they should be with respect to the non-centered functional ξ. This requires corrections to the notation, which we provide here, but since higher order cumulants for centered and non-centered measures coincide, it does not change the arguments of [57], which we include for completeness and which go as follows.

We have

$$
dM_\lambda^k(v_1,\ldots,v_k) = m_\lambda(v_1,\ldots,v_k)\prod_{i=1}^k \kappa(v_i)\,d(\lambda^{1/d}v_i),
$$

where $m_\lambda(v_1,\ldots,v_k)$ is given by mixed moment

$$
m_\lambda(v_1,\ldots,v_k) := \mathbf{E}\left[\prod_{i=1}^k \xi_\lambda(v_i;\Pi_{\lambda\kappa}\cup\{v_j\}_{j=1}^k)\right]. \qquad (8.53)
$$

Due to the behaviour of M_λ^k on the diagonal subspaces we make the standing assumption that if the differential $d(\lambda_1^{1/d}v_1)\cdots d(\lambda_1^{1/d}v_k)$ involves repetition of certain coordinates, then it collapses into the corresponding lower order differential in which each coordinate occurs only once. For each $k \in \mathbb{N}$, by the assumed moment bounds (8.35), the mixed moment on the right hand side of (8.53) is bounded uniformly in λ by a constant $c(\xi,k)$. Likewise, the k-th summand in (8.52) is finite.

For all $i = 1,2,\ldots$ we let K_i denote the i-th copy of K. For any subset T of the positive integers, we let

$$K^T := \prod_{i \in T} K_i.$$

If $|T| = l$, then for all $\lambda \geq 1$, by M_λ^T we mean a copy of the l-th moment measure on the l-fold product space K_λ^T. M_λ^T is equal to M_λ^l as defined above.

When the series (8.52) is convergent, the logarithm of the Laplace functional gives

$$\log\left[1 + \sum_{k=1}^{\infty} \frac{1}{k!}\lambda^{-k/2}\langle(-f)^k, M_\lambda^k\rangle\right] = \sum_{l=1}^{\infty} \frac{1}{l!}\lambda^{-l/2}\langle(-f)^l, c_\lambda^l\rangle; \tag{8.54}$$

the signed measures c_λ^l are cumulant measures. Regardless of the validity of (8.52), the existence of all cumulants c_λ^l, $l = 1, 2, \ldots$ follows from the existence of all moments in view of the representation

$$c_\lambda^l = \sum_{T_1, \ldots, T_p} (-1)^{p-1}(p-1)! M_\lambda^{T_1} \cdots M_\lambda^{T_p},$$

where $T_1, \ldots, T_p$ ranges over all unordered partitions of the set $1, \ldots, l$ (see p. 30 of [341]). The first cumulant measure coincides with the expectation measure and the second cumulant measure coincides with the variance measure.

We follow the proof of Theorem 2.4 of [57], with these small changes: (a) replace the centered functional $\bar{\xi}$ with the non-centered ξ (b) correspondingly, let all cumulants c_λ^l, $l = 1, 2, \ldots$ be the cumulant measures for the *non-centered* moment measures M_λ^k, $k = 1, 2, \ldots$. Since c_λ^1 coincides with the expectation measure, Theorem 8.4 gives for all $f \in C(K)$

$$\lim_{\lambda \to \infty} \lambda^{-1}\langle f, c_\lambda^1\rangle = \lim_{\lambda \to \infty} \lambda^{-1}\mathbf{E}[\langle f, \mu_\lambda^\xi\rangle] = \int_K f(x)\mathbf{E}[\xi(o, \Pi_{\kappa(x)})]\kappa(x)dx.$$

We already know from the variance convergence that

$$\lim_{\lambda \to \infty} \lambda^{-1}\langle f^2, c_\lambda^2\rangle = \lim_{\lambda \to \infty} \lambda^{-1}\, \mathbf{var}[\langle f, \mu_{\lambda\kappa}^\xi\rangle] = \int_K f(x)^2 V^\xi(\kappa(x))\kappa(x)dx.$$

Thus, to prove (8.51), it will be enough to show for all $k \geq 3$ and all $f \in C(K)$ that $\lambda^{-k/2}\langle f^k, c_\lambda^k\rangle \to 0$ as $\lambda \to \infty$. This will be done in Lemma 8.4 below, but first we recall some terminology from [57].

A cluster measure $U_\lambda^{S,T}$ on $K^S \times K^T$ for non-empty $S, T \subset \{1, 2, \ldots\}$ is defined by

$$U_\lambda^{S,T}(B \times D) = M_\lambda^{S \cup T}(B \times D) - M_\lambda^S(B) M_\lambda^T(D)$$

for all Borel B and D in K^S and K^T, respectively.

Let S_1, S_2 be a partition of S and let T_1, T_2 be a partition of T. A product of a cluster measure $U_\lambda^{S_1,T_1}$ on $K^{S_1} \times K^{T_1}$ with products of moment measures $M^{|S_2|}$ and $M^{|T_2|}$ on $K^{S_2} \times K^{T_2}$ will be called a (S,T) semi-cluster measure.

For each non-trivial partition (S,T) of $\{1,\dots,k\}$ the k-th cumulant c^k is represented as

$$c^k = \sum_{(S_1,T_1),(S_2,T_2)} \alpha((S_1,T_1),(S_2,T_2))U^{S_1,T_1} M^{|S_2|} M^{|T_2|}, \tag{8.55}$$

where the sum ranges over partitions of $\{1,\dots,k\}$ consisting of pairings (S_1,T_1), (S_2,T_2), where $S_1, S_2 \subset S$ and $T_1, T_2 \subset T$, and where $\alpha((S_1,T_1),(S_2,T_2))$ are integer valued pre-factors. In other words, for any non-trivial partition (S,T) of $\{1,\dots,k\}$, c^k is a linear combination of (S,T) semi-cluster measures; see Lemma 5.1 of [57].

The following bound is critical for showing that $\lambda^{-k/2}\langle f, c_\lambda^k\rangle \to 0$ for $k \geq 3$ as $\lambda \to \infty$. This lemma appears as Lemma 5.2 in [57].

Lemma 8.3. *If ξ is exponentially stabilizing as in (8.32), then the functions m_λ cluster exponentially, that is there are positive constants $a_{j,l}$ and $c_{j,l}$ such that uniformly*

$$|m_\lambda(x_1,\dots x_j, y_1,\dots, y_l) - m_\lambda(x_1,\dots,x_j)m_\lambda(y_1,\dots,y_l)| \leq a_{j,l}\exp(-c_{j,l}\delta\lambda^{1/d}),$$

where $\delta := \min_{1\leq i\leq j, 1\leq p\leq l} |x_i - y_p|$ is the separation between the sets $\{x_i\}_{i=1}^j$ and $\{y_p\}_{p=1}^l$ of points in K.

The constants $a_{j,l}$, while independent of λ, may grow quickly in j and l, but this will not affect the decay of the cumulant measures in the scale parameter λ. The next lemma provides the desired decay of the cumulant measures; we provide a proof which is slightly different from that given for Lemma 5.3 of [57].

Lemma 8.4. *For all $f \in C(K)$ and $k = 2,3,\dots$ we have $\lambda^{-1}\langle f^k, c_\lambda^k\rangle \in O(\|f\|_\infty^k)$.*

Proof. We need to estimate

$$\int_{K^k} f(v_1)\dots f(v_k)\, dc_\lambda^k(v_1,\dots,v_k).$$

We will modify the arguments in [57]. Given $v := (v_1,\dots,v_k) \in K^k$, let $D_k(v) := D_k(v_1,\dots,v_k) := \max_{i\leq k}(\|v_1 - v_i\| + \dots + \|v_k - v_i\|)$ be the l^1 diameter for v. Let $\Xi(k)$ be the collection of all partitions of $\{1,\dots,k\}$ into exactly two subsets S and T. For all such partitions consider the subset $\sigma(S,T)$ of $K^S \times K^T$ having the property that $v \in \sigma(S,T)$ implies $d(x(v),y(v)) \geq D_k(v)/k^2$, where $x(v)$ and $y(v)$ are the projections of v onto K^S and K^T, respectively, and where $d(x(v),y(v))$ is the minimal Euclidean distance between pairs of points from $x(v)$ and $y(v)$. It is easy to see that for every $v := (v_1,\dots,v_k) \in K^k$, there is a splitting of v, say

$x := x(v)$ and $y := y(v)$, such that $d(x,y) \geq D_k(v)/k^2$; if this were not the case then a simple argument shows that, given $v := (v_1,\ldots,v_k)$ the distance between any pair of constituent components must be strictly less than $D_k(v)/k$, contradicting the definition of D_k. It follows that K^k is the union of the sets $\sigma(S,T)$, $(S,T) \in \Xi(k)$. The key to the proof of Lemma 8.4 is to evaluate the cumulant c_λ^k over each $\sigma(S,T) \in \Xi(k)$, that is to write $\langle f, c_\lambda^k \rangle$ as a finite sum of integrals

$$\langle f, c_\lambda^k \rangle = \sum_{\sigma(S,T)\in\Xi(k)} \int_{\sigma(S,T)} f(v_1)\cdots f(v_k)\, dc_\lambda^k(v_1,\ldots,v_k),$$

then appeal to the representation (8.55) to write the cumulant measure $dc_\lambda^k(v_1,\ldots,v_k)$ on each $\sigma(S,T)$ as a linear combination of (S,T) semi-cluster measures, and finally to appeal to Lemma 8.3 to control the constituent cluster measures U^{S_1,T_1} by an exponentially decaying function of $\lambda^{1/d} D_k(v) := \lambda^{1/d} D_k(v_1,\ldots,v_k)$.

Given $\sigma(S,T)$, $S_1 \subset S$ and $T_1 \subset T$, this goes as follows. Let $x \in K^S$ and $y \in K^T$ denote elements of K^S and K^T, respectively; likewise we let $\tilde{x}$ and $\tilde{y}$ denote elements of K^{S_1} and K^{T_1}, respectively. Let $\tilde{x}^c$ denote the complement of $\tilde{x}$ with respect to x and likewise with $\tilde{y}^c$. The integral of f against one of the (S,T) semi-cluster measures in (8.55), induced by the partitions (S_1,S_2) and (T_1,T_2) of S and T respectively, has the form

$$\int_{\sigma(S,T)} f(v_1)\cdots f(v_k)\, d\left(M_\lambda^{|S_2|}(\tilde{x}^c) U_\lambda^{i+j}(\tilde{x},\tilde{y}) M_\lambda^{|T_2|}(\tilde{y}^c)\right).$$

Letting $u_\lambda(\tilde{x},\tilde{y}) := m_\lambda(\tilde{x},\tilde{y}) - m_\lambda(\tilde{x})m_\lambda(\tilde{y})$, the above equals

$$\int_{\sigma(S,T)} f(v_1)\cdots f(v_k) m_\lambda(\tilde{x}^c) u_\lambda(\tilde{x},\tilde{y}) m_\lambda(\tilde{y}^c) \prod_{i=1}^k \kappa(v_i)\, d(\lambda^{1/d} v_i). \tag{8.56}$$

We use Lemma 8.3 to control $u_\lambda(\tilde{x},\tilde{y}) := m_\lambda(\tilde{x},\tilde{y}) - m_\lambda(\tilde{x})m_\lambda(\tilde{y})$, we bound f and κ by their respective sup norms, we bound each mixed moment by $c(\xi,k)$, and we use $\sigma(S,T) \subset K^k$ to show that

$$\int_{\sigma(S,T)} f(v_1)\cdots f(v_k)\, d\left(M_\lambda^{|S_2|}(\tilde{x}^c) U_\lambda^{i+j}(\tilde{x},\tilde{y}) M_\lambda^{|T_2|}(\tilde{y}^c)\right)$$

$$\leq D(k) c(\xi,k)^2 \|f\|_\infty^k \|\kappa\|_\infty^k \int_{K^k} \exp(-c\lambda^{1/d} D_k(v)/k^2)\, d(\lambda^{1/d} v_1)\cdots d(\lambda^{1/d} v_k).$$

Letting $z_i := \lambda^{1/d} v_i$ the above bound becomes

$$\lambda D(k)c(\xi,k)^2\|f\|_\infty^k\|\kappa\|_\infty^k \int_{(\lambda^{1/d}K)^k} \exp(-cD_k(z)/k^2)\,dz_1\cdots dz_k$$

$$\leq \lambda D(k)c(\xi,k)^2\|f\|_\infty^k\|\kappa\|_\infty^k \int_{(\mathbb{R}^d)^{k-1}} \exp(-cD_k(0,z_1,\ldots,z_{k-1})/k^2)\,dz_1\cdots dz_k$$

where we use the translation invariance of $D_k(\cdot)$. Upon a further change of variable $w := z/k$ we have

$$\int_{\sigma(S,T)} f(v_1)\cdots f(v_k)\,d\left(M_\lambda^{|S_2|}(\tilde{x}^c)U_\lambda^{i+j}(\tilde{x},\tilde{y})M_\lambda^{|T_2|}(\tilde{y}^c)\right)$$

$$\leq \lambda\tilde{D}(k)c(\xi,k)^2\|f\|_\infty^k\|\kappa\|_\infty^k \int_{(\mathbb{R}^d)^{k-1}} \exp(-cD_k(0,w_1,\ldots,w_{k-1}))dw_1\cdots dw_{k-1}.$$

Finally, since $D_k(0,w_1,\ldots,w_{k-1}) \geq \|w_1\| + \ldots + \|w_{k-1}\|$ we obtain

$$\int_{\sigma(S,T)} f(v_1)\cdots f(v_k)d\left(M_\lambda^{|S_2|}(\tilde{x}^c)U_\lambda^{i+j}(\tilde{x},\tilde{y})M_\lambda^{|T_2|}(\tilde{y}^c)\right)$$

$$\leq \lambda\tilde{D}(k)c(\xi,k)^2\|f\|_\infty^k\|\kappa\|_\infty^k\left(\int_{\mathbb{R}^d}\exp(-\|w\|)\,dw\right)^{k-1} = O(\lambda)$$

as desired. □

8.3.6 Central Limit Theorem for Functionals of Binomial Input

To obtain central limit theorems for functionals over binomial input $\mathcal{X}_n := \{\eta_i\}_{i=1}^n$ we need some more definitions. For all functionals ξ and $\tau \in (0,\infty)$, recall the "add one cost" $\Delta^\xi(\tau)$ defined at (8.40). For all $j = 1,2,\ldots$, let $\mathcal{S}_j$ be the collection of all subsets of $\mathbb{R}^d$ of cardinality at most j.

Definition 8.6. Say that ξ has a moment of order $p > 0$ (with respect to binomial input $\mathcal{X}_n$) if

$$\sup_{n\geq 1, x\in\mathbb{R}^d, \mathcal{D}\in\mathcal{S}_3}\ \sup_{(n/2)\leq m\leq(3n/2)} \mathbf{E}[|\xi_n(x,\mathcal{X}_m\cup\mathcal{D})|^p] < \infty. \tag{8.57}$$

Definition 8.7. ξ is *binomially exponentially stabilizing* for κ if for all $x\in\mathbb{R}^d$, $\lambda\geq 1$, and $\mathcal{D}\subset\mathcal{S}_2$ almost surely there exists $R := R_{\lambda,n}(x,\mathcal{D}) < \infty$ such that for all finite $\mathcal{A}\subset(\mathbb{R}^d \setminus B_{\lambda^{-1/d}R}(x))$, we have

$$\xi_\lambda\left(x,([\mathcal{X}_n\cup\mathcal{D}]\cap B_{\lambda^{-1/d}R}(x))\cup\mathcal{A}\right) = \xi_\lambda\left(x,[\mathcal{X}_n\cup\mathcal{D}]\cap B_{\lambda^{-1/d}R}(x)\right), \tag{8.58}$$

and moreover there is an $\varepsilon > 0$ such that the tail probability $\tau_\varepsilon(t)$ defined for $t > 0$ by

$$\tau_\varepsilon(t) := \sup_{\lambda \geq 1,\, n \in \mathbb{N} \cap ((1-\varepsilon)\lambda, (1+\varepsilon)\lambda)} \ \sup_{x \in \mathbb{R}^d,\, \mathcal{D} \subset \mathcal{S}_2} \mathbf{P}(R_{\lambda,n}(x, \mathcal{D}) > t)$$

satisfies $\limsup_{t \to \infty} t^{-1} \log \tau_\varepsilon(t) < 0$.

If ξ is homogeneously stabilizing then in most examples of interest, similar methods can be used to show that ξ is binomially exponentially stabilizing whenever κ is bounded away from zero.

Exercise 8.7. Show that the interaction function ξ from Examples 1 and 2 is binomially exponentially stabilizing whenever κ is bounded away from zero on its support, assumed compact and convex.

Theorem 8.6 (CLT for binomial input). *Assume that κ is Lebesgue-almost everywhere continuous. Let ξ be a homogeneously stabilizing (8.30) translation invariant functional satisfying the moment conditions (8.35) and (8.57) for some $p > 2$. Suppose further that K is bounded and that ξ is exponentially stabilizing with respect to κ and K as in (8.32) and binomially exponentially stabilizing with respect to κ and K as in (8.58). Then for all $f \in \mathbb{B}(K)$ we have*

$$\lim_{n \to \infty} n^{-1} \mathbf{var}[\langle f, \rho_n \rangle] = \tau^2(f)$$
$$:= \int_K f(x)^2 V^\xi(\kappa(x))\kappa(x)\, dx - \left(\int_K f(x) \Delta^\xi(\kappa(x))\kappa(x) dx \right)^2 \tag{8.59}$$

as well as convergence of the finite-dimensional distributions

$$(\langle f_1, n^{-1/2}\overline{\rho}_n \rangle, \ldots, \langle f_k, n^{-1/2}\overline{\rho}_n \rangle),$$

$f_1, \ldots, f_k \in \mathbb{B}(K)$, to a mean zero Gaussian field with covariance kernel

$$(f, g) \mapsto \int_K f(x) g(x) V^\xi(\kappa(x))\kappa(x)\, dx$$
$$- \int_K f(x) \Delta^\xi(\kappa(x))\kappa(x)\, dx \int_K g(x) \Delta^\xi(\kappa(x))\kappa(x)\, dx. \tag{8.60}$$

Proof. We sketch the proof, borrowing heavily from coupling arguments appearing in the complete proofs given in [57, 395, 398]. Fix $f \in \mathbb{B}(K)$. Put $H_n := \langle f, \rho_n \rangle$, $H_n' := \langle f, \mu_n \rangle$, where μ_n is defined at (8.34) and assume that $\Pi_{n\kappa}$ is coupled to $\mathcal{X}_n$ by setting $\Pi_{n\kappa} = \bigcup_{i=1}^{N(n)} \eta_i$, where $N(n)$ is an independent Poisson random variable with mean n. Put

$$\alpha := \alpha(f) := \int_K f(x) \Delta^\xi(\kappa(x))\kappa(x)\, dx.$$

Conditioning on the random variable $N := N(n)$ and using that N is concentrated around its mean, it can be shown that as $n \to \infty$ we have

$$\mathbf{E}[(n^{-1/2}(H_n' - H_n - (N(n) - n)\alpha))^2] \to 0. \tag{8.61}$$

The arguments are long and technical (cf. Sect. 5 of [395], Sect. 4 of [398]).

Let $\sigma^2(f)$ be as at (8.41) and let $\tau^2(f)$ be as at (8.59), so that $\tau^2(f) = \sigma^2(f) - \alpha^2$.

By Theorem 8.5 we have as $n \to \infty$ that $n^{-1}\,\mathbf{var}[H_n'] \to \sigma^2(f)$ and $n^{-1/2}(H_n' - \mathbf{E}H_n') \xrightarrow{d} N(0, \sigma^2(f))$. We now deduce Theorem 8.6, following verbatim by now standard arguments (see for example p. 1020 of [398], p. 251 of [57]), included here for sake of completeness.

To prove convergence of $n^{-1}\,\mathbf{var}[H_n]$, we use the identity

$$n^{-1/2} H_n' = n^{-1/2} H_n + n^{-1/2}(N(n) - n)\alpha + n^{-1/2}[H_n' - H_n - (N(n) - n)\alpha]. \tag{8.62}$$

The variance of the third term on the right-hand side of (8.62) goes to zero by (8.61), whereas the second term has variance α^2 and is independent of the first term. It follows that with $\sigma^2(f)$ defined at (8.41), we have

$$\sigma^2(f) = \lim_{n\to\infty} n^{-1}\,\mathbf{var}[H_n'] = \lim_{n\to\infty} n^{-1}\,\mathbf{var}[H_n] + \alpha^2,$$

so that $\sigma^2(f) \geq \alpha^2$ and $n^{-1}\,\mathbf{var}[H_n] \to \tau^2(f)$. This gives (8.59).

Now to prove Theorem 8.6 we argue as follows. By Theorem 8.5, we have $n^{-1/2}(H_n' - \mathbf{E}H_n') \xrightarrow{d} N(0, \sigma^2(f))$. Together with (8.61), this yields

$$n^{-1/2}[H_n - \mathbf{E}H_n' + (N(n) - n)\alpha] \xrightarrow{d} N(0, \sigma^2(f)).$$

However, since $n^{-1/2}(N(n) - n)\alpha$ is independent of H_n and is asymptotically normal with mean zero and variance α^2, it follows by considering characteristic functions that

$$n^{-1/2}(H_n - \mathbf{E}H_n') \xrightarrow{d} N(0, \sigma^2(f) - \alpha^2). \tag{8.63}$$

By (8.61), the expectation of $n^{-1/2}(H_n' - H_n - (N(n) - n)\alpha)$ tends to zero, so in (8.63) we can replace $\mathbf{E}H_n'$ by $\mathbf{E}H_n$, which gives us

$$n^{-1/2}(H_n - \mathbf{E}H_n) \xrightarrow{d} N(0, \tau^2(f)).$$

To obtain convergence of finite-dimensional distributions (8.60) we use the Cramér–Wold device.

□

8.4 Applications

Consider a linear statistic $H^\xi(\mathcal{X})$ of a large geometric structure on $\mathcal{X}$. If we are interested in the limit behavior of H^ξ on random point sets, then the results of the previous section suggest checking whether the interaction function ξ is stabilizing. Verifying the stabilization of ξ is sometimes non-trivial and may involve discretization methods. Here we describe four non-trivial statistics H^ξ for which one may show stabilization/localization of ξ. Our list is non-exhaustive and primarily focusses on the problems described in Sect. 8.1.1.

8.4.1 Random Packing

Given $d \in \mathbb{N}$ and $\lambda \geq 1$, let $\eta_{1,\lambda}, \eta_{2,\lambda}, \ldots$ be a sequence of independent random d-vectors uniformly distributed on the cube $Q_\lambda := [0, \lambda^{1/d})^d$. Let $\tau_i, i \geq 1$, be i.i.d. time marks, independent of $\eta_i, i \geq 1$, and uniformly distributed on $[0, 1]$. Equip each vector η_i with the time mark τ_i and re-order the indices so that τ_i are increasing. Let S be a fixed bounded closed convex set in $\mathbb{R}^d$ with non-empty interior (i.e., a "solid") with centroid at the origin o of $\mathbb{R}^d$ (for example, the unit ball), and for $i \in \mathbb{N}$, let $S_{i,\lambda}$ be the translate of S having centroid at $\eta_{i,\lambda}$ and arrival time τ_i. Thus $\mathcal{S}_\lambda := (S_{i,\lambda})_{i \geq 1}$ is an infinite sequence of solids sequentially arriving at uniform random positions in Q_λ at arrival times $\tau_i, i \geq 1$ (the centroids lie in Q_λ but the solids themselves need not lie wholly inside Q_λ).

Let the first solid $S_{1,\lambda}$ be packed (i.e., accepted), and recursively for $i = 2, 3, \ldots$, let the i-th solid $S_{i,\lambda}$ be packed if it does not overlap any solid in $\{S_{1,\lambda}, \ldots, S_{i-1,\lambda}\}$ which has already been packed. If not packed, the i-th solid is discarded. This process, known as *random sequential adsorption (RSA) with infinite input*, is irreversible and terminates when it is not possible to accept additional solids. At termination, we say that the sequence of solids $\mathcal{S}_\lambda$ *jams* Q_λ or *saturates* Q_λ. The number of solids accepted in Q_λ at termination is denoted by the *jamming* number $N_\lambda := N_{\lambda,d} := N_{\lambda,d}(\mathcal{S}_\lambda)$.

There is a large literature of experimental results concerning the jamming numbers, but a limited collection of rigorous mathematical results, especially in $d \geq 2$. The short range interactions of arriving particles lead to complicated long range spatial dependence between the status of particles. Dvoretzky and Robbins [163] show in $d = 1$ that the jamming numbers $N_{\lambda,1}$ are asymptotically normal.

By writing the jamming number as a linear statistic involving a stabilizing interaction ξ on marked point sets, and recalling Remark 1 following Theorem 8.5, one may establish [458] that $N_{\lambda,d}$ are asymptotically normal for all $d \geq 1$. This puts the experimental results and Monte Carlo simulations of Quintanilla and Torquato [410] and Torquato (Chap. 11.4 of [494]) on rigorous footing.

Theorem 8.7. *Let $\mathcal{S}_\lambda$ and $N_\lambda := N_\lambda(\mathcal{S}_\lambda)$ be as above. There are constants $\mu := \mu(S,d) \in (0,\infty)$ and $\sigma^2 := \sigma^2(S,d) \in (0,\infty)$ such that as $\lambda \to \infty$ we have*

$$\left|\lambda^{-1}\mathbf{E}N_\lambda - \mu\right| = O(\lambda^{-1/d}) \tag{8.64}$$

and $\lambda^{-1}\,\mathbf{var}[N_\lambda] \to \sigma^2$ *with*

$$\sup_{t\in\mathbb{R}}\left|\mathbf{P}\left(\frac{N_\lambda - \mathbf{E}N_\lambda}{\sqrt{\mathbf{var}[N_\lambda]}} \le t\right) - \mathbf{P}(N(0,1)\le t)\right| = O((\log\lambda)^{3d}\lambda^{-1/2}). \tag{8.65}$$

To prove this, one could enumerate the arriving solids in $\mathcal{S}_\lambda$, by (x_i, t_i), where $x_i \in \mathbb{R}^d$ is the spatial coordinate of the i-th solid and $t_i \in [0,\infty)$ is its temporal coordinate, i.e. the arrival time. Furthermore, letting $\mathcal{X} := \{(x_i, t_i)\}_{i=1}^{\infty}$ be a marked point process, one could set $\xi((x,t),\mathcal{X})$ to be one or zero depending on whether the solid with center at $x \in \mathcal{S}_\lambda$ is accepted or not; $H^\xi(\mathcal{X})$ is the total number of solids accepted. Thus ξ is defined on elements of the marked point process $\mathcal{X}$. A natural way to prove Theorem 8.7 would then be to show that ξ satisfies the conditions of Theorem 8.5. The moment conditions (8.35) are clearly satisfied as ξ is bounded by 1. To show stabilization it turns out that it is easier to *discretize* as follows.

For any $A \subset \mathbb{R}^d$, let $A_+ := A \times \mathbb{R}_+$. Let $\zeta(\mathcal{X}, A)$ be the number of solids with centers in $\mathcal{X} \cap A$ which are packed according to the packing rules. Abusing notation, let Π denote a homogeneous Poisson point process in $\mathbb{R}^d \times \mathbb{R}_+$ with intensity $dx \times ds$, with dx denoting Lebesgue measure on $\mathbb{R}^d$ and ds denoting Lebesgue measure on $\mathbb{R}_+$. Abusing the terminology at (8.30), ζ is *homogeneously stabilizing* since it may be shown that almost surely there exists $R < \infty$ (a radius of homogeneous stabilization for ζ) such that for all $\mathcal{X} \subset (\mathbb{R}^d \setminus B_R)_+$ we have

$$\zeta((\Pi \cap (B_R)_+) \cup \mathcal{X}, Q_1) = \zeta(\Pi \cap (B_R)_+, Q_1). \tag{8.66}$$

Since ζ is homogeneously stabilizing it follows that the limit

$$\zeta(\Pi, i + Q_1) := \lim_{r\to\infty} \zeta(\Pi \cap (B_r(i))_+, i + Q_1)$$

exists almost surely for all $i \in \mathbb{Z}^d$. The random variables $(\zeta(\Pi, i + Q_1),\ i \in \mathbb{Z}^d)$ form a stationary random field. It may be shown that the tail probability for R decays exponentially fast.

Given ζ, for all $\lambda > 0$, all $\mathcal{X} \subset \mathbb{R}^d \times \mathbb{R}_+$, and all Borel $A \subset \mathbb{R}^d$ we let $\zeta_\lambda(\mathcal{X}, A) := \zeta(\lambda^{1/d}\mathcal{X}, \lambda^{1/d}A)$. Let Π_λ, $\lambda \ge 1$, denote a homogeneous Poisson point process in $\mathbb{R}^d \times \mathbb{R}_+$ with intensity measure $\lambda\, dx \times ds$. Define the random measure μ_λ^ζ on $\mathbb{R}^d$ by

$$\mu_\lambda^\zeta(\cdot) := \zeta_\lambda(\Pi_\lambda \cap Q_1, \cdot) \tag{8.67}$$

and the centered version $\overline{\mu}_\lambda^\zeta := \mu_\lambda^\zeta - \mathbf{E}[\mu_\lambda^\zeta]$. Modification of the stabilization methods of Sect. 8.3 then yield Theorem 8.7; this is spelled out in [458].

For companion results for RSA packing with *finite input per unit volume* we refer to [400].

8.4.2 Convex Hulls

Let $K \subset \mathbb{R}^d$ be a compact convex body with non-empty interior and having a C^3 boundary of positive Gaussian curvature $x \mapsto H_{d-1}(x)$, with $x \in \partial K$. Letting Π_λ be a Poisson point process in $\mathbb{R}^d$ of intensity λ we let K_λ be the convex hull of $K \cap \Pi_\lambda$.

The random polytope K_λ, together with the analogous polytope K_n obtained by considering n i.i.d. uniformly distributed points in $B_1(o)$, are well-studied objects in stochastic geometry, with a long history originating with the work of Rényi and Sulanke [421]. See the surveys of Affentranger [3], Buchta [88], Gruber [207], Schneider [444, 446], and Weil and Wieacker [513], together with Chap. 8.2 in Schneider and Weil [451]. See the overview in Sect. 7.1.

Functionals of K_λ of interest include its volume, here denoted $V_d(K_\lambda)$ and the number of k-dimensional faces of K_λ, here denoted $f_k(K_\lambda)$, $k \in \{0, 1, \dots, d-1\}$. Note that $f_0(K_\lambda)$ is the number of vertices of K_λ. The k-th intrinsic volumes of K_λ are denoted by $V_k(K_\lambda)$, $k \in \{1, \dots, d-1\}$.

As seen in Sect. 7.1, we have for all $d \geq 2$ and all $k \in \{0, \dots, d-1\}$ that there are constants $D_{k,d}$ such that

$$\lim_{\lambda \to \infty} \lambda^{-(d-1)/(d+1)} \mathbf{E} f_k(K_\lambda) = D_{k,d} \int_{\partial K} H_{d-1}(x)^{1/(d+1)} dx.$$

and one may wonder whether there exist similar asymptotics for limiting variances. This is indeed the case, which may be seen as follows.

Define the functional $\xi(x, \mathcal{X})$ to be one or zero, depending on whether $x \in \mathcal{X}$ is a vertex in the convex hull of $\mathcal{X}$. When $K = B_1(o)$ the unit ball in $\mathbb{R}^d$, by reformulating functionals of convex hulls in terms of functionals of re-scaled parabolic growth processes in space and time, it may be shown that ξ is exponentially localizing [111]. The arguments are non-trivial and we refer to [111] for details. Taking into account the proper scaling in space-time, a modification of Theorem 8.5 yields variance asymptotics for $V_d(K_\lambda)$, namely

$$\lim_{\lambda \to \infty} \lambda^{(d+3)/(d+1)} \mathbf{var}[V_d(K_\lambda)] = \sigma_V^2, \tag{8.68}$$

where $\sigma_V^2 \in (0, \infty)$ is a constant. This adds to Reitzner's central limit theorem (Theorem 1 of [419]), his variance approximation $\mathbf{var}[V_d(K_\lambda)] \approx \lambda^{-(d+3)/(d+1)}$ (Theorem 3 and Lemma 1 of [419]), and Hsing [248], which is confined to $d = 2$. The stabilization methods of Theorem 8.5 yield a central limit theorem for $V_d(K_\lambda)$.

Let $k \in \{0, 1, \dots, d-1\}$. Consider the functional $\xi_k(x, \mathcal{X})$, defined to be zero if x is not a vertex in the convex hull of $\mathcal{X}$ and otherwise defined to be the product of $(k+1)^{-1}$ and the number of k-dimensional faces containing x. Consideration of the parabolic growth processes and the stabilization of ξ_k in the context of such processes (cf. [111]) yield variance asymptotics and a central limit theorem for the number of k-dimensional faces of K_λ, yielding for all $k \in \{0, 1, \dots, d-1\}$

$$\lim_{\lambda \to \infty} \lambda^{-(d-1)/(d+1)} \mathbf{var}[f_k(K_\lambda)] = \sigma^2_{f_k}, \tag{8.69}$$

where $\sigma^2_{f_k} \in (0, \infty)$ is given as a closed form expression described in terms of paraboloid growth processes. For the case $k = 0$, this is proved in [459], whereas [111] handles the cases $k > 0$. This adds to Reitzner (Lemma 2 of [419]), whose breakthrough paper showed $\mathbf{var}[f_k(K_\lambda)] \approx \lambda^{(d-1)/(d+1)}$.

Theorem 8.5 also yields variance asymptotics for the intrinsic volumes $V_k(K_\lambda)$ of K_λ for all $k \in \{1, \ldots, d-1\}$, namely

$$\lim_{\lambda \to \infty} \lambda^{(d+3)/(d+1)} \mathbf{var}[V_k(K_\lambda)] = \sigma^2_{V_k}, \tag{8.70}$$

where again $\sigma^2_{V_k}$ is explicitly described in terms of paraboloid growth processes. This adds to Bárány et al. (Theorem 1 of [46]), which shows $\mathbf{var}[V_k(K_n)] \approx n^{-(d+3)/(d+1)}$.

8.4.3 *Intrinsic Dimension of High Dimensional Data Sets*

Given a finite set of samples taken from a multivariate distribution in $\mathbb{R}^d$, a fundamental problem in learning theory involves determining the intrinsic dimension of the sample [156, 299, 427, 492]. Multidimensional data ostensibly belonging to a high-dimensional space $\mathbb{R}^d$ often are concentrated on a smooth submanifold $\mathcal{M}$ or hypersurface with intrinsic dimension m, where $m < d$. The problem of determining the intrinsic dimension of a data set is of fundamental interest in machine learning, signal processing, and statistics and it can also be handled via analysis of the sums (8.1).

Discerning the intrinsic dimension m allows one to reduce dimension with minimal loss of information and to consequently avoid difficulties associated with the "curse of dimensionality". When the data structure is linear there are several methods available for dimensionality reduction, including principal component analysis and multidimensional scaling, but for non-linear data structures, mathematically rigorous dimensionality reduction is more difficult. One approach to dimension estimation, inspired by Levina and Bickel [328] uses probabilistic methods involving the k-nearest neighbour graph $G^N(k, \mathcal{X})$ defined in Sect. 8.1.2.

For all $k = 3, 4, \ldots$, the Levina and Bickel estimator of the dimension of a data cloud $\mathcal{X} \subset \mathcal{M}$, is given by

$$\hat{m}_k(\mathcal{X}) := (\mathrm{card}(\mathcal{X}))^{-1} \sum_{x \in \mathcal{X}} \xi_k(x, \mathcal{X}),$$

where for all $x \in \mathcal{X}$ we have

$$\xi_k(x, \mathcal{X}) := (k-2)\left(\sum_{j=1}^{k-1} \log \frac{D_k(x)}{D_j(x)}\right)^{-1},$$

where $D_j(x) := D_j(x, \mathcal{X})$, $1 \leq j \leq k$, are the distances between x and its j-th nearest neighbour in $\mathcal{X}$. We also define for all $\rho > 0$ the functionals

$$\xi_{k,\rho}(x, \mathcal{X}) := (k-2)\left(\sum_{j=1}^{k-1} \log \frac{D_k(x)}{D_j(x)}\right)^{-1} \mathbf{1}(D_k(x) < \rho)$$

and we put

$$\hat{m}_{k,\rho}(\mathcal{X}) := (\mathrm{card}(\mathcal{X}))^{-1} \sum_{x \in \mathcal{X}} \xi_{k,\rho}(x, \mathcal{X}).$$

Let $\{\eta_i\}_{i=1}^n$ be i.i.d. random variables with values in a submanifold $\mathcal{M}$ and put $\mathcal{X}_n := \{\eta_i\}_{i=1}^n$. Levina and Bickel [328] argue that $\hat{m}_k(\mathcal{X}_n)$ approximates the intrinsic dimension of $\mathcal{X}_n$, i.e., the dimension of $\mathcal{M}$. Indeed, $\hat{m}_k$ is an unbiased estimator when the underlying sample is a homogeneous Poisson point process on $\mathbb{R}^m$, as seen by the next exercise.

Exercise 8.8. Recall that Π_1 is a homogeneous Poisson point process on $\mathbb{R}^m$ of intensity 1. Conditional on D_k, the collection $\{(\frac{D_j(o,\Pi_1)}{D_k(o,\Pi_1)})^m\}_{j=1}^{k-1}$ is a sample from a Unif[0, 1]-distribution. Deduce that

$$\mathbf{E}\xi_k(o, \Pi_1) = m(k-2)\mathbf{E}\left(\sum_{j=1}^{k-1} \log(1/U_j)\right)^{-1} = m.$$

Subject to regularity conditions on $\mathcal{M}$ and the density κ, the papers [403, 526] substantiate the arguments of Levina and Bickel and show (a) consistency of the dimension estimator $\hat{m}_k(\mathcal{X}_n)$ and (b) a central limit theorem for $\hat{m}_{k,\rho}(\mathcal{X}_n)$, ρ fixed and small, together with a rate of convergence. This goes as follows.

For all $\tau > 0$, recall that Π_τ is a homogeneous Poisson point process on $\mathbb{R}^m$ of intensity τ. Recalling the notation (8.39) and (8.40), we put

$$\begin{aligned} V^{\xi_k}(\tau, m) &:= \mathbf{E}[\xi_k(o, \Pi_\tau)^2] \\ &\quad + \tau \int_{\mathbb{R}^m} \left[\mathbf{E}[\xi_k(o, \Pi_\tau \cup \{u\})\xi_k(u, \Pi_\tau \cup o)] - (\mathbf{E}[\xi_k(o, \Pi_\tau)])^2\right] du \end{aligned} \tag{8.71}$$

and

$$\delta^{\xi_k}(\tau, m) := \mathbf{E}[\xi_k(o, \Pi_\tau)] + \tau \int_{\mathbb{R}^m} \mathbf{E}[\xi_k(o, \Pi_\tau \cup \{u\}) - \xi_k(o, \Pi_\tau)]\, du. \tag{8.72}$$

We put $\delta^{\xi_k}(m) := \delta^{\xi_k}(1, m)$. Let $\Pi_{\lambda\kappa}$ be the collection $\{\eta_1, \ldots, \eta_{N(\lambda)}\}$, where η_i are i.i.d. with density κ and $N(\lambda)$ is an independent Poisson random variable with parameter λ. Thus $\Pi_{\lambda\kappa}$ is a Poisson point process on $\mathcal{M}$ with intensity $\lambda\kappa$. By extending Theorems 8.4 and 8.5 to C^1 submanifolds $\mathcal{M}$ as in [403], we obtain the following limit theory for the Levina and Bickel estimator.

Theorem 8.8. *Let κ be bounded away from zero and infinity on $\mathcal{M}$. We have for all $k \geq 4$*

$$\lim_{\lambda \to \infty} |\hat{m}_k(\Pi_{\lambda\kappa}) - m| = \lim_{n \to \infty} |\hat{m}_k(\mathcal{X}_n) - m| = 0, \tag{8.73}$$

where $m = \dim(\mathcal{M})$ and where the convergence holds in probability. If κ is a.e. continuous then there exists $\rho_1 > 0$ such that if $\rho \in (0, \rho_1)$ and $k \geq 7$, then

$$\lim_{n \to \infty} n\, \mathbf{var}[\hat{m}_{k,\rho}(\mathcal{X}_n)] = \sigma_k^2(m) := \frac{m^2}{k-3} - (\delta^{\xi_k}(m))^2 \tag{8.74}$$

and as $n \to \infty$,

$$n^{1/2}(\hat{m}_{k,\rho}(\mathcal{X}_n) - \mathbf{E}\hat{m}_{k,\rho}(\mathcal{X}_n)) \xrightarrow{d} N(0, \sigma_k^2(m)). \tag{8.75}$$

Remark. Theorem 8.8 adds to Chatterjee [116], who does not provide variance asymptotics (8.74) and who considers convergence rates with respect to the weaker Kantorovich–Wasserstein distance. Bickel and Yan (Theorems 1 and 3 of Sect. 4 of [67]) establish a central limit theorem for $\hat{m}_k(\mathcal{X}_n)$ for linear $\mathcal{M}$.

8.4.4 Clique Counts, Vietoris–Rips Complex

A central problem in data analysis involves discerning and counting clusters. Geometric graphs and the Vietoris–Rips complex play a central role and both are amenable to asymptotic analysis via stabilization techniques. The Vietoris–Rips complex is studied in connection with the statistical analysis of high-dimensional data sets [118], manifold reconstruction [119], and it has also received attention amongst topologists in connection with clustering and connectivity questions of data sets [112].

If $\mathcal{X} \subset \mathbb{R}^d$ is finite and $\beta > 0$, then the *Vietoris–Rips complex* $\mathcal{R}^\beta(\mathcal{X})$ is the abstract simplicial complex whose k-simplices (cliques of order $k + 1$) correspond to unordered $(k + 1)$ tuples of points of $\mathcal{X}$ which are pairwise within Euclidean distance β of each other. Thus, if there is a subset S of $\mathcal{X}$ of size $k + 1$ with all points of S distant at most β from each other, then S is a k-simplex in the complex.

Given $\mathcal{R}^\beta(\mathcal{X})$ and $k \in \mathbb{N}$, let $N_k^\beta(\mathcal{X})$ be the cardinality of k-simplices in $\mathcal{R}^\beta(\mathcal{X})$. Let $\xi_k^\beta(x, \mathcal{X})$ be the product of $(k + 1)^{-1}$ and the cardinality of k-simplices containing x in $\mathcal{R}^\beta(\mathcal{X})$. Thus $N_k^\beta(\mathcal{X}) = \sum_{x \in \mathcal{X}} \xi_k^\beta(x, \mathcal{X})$. The value of $\xi_k^\beta(x, \mathcal{X})$ depends only on points distant at most β from x, showing that β is a radius of stabilization for ξ_k^β and thus ξ_k^β is trivially exponentially stabilizing (8.32) and binomially exponentially stabilizing (8.58).

The next scaling result, which holds for C^1 submanifolds $\mathcal{M}$, links the large scale behavior of the clique count with the density κ of the underlying point set. Let η_i be i.i.d. with density κ on the manifold $\mathcal{M}$. Put $\mathcal{X}_n := \{\eta_i\}_{i=1}^n$. Let Π_τ be a homogeneous Poisson point process on $\mathbb{R}^m$ of constant intensity τ, dx the volume measure on $\mathcal{M}$, and let $V^{\xi_k^\beta}$ and $\delta^{\xi_k^\beta}$ be defined as in (8.39) and (8.40), respectively, with ξ replaced by ξ_k^β. It is shown in [403] that a generalization of Theorems 8.4 and 8.6 to binomial input on manifolds yields:

Theorem 8.9. *Let κ be bounded on $\mathcal{M}$; $\dim \mathcal{M} = m$. For all $k \in \mathbb{N}$ and all $\beta > 0$ we have*

$$\lim_{n\to\infty} n^{-1} N_k^\beta(n^{1/m}\mathcal{X}_n) = \int_{\mathcal{M}} \mathbf{E}[\xi_k^\beta(o, \Pi_{\kappa(x)})]\kappa(x)\,dx \;\; \text{in } L^2. \tag{8.76}$$

If κ is a.e. continuous and bounded away from zero on its support, assumed to be a compact subset of $\mathcal{M}$, then

$$\lim_{n\to\infty} n^{-1}\,\mathbf{var}[N_k^\beta(n^{1/m}\mathcal{X}_n)]$$
$$= \sigma_k^2(m) := \int_{\mathcal{M}} V^{\xi_k^\beta}(\kappa(x))\kappa(x)\,dx - \left(\int_{\mathcal{M}} \delta^{\xi_k^\beta}(\kappa(x))\kappa(x)\,dx\right)^2 \tag{8.77}$$

and, as $n \to \infty$

$$n^{-1/2}(N_k^\beta(n^{1/m}\mathcal{X}_n) - \mathbf{E}N_k^\beta(n^{1/m}\mathcal{X}_n)) \xrightarrow{d} N(0, \sigma_k^2(m)). \tag{8.78}$$

This result extends Proposition 3.1, Theorems 3.13 and 3.17 of [394]. For more details and for further simplification of the limits (8.76) and (8.77) we refer to [403].

Chapter 9
Introduction to Random Fields

Alexander Bulinski and Evgeny Spodarev

Abstract This chapter gives preliminaries on random fields necessary for understanding of the next two chapters on limit theorems. Basic classes of random fields (Gaussian, stable, infinitely divisible, Markov and Gibbs fields, etc.) are considered. Correlation theory of stationary random functions as well as elementary nonparametric statistics and an overview of simulation techniques are discussed in more detail.

9.1 Random Functions

Let $(\Omega, \mathcal{A}, \mathbf{P})$ be a probability space and $(S, \mathcal{B})$ be a *measurable space* (i.e. we consider an arbitrary set S endowed with a sigma-algebra $\mathcal{B}$). We always assume that $\Omega \neq \emptyset$ and $S \neq \emptyset$.

Definition 9.1. A *random element* $\xi : \Omega \to S$ is an $\mathcal{A}|\mathcal{B}$*-measurable mapping* (one writes $\xi \in \mathcal{A}|\mathcal{B}$), that is,

$$\xi^{-1}(B) := \{\omega \in \Omega : \xi(\omega) \in B\} \in \mathcal{A} \text{ for all } B \in \mathcal{B}. \tag{9.1}$$

If ξ is a random element, then for a given $\omega \in \Omega$, the value $\xi(\omega)$ is called a *realization* of ξ.

We say that the sigma-algebra $\mathcal{B}$ consisting of some subsets of S is *generated by a system* $\mathcal{M}$ of subsets of S if $\mathcal{B}$ is the intersection of all σ-algebras (of subsets of S)

A. Bulinski (✉)
Moscow State University, Moscow, Russia
e-mail: bulinski@mech.math.msu.su

E. Spodarev
Ulm University, Ulm, Germany
e-mail: evgeny.spodarev@uni-ulm.de

E. Spodarev (ed.), *Stochastic Geometry, Spatial Statistics and Random Fields*,
Lecture Notes in Mathematics 2068, DOI 10.1007/978-3-642-33305-7_9,

containing $\mathcal{M} \cup \{S\}$. One uses the notation $\mathcal{B} = \sigma\{\mathcal{M}\}$. For topological (or metric) space S one usually takes $\mathcal{B} = \mathcal{B}(S)$ where $\mathcal{B}(S)$ is the *Borel σ-algebra*. Recall that $\mathcal{B}(S)$ is generated by all open subsets of S.

If $S = \mathbb{R}^n$ and $\mathcal{B} = \mathcal{B}(\mathbb{R}^n)$ where $n \in \mathbb{N}$, then random element ξ is called a *random variable* when $n = 1$ and a *random vector* when $n > 1$. One also says that ξ is a random variable with values in a state-space S (endowed with σ-algebra $\mathcal{B}$) if (9.1) holds.

Exercise 9.1. Let $(\Omega, \mathcal{A})$ and $(S, \mathcal{B})$ be measurable spaces and $\mathcal{B} = \sigma\{\mathcal{M}\}$ with $\mathcal{M}$ being a family of subsets of S. Prove that a mapping $\xi : \Omega \to S$ is $\mathcal{A}|\mathcal{B}$-measurable iff $\xi^{-1}(C) \in \mathcal{A}$ for any $C \in \mathcal{M}$.

Example 9.1 (Point process). Let $\mathcal{N}$ be the set of all *locally finite simple point patterns* $\varphi = \{x_i\}_{i=1}^{\infty} \subset \mathbb{R}^d$, cf. Sect. 3.1.1. It means $\varphi(B) := |\varphi \cap B| < \infty$ for any bounded set $B \in \mathcal{B}(\mathbb{R}^d)$ (one writes $B \in \mathcal{B}_0(\mathbb{R}^d)$) where $|A|$ stands for the cardinality of A and we assume that $x_i \neq x_j$ for $i \neq j$. Let $\mathfrak{N}$ be the minimal σ-algebra generated in $\mathcal{N}$ by all sets of the form $\{\varphi \in \mathcal{N} : \varphi(B) = k\}$ for $k \in \mathbb{Z}_+$ and $B \in \mathcal{B}_0(\mathbb{R}^d)$. Take $(S, \mathcal{B}) = (\mathcal{N}, \mathfrak{N})$. The *point process* $\Psi : \Omega \to \mathcal{N}$ is an $\mathcal{A}|\mathfrak{N}$-measurable random element. Another possibility to define Ψ is to use a *random counting measure*

$$\Psi(\omega, B) = \sum_{i=1}^{\infty} \delta_{x_i(\omega)}(B), \quad \omega \in \Omega, \; B \in \mathcal{B}_0(\mathbb{R}^d), \tag{9.2}$$

where δ_x is the Dirac measure concentrated at a point x and a point process $\{x_i(\omega)\}$ can be viewed as a support of this measure, see Sect. 4.1.1 (Fig. 9.1).

Example 9.2 (Random closed sets). Let $\mathcal{F}$ be the family of all closed sets in $\mathbb{R}^d$. Introduce σ-algebra $\mathfrak{F}$ generated by the classes of sets $\mathfrak{F}_B = \{A \in \mathcal{F} : A \cap B \neq \emptyset\}$ where $B \subset \mathbb{R}^d$ is any compact. The *random closed set* (RACS) is a random element $X : \Omega \to \mathcal{F}, \; X \in \mathcal{A}|\mathfrak{F}$, cf. Sect. 1.2.1. In particular one can take $X = \cup_{i=1}^{\infty} B_r(x_i)$ for fixed $r > 0$. Here $B_r(x)$ is a closed ball in $\mathbb{R}^d$ of radius $r > 0$, centered at x, and $\{x_i\}_{i=1}^{\infty}$ is a point process ($x_i = x_i(\omega)$). This corresponds to the special case of the so called *germ-grain models*, cf. Example 1 of Sect. 4.1.3.

Now we consider a general definition of random functions.

Definition 9.2. Let T be an arbitrary index set and $(S_t, \mathcal{B}_t)_{t \in T}$ a collection of measurable spaces. A family $\xi = \{\xi(t), \, t \in T\}$ of random elements $\xi(t) : \Omega \to S_t$ defined on a probability space $(\Omega, \mathcal{A}, \mathbf{P})$ and $\mathcal{A}|\mathcal{B}_t$-measurable for each $t \in T$ is called a *random function* (associated with $(S_t, \mathcal{B}_t)_{t \in T}$).

In other words, $\xi = \xi(\omega, t)$ is defined on $\Omega \times T$, $\xi(\omega, t) \in S_t$ for any $\omega \in \Omega$ and $t \in T$, moreover, $\xi(\cdot, t) \in \mathcal{A}|\mathcal{B}_t$ for each $t \in T$. Usually $(S_t, \mathcal{B}_t) = (S, \mathcal{B})$ for any $t \in T$. The argument ω is traditionally omitted and one writes $\xi(t)$ instead of

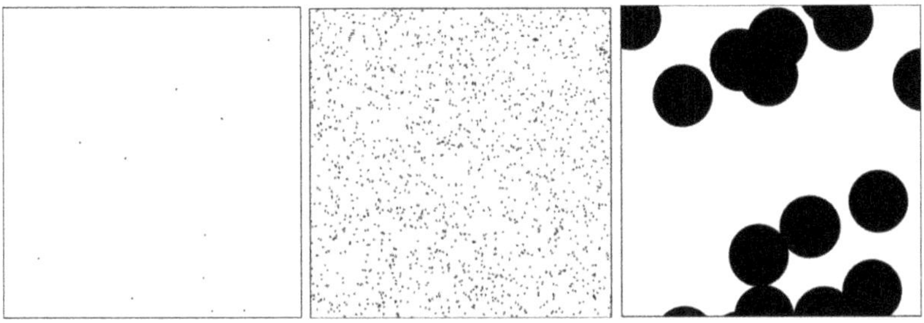

(a) $\lambda = 0.0001$ (b) $\lambda = 0.01$ (c) $\lambda = 0.0001, r = 40$

Fig. 9.1 A realization of a homogeneous Poisson point process with intensity $\lambda = 0.0001$ (*left*), $\lambda = 0.01$ (*middle*) and a germ-grain model of equal discs based on a homogeneous Poisson point process with intensity $\lambda = 0.0001$

$\xi(\omega, t)$. A random function $\xi = \{\xi(t),\ t \in T\}$ is called a *stochastic process*[1] if $T \subset \mathbb{R}$ and a *random field* if $T \subset \mathbb{R}^d$ where $d > 1$. If $T = \{x \in \mathbb{R}^d : a_i \le x_i \le b_i,\ i = 1, \dots, d\}$ (where $a_i < b_i$, $i = 1, \dots, d$), $T = \mathbb{R}^d$ or $T = \mathbb{R}^d_+ = [0, \infty)^d$, then $\xi = \{\xi(t),\ t \in T\}$ is called a *random field (or process) with continuous parameter (time)*. Whenever $T = \mathbb{Z}^d$, $T = \mathbb{Z}^d_+$ or $T = \mathbb{N}^d$ one calls ξ a *random field (or process) with discrete parameter (time)*.

In fact a random function $\xi = \{\xi(t),\ t \in T\}$ can be viewed as a random element with values in some functional space endowed with specified σ-algebra. Set $S_T = \prod_{t \in T} S_t$, i.e. we consider the Cartesian product of spaces S_t, $t \in T$. Thus $x \in S_T$ is a function such that $x(t) \in S_t$ for each $t \in T$. Introduce, for $t \in T$ and $B_t \in \mathcal{B}_t$, the *elementary cylinder* in S_T as follows

$$C_T(B_t) := \{x \in S_T :\ x(t) \in B_t\}.$$

This set contains all functions $x \in S_T$ that go through the "gate" B_t (see Fig. 9.2).

Definition 9.3. A *cylindric σ-algebra* $\mathcal{B}_T$ is a σ-algebra generated in S_T by the collection of all elementary cylinders. One writes $\mathcal{B}_T = \bigotimes_{t \in T} \mathcal{B}_t$ and if $\mathcal{B}_t = \mathcal{B}$ for all $t \in T$ then one uses the notation $\mathcal{B}^T$.

Exercise 9.2. A family $\xi = \{\xi(t),\ t \in T\}$ is a random function defined on a probability space $(\Omega, \mathcal{A}, \mathbf{P})$ and associated with a collection of measurable spaces $(S_t, \mathcal{B}_t)_{t \in T}$ iff, for $\omega \in \Omega$, the mapping $\omega \longmapsto \xi(\omega, \cdot)$ is $\mathcal{A}|\mathcal{B}_T$-measurable. *Hint:* use Exercise 9.1.

For any fixed $\omega \in \Omega$, the function $\xi(\omega, t)$, $t \in T$, is called a *trajectory* of ξ. Thus in view of Exercise 9.2 the trajectory is a realization of the random element ξ with values in a space $(S_T, \mathcal{B}_T)$.

---

[1]The notation T comes from "time", since for random processes $t \in T$ is often interpreted as the time parameter.

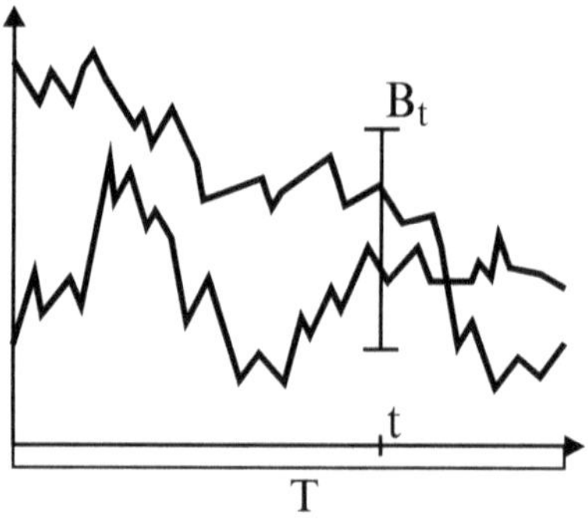

Fig. 9.2 Trajectories going through a "gate" B_t

Definition 9.4. Let $(S, \mathcal{B})$ be a measurable space and $\xi : \Omega \to S$ a random element defined on a probability space $(\Omega, \mathcal{A}, \mathbf{P})$. The *distribution* (or *law*) of ξ is a probability measure $\mathbf{P}_\xi$ on $(S, \mathcal{B})$ such that $\mathbf{P}_\xi(B) = \mathbf{P}(\xi^{-1}(B))$, $B \in \mathcal{B}$.

Alternatively the notation $\text{Law}(\xi)$ or $\mathbf{P}\xi^{-1}$ can be used.

Lemma 9.1. *Any probability measure μ on $(S, \mathcal{B})$ can be considered as a distribution of some random element ξ.*

Proof. Take $\Omega = S$, $\mathcal{A} = \mathcal{B}$, $\mathbf{P} = \mu$ and set $\xi(\omega) = \omega$ for $\omega \in \Omega$. □

Recall the following basic concept.

Definition 9.5. Let $(S_t, \mathcal{B}_t)_{t \in T}$ be a collection of measurable spaces. A family $\xi = \{\xi(t), t \in T\}$ consists of *independent random elements* $\xi(t) : \Omega \to S_t$ defined on a probability space $(\Omega, \mathcal{A}, \mathbf{P})$ if, for any $n \in \mathbb{N}$, $\{t_1, \dots, t_n\} \subset T$, $B_k \in \mathcal{B}_{t_k}$ where $k = 1, \dots, n$, one has

$$\mathbf{P}(\xi(t_1) \in B_1, \dots, \xi(t_n) \in B_n) = \prod_{k=1}^{n} \mathbf{P}(\xi(t_k) \in B_k). \tag{9.3}$$

Otherwise one says that ξ consists of *dependent* random elements.

Independence of $\xi(t)$, $t \in T$, is equivalent to the statement that σ-algebras $\mathcal{A}_t := \{(\xi(t))^{-1}(\mathcal{B}_t)\}$ are independent. For the sake of simplicity we assume that a single random element forms an "independent family" of random elements. We use the standard notation $\{C_1, \dots, C_n\} := \cap_{k=1}^{n} C_k$ in (9.3) for events $C_k = \{\xi(t_k) \in B_k\}$.

Let us consider the problem of existence of random functions with some predefined properties. We start with the following result, see for example [283, p. 93].

Theorem 9.1 (Lomnicki, Ulam). *Let $(S_t, \mathcal{B}_t)_{t \in T}$ be an arbitrary family of measurable spaces and μ_t be a probability measure on $(S_t, \mathcal{B}_t)$ for every $t \in T$. Then there exists a random function $\xi = \{\xi(t), t \in T\}$ (associated with $(S_t, \mathcal{B}_t)_{t \in T}$) defined on a probability space $(\Omega, \mathcal{A}, \mathbf{P})$ and such that*

1. ξ *consists of independent random elements,*
2. $\mathbf{P}_{\xi(t)} = \mu_t$ *on* $(S_t, \mathcal{B}_t)$ *for each* $t \in T$.

Many useful classes of random functions are constructed on the basis of independent random elements, see Examples in Sect. 9.2.

Exercise 9.3. Let $\xi = \{\xi(t),\ t \in T\}$ be a random function defined on $(\Omega, \mathcal{A}, \mathbf{P})$ and associated with a family of measurable spaces $(S_t, \mathcal{B}_t)_{t\in T}$. For any finite set $\{t_1, \ldots, t_n\} \subset T$ consider a column vector $\zeta = (\xi(t_1), \ldots, \xi(t_n))^\top$ where $\top$ stands for transposition. Prove that ζ is an $\mathcal{A}|\mathcal{B}_{t_1,\ldots,t_n}$-measurable random element with values in the space $S_{t_1,\ldots,t_n} := S_{t_1} \times \ldots \times S_{t_n}$ endowed with σ-algebra $\mathcal{B}_{t_1,\ldots,t_n}$ generated by "rectangles" $B_1 \times \ldots \times B_n$ where $B_k \in \mathcal{B}_{t_k}$, $k = 1, \ldots, n$. *H*int: use Exercise 9.1.

Taking into account the last exercise we can give

Definition 9.6. Let $\xi = \{\xi(t),\ t \in T\}$ be a random function defined on $(\Omega, \mathcal{A}, \mathbf{P})$ and associated with a family of measurable spaces $(S_t, \mathcal{B}_t)_{t\in T}$. For $n \in \mathbb{N}$ and $\{t_1, \ldots, t_n\} \subset T$ we call the *finite-dimensional distribution* of a random function ξ the law $P_{t_1,\ldots,t_n}$ of the random vector $(\xi(t_1), \ldots, \xi(t_n))^\top$ on $(S_{t_1,\ldots,t_n}, \mathcal{B}_{t_1,\ldots,t_n})$. Thus

$$P_{t_1,\ldots,t_n}(C) = \mathbf{P}((\xi(t_1), \ldots, \xi(t_n))^\top \in C), \quad C \in \mathcal{B}_{t_1,\ldots,t_n}.$$

In particular for $C = B_{t_1} \times \ldots \times B_{t_n}$ where $B_k \in \mathcal{B}_{t_k}$, $k = 1, \ldots, n$, we have

$$P_{t_1,\ldots,t_n}(B_1 \times \ldots \times B_n) = \mathbf{P}(\xi(t_1) \in B_1, \ldots, \xi(t_n) \in B_n). \tag{9.4}$$

Theorem 9.1 shows that one can construct a random function (consisting of independent random elements) with finite-dimensional distributions $P_{t_1,\ldots,t_n}$ for which expressions in (9.4) are determined by (9.3). We also mention in passing the important Ionescu–Tulcea Theorem (see for example [472, p. 249]) permitting to define a sequence of random elements $(X_n)_{n\in\mathbb{N}}$ taking values in arbitrary measurable spaces $(S_n, \mathcal{B}_n)$ and having the finite-dimensional distributions specified by means of probability kernels.

Under wide conditions, one can also ensure the existence of a family of dependent random elements. Observe that the finite-dimensional distributions $P_{t_1,\ldots,t_n}$ of a random function $\xi = \{\xi(t),\ t \in T\}$ associated with a family of measurable spaces $(S_t, \mathcal{B}_t)_{t\in T}$ possess (in nontrivial case $|T| > 1$) the important properties listed in

Exercise 9.4. Show that for any integer $n \geq 2$, $\{t_1, \ldots, t_n\} \subset T$, $B_k \in \mathcal{B}_{t_k}$, $k = 1, \ldots, n$, and an arbitrary permutation $(i_1, \ldots, i_n)$ of $(1, \ldots, n)$ the *consistency conditions* are satisfied:

$$P_{t_1,\ldots,t_n}(B_1 \times \ldots \times B_n) = P_{t_{i_1}\times\ldots\times t_{i_n}}(B_{i_1} \times \ldots \times B_{i_n}), \tag{9.5}$$

$$P_{t_1,\ldots,t_n}(B_1 \times \ldots \times B_{n-1} \times S_{t_n}) = P_{t_1,\ldots,t_{n-1}}(B_1, \ldots, B_{n-1}). \tag{9.6}$$

To formulate the fundamental theorem on the existence of a random function with given finite-dimensional distributions, we restrict ourselves to the case of $(S_t, \mathcal{B}_t) = (\mathbb{R}^m, \mathcal{B}(\mathbb{R}^m))$ for some $m \in \mathbb{N}$. A more general version of Theorem 9.2 (for the so-called *Borel spaces*) can be found, e.g., in [102, p. 26] and [283, p. 92].

Theorem 9.2 (Kolmogorov). *Let probability measures $\mathbf{P}_{t_1,\dots,t_n}$ be given on spaces $(\mathbb{R}^m \times \ldots \times \mathbb{R}^m, \mathcal{B}(\mathbb{R}^m) \otimes \ldots \otimes \mathcal{B}(\mathbb{R}^m))$ for any $\{t_1, \dots, t_n\} \subset T$ and $n \in \mathbb{N}$. Assume that they satisfy the consistency conditions. Then there exists a random function $\xi = \{\xi(t),\ t \in T\}$ defined on a probability space $(\Omega, \mathcal{A}, \mathbf{P})$ and such that its finite-dimensional distributions coincide with $\mathbf{P}_{t_1,\dots,t_n}$.*

Note that Theorem 9.2 provides the necessary and sufficient conditions for existence of a random function (defined on an arbitrary T and taking values in a Euclidean space for each $t \in T$) with given finite-dimensional distributions. Moreover, in this case it can be convenient to verify the consistency conditions by means of the characteristic functions of probability measures on Euclidean spaces. Recall that for any $n \in \mathbb{N}$ and $\{t_1, \dots, t_n\} \subset T$, the characteristic function of the random vector $\xi_{t_1,\dots,t_n} = (\xi(t_1), \dots, \xi(t_n))^\top$ (or, equivalently, of the finite-dimensional distribution $P_{t_1,\dots,t_n}$ of ξ) is

$$\varphi_{t_1,\dots,t_n}(\lambda) = \mathbf{E}\exp\{i\,\langle \lambda, \xi_{t_1,\dots,t_n}\rangle\}, \quad \lambda = (\lambda_1, \dots, \lambda_n)^\top \in \mathbb{R}^{mn}, \quad i^2 = -1, \quad (9.7)$$

$\langle \cdot, \cdot \rangle$ being a scalar product in $\mathbb{R}^{mn}$.

Exercise 9.5. Let $\varphi_{t_1,\dots,t_n}(\lambda_1, \dots, \lambda_n)$ be the characteristic function of the probability measure $P_{t_1,\dots,t_n}$ on $(\mathbb{R}^{mn}, \mathcal{B}(\mathbb{R}^{mn}))$ where $\{t_1, \dots, t_n\} \subset T$ and $\lambda_k \in \mathbb{R}^m$, $k = 1, \dots, n$. Prove that consistency conditions (9.5) and (9.6) are equivalent to the following ones: given integer $n \geq 2$, for each $\lambda = (\lambda_1, \dots, \lambda_n)^\top \in \mathbb{R}^{mn}$ and any permutation $(i_1, \dots, i_n)$ of $(1, \dots, n)$ one has

$$\varphi_{t_{i_1},\dots,t_{i_n}}(\lambda_{i_1}, \dots, \lambda_{i_n}) = \varphi_{t_1,\dots,t_n}(\lambda_1, \dots, \lambda_n),$$

$$\varphi_{t_1,\dots,t_n}(\lambda_1, \dots, \lambda_{n-1}, o) = \varphi_{t_1,\dots,t_{n-1}}(\lambda_1, \dots, \lambda_{n-1}), \quad o = (0, \dots, 0)^\top \in \mathbb{R}^m.$$

Let $(S, \mathcal{B})$ be a measurable space where S is a group with addition operation. Assume that $-A = \{-x : x \in A\} \in \mathcal{B}$ for any $A \in \mathcal{B}$. This holds, for instance, if $\mathcal{B} = \mathcal{B}(\mathbb{R}^d)$.

Definition 9.7. One says that a measure μ on $(S, \mathcal{B})$ is *symmetric* if $\mu(-A) = \mu(A)$ for all $A \in \mathcal{B}$. A random element ξ with values in S is *symmetric* if $\mathbf{P}_\xi$ is symmetric. In other words, $\mathbf{P}_{-\xi} = \mathbf{P}_\xi$.

Exercise 9.6. Prove that a real-valued random function $\xi = \{\xi(t),\ t \in T\}$ is *symmetric* iff all its finite-dimensional distributions are symmetric.

Let T be a metric space. For Sect. 9.7.4, we need the following

Definition 9.8. Let $\xi = \{\xi(t),\ t \in T\}$ be a real-valued random function such that $\mathbf{E}|\xi(t)|^p < \infty$ for all $t \in T$ and some $p \geq 1$. One says that ξ is *L^p-continuous* if $\xi(s) \xrightarrow{L^p} \xi(t)$ as $s \to t$ for each $t \in T$, i.e., $\mathbf{E}|\xi(s) - \xi(t)|^p \to 0$. It is called *$L^p$-separable* if there exists a countable subset $T_0 \subset T$ such that for any $t \in T$ one can find a sequence $(t_n) \subset T_0$ with $\xi(t_n) \xrightarrow{L^p} \xi(t)$ as $n \to \infty$.

Remark 9.1. If T is a separable metric space and ξ is L^p-continuous, then ξ is L^p-separable.

9.2 Some Basic Examples

In general the construction of a random function involving the Kolmogorov theorem is not easy. Instead one can employ, for example, the representations of the form $\xi(t) = g(t, \eta_1, \eta_2, \ldots), t \in T$, where g is an appropriately measurable function and $\eta_1, \eta_2, \ldots$ are random elements already known to exist.

9.2.1 *White Noise*

Definition 9.9. A random function $\xi = \{\xi(t),\ t \in T\}$ defined on $(\Omega, \mathcal{A}, \mathbf{P})$ is called *white noise* if $\xi(t), t \in T$, are independent and identically distributed (i.i.d.) random variables.

The white noise exists due to Theorem 9.1. Alternatively, one can easily verify consistency conditions, as all finite-dimensional distributions are products of the marginal distributions, and apply Theorem 9.2. Note that white noise is employed to model noise in images, such as *salt-and-pepper noise* ($\xi(t) \sim \text{Ber}(p),\ t \in T$) for binary images or *Gaussian white noise* ($\xi(t) \sim N(0, \sigma^2), \sigma^2 > 0$) for *greyscale images*, see for example [520, pp. 16–17].

9.2.2 *Gaussian Random Functions*

The famous simple (but important) example of a random function with finite-dimensional distributions given explicitly is that of a Gaussian one.

Definition 9.10. A real-valued random function $\xi = \{\xi(t),\ t \in T\}$ is called *Gaussian* if all its finite-dimensional distributions $P_{t_1,\ldots,t_n}$ are Gaussian, i.e. for any $n \in \mathbb{N}$ and $\{t_1, \ldots, t_n\} \subset T$ the distribution of the random vector $\xi_{t_1,\ldots,t_n} = (\xi(t_1), \ldots, \xi(t_n))^\top$ is normal law in $\mathbb{R}^n$ with mean $\mu_{t_1,\ldots,t_n}$ and covariance matrix $\Sigma_{t_1,\ldots,t_n}$. In other words, $\xi_\tau \sim N(\mu_\tau, \Sigma_\tau)$ where $\tau = (t_1, \ldots, t_n)^\top$. Here we use the vector τ instead of indices $t_1, \dots, t_n$. This means (see (9.7) with $m = 1$) that the characteristic function of ξ_τ is provided by formula

$$\varphi_\tau(\lambda) = \exp\left\{ i\langle\lambda, \mu_\tau\rangle - \frac{1}{2}\langle\Sigma_\tau\lambda, \lambda\rangle \right\}, \quad \lambda \in \mathbb{R}^n, \quad i^2 = -1. \tag{9.8}$$

One can verify that $\mu_\tau = (\mathbf{E}\xi(t_1), \dots, \mathbf{E}\xi(t_n))^\top$ and $\Sigma_\tau = (\mathbf{cov}(\xi(t_i), \xi(t_j)))_{i,j=1}^n$. If Σ_τ is nondegenerate, then P_τ has the density

$$f_\tau(x) = \frac{1}{(2\pi)^{n/2}\sqrt{\det \Sigma_\tau}} \exp\left\{ -\frac{1}{2}(x - \mu_\tau)^\top \Sigma_\tau^{-1}(x - \mu_\tau) \right\}, \quad x \in \mathbb{R}^n. \tag{9.9}$$

Exercise 9.7. Show that η is a Gaussian vector in $\mathbb{R}^n$ iff for any nonrandom vector $c \in \mathbb{R}^n$ the random variable $\langle c, \eta\rangle$ is normally distributed.

As an example, we mention a *Gaussian field with generalized Cauchy covariance* recently studied in [333]. It is a translation invariant (i.e., stationary, see Sect. 9.5) Gaussian field $\xi = \{\xi(t),\ t \in \mathbb{R}^d\}$ with zero mean ($\mathbf{E}\,\xi(t) = 0,\ t \in \mathbb{R}^d$) and covariance $C(t) = \mathbf{E}\,(\xi(o)\xi(t))$ given by

$$C(t) = (1 + \| t \|_2^\alpha)^{-\beta}, \qquad t \in \mathbb{R}^d, \tag{9.10}$$

where $\alpha \in (0, 2]$, $\beta > 0$ and $\| \cdot \|_2$ is the Euclidean norm in $\mathbb{R}^d$. Note that (9.10) has the same functional form as the characteristic function of the *generalized multivariate Linnik distribution* first studied in [13]; ξ becomes a Gaussian field with usual Cauchy covariance when $\alpha = 2$ and $\beta = 1$.

Gaussian random functions are widely used in applications ranging from modelling the microstructure of surfaces in material science (for example the surfaces of metal or paper, see Fig. 9.3) to models of fluctuations of cosmic microwave background radiation (see Fig. 9.4, [271]).

In a way similar to (9.8) we can introduce a Gaussian random field with values in $\mathbb{R}^m$. In this case, the distribution of $\xi_{t_1,\dots,t_n}$ is nm-dimensional Gaussian. A random field $\xi = \{\xi(t),\ t \in T\}$ where $\xi(t)$ takes values in $\mathbb{C}$ is called *Gaussian complex-valued* if $\eta = \{(\mathrm{Re}\,\xi(t), \mathrm{Im}\,\xi(t))^\top,\ t \in T\}$ is a Gaussian field with values in $\mathbb{R}^2$, i.e. the vector $(\mathrm{Re}\,\xi(t_1), \mathrm{Im}\,\xi(t_1), \dots, \mathrm{Re}\,\xi(t_n), \mathrm{Im}\,\xi(t_n))^\top$ has the normal distribution in $\mathbb{R}^{2n}$ for any $n \in \mathbb{N}$ and $\{t_1, \dots, t_n\} \subset T$.

9.2.3 Lognormal and χ^2 Random Functions

A random function $\xi = \{\xi(t),\ t \in T\}$ is called *lognormal* if $\xi(t) = \mathrm{e}^{\eta(t)}$ where $\eta = \{\eta(t),\ t \in T\}$ is a Gaussian random field. A random function $\xi = \{\xi(t),\ t \in T\}$ is called χ^2 if $\xi(t) = \|\eta(t)\|_2^2$, $t \in T$, where $\eta = \{\eta(t),\ t \in T\}$ is a Gaussian random field with values in $\mathbb{R}^n$ such that $\eta(t) \sim N(o, \mathrm{I})$, $t \in T$, I denotes the $(n \times n)$-identity matrix. Evidently, $\xi(t)$ is χ_n^2-distributed for all $t \in T$.

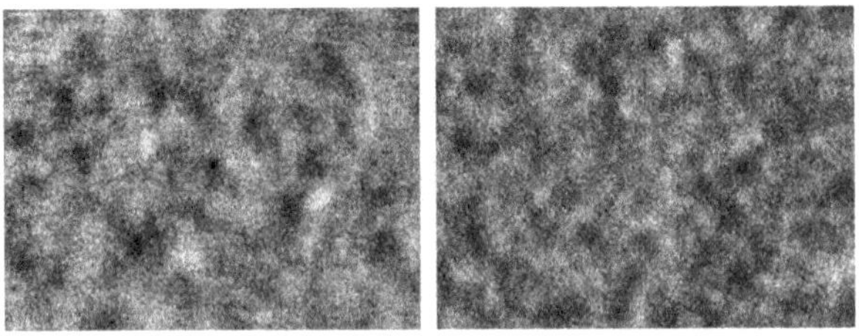

Fig. 9.3 Paper surface (*left*) and simulated Gaussian random field (*right*) based on the estimated data (courtesy of Voith Paper, Heidenheim) $\mathbf{E}\xi(t) = 126$, $\mathbf{cov}(\xi(0), \xi(t)) = 491\exp\{-\|t\|_2/56\}$

Fig. 9.4 Fluctuations of cosmic microwave background (CMB) radiation (measured in 10^{-6} K) around the mean value of 2.725 K. The image covers the whole sky with 5 years of WMAP (Wilkinson Microwave Anisotropy Probe, 2007) data (courtesy of H. Janzer)

9.2.4 Cosine Fields

Let η be a random variable uniformly distributed on $[0, 1]$. Consider a random field $\xi = \{\xi(t), t \in \mathbb{R}^d\}$ where $\xi(t) = \sqrt{2}\cos(2\pi\eta + \langle t, \zeta\rangle)$, ζ being an $\mathbb{R}^d$-valued random vector independent of η, $d \geq 1$. Each realization of ξ is a cosine wave surface.

Exercise 9.8. Let $\xi_1, \xi_2, \dots$ be independent cosine waves. Find the weak limits for finite-dimensional distributions of the fields $\{\frac{1}{\sqrt{n}}\sum_{k=1}^{n}\xi_k(t), t \in \mathbb{R}^d\}$ as $n \to \infty$.

9.2.5 Shot-Noise Random Fields and Moving Averages

Let $\Pi_\lambda = \{x_i\}$ be a homogeneous Poisson point process with intensity $\lambda > 0$ (see Sect. 3.1.2). Introduce a *shot-noise field*

$$\xi(t) = \sum_{x_i \in \Pi_\lambda} g(t - x_i), \quad t \in \mathbb{R}^d, \tag{9.11}$$

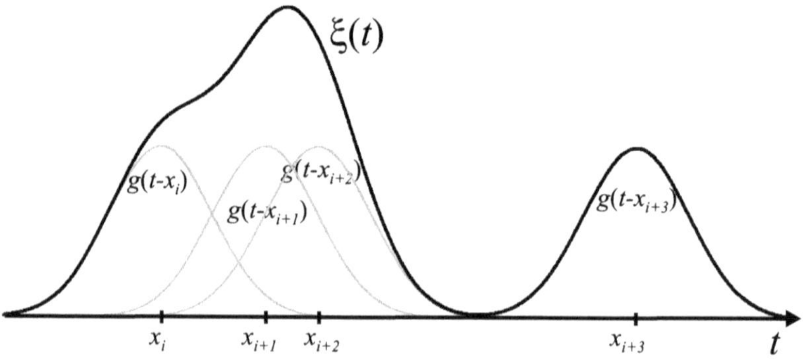

Fig. 9.5 Construction of a shot-noise process in $\mathbb{R}$

(see Fig. 9.5 and Fig. 9.6 (left)) where the series converges in $L^1(\Omega, \mathcal{A}, \mathbf{P})$ and $g : \mathbb{R}^d \to \mathbb{R}$ is a deterministic *response function* such that

$$\int_{\mathbb{R}^d} |g(x)| \, dx < \infty. \tag{9.12}$$

Exercise 9.9. Show that ξ in (9.11) is a well-defined random field. Prove that $\mathbf{E}\xi(t)^2 < \infty$ for any $t \in \mathbb{R}^d$ if

$$\int_{\mathbb{R}^d} g^2(x) \, dx < \infty. \tag{9.13}$$

A large class of response functions can be constructed as follows. Take $g(x) = K(\|x\|_2/a)$ where the *kernel* $K : \mathbb{R} \to \mathbb{R}_+$ is a probability density function with compact support supp K, i.e. the closure of the set $\{x \in \mathbb{R} : K(x) > 0\}$. For instance, one can take for K the *Epanechnikov kernel* (Fig. 9.7)

$$K(x) = \frac{3}{4}(1 - x^2)\mathbf{1}(x \in [-1, 1]), \quad x \in \mathbb{R},$$

or the *bisquare kernel*

$$K(x) = \frac{15}{16}(1 - x^2)^2\mathbf{1}(x \in [-1, 1]), \quad x \in \mathbb{R}.$$

However, kernels with unbounded support such as the *Gaussian kernel*

$$K(x) = \frac{1}{\sqrt{2\pi}} \mathrm{e}^{-x^2/2}, \quad x \in \mathbb{R},$$

can be used as well.

Fig. 9.6 Examples of simulated realizations of a shot-noise random field (*left*), Gaussian white noise (*middle*) and a Gaussian random field with a non-trivial covariance function (*right*)

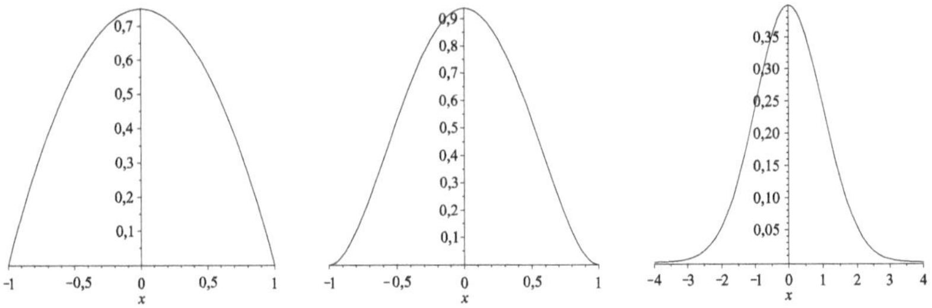

Fig. 9.7 Epanechnikov kernel (*left*), bisquare kernel (*middle*) and Gaussian kernel (*right*)

Exercise 9.10. Let ξ be a shot-noise field (9.11). Show that the characteristic function of $\xi_{t_1,\dots,t_n} = (\xi(t_1),\dots,\xi(t_n))^\top$, $t_1,\dots,t_n \in \mathbb{R}^d$, $n \in \mathbb{N}$ is given by

$$\varphi_{\xi_{t_1,\dots,t_n}}(s) = \exp\left\{\lambda \int_{\mathbb{R}^d} (\mathrm{e}^{i\langle s, g_{t_1,\dots,t_n}(u)\rangle} - 1)\, du\right\}, \quad s \in \mathbb{R}^n,$$

where $g_{t_1,\dots,t_n}(u) = (g(t_1 - u),\dots,g(t_n - u))^\top$.

Definition by formula (9.11) extends to comprise random response functions (see for example [96, pp. 39–43] or [343, p. 31]) and (non-homogeneous) point processes $\{x_i\}$ more general than Π_λ. Note that a shot-noise field (9.11) can be written as a stochastic integral

$$\xi(t) = \int_{\mathbb{R}^d} g(t - x)\Pi_\lambda(dx), \quad t \in \mathbb{R}^d,$$

where $\Pi_\lambda(\cdot)$ is interpreted as a random Poisson counting measure defined in (9.2) (see [489, Chap. 7]). Therefore it is a special case of *moving averages*:

$$\xi(t) = \int_E g(t, x)\mu(dx), \quad t \in T. \tag{9.14}$$

Here $\mu(\cdot)$ is a random *independently scattered measure* on a measurable space $(E, \mathcal{E})$, i.e., for each $n \in \mathbb{N}$ and any pairwise disjoint $B_1, \dots, B_n \in \mathcal{E}$, the random variables $\mu(B_1), \dots, \mu(B_n)$ are independent, and $g : T \times E \to \mathbb{R}$ is a deterministic function. In general μ can take the value ∞. Thus we only consider $B_1, \dots, B_n$ with $\mu(B_i) < \infty$ a.s., $i = 1, \dots, n$. The integral in (9.14) is understood to converge (for certain functions g) in probability if the measure μ is infinitely divisible; for more details see Sect. 9.7.3. In particular, one can use the *Gaussian white noise measure* μ in (9.14), that is, an independently scattered random measure defined on $(E, \mathcal{E}) = (\mathbb{R}^d, \mathcal{B}(\mathbb{R}^d))$ and such that $\mu(B) \sim N(o, \nu_d(B))$ for $B \in \mathcal{B}_o(\mathbb{R}^d)$ where $\nu_d(B)$ denotes the Lebesgue measure (or volume) of B.

Moving averages (MA-processes) with discrete parameter space $T = \mathbb{Z}$ are widely used in econometrics, see for example [81, 87, 277]. The case of $T = \mathbb{Z}^d$, $d > 1$, is considered, for example in [391]; see also references therein.

9.2.6 Random Cluster Measures

Let $\{\mu, \mu_i, i \in \mathbb{N}\}$ be a family of i.i.d. random measures defined on $(\mathbb{R}^d, \mathcal{B}(\mathbb{R}^d))$ and independent of a Poisson spatial process $\Pi_\Lambda = \{x_i\}$ with intensity measure Λ in $\mathbb{R}^d$. Introduce a *random cluster measure*

$$X(\omega, B) := \sum_i \mu_i(\omega, B + x_i(\omega)), \quad \omega \in \Omega, \quad B \in \mathcal{B}(\mathbb{R}^d), \tag{9.15}$$

taking values in $\overline{\mathbb{R}}_+$. Thus we have a random function defined on a set $T = \mathcal{B}(\mathbb{R}^d)$. Due to the explicit construction of the process Π_Λ (see Theorem 4.3) we can consider (9.15) as the convenient notation for the following series

$$\sum_{m=1}^{\infty} \sum_{j=1}^{\tau_m} \mu_{mj}(\omega, B + \xi_{mj}), \quad \omega \in \Omega, \quad B \in \mathcal{B}(\mathbb{R}^d). \tag{9.16}$$

Here μ_{mj} are independent copies of μ which are independent of the array of random vectors $(\tau_m, \xi_{mj})_{m,j \in \mathbb{N}}$ used in (4.4) to construct Π_Λ. Cluster random measures find various applications in astronomy, ecology, etc., see for example [325]. Simulation problems of such measures are discussed for example in [86].

9.2.7 Stable Random Functions

Definition 9.11. A nondegenerate random vector $\eta = (\eta_1, \dots, \eta_n)^\top$ taking values in $\mathbb{R}^n$ is called *stable* if for any $k \in \mathbb{N}$ there are some $c(k) \geq 0$ and $d(k) \in \mathbb{R}^n$ such that

$$\text{Law}(\eta^1 + \ldots + \eta^k) = \text{Law}(c(k)\eta + d(k))$$

where $\eta^1, \ldots, \eta^k$ are independent copies of η.

In this case $c(k) = k^{1/\alpha}$ for some $\alpha \in (0, 2]$, see for example [432, Corollary 2.1.3]. Then vector $\eta = (\eta_1, \ldots, \eta_n)^\top$ is said to have an *α-stable distribution*. It is known (see for example [432, Theorem 2.3.1]) that η is an α-stable random vector, $0 < \alpha < 2$, iff its characteristic function $\varphi_\eta(s) = \mathbf{E}\exp\{i\langle s, \eta\rangle\}$, $s \in \mathbb{R}^n$, has the form

$$\varphi_\eta(s) = \exp\left\{i\langle b, s\rangle - \int_{S^{n-1}} |\langle s, u\rangle|^\alpha \left(1 - i\ \text{sgn}(\langle s, u\rangle)\varkappa_n(\alpha, s, u)\right) \Gamma(du)\right\}$$

where $\mathbb{S}^{n-1}$ is the unit sphere in $\mathbb{R}^n$, $b \in \mathbb{R}^n$, Γ (the *spectral measure* of η) is a finite measure on $\mathcal{B}(\mathbb{S}^{n-1})$ and

$$\varkappa_n(\alpha, s, u) = \begin{cases} \tan\frac{\pi\alpha}{2}, & \alpha \neq 1, \\ -\frac{2}{\pi}\log|\langle s, u\rangle|, & \alpha = 1. \end{cases}$$

The pair (b, Γ) is unique. For $\alpha > 1$, the distribution of η is *centered* if $b = 0$. For $\alpha \in (0, 2)$ the distribution of η is *symmetric* iff there exists a unique symmetric measure Γ on $\mathbb{S}^{n-1}$ such that

$$\varphi_\eta(s) = \exp\left\{-\int_{\mathbb{S}^{n-1}} |\langle s, u\rangle|^\alpha \Gamma(du)\right\}, \quad s \in \mathbb{R}^n,$$

see [432, Theorem 2.4.3]. For $n = 1$, η is an α-stable random variable with characteristic function $\varphi_\eta(s) = \exp\{ibs - \sigma^\alpha |s|^\alpha (1 - i\beta\,\text{sgn}(s)\varkappa_1(\alpha, s))\}$ and

$$\varkappa_1(\alpha, s) = \begin{cases} \tan\frac{\pi\alpha}{2}, & \alpha \neq 1, \\ -\frac{2}{\pi}\log|s|, & \alpha = 1. \end{cases}$$

Here $\sigma > 0$, $\beta \in [-1, 1]$, $b \in \mathbb{R}$ are the parameters of *scale*, *skewness* and *shift*, respectively. For short, one writes $\eta \in S_\alpha(\sigma, \beta, b)$. A geometric approach to the study of stable laws is given in recent papers [143–145, 364, 365].

Exercise 9.11. Show that 2-stable distributions (i.e. $\alpha = 2$) are Gaussian.

A random function $\xi = \{\xi(t), t \in T\}$ is called *α-stable* if all its finite-dimensional distributions are α-stable. Applications of such random functions range from physics to finance and insurance; see for example [287, 411, 496]. Since α-stable laws are heavy-tailed and thus have finite absolute moments of order $p < \alpha$, their variance does not exist (here we exclude the Gaussian case $\alpha = 2$). This explains the fact that they are often used to model random phenomena with very irregular trajectories (in time and/or in space) and very high volatility such as stock prices, (total) claim amounts in insurance portfolios with dangerous risks, etc.

For various aspects of the theory of α-stable random functions we refer to [432, 496]. Apart from α-stability, other notions of stability (e.g., max-stability and operator-stability) can be used to generate further classes of stable random fields, see for example [68, 217–219, 278, 422, 439, 486, 487, 506].

9.2.8 *Random Fields Related to Random Closed Sets*

Let X be a random closed set in $\mathbb{R}^d$ (cf. Chap. 1.2.1). Put $\xi(t) = g(X \cap (W - t))$, $t \in \mathbb{R}^d$, where $W \in \mathcal{B}_0(\mathbb{R}^d)$ is a *scanning observation window* and $g : \mathcal{B}_0(\mathbb{R}^d) \to \mathbb{R}$ is a measurable geometric functional, $W \ominus t = \{s - t, s \in W\}$. Then $\xi = \{\xi(t), t \in \mathbb{R}^d\}$ is a random field describing the geometric properties of X. For instance, one can take $W = B_r(0)$, $g(\cdot) = \nu_d(\cdot)$. Then $\xi(t)$ is the volume of a part of X observed within the r-neighbourhood of $t \in \mathbb{R}^d$. For random sets X with realizations belonging to a certain class of sets (say, to an *extended convex ring*, see for example [451, p. 12]), other choices of g are possible, such as intrinsic volumes (Minkowski functionals) or their linear combinations (e.g., *Wills functional*). In the latter case, g is defined on the *convex ring* of subsets of $\mathbb{R}^d$ and not on the whole $\mathcal{B}_0(\mathbb{R}^d)$.

9.3 Markov and Gibbs Random Fields

This section deals with a class of random fields on finite graphs which are widely used in applications, for instance, in image processing. These are Markov or Gibbs fields that allow for a complex dependence structure between neighbouring nodes of the graph. After defining the Markov property of random fields on graphs, energy and potential are introduced which are essential to Gibbs fields. Some basic results such as the Averintsev–Clifford–Hammersley Theorem are given. An example of Gaussian Markov fields is considered in more detail.

9.3.1 *Preliminaries*

There are different approaches to adjusting the techniques of conditional probabilities to random fields. It was developed for stochastic processes and is indispensible, e.g., for the theory of Markov processes and martingales (submartingales, etc.). The latter have important applications, say, in stochastic calculus and financial mathematics. The multiindex generalizations of martingales are considered, e.g. in [296].

Extensions of the Markov property to the multiparameter case coming from statistical physics were studied intensively starting from 1960s. It is worthwhile to mention the powerful Dobrushin–Lanford–Ruelle construction of probability measures on a space of configurations (on a functional space) using the family of *conditional distributions* subject to the corresponding consistency conditions. This is a special interesting branch of the modern probability theory leading to the mathematical study of Gibbs random fields with numerous applications in statistical physics, see, for example [184]. As an elementary introduction to this topic we provide a simple proof of the classical Averintsev–Clifford–Hammersley Theorem[2] clarifying the relationship between Markov and Gibbs random fields.

Let $\mathbb{G} = (\mathbb{V}, \mathbb{E})$ be a finite undirected graph with the set of vertices (*nodes* or *sites*) $\mathbb{V}$ and the set of edges $\mathbb{E}$. We say that two vertices are *neighbours* if there exists an edge $e \in \mathbb{E}$ which connects them (the case $|\mathbb{V}| = 1$ is trivial, $|\mathbb{V}|$ stands for the cardinality of the set of vertices $\mathbb{V}$). For nonempty set $A \subset \mathbb{V}$ let its *boundary* be

$$\partial A = \{\text{all vertices } v \in \mathbb{V} \setminus A \text{ that have a neighbour in } A\}.$$

Thus $\partial \mathbb{V} = \emptyset$. It is convenient to put $\partial \emptyset = \mathbb{V}$. Obviously $\partial A = \cup_{t \in A} \partial\{t\} \setminus A$ (see Fig. 9.8 for examples).

Let $\xi = \{\xi(t),\ t \in \mathbb{V}\}$ be a random field defined on a probability space $(\Omega, \mathcal{A}, \mathbf{P})$ such that each random variable $\xi(t)$ takes values in a finite or countable set S. Further on we assume that $\xi(t) \in \mathcal{A} \mid \mathcal{B}$ where $\mathcal{B} = 2^S$, i.e. $\mathcal{B}$ is the collection of all subsets of S (equivalently $\{\xi(t) = x\} \in \mathcal{A}$ for any $x \in S$ and each $t \in \mathbb{V}$).

Definition 9.12. A random field $\xi = \{\xi(t),\ t \in \mathbb{V}\}$ is called *Markov* (corresponding to the graph $\mathbb{G}$) if for each $t \in \mathbb{V}$ and arbitrary $x = \{x_t,\ t \in \mathbb{V}\} \in S^{\mathbb{V}}$ one has

$$\begin{aligned}\mathbf{P}(\xi(t) = x_t \,|\, \xi(s) = x_s \text{ for } s \in \mathbb{V} \setminus \{t\}) \\ = \mathbf{P}(\xi(t) = x_t \,|\, \xi(s) = x_s \text{ for } s \in \partial\{t\})\end{aligned} \tag{9.17}$$

whenever $\mathbf{P}(\xi(s) = x_s \text{ for } s \in \mathbb{V} \setminus \{t\}) \neq 0$.

Thus, to calculate the left-hand side of (9.17) one can specify the values of the field ξ only in the neighbourhood of the site t. For $x = \{x_t,\ t \in \mathbb{V}\} \in S^{\mathbb{V}}$ and nonempty $T \subset \mathbb{V}$, set $x_T := \{x_t,\ t \in T\} \in S^T$. If $\xi = \{\xi(t),\ t \in \mathbb{V}\}$ is a random field then we write $\xi_T := \{\xi(t),\ t \in T\}$ and $\xi_T = x_T$ means $\xi(t) = x(t)$ for each $t \in T$ when $T \neq \emptyset$. We put $\{\xi_T = x_T\} = \Omega$ if $T = \emptyset$. Thus for $\partial\{t\} = \emptyset$ relation (9.17) implies that $\xi(t)$ is independent of $\{\xi(s),\ s \in \mathbb{V} \setminus \{t\}\}$. Moreover, for $\mathbb{V} = \{t\}$, a single random variable $\xi(t)$ can be viewed as a Markov random field.

Formula (9.17) can be written in a more convenient form, namely,

$$\mathbf{P}(\xi(t) = x_t \mid \xi_{\mathbb{V}\setminus\{t\}} = x_{\mathbb{V}\setminus\{t\}}) = \mathbf{P}(\xi(t) = x_t \mid \xi_{\partial\{t\}} = x_{\partial\{t\}}). \tag{9.18}$$

---

[2]The important contributions of other researchers to establishing this result are discussed in [520, pp. 69–70].

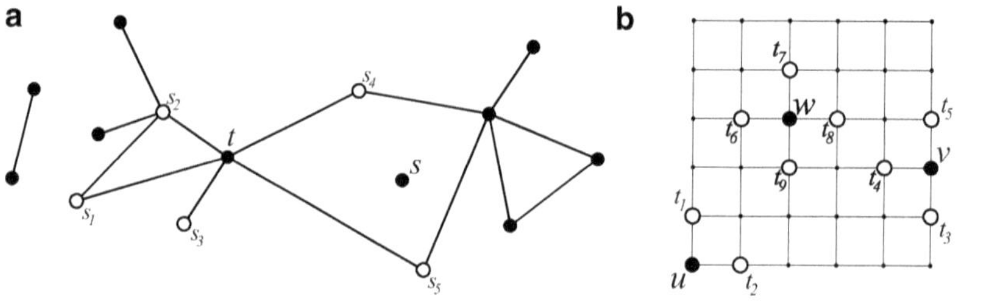

Fig. 9.8 (**a**) $\partial\{t\} = \{s_1, s_2, s_3, s_4, s_5\}$, $\partial\{s\} = \emptyset$; (**b**) $\partial\{u\} = \{t_1, t_2\}$, $\partial\{v\} = \{t_3, t_4, t_5\}$, $\partial\{w\} = \{t_6, t_7, t_8, t_9\}$

For a Markov random field ξ introduce the *local characteristic* $\pi^t(x)$ as the right-hand side of (9.18). A family $\{\pi^t(x)\}_{t\in\mathbb{V}})$ is called the *local specification*.

Assume that $\sigma\{\xi_T\} = \{\emptyset, \Omega\}$ if $T = \emptyset$.

Exercise 9.12. Show that (9.18) is equivalent to the following relation:

$$\mathbf{E}(f(\xi(t)) \mid \xi_{\mathbb{V}\setminus\{t\}}) = \mathbf{E}(f(\xi(t)) \mid \xi_{\partial\{t\}}) \tag{9.19}$$

for any bounded function $f : S \to \mathbb{R}$.

Exercise 9.13. Let $\mathbb{G}_A = (A, \mathbb{E}_A)$ be a subgraph of $\mathbb{G} = (\mathbb{V}, \mathbb{E})$, i.e. $A \subset \mathbb{V}$ and $\mathbb{E}_A$ consists of edges belonging to $\mathbb{E}$ and connecting the vertices of A only. Prove the following statement: if $\xi = \{\xi(t),\ t \in \mathbb{V}\}$ is a Markov random field corresponding to $\mathbb{G}$ and $\partial A = \emptyset$ then $\xi_A = \{\xi(t),\ t \in A\}$ is a Markov random field corresponding to $\mathbb{G}_A$.

Exercise 9.14. Let $\xi = \{\xi(t),\ t \in \mathbb{V}\}$ be a Markov random field (corresponding to a graph $\mathbb{G}$) and $M = \{t \in \mathbb{V} : \partial\{t\} = \emptyset\}$. Show that $\xi_M = \{\xi(t),\ t \in M\}$ is a collection of independent random variables and ξ_M is independent of the family $\xi_{\mathbb{V}\setminus M}$.

Exercise 9.15. Let $\xi = \{\xi(t),\ t \in \mathbb{V}\}$ be a Markov random field (corresponding to a graph $\mathbb{G}$) defined on a probability space $(\Omega, \mathcal{A}, \mathbf{P})$. Let $\{\xi(s),\ s \in \mathbb{U}\}$ be a family of independent random variables taking values in S and defined on the same probability space. Assume also that $\{\xi(t),\ t \in \mathbb{V}\}$ and $\{\xi(s),\ s \in \mathbb{U}\}$ are independent. Prove that $\{\xi(t),\ t \in \mathbb{V} \cup \mathbb{U}\}$ is a Markov random field corresponding to the enlarged graph $(\mathbb{V} \cup \mathbb{U}, \mathbb{E})$, that is, we add the vertices $\mathbb{U}$ as singletons to the collection $\mathbb{V}$ and do not introduce the new edges.

Exercise 9.16. Let $\mathbb{G} = (\mathbb{V}, \mathbb{E})$ and $\tilde{\mathbb{G}} = (\mathbb{V}, \tilde{\mathbb{E}})$ be graphs with the same sets of vertices but different sets of edges. Let $\xi = \{\xi(t),\ t \in \mathbb{V}\}$ be a Markov random field corresponding to $\mathbb{G}$. Is ξ a Markov random field corresponding to $\tilde{\mathbb{G}}$?

Remark 9.2. One uses also the system of neighbourhoods $\mathcal{U}_t$ for $t \in \mathbb{V}$ (such that $t \notin \mathcal{U}_t$ and if $s \in \mathcal{U}_t$ then $t \in \mathcal{U}_s$) to define a Markov field, see, for example [84,520]. Namely, the analogue of (9.18) is employed with $\partial\{t\}$ replaced by $\mathcal{U}_t$. Evidently we obtain an equivalent definition if we introduce the graph such that s is a neighbour of t iff $s \in \mathcal{U}_t$. Any field $\xi = \{\xi(t),\, t \in \mathbb{V}\}$ is Markovian when $\mathcal{U}_t = \mathbb{V}$ for each $t \in \mathbb{V}$. However, interesting (e.g., for modelling) Markov fields are those with relatively small neighbourhoods for each site.

Any element $\omega \in S^{\mathbb{V}}$ will be called a *configuration*. Let ξ be a random field defined on some probability space and such that $\mathrm{Law}(\xi) = Q$ on $(S^{\mathbb{V}}, \mathcal{A})$. Introduce $\tilde{\xi}(s,\omega) := \omega(s)$ for $s \in \mathbb{V}$ and $\omega \in S^{\mathbb{V}}$ (i.e. $\tilde{\xi}$ is the identical map on $S^{\mathbb{V}}$), then $\mathrm{Law}(\tilde{\xi}) = \mathrm{Law}(\xi) = Q$. Thus the study of the random field ξ and the configurations is in a sense equivalent.

9.3.2 Energy and Potential

Let the *energy* $\mathsf{E} : S^{\mathbb{V}} \to \mathbb{R}$ be an arbitrary function.

Definition 9.13. A *Gibbs measure* (or *Gibbs state*) $\mathbb{P}$ is defined on a space $(S^{\mathbb{V}}, \mathcal{A})$ as follows: $\mathbb{P}(\emptyset) = 0$ and

$$\mathbb{P}(B) = \frac{1}{Z} \sum_{\omega \in B} \exp\{-\mathsf{E}(\omega)\}, \quad B \in \mathcal{A}, \quad B \neq \emptyset, \tag{9.20}$$

where the normalizer (*partition function*)

$$Z = \sum_{\omega \in S^{\mathbb{V}}} \exp\{-\mathsf{E}(\omega)\} \tag{9.21}$$

is assumed finite.[3]

Note that in 1902 Gibbs introduced the probability distribution on the configuration space $S^{\mathbb{V}}$ by formula

$$\mathbb{P}_{\mathsf{T}}(\omega) = \frac{1}{Z_{\mathsf{T}}} \exp\left\{-\frac{\mathsf{E}(\omega)}{\mathsf{T}}\right\} \tag{9.22}$$

where $\mathsf{T} > 0$ is the *temperature*, E is the *energy* of configuration ω and Z_{T} is the normalizing constant. To simplify the notation we omit T in (9.20), that is we set the energy to be $\frac{1}{\mathsf{T}}\mathsf{E}$. One also writes the *Hamiltonian* H instead of E.

Let Q be a probability measure on $(S^{\mathbb{V}}, \mathcal{A})$. Introduce the *entropy*

$$\mathcal{H}(Q) := -\sum_{\omega \in S^{\mathbb{V}}} Q(\omega) \log Q(\omega)$$

[3]Clearly Z is finite if S is a finite set.

and *mean energy* $\mathbf{E}(\mathsf{E}, Q) := \sum_{\omega \in S^{\mathbb{V}}} \mathsf{E}(\omega) Q(\omega)$, where $0 \log 0 := 0$. If S is countable we assume that the series are absolutely convergent.

Exercise 9.17 (Gibbs variational principle). Show that the *free energy* $\mathbf{E}(\mathsf{E}, Q) - \mathcal{H}(Q)$ satisfies the inequality

$$\mathbf{E}(\mathsf{E}, Q) - \mathcal{H}(Q) \geq -\log Z$$

with equality achieved only for $Q = \mathbb{P}$ where $\mathbb{P}$ and Z are given in (9.20) and (9.21), respectively.

Usually the energy is described by means of the potential function expressing the interaction in subsystems.

Definition 9.14. A *potential* is a real-valued function $\mathsf{V}_A(\omega)$ defined for all $A \subset \mathbb{V}$ and $\omega \in S^{\mathbb{V}}$, such that $\mathsf{V}_\emptyset(\cdot) = 0$. The *energy* of a configuration ω is given by

$$\mathsf{E}(\omega) := \sum_{A \subset \mathbb{V}} \mathsf{V}_A(\omega) \tag{9.23}$$

where "$\subset$" always denotes non-strict inclusion.

Thus the Gibbs measure corresponding to energy E is defined by means of its potential. To see that arbitrary energy can be represented according to (9.23) with appropriate potential we use (as in [201]) the following well-known result.

Lemma 9.2 (Möbius formula). *Let f and g be two functions defined on all subsets of a finite set C. Then the following formulae are equivalent:*

$$f(A) = \sum_{B \subset A} (-1)^{|A \setminus B|} g(B) \;\; \textit{for all} \;\; A \subset C, \tag{9.24}$$

$$g(A) = \sum_{B \subset A} f(B) \;\; \textit{for all} \;\; A \subset C. \tag{9.25}$$

Proof. Let (9.24) hold. The change of the order of summation leads to the relations

$$\sum_{B \subset A} f(B) = \sum_{B \subset A} \sum_{D \subset B} (-1)^{|B \setminus D|} g(D) = \sum_{D \subset A} g(D) \sum_{B : D \subset B \subset A} (-1)^{|B \setminus D|}$$

$$= \sum_{D \subset A} g(D) \sum_{F \subset A \setminus D} (-1)^{|F|} = \sum_{D \subset A} g(D) \sum_{k=0}^{|A \setminus D|} \binom{|A \setminus D|}{k} (-1)^k = g(A)$$

as $\sum_{k=0}^{m} \binom{m}{k} (-1)^k = (1-1)^m = 0$ for $m \in \mathbb{N}$ and $\binom{0}{0}(-1)^0 = 1$ (in the case $D = A$).

We used $F := B \setminus D$ and took into account that there exist $\binom{|A \setminus D|}{k}$ sets $F \subset A \setminus D$ such that $|F| = k$ for $k = 0, \ldots, |A \setminus D|$, see Fig. 9.9.

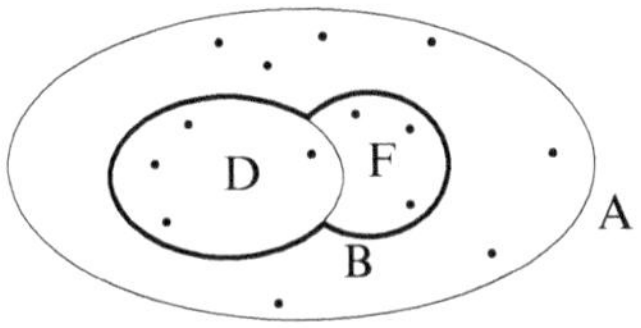

Fig. 9.9 Illustration to the proof of Lemma 9.2 ($F = B \setminus D, |F| = k$)

In the same manner (9.25) implies (9.24). □

Now we turn to (9.23). Fix an arbitrary element $o \in S$ and introduce the *vacuum configuration* $\mathbf{o} = \{\omega(t),\ t \in \mathbb{V}\} \in S^{\mathbb{V}}$ such that $\omega(t) = o$ for all $t \in \mathbb{V}$. For a set $B \subset \mathbb{V}$ and $\omega \in S^{\mathbb{V}}$ define (see Fig. 9.10) the configuration ω^B letting

$$\omega^B(t) = \begin{cases} \omega(t), & t \in B, \\ o, & t \in \mathbb{V} \setminus B. \end{cases} \tag{9.26}$$

In particular $\omega^{\mathbb{V}} = \omega$ and $\omega^{\emptyset} = \mathbf{o}$.

Notice that if we take $\mathsf{E}+c$ in (9.20) instead of E for constant c, then the measure $\mathbb{P}$ will be the same (the new normalizing factor $e^c Z$ arises instead of Z). Therefore w.l.g. we can assume that $\mathsf{E}(\mathbf{o}) = 0$, that is, "the energy of the vacuum state is zero". Now introduce the *canonical potential*[4]

$$\mathsf{V}_A(\omega) := \sum_{B \subset A} (-1)^{|A \setminus B|} \mathsf{E}(\omega^B). \tag{9.27}$$

Corollary 9.1. *Let* $\mathsf{E} : S^{\mathbb{V}} \to \mathbb{R}$ *be a function such that* $\mathsf{E}(\mathbf{o}) = 0$ *where* $\mathbf{o} \in S^{\mathbb{V}}$. *Then, for the canonical potential defined by* (9.27), *relation* (9.23) *holds.*

Proof. For each (fixed) $\omega \in S^{\mathbb{V}}$ consider the set function $g(B) := \mathsf{E}(\omega^B)$ where $B \in \mathcal{A}$. Note that $\mathsf{V}_{\emptyset}(\omega) = (-1)^0 \mathsf{E}(\omega^{\emptyset}) = \mathsf{E}(\mathbf{o}) = 0$. As $\omega^{\mathbb{V}} = \omega$ we come to (9.23) by virtue of Lemma 9.2. □

Lemma 9.3. *A probability measure* $\mathbb{P}$ *on a space* $(S^{\mathbb{V}}, \mathcal{A})$ *can be viewed as the Gibbs measure corresponding to some (canonical) potential iff*

$$\mathbb{P}(A) > 0 \text{ for any } A \in \mathcal{A}, \quad A \neq \emptyset. \tag{9.28}$$

Proof. The necessity is obvious. To prove the sufficiency introduce

$$\mathsf{E}(\omega) := -\log \frac{\mathbb{P}(\{\omega\})}{\mathbb{P}(\mathbf{o})}, \quad \omega \in S^{\mathbb{V}}. \tag{9.29}$$

[4]Note that the canonical potential depends on $o \in S$.

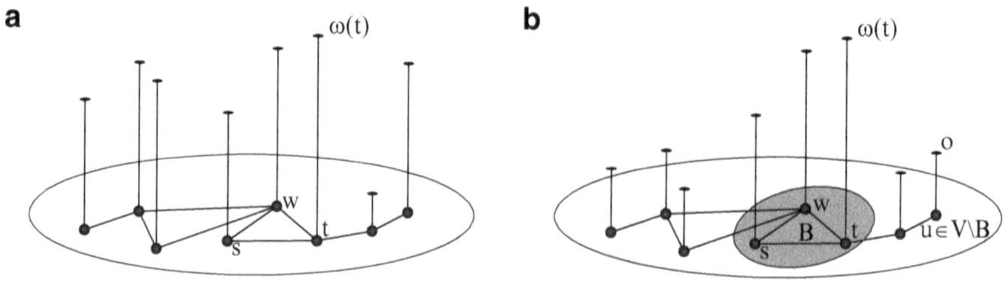

Fig. 9.10 Visualization of the configurations ω and ω^B, $B = \{s, t, w\}$

Corollary 9.1 yields (9.20) with potential given by (9.27). Clearly $Z = 1/\mathbb{P}(\mathbf{o})$. □

One says that probability measure $\mathbb{P}$ is *strictly positive* if (9.28) holds.

Definition 9.15. A set $A \subset \mathbb{V}$ with $|A| > 1$ is called a *clique*[5] when a subgraph G_A induced by A is *complete*, i.e. any two distinct vertices of A are neighbours, see Fig. 9.11. It is convenient to say that any singleton $\{t\} \subset \mathbb{V}$ is a clique. The *nearest neighbour potential* is defined by the relation

$$\mathsf{V}_A(\omega) = 0 \ \text{ for all } \omega \text{ if } A \text{ is not a clique.} \tag{9.30}$$

Obviously if the vertices s and t are neighbours then $\{s, t\}$ is a clique.

Example 9.3 (Ising model). In 1925 Ising introduced the model to describe the phenomenon of phase transition in ferromagnetic materials. In Ising's finite model

$$\mathbb{V} = \mathbb{Z}_m^2 = \{(i, j) \in \mathbb{Z}^2 \cap [1, m]^2\},$$

sites s, t are neighbours if the Euclidean distance between them is equal to 1, the state space $S = \{-1, 1\}$ where ± 1 is the orientation of the spin (intrinsic magnetic moment) at a given site. The non-zero values of the potential are given by the formula

$$\mathsf{V}_{\{t\}}(\omega) = -\frac{H}{k}\omega(t), \ \ \mathsf{V}_{\langle s,t\rangle}(\omega) = -\frac{J}{k}\omega(s)\omega(t)$$

where k is the Boltzmann constant, H is the external magnetic field, J is the energy of an elementary magnetic dipole and here $\langle s, t\rangle$ is the two-element clique formed by sites s and t. Thus

$$\mathsf{E}(\omega) = -\frac{J}{k}\sum_{\langle s,t\rangle}\omega(s)\omega(t) - \frac{H}{k}\sum_{t\in\mathbb{V}}\omega(t),$$

and we obtain the Gibbs measure using (9.22).

[5]Because in ordinary language a clique is a group of people who know and favour each other.

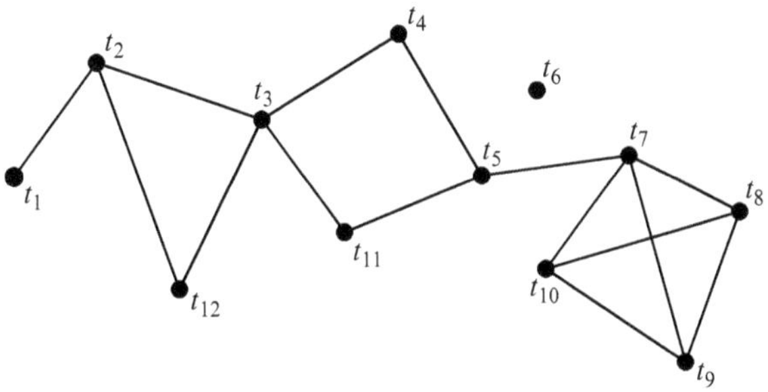

Fig. 9.11 $A = \{t_2, t_3, t_{12}\}$, $B = \{t_7, t_8, t_9, t_{10}\}$, $C = \{t_6\}$ are the cliques, $D = \{t_3, t_4, t_5, t_{11}\}$ is not a clique

Definition 9.16. The events B and C are conditionally independent given event D with $\mathbf{P}(D) > 0$ if

$$\mathbf{P}(B \cap C \mid D) = \mathbf{P}(B \mid D)\mathbf{P}(C \mid D), \tag{9.31}$$

i.e. B and C are independent in the space $(\Omega, \mathcal{A}, \mathbf{P}_D)$ where $\mathbf{P}_D$ is the conditional probability given D.

Exercise 9.18 ([84, p. 10]). Show that, for events B, C, D such that $\mathbf{P}(B \cap D) > 0$, (9.31) is equivalent to the relation

$$\mathbf{P}(C \mid B \cap D) = \mathbf{P}(C \mid D).$$

We shall apply the simple but useful result established in [84, p. 12].

Lemma 9.4. *Let α, β, γ be three discrete random variables (or vectors) taking values in finite or countable spaces F, H, K respectively. Assume that for any $a \in F$, $b \in H$ and $c \in K$ one has $\mathbf{P}(\beta = b,\ \gamma = c) > 0$ and*

$$\mathbf{P}(\alpha = a \mid \beta = b,\ \gamma = c) = g(a, b).$$

Then for all $a \in F$ and $b \in H$

$$\mathbf{P}(\alpha = a \mid \beta = b) = g(a, b). \tag{9.32}$$

Proof. Obviously for any $a \in F$ and $b \in H$

$$\begin{aligned}\mathbf{P}(\alpha = a,\ \beta = b) &= \sum_{c \in K} \mathbf{P}(\alpha = a,\ \beta = b,\ \gamma = c)\\ &= \sum_{c \in K} \mathbf{P}(\alpha = a \mid \beta = b,\ \gamma = c)\mathbf{P}(\beta = b,\ \gamma = c)\\ &= g(a, b) \sum_{c \in K} \mathbf{P}(\beta = b,\ \gamma = c) = g(a, b)\mathbf{P}(\beta = b).\end{aligned}$$

Observe that $\mathbf{P}(\beta = b) = \sum_{c \in K} \mathbf{P}(\beta = b, \gamma = c) > 0$. Hence (9.32) holds. The proof is complete. □

9.3.3 Averintsev–Clifford–Hammersley Theorem

We start with auxiliary results.

Lemma 9.5. *Let $\mathbb{G} = (\mathbb{V}, \mathbb{E})$ be a finite graph. Let $\mathbb{P}$ be a Gibbs measure defined on $(S^{\mathbb{V}}, \mathcal{A})$, and the canonical potential corresponding to an energy E (and some element $o \in S$) be that of the nearest neighbour. Then $\mathbb{P} = \mathbf{P}_\xi$ where the Markov random field $\xi = \{\xi(t),\ t \in \mathbb{V}\}$ corresponds to $\mathbb{G}$ ($\mathbf{P}_\xi$ is the law of ξ on $(S^{\mathbb{V}}, \mathcal{A})$).*

Proof. Note that if $B \subset A \subset \mathbb{V}$ then $(\omega^A)^B = \omega^B$ in view of (9.26). Consequently due to (9.27) for the canonical potential $\mathsf{V}_A(\omega)$ one has

$$\mathsf{V}_A(\omega) = \mathsf{V}_A(\omega^A), \quad \omega \in S^{\mathbb{V}}, \quad A \subset \mathbb{V}. \tag{9.33}$$

Therefore for any $A \subset \mathbb{V}$, given configurations ω and ν, we obtain

$$\mathsf{V}_A(\omega) = \mathsf{V}_A(\nu) \ \text{ if } \ \omega(t) = \nu(t), \quad t \in A. \tag{9.34}$$

Consider the probability space $(S^{\mathbb{V}}, \mathcal{A}, \mathbb{P})$ and let the random field ξ be identical mapping of $S^{\mathbb{V}}$ ($\xi(s,\omega) := \omega(s)$ for $s \in \mathbb{V}$, $\omega \in S^{\mathbb{V}}$). Then $\mathbf{P}_\xi = \mathbb{P}$. Fix any $t \in \mathbb{V}$ and introduce $\mathbb{U} = \mathbb{V} \setminus \{t\}$. Consider $\omega = \{\omega(s),\ s \in \mathbb{V}\} \in S^{\mathbb{V}}$ where $\omega(t) = x_t$ and $\omega(s) = x_s$ for $s \in \mathbb{U}$. Let $\mathcal{M} = \{A \subset \mathbb{V} :\ A \text{ is a clique and } t \notin A\}$ and $\mathcal{N} = \{A \subset \mathbb{V} :\ A \text{ is a clique and } t \in A\}$, see Fig. 9.12.

Taking into account (9.20), (9.23) and (9.30) the left-hand side L of (9.17) can be written as follows

$$\frac{\mathbb{P}(\omega)}{\sum_{\nu \in S^{\mathbb{V}}:\ \nu_{\mathbb{U}} = \omega_{\mathbb{U}}} \mathbb{P}(\nu)} = \frac{\exp\{-\sum_{A \in \mathcal{M}} \mathsf{V}_A(\omega^A)\} \exp\{-\sum_{A \in \mathcal{N}} \mathsf{V}_A(\omega^A)\}}{\sum_{\nu \in S^{\mathbb{V}}:\ \nu_{\mathbb{U}} = \omega_{\mathbb{U}}} \exp\{-\sum_{A \in \mathcal{M}} \mathsf{V}_A(\nu^A)\} \exp\{-\sum_{A \in \mathcal{N}} \mathsf{V}_A(\nu^A)\}}.$$

If $\nu(u) = \omega(u)$ for $u \in \mathbb{U}$ and $A \in \mathcal{M}$ then $A \subset \mathbb{U}$ and $\nu^A = \omega^A$. Thus,

$$\begin{aligned} L &= \frac{\exp\{-\sum_{A \in \mathcal{N}} \mathsf{V}_A(\omega^A)\}}{\sum_{\nu \in S^{\mathbb{V}}:\ \nu_{\mathbb{U}} = \omega_{\mathbb{U}}} \exp\{-\sum_{A \in \mathcal{N}} \mathsf{V}_A(\nu^A)\}} \\ &= \frac{\exp\{-\sum_{A \in \mathcal{N}} \mathsf{V}_A(\omega^A)\}}{\sum_{z \in S} \exp\{-\sum_{A \in \mathcal{N}} \mathsf{V}_A(\omega_z^A)\}} \end{aligned} \tag{9.35}$$

where $\omega_z(t) = z$ and $\omega_z(u) = \omega(u)$ for all $u \in \mathbb{U}$. Let $\partial\{t\} \neq \emptyset$. Note that if A is a clique containing $\{t\}$ then any vertex $s \in A \setminus \{t\}$ is a neighbour of t. Therefore $A \subset \{t\} \cup \partial\{t\}$ if $A \in \mathcal{N}$. Hence the right-hand side of (9.35) does not depend on $\omega^{\mathbb{V} \setminus (\{t\} \cup \partial\{t\})}$. Set $\alpha = \xi(t)$, $\beta = \xi_{\partial\{t\}}$ and $\gamma = \xi_{\mathbb{V} \setminus (\{t\} \cup \partial\{t\})}$ and note that

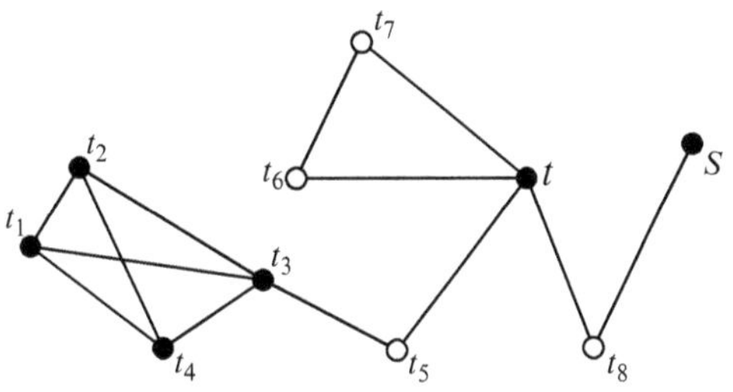

Fig. 9.12 The clique $A = \{t_1, t_2, t_3, t_4\} \in \mathcal{M}$ and the clique $C = \{t, t_6, t_7\} \in \mathcal{N}$, $C \subset \partial\{t\} = \{t_5, t_6, t_7, t_8\}$

$$L = \mathbf{P}(\alpha = \omega(t) \mid \beta = \omega_{\partial\{t\}}, \gamma = \omega_{\mathbb{V}\setminus(\{t\}\cup\partial\{t\})}).$$

Now Lemma 9.4 leads to the desired relation (9.18). In case of $\partial\{t\} = \emptyset$ we have $\{\beta = \omega_{\partial\{t\}}\} = \Omega$ (or one can assume that in Lemma 9.4 the random variable β takes values in a set H consisting of a single element) and the reasoning is the same as before. The proof is complete. □

Lemma 9.6. *Let $\xi = \{\xi(t), t \in \mathbb{V}\}$ be a Markov random field corresponding to a finite graph $\mathbb{G} = (\mathbb{V}, \mathbb{E})$ and such that values of $\xi(t)$ belong to a finite or countable set S for each $t \in \mathbb{V}$. Assume that the measure $\mathbb{P} := \mathbf{P}_\xi$ is strictly positive. Then $\mathbb{P}$ is a Gibbs measure corresponding to the nearest neighbour canonical potential.*

Proof. The case $|\mathbb{V}| = 1$ is trivial. So we consider $|\mathbb{V}| \geq 2$. Let $\mathsf{E}(\omega)$ and $\mathsf{V}_A(\omega)$ be defined according to (9.27) and (9.29), respectively (with any fixed $o \in S$). In view of Lemma 9.3 we only need to verify that the canonical potential for energy E is a nearest neighbour one, i.e. (9.30) holds. If A is not a clique then there exist $s, t \in A$ such that they are not neighbours. Introduce $C = A \setminus \{s, t\}$. Then (9.27) reads

$$\begin{aligned}\mathsf{V}_A(\omega) &= \sum_{B\subset A} (-1)^{|A\setminus B|} \mathsf{E}(\omega^B) \\ &= \sum_{D\subset C} (-1)^{|C\setminus D|} (\mathsf{E}(\omega^{D\cup\{s,t\}}) - \mathsf{E}(\omega^{D\cup\{s\}}) - \mathsf{E}(\omega^{D\cup\{t\}}) + \mathsf{E}(\omega^D)).\end{aligned}$$

To complete the proof of the Lemma it suffices to show that for each $D \subset C$ one has

$$\mathsf{E}(\omega^{D\cup\{s,t\}}) - \mathsf{E}(\omega^{D\cup\{s\}}) - \mathsf{E}(\omega^{D\cup\{t\}}) + \mathsf{E}(\omega^D) = 0.$$

In view of (9.29) the last relation is tantamount to

$$\frac{\mathbb{P}(\omega^{D\cup\{s,t\}})}{\mathbb{P}(\omega^{D\cup\{s\}})} = \frac{\mathbb{P}(\omega^{D\cup\{t\}})}{\mathbb{P}(\omega^D)}. \tag{9.36}$$

Set $M = \{\xi_D = \omega_D, \xi_T = \mathbf{o}_T\}$ where $T = (C \setminus D) \cup (\mathbb{V} \setminus A)$ (Fig. 9.13). Then (9.36) can be rewritten in the equivalent form

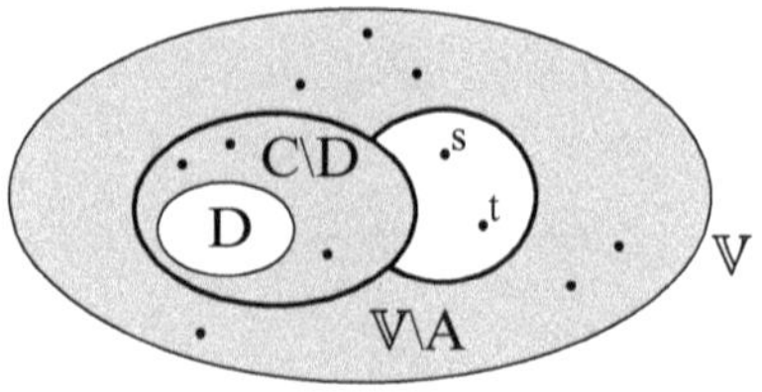

Fig. 9.13 $A = C \cup \{s,t\}$, $T = (C \setminus D) \cup (\mathbb{V} \setminus A)$

$$\frac{\mathbf{P}(\xi(t) = \omega(t), \xi(s) = \omega(s), M)}{\mathbf{P}(\xi(t) = o, \xi(s) = \omega(s), M)} = \frac{\mathbf{P}(\xi(t) = \omega(t), \xi(s) = o, M)}{\mathbf{P}(\xi(t) = o, \xi(s) = o, M)}. \qquad (9.37)$$

Due to the Markov property and strict positivity of $\mathbf{P}_\xi$ the left-hand side of (9.37) is equal to the following expression

$$\frac{\mathbf{P}(\xi(t) = \omega(t)|\xi_{\partial\{t\}} = \tau_{\partial\{t\}})\mathbf{P}(\xi(s) = \omega(s), M)}{\mathbf{P}(\xi(t) = o|\xi_{\partial\{t\}} = \tau_{\partial\{t\}})\mathbf{P}(\xi(s) = \omega(s), M)} = \frac{\mathbf{P}(\xi(t) = \omega(t) \mid \xi_{\partial\{t\}} = \tau_{\partial\{t\}})}{\mathbf{P}(\xi(t) = o \mid \xi_{\partial\{t\}} = \tau_{\partial\{t\}})}$$

where $\tau(u) = \omega(u)$ for $u \in D \cup \{s,t\}$ and $\tau(u) = o$ for $u \in T$. Analogously the right-hand side of (9.37) has the form

$$\frac{\mathbf{P}(\xi(t) = \omega(t) \mid \xi_{\partial\{t\}} = \mu_{\partial\{t\}})}{\mathbf{P}(\xi(t) = o \mid \xi_{\partial\{t\}} = \mu_{\partial\{t\}})}$$

where $\mu(u) = \omega(u)$ for $u \in D \cup \{t\}$ and $\mu(u) = o$ for $u \in T \cup \{s\}$.

Note that $\tau = \mu$ on $\mathbb{V} \setminus \{s\}$. Thus $\{\xi_{\partial\{t\}} = \tau_{\partial\{t\}}\} = \{\xi_{\partial\{t\}} = \mu_{\partial\{t\}}\}$ as $s \notin \partial\{t\}$. The proof is complete. □

Thus Lemmas 9.5 and 9.6 imply

Theorem 9.3 (Averintsev, Clifford and Hammersley). *Let $\mathbb{G} = (\mathbb{V}, \mathbb{E})$ be a finite graph. Let $\mathbb{P}$ be a probability measure on the space $(S^{\mathbb{V}}, \mathcal{A})$, $\mathcal{A}$ contains all subsets of $S^{\mathbb{V}}$ and a set S is finite or countable. Then the following statements are equivalent:*

1. *$\mathbb{P} = \mathbf{P}_\xi$ where $\xi = \{\xi(t), t \in \mathbb{V}\}$ is a Markov random field corresponding to the graph $\mathbb{G}$ and having strictly positive law,*
2. *$\mathbb{P}$ is the Gibbs measure on $(S^{\mathbb{V}}, \mathcal{A})$ corresponding to the nearest neighbour canonical potential.*

Exercise 9.19. Let $\xi = \{\xi(t), t \in \mathbb{V}\}$ be a random field (corresponding to a finite graph $\mathbb{G}$). Prove that it is Markov with strictly positive law iff, for any $x = \{x_t, t \in V\} \in S^{\mathbb{V}}$ and each $T \subset \mathbb{V}$, the following relation holds

$$\mathbf{P}(\xi_T = x_T \mid \xi_{\mathbb{V}\setminus T} = x_{\mathbb{V}\setminus T}) = \mathbf{P}(\xi_T = x_T \mid \xi_{\partial T} = x_{\partial T}).$$

Exercise 9.20. Give an example of the Markov random field which is not the Gibbs one. Give an example of the Gibbs random field which is not the Markov one.

Remark 9.3. Exercises 9.18 and 9.19 show that the random field $\xi = \{\xi(t),\ t \in \mathbb{V}\}$ with strictly positive law is Markov iff, for any $x = \{x_t,\ t \in \mathbb{V}\} \in S^{\mathbb{V}}$ and each $T \subset \mathbb{V}$, the events $\{\xi_T = x_T\}$ and $\{\xi_{\mathbb{V}\setminus(T\cup\partial T)} = x_{\mathbb{V}\setminus(T\cup\partial T)}\}$ are conditionally independent given $\{\xi_{\partial T} = x_{\partial T}\}$.

9.3.4 Gaussian Markov Random Fields

To illustrate possible generalizations of the Markov random fields defined on a finite graph and taking non discrete values, we discuss some elementary facts related to *Gaussian Markov fields.*

Let ζ, η be random vectors with values in $\mathbb{R}^k$ and $\mathbb{R}^m$, respectively, such that there exists the joint density[6] $p_{\zeta,\eta}$ and the density p_η of η is strictly positive (i.e., one can find such version of the density). Then it is possible to introduce the *conditional density* $p_{\eta|\zeta}$ by formula

$$p_{\zeta|\eta}(z \mid y) := \frac{p_{\zeta,\eta}(z, y)}{p_\eta(y)}, \quad z \in \mathbb{R}^k, \quad y \in \mathbb{R}^m. \tag{9.38}$$

Definition 9.17. Let γ, η, ζ be random vectors (with values in $\mathbb{R}^n$, $\mathbb{R}^m$ and $\mathbb{R}^k$, respectively) having joint density $p_{\gamma,\eta,\zeta}$ and such that the density of η is strictly positive. Then γ and ζ are *conditionally independent* given η (one writes $\gamma \perp \zeta \mid \eta$) if for all values x, y, z one has

$$p_{\gamma,\zeta|\eta}(x, z \mid y) = p_{\gamma|\eta}(x \mid y)\, p_{\zeta|\eta}(z \mid y). \tag{9.39}$$

The following result can be viewed as a counterpart of Lemma 9.4.

Lemma 9.7. *The relation $\gamma \perp \zeta \mid \eta$ is equivalent to the* factorization condition*:*

$$p_{\gamma,\zeta,\eta}(x, z, y) = f(x, y)g(z, y) \tag{9.40}$$

for some nonnegative functions f, g and all values of x, y, z whenever p_η is strictly positive.

Proof. Obviously (9.39) yields (9.40) as $p_{\gamma,\zeta,\eta}(x, z, y) = p_{\gamma,\zeta|\eta}(x, z \mid y)p_\eta(y)$ in view of (9.38) where instead of ζ we use now the vector (γ, ζ).

Let (9.40) hold. Taking into account strict positivity of p_η we see that (9.39) is equivalent to the following relation (for all x, y, z)

[6] All densities are considered with respect to the corresponding Lebesgue measures.

$$p_{\gamma,\zeta,\eta}(x,z,y)p_\eta(y) = p_{\gamma,\eta}(x,y)p_{\zeta,\eta}(z,y). \tag{9.41}$$

Using the formula for marginal densities of random vectors we have

$$p_{\gamma,\eta}(x,y) = \int_{\mathbb{R}^k} p_{\gamma,\zeta,\eta}(x,z,y)dz, \quad p_{\zeta,\eta}(z,y) = \int_{\mathbb{R}^n} p_{\gamma,\zeta,\eta}(x,z,y)dx,$$

$$p_\eta(y) = \int_{\mathbb{R}^{k+n}} p_{\gamma,\zeta,\eta}(x,z,y)dxdz.$$

Applying (9.40) and the Fubini theorem one infers that (9.41) is satisfied. The proof is complete. □

Let $\mathbb{G} = (\mathbb{V},\mathbb{E})$ be a finite *labeled graph*, i.e. we enumerate the points of $\mathbb{V}$ to have the collection $\{t_1,\dots,t_N\}$. Moreover, we can identify t_i with i for $i = 1,\dots,n$ and now instead of a random field $\xi = \{\xi(t),\ t \in \mathbb{V}\}$ it is convenient to consider the random vector $\xi = (\xi_1,\dots,\xi_n)^\top$ where $\xi_i := \xi(t_i)$. Assume that $\xi \sim N(a,C)$ where the covariance matrix C is positive definite ($C > 0$). Then there exists the *precision matrix* $Q = C^{-1}$ and the vector ξ has a density p_ξ which is provided by formula (9.9). It is easily seen that $Q^\top = Q$ and $Q > 0$. Note that any such matrix $Q = (Q_{ij})$ produces a matrix $C = Q^{-1}$ which can be considered as the covariance matrix of a Gaussian vector ξ. For a set $A \subset \{1,\dots,n\}$ introduce ξ_{-A} as a vector ξ without components belonging to A. Thus $\xi_{-i} = (\xi_1,\dots,\xi_{i-1},\xi_{i+1},\dots,\xi_n)^\top$ when $1 < i < n$ ($\xi_{-1} = (\xi_2,\dots,\xi_n)^\top$ and $\xi_{-n} = (\xi_1,\dots,\xi_{n-1})^\top$). The similar notation x_{-A} will be used for a nonrandom vector $x \in \mathbb{R}^n$ and $A \subset \{1,\dots,n\}$.

Theorem 9.4 ([430, p. 21]). *Let $\xi \sim N(a,C)$ where $C > 0$. Then for $i,j \in \{1,\dots,n\}$, $i \neq j$, one has*

$$\xi_i \perp \xi_j \mid \xi_{-\{i,j\}} \iff Q_{ij} = 0.$$

Proof. Let $Q_{ij} = 0$. Then obviously

$$p_\xi(x) = c_n(Q)\exp\left\{-\frac{1}{2}\sum_{k=1}^{n}\sum_{k=1}^{n} Q_{km}(x_k - a_k)(x_m - a_m)\right\} = f(x_{-\{i\}})g(x_{-\{j\}})$$

with explicit formulae for f and g, here $x \in \mathbb{R}^n$ and $c_n(Q) = (2\pi)^{-n/2}(\det Q)^{1/2}$. Therefore Lemma 9.7 implies the conditional independence of ξ_i and ξ_j given $\xi_{-\{i,j\}}$.

Assume that $\xi_i \perp \xi_j \mid \xi_{-\{i,j\}}$. Then employment of Lemma 9.7 leads to the relation

$$\exp\{-Q_{ij}(x_i - a_i)(x_j - a_j)\} = h(x_{-\{i\}})r(x_{-\{j\}})$$

for some positive functions h and r. Consequently taking logarithm of both sides of the last relation one can easily show that $Q_{ij} = 0$. The proof is complete. □

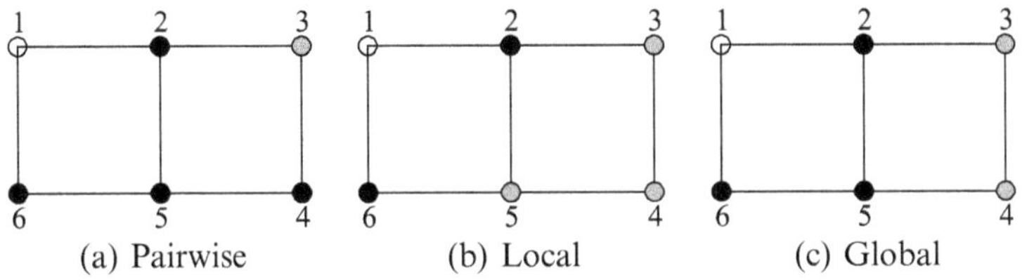

Fig. 9.14 $\xi_A \perp \xi_B \mid \xi_D$. **(a)** $A = \{1\}$, $B = \{3\}$, $D = \mathbb{V}_{-\{1,3\}} = \{2,4,5,6\}$, **(b)** $A = \{1\}$, $B = \{3,4,5\}$, $D = \partial\{1\} = \{2,6\}$, **(c)** $A = \{1\}$, $B = \{3,4\}$, $D = \{2,5,6\}$

Exercise 9.21. Assume that $\zeta = (\zeta_1, \dots, \zeta_n)^\top$ is a Gaussian random vector and sets $I, J \subset \{1, \dots, n\}$ are disjoint. Prove that ζ_I and ζ_J are independent iff $\mathbf{cov}(\zeta_i, \zeta_j) = 0$ for all $i \in I$ and $j \in J$.

Consider a Gaussian random field $\xi = \{\xi_t,\ t \in \mathbb{V}\}$ defined on a finite set $\mathbb{V} = \{t_1, \dots, t_n\}$. We suppose w.l.g. that $\mathbb{V} = \{1, \dots, n\}$. Let $(\xi_1, \dots, \xi_n) \sim N(a, C)$ with $C > 0$. Put $Q = C^{-1}$ and introduce the graph $\mathbb{G} = (\mathbb{V}, \mathbb{E})$ where the vertices i and j are neighbours if $Q_{ij} \neq 0$. Recall that a *path* from i to j $(i, j \in \mathbb{V})$ is a collection $i_1, \dots, i_m$ of distinct vertices $(m \geq 2)$ such that $\langle i_k, i_{k+1} \rangle \in \mathbb{E}$ (i.e. i_k and i_{k+1} are neighbours for $k = 1, \dots, m-1$) and $i_1 = i$, $i_m = j$. Let $D \subset \mathbb{V}$. One says that disjoint sets $A, B \subset \mathbb{V} \setminus D$ are *separated* by D if there is no path starting at a point of A and coming to a point of B without passing through D (i.e. any path $i_1, \dots, i_m$ from a site $i_1 \in A$ to a site $i_m \in B$ contains a site belonging to D). The following result explains that the graph $\mathbb{G}$ introduced above is the appropriate tool to describe the *conditional dependence structure* of the components of a nondegenerate Gaussian vector or random field (cf. Exercise 9.21).

Theorem 9.5 ([478]). *Let ξ and $\mathbb{G}$ be a Gaussian random field and a graph introduced above. Then the following statements hold.*

1. Pairwise Markov property*:*

$$\xi_i \perp \xi_j \mid \xi_{-\{i,j\}} \quad \text{if } i \neq j,\ \langle i, j \rangle \notin \mathbb{E}.$$

2. Local Markov property*:*

$$\xi_i \perp \xi_{-\{i\} \cup \partial\{i\}} \mid \xi_{\partial\{i\}} \quad \text{for each } i \in \mathbb{V}.$$

3. Global Markov property*:*

$$\xi_A \perp \xi_B \mid \xi_D$$

for all nonempty sets $A, B \subset \mathbb{V}$ such that D separates them, see Fig. 9.14.

The Gaussian Markov random fields play an important role in various applications. In this regard we refer to the monograph [430].

9.3.5 Final Remarks

It should be noted that the deep theory of Gibbs measures begins when the set of sites $\mathbb{V}$ is infinite. In this case one cannot use the simple formula (9.20). The interesting effect of the existence of different measures with given family of local specifications is interpreted in terms of phase transitions known in statistical physics. We recommend to start with a simple example of phase transition considered in [310, Chap. 22], see also [84, 184, 202, 297]. Note in passing that critical phenomena depend essentially on the temperature T (i.e., whether $\mathsf{T} < \mathsf{T}_c$ or $\mathsf{T} > \mathsf{T}_c$ for specified threshold T_c). Markov random fields are used in image processing and texture analysis, see, for example [330]. We refer to [225] where the random field is defined on the subset D of a plane divided by several regions $D_1, \ldots, D_N$. In that paper the authors propose the (non-Gibbs Markov random field) model where the mutual location of these regions and distances $\mathrm{dist}(D_i, D_j)$ play an essential role in dependence structure. There are various generalizations of the Markov property, see for example [429]. Gaussian Markov random fields are studied in [430] where the applications to geostatistics are provided. The region based hidden Markov random field (RBHMRF) model is used to encode the characteristics of different brain regions into a probabilistic framework for brain MR image segmentation, see [121]. The statistical analysis of Markov random fields and related models can be found e.g. in [177, 373].

9.4 Moments and Covariance

Let $\xi = \{\xi(t),\ t \in T\}$ be a real-valued random function.

Definition 9.18. The *(mixed) moment*

$$\mu^{(j_1,\ldots,j_n)}(t_1,\ldots,t_n) := \mathbf{E}(\xi^{j_1}(t_1)\cdot\ldots\cdot\xi^{j_n}(t_n)),$$

provided that this expectation is finite. Here $j_1,\ldots,j_n \in \mathbb{N}$, $t_1,\ldots,t_n \in T$ and $n \in \mathbb{N}$.

Exercise 9.22. Show that $\mu^{(j_1,\ldots,j_n)}(t_1,\ldots,t_n)$ exists if $\mathbf{E}|\xi(t)|^j < \infty$ for all $t \in T$ and $j = j_1 + \ldots + j_n$.

Introduce now

1. *Mean value function* $\mu(t) = \mu^{(1)}(t) = \mathbf{E}(\xi(t)), t \in T$.
2. *Covariance function*

$$C(s,t) = \mathbf{cov}(\xi(s), \xi(t)) = \mu^{(1,1)}(s,t) - \mu^{(1)}(s)\mu^{(1)}(t), \quad s,t \in T.$$

Obviously the covariance function is *symmetric*, i.e. $C(s,t) = C(t,s)$ for all $s,t \in T$ and $C(t,t) = \mathbf{var}\,\xi(t)$, $t \in T$.

Definition 9.19. A function $C : T \times T \to \mathbb{C}$ is called *positive semi-definite* if for any $n \in \mathbb{N}$, $t_k \in T$ and $z_k \in \mathbb{C}$ $(k = 1,\ldots,n)$ one has

$$\sum_{k,m=1}^{n} C(t_k, t_m) z_k \overline{z_m} \geq 0. \tag{9.42}$$

Here $\overline{z_m}$ means the complex conjugate of z_m.

Exercise 9.23. Prove that covariance function C of a real-valued random field $\xi = \{\xi(t),\ t \in T\}$ with a finite second moment is positive semi-definite.

Note that (contrary to the covariance function properties) arbitrary real-valued function $f(t)$, $t \in T$, can be considered as a mean value function of some random function $\xi = \{\xi(t),\ t \in T\}$ (e.g., the Gaussian white noise plus f). It shows the (deterministic) *trend* of the random function ξ. The correlation coefficient

$$R(s,t) = \frac{C(s,t)}{\sqrt{C(s,s)C(t,t)}}$$

is sometimes called the *correlation function* of ξ. Clearly, it is well defined if $\mathbf{var}\,\xi(s) > 0$ and $\mathbf{var}\,\xi(t) > 0$. For such $s,t \in T$ we have $|R(s,t)| \leq 1$ by the Cauchy–Schwarz inequality.

Exercise 9.24. Give an example of a function that is not positive semi-definite.

Exercise 9.25. Let $\mu : T \to \mathbb{R}$ be an arbitrary function and $C : T \times T \to \mathbb{R}$ be a positive semi-definite symmetric function. Show that there exists a random function $\xi : \Omega \times T \to \mathbb{R}$ such that its mean value function and covariance function are $\mu(\cdot)$ and $C(\cdot,\cdot)$, respectively. *H*int: use a Gaussian random function.

Exercise 9.26. Give an example of two different random functions ξ and η with $\mathbf{E}\xi(t) = \mathbf{E}\eta(t)$ and $\mathbf{E}(\xi(s)\xi(t)) = \mathbf{E}(\eta(s)\eta(t))$ for all $s,t \in T$.

Exercise 9.27. Show that there exists such random function $\xi = \{\xi(t),\ t \in T\}$ that any collection of its mixed moments does not determine the distribution of ξ in $(S_T, \mathcal{B}_T)$ uniquely.

Let $\{\xi(t),\ t \in T\}$ be a real-valued random function such that $\mathbf{E}|\xi(t)|^k < \infty$ for some $k \in \mathbb{N}$ and all $t \in T$. For any $s,t \in T$ consider the increment $\xi(t) - \xi(s)$. The function $\gamma_k(s,t) := \mathbf{E}(\xi(t) - \xi(s))^k$ is called the *mean increment of order* $k \in \mathbb{N}$. In particular,

$$\gamma(s,t) := \frac{1}{2}\gamma_2(s,t) = \frac{1}{2}\mathbf{E}(\xi(t) - \xi(s))^2$$

is called the *variogram* of ξ. Variograms are frequently used in geostatistics.

Exercise 9.28. Verify that the following relationship holds

$$\gamma(s,t) = \frac{C(s,s) + C(t,t)}{2} - C(s,t) + \frac{1}{2}(\mu(s) - \mu(t))^2, \quad s,t \in T. \tag{9.43}$$

Notice that *cumulants*, i.e. the coefficients of the Taylor expansion at the origin for the logarithm of characteristic function of the vector $(\xi(t_1), \dots, \xi(t_n))^\top$ (under appropriate moment conditions), are also useful tools to study random functions. For the relationship between moments and cumulants see for example [472, pp. 289–293]. A recent application of cumulants in geostatistics can be found in [155].

To finish this section, we introduce the covariance function for complex-valued process $\xi = \{\xi(t),\ t \in T\}$ with arbitrary index set T. Let $\mathbf{E}|\xi(t)|^2 < \infty$ for any $t \in T$. Put

$$C(s,t) := \mathbf{E}(\xi(s) - \mathbf{E}\xi(s))\overline{(\xi(t) - \mathbf{E}\xi(t))}, \quad s,t \in T. \tag{9.44}$$

Then $C(s,t) = \overline{C(t,s)}$ and (9.42) holds.

9.5 Stationarity, Isotropy and Scaling Invariance

In this section, we introduce the notions of spatial homogeneity of the distribution of random functions. Let the index set T be a linear vector space[7] with addition "+", subtraction "–" and multiplication "·" of vectors by real numbers.

Definition 9.20. A random function $\xi = \{\xi(t),\ t \in T\}$ is called (*strictly*) *stationary* if for any $n \in \mathbb{N}$ and $h, t_1, \dots, t_n \in T$ one has

$$\text{Law}(\xi(t_1), \dots, \xi(t_n)) = \text{Law}(\xi(t_1 + h), \dots, \xi(t_n + h)),$$

i.e. all finite-dimensional distributions of ξ are invariant under shifts in T.

Definition 9.21. A (complex-valued) random function $\xi = \{\xi(t),\ t \in T\}$ is called *wide sense stationary* if $\mathbf{E}|\xi(t)|^2 < \infty$ for each $t \in T$,

$$\mu(t) \equiv \mu \quad \text{and} \quad C(s,t) = C(s+h, t+h), \quad h,s,t \in T,$$

where μ is a constant and C appeared in (9.44).

One writes $C(t) := C(t,0)$ where $t \in T$ and 0 is the zero in T. Therefore, $C(s,t) = C(s-t)$ for $s,t \in T$. Note that both definitions of stationarity do not imply each other. It is clear that if ξ is strictly stationary with $\mathbf{E}|\xi(t)|^2 < \infty$, $t \in T$, then it is stationary in the wide sense.

[7] For the sake of simplicity, we do not consider the case when T is a subset of a linear vector space.

Definition 9.22. A random function $\xi = \{\xi(t),\ t \in T\}$ is *intrinsically stationary of order two* if its mean increments $\gamma_i(s,t)$, $s,t \in T$, exist up to the order 2, and for all $s,t,h \in T$

$$\gamma_1(s,t) = 0, \qquad \gamma_2(s,t) = \gamma_2(s+h,t+h).$$

It is clear that intrinsic stationarity of order two (which is often used in practice) is a little bit more general than stationarity in the wide sense, because we require the existence of moments of increments of ξ and not of $\xi(t)$ itself. However, this distinction is of superficial nature, as most random functions which are of practical interest are wide sense stationary (and hence intrinsically stationary of order two).

The notion of isotropy can be introduced in the strict or wide sense as the notion of stationarity. To define it we assume that $T = \mathbb{R}^d$, $d \geq 2$. It is often required that isotropic processes are also stationary. However, we shall not do it (Fig. 9.15).

Definition 9.23. A random field $\xi = \{\xi(t),\ t \in \mathbb{R}^d\}$ is called *isotropic in the strict sense* or *in the wide sense* if for any $n \in \mathbb{N}$, $t_1, \dots, t_n \in \mathbb{R}^d$ and $A \in \mathsf{SO}$ one has

$$\mathrm{Law}(\xi(At_1), \dots, \xi(At_n)) = \mathrm{Law}(\xi(t_1), \dots, \xi(t_n))$$

or

$$\mu(At) = \mu(t), \quad C(As, At) = C(s,t), \quad s,t \in \mathbb{R}^d,$$

respectively.

Further on we suppose that $\xi = \{\xi(t),\ t \in T\}$ is a *centered* complex-valued random function (i.e. $\mathbf{E}\xi(t) = o,\ t \in T$) defined on a linear space T. If ξ is not centered, one can consider the random function $\eta = \{\eta(t) = \xi(t) - \mu(t),\ t \in T\}$. Sometimes we shall assume that ξ is wide sense stationary. In this case its covariance function $C(h) = \mathbf{E}\{\xi(t)\xi(t+h)\}$, $h \in T$, posseses the following obvious properties:

$$C(0) = \mathbf{var}(\xi(h)) \geq 0, \quad C(h) = \overline{C(-h)} \text{ and } |C(h)| \leq C(0) \text{ for any } h \in T.$$

Exercise 9.29. Let $\xi = \{\xi(t),\ t \in \mathbb{R}^d\}$ be a shot-noise field introduced in (9.11) with response function g satisfying (9.12) and (9.13). Prove that for any $s,t \in \mathbb{R}^d$

1. $\mathbf{E}\xi(t) = \lambda \int_{\mathbb{R}^d} g(t-z)\,dz$
2. $\mathbf{cov}(\xi(s), \xi(t)) = \lambda \int_{\mathbb{R}^d} g(t-z)g(s-z)\,dz.$

*H*int: use the Campbell–Mecke formula, cf. Sect. 4.1.2, [489, pp. 36–39] and [451, Theorems 3.1.2 and 3.1.3].

Let us define operator scaling stable random fields according to [68]. Consider a real $(d \times d)$-matrix A whose eigenvalues have positive real parts.

Definition 9.24. A real-valued random field $\{\xi(t),\ t \in \mathbb{R}^d\}$ is called *operator scaling* for A and $H > 0$ if for any $c > 0$

$$Law\{\xi(c^A t),\ t \in \mathbb{R}^d\} = Law\{c^H \xi(t),\ t \in \mathbb{R}^d\},$$

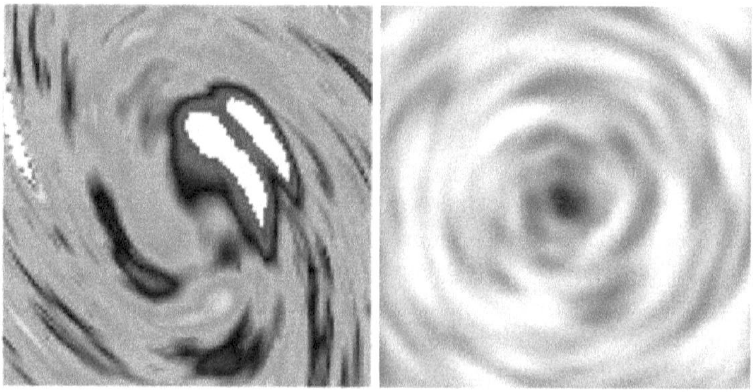

Fig. 9.15 Two typical realizations of Gaussian isotropic non-stationary random fields with covariance function from [441, Example 16]. Courtesy of M. Schlather

that is, all finite-dimensional distributions of these fields coincide. As usual we suppose $c^A = \exp\{A \log c\}$ with $\exp\{B\} = \sum_{k=0}^{\infty} \frac{B^k}{k!}$ for a matrix B.

These fields can be seen as anisotropic generalizations of self-similar random fields. Let us recall that a real-valued random field $\{\xi(t),\ t \in \mathbb{R}^d\}$ is said to be H*-self-similar* with $H > 0$ if for any $c > 0$

$$Law\{\xi(ct),\ t \in \mathbb{R}^d\} = Law\{c^H \xi(t),\ t \in \mathbb{R}^d\}.$$

Then a H-self-similar field is also an operator scaling field for the identity matrix $A = I_d$ of size $d \times d$.

Numerous natural phenomena have been shown to be self-similar. For instance, self-similar random fields are used to model persistent phenomena in internet traffic, hydrology, geophysics or financial markets, see for example [432, 497, 518]. A very important class of such fields is given by Gaussian random fields and especially by fractional Brownian fields.

The *fractional Brownian field* ξ_H where $H \in (0,1)$ is the so-called *Hurst parameter*, is H-self-similar and has stationary increments, i.e.

$$Law\{\xi_H(t+h) - \xi_H(h),\ t \in \mathbb{R}^d\} = Law\{\xi_H(t),\ t \in \mathbb{R}^d\}, \qquad h \in \mathbb{R}^d.$$

It is an isotropic generalization of the *fractional Brownian motion* introduced in [308] and studied in [342]. A robust method to simultaneously estimate the Hurst parameter in each scaling direction (for anisotropic stationary scalar random fields with spatial long memory) by means of a local Whittle estimator is given in [209]. It is shown that this estimator is consistent and asymptotically normal under specified conditions.

A weaker self-similarity property known as *local self-similarity* was studied in [293].

Definition 9.25. Let $\alpha \in (0, 2]$. A centered stationary Gaussian field is *locally self-similar (lss) of order* $\alpha/2$ if its covariance satisfies

$$C(h) = A - B\|h\|_2^{\alpha}\Big(1 + O(\|h\|_2^{\delta})\Big), \qquad \|h\|_2 \to 0$$

for some positive constants A, B and δ.

This class includes the centered Gaussian field with covariance function $C(t) = \exp\{-\beta\|t\|_2^{\alpha}\}$, $t \in \mathbb{R}^d$ ($\alpha \in (0, 2]$ and $\beta > 0$), which has the same functional form as the characteristic function of the multivariate symmetric stable distribution (see [311]).

Exercise 9.30. Show that the Gaussian random field with generalized Cauchy covariance introduced in Sect. 9.2.2 is not self-similar, but locally self-similar of order $\alpha/2$ with $A = 1$, $B = \beta$ and $\delta = \alpha$.

A nice property of $\alpha/2$-lss fields is that their fractal dimension is determined by α (see [333, Proposition 2.6]).

9.6 Positive Semi-definite Functions

Which function can be a covariance function of a stationary continuous (in mean square sense) random field? The answer to this question yields the famous Bochner–Khinchin Theorem: it must be positive semi-definite. In this section we provide some criteria for positive semi-definiteness as well as principles of construction of new covariance structures out of known "building blocks".

9.6.1 General Results

To formulate the important results for stationary random fields we recall the following concept.

Definition 9.26. Let T be an Abelian group (with addition). A function $f : T \to \mathbb{C}$ is *positive semi-definite* if $C(s, t) = f(s - t)$, $s, t \in T$ satisfies (9.42).

The following classical result was established (for $d = 1$) in 1932–1934 independently by Bochner and Khinchin.

Theorem 9.6 (Bochner–Khinchin). *A function* $f : \mathbb{R}^d \to \mathbb{C}$*, continuous at the origin* $o \in \mathbb{R}^d$*, is positive semi-definite iff it can be represented as the characteristic function of a finite measure* μ_f *on* $\mathbb{R}^d$*, that is,*

$$f(t) = \varphi_{\mu_f}(t) = \int_{\mathbb{R}^d} \exp\{i\,\langle t, x\rangle\}\mu_f(dx), \quad t \in \mathbb{R}^d, \quad i^2 = -1. \tag{9.45}$$

Definition 9.27. The measure μ_f appearing in (9.45) is called a *spectral measure* of f. If μ_f has a density h with respect to the Lebesgue measure on $\mathbb{R}^d$ then h is called the *spectral density* of f.

Remark 9.4. A function f satisfying (9.45) is real-valued iff the measure μ_f is symmetric.

Indeed, function f is real-valued iff $f(t) = \overline{f(t)}$ for all $t \in \mathbb{R}^d$. We have

$$\overline{f(t)} = f(-t) = \int_{\mathbb{R}^d} e^{i\langle t,-x\rangle}\mu_f(dx) = \int_{\mathbb{R}^d} e^{i\langle t,x\rangle}\mu_f(-dx), \quad t \in \mathbb{R}^d.$$

Using the one to one correspondence between finite measures and their characteristic functions and in view of (9.45) the relation $\mu_f(A) = \mu_f(-A)$ is equivalent to the symmetry property of f.

Note also that any semi-definite function $f : \mathbb{R}^d \to \mathbb{C}$ which is continuous at the origin $o \in \mathbb{R}^d$ is automatically continuous on $\mathbb{R}^d$ in view of (9.45).

The next theorem characterizes all *measurable* positive semi-definite functions.

Theorem 9.7 (Riesz, [435, p. 81]). *A function $f : \mathbb{R}^d \to \mathbb{C}$ is positive semi-definite and measurable iff $f = f_c + f_0$ where $f_c : \mathbb{R}^d \to \mathbb{C}$ and $f_0 : \mathbb{R}^d \to \mathbb{C}$ are some positive semi-definite functions such that f_c is continuous on $\mathbb{R}^d$ (and hence the Bochner–Khinchin theorem can be applied) and f_0 equals zero almost everywhere on $\mathbb{R}^d$ with respect to the Lebesgue measure.*

Remark 9.5. The discontinuous semi-definite function f_0 widely used in applications is the so-called *nugget effect function*: $f_0(t) = a\mathbf{1}(t = 0)$ where $a > 0$. This function is a covariance function of a white noise random field with variance $a > 0$. It is employed in geostatistics to model the discontinuity at zero of the variogram of the data. This allows to consider random field models with L^2-discontinuous realizations.

The full description of the positive semi-definite functions defined on $T = Z^d$ is provided by the following result proved for $d = 1$ by Herglotz [241].

Theorem 9.8. *A function $f : \mathbb{Z}^d \to \mathbb{C}$ is positive semi-definite iff there exists a finite* (spectral) measure μ_f *on* $([-\pi,\pi]^d, \mathcal{B}([-\pi,\pi]^d))$ *such that*

$$f(t) = \int_{[-\pi,\pi]^d} \exp\{i\,\langle t,x\rangle\}\mu_f(dx), \quad t \in \mathbb{Z}^d.$$

Exercise 9.31. Show that the function $f_0(t) = a$, $t \in \mathbb{Z}^d$, is positive semi-definite for any $a \geq 0$. Find the corresponding spectral measure.

In general, it is not so easy to verify the conditions of Theorem 9.6. Because of that, we give sufficient conditions for positive semi-definiteness.

Theorem 9.9 (Pólya–Askey, [18]). *A function $f(t) = g(\|t\|_2)$ defined for $t \in \mathbb{R}^d$ is positive semi-definite if $g : \mathbb{R}_+ \to \mathbb{R}$ satisfies the conditions:*

1. *g is continuous on $\mathbb{R}_+$,*
2. *$g(0) = 1$,*
3. *$\lim_{t\to\infty} g(t) = 0$,*
4. *$(-1)^k g^{(k)}(t)$ is convex for $k = [d/2]$ ($[a]$ stands for the integer part of a).*

Theorem 9.10 (Gneiting, [191]). *If $d \geq 2$ then the previous theorem holds with replacement of condition 4 by the following one*

$$(-1)^{k+1}\frac{d^k}{dt^k}g'(\sqrt{t}) \text{ is convex for } k = \left[\frac{d-2}{2}\right].$$

9.6.2 Isotropic Case

Following [437, 440], consider covariance functions of stationary isotropic random fields with index space $\mathbb{R}^d$. They have the form $C(x, y) = C_0(\|x - y\|_2)$ where $x, y \in \mathbb{R}^d$.

The characterization result below can be obtained by passing to polar coordinates in Bochner's theorem.

Theorem 9.11 (Schoenberg, [453]). *A function $f : \mathbb{R}^d \to \mathbb{R}$, $d \geq 2$, continuous and rotation-invariant, is positive semi-definite iff $f(t) = f_0(\|t\|_2)$, $t \in \mathbb{R}^d$, with $f_0 : \mathbb{R}_+ \to \mathbb{R}$ being equal to*

$$f_0(s) = \Gamma\left(\frac{d}{2}\right)\left(\frac{2}{s}\right)^{\nu}\int_0^{\infty} r^{-\nu} J_{\nu}(rs)\mu(dr), \quad s \geq 0, \tag{9.46}$$

where μ is a finite measure on $[0, \infty)$, $\nu = d/2 - 1$ and

$$J_{\nu}(r) = \sum_{j=0}^{\infty}\frac{(-1)^j}{j!\,\Gamma(\nu + j + 1)}\left(\frac{r}{2}\right)^{\nu+2j}, \quad r \geq 0,$$

is the Bessel function of the first kind of order $\nu \geq -1/2$.

Example 9.4. Let us give Schoenberg's representation of a covariance function C of a wide-sense stationary isotropic random field for $d = 3$. Since $J_{1/2}(r) = \sqrt{\frac{2}{\pi r}}\sin(r)$, $r \in \mathbb{R}_+$, we get

$$C(x, y) = C_0(\|x - y\|_2) = \int_0^{\infty}\frac{\sin(sr)}{sr}\mu(dr)\Big|_{s=\|x-y\|_2}$$

from (9.46) because

$$C_0(s) = \Gamma(3/2)\int_0^{\infty}\sqrt{\frac{2}{sr}}\sqrt{\frac{2}{\pi sr}}\sin(sr)\mu(dr) = \int_0^{\infty}\frac{\sin(sr)}{sr}\mu(dr).$$

For isotropic measurable positive semi-definite functions, a stronger version of the Riesz theorem is available:

Theorem 9.12 ([194]). *If f is a rotation-invariant, measurable positive semi-definite function $f : \mathbb{R}^d \to \mathbb{C}$, $d \geq 2$, then*

$$f(t) = f_c(t) + a \cdot \mathbf{1}(t = 0)$$

where $a \geq 0$ and f_c is positive semi-definite and continuous.

It follows from Theorem 9.12 that measurable isotropic covariance functions are discontinuous at most at the origin.

9.6.3 Construction of Positive Semi-definite Functions

Theorem 9.13. *Let $f_k : \mathbb{R}^d \to \mathbb{C}$ be positive semi-definite functions and $a_k \geq 0$ for $k \in \mathbb{N}$. The function $f : \mathbb{R}^d \to \mathbb{C}$ is positive semi-definite if the following operations are used.*

1. Scaling: $f(t) = f_1(at)$, $a \in \mathbb{R}$, $t \in \mathbb{R}^d$.
2. Linear combination: $f(t) = \sum_{k=1}^n a_k f_k(t)$, $t \in \mathbb{R}^d$.
3. Multiplication: $f(t) = \prod_{k=1}^n f_k(t)$, $t \in \mathbb{R}^d$.
4. Pointwise limit: $f(t) = \lim_{k\to\infty} f_k(t)$, $t \in \mathbb{R}^d$, *whenever the limit exists.*
5. Convolution: $f(t) = \int_{\mathbb{R}^d} f_1(t - y)\overline{f_2(y)}\, dy$, $t \in \mathbb{R}^d$, *if additionally f_1 and f_2 are continuous and the integral exists.*

Proof. 1. This follows directly from Definition 9.26.
2. Consider a random field $\xi(t) = \sum_{k=1}^n \sqrt{a_k}\xi_k(t)$ where $\xi_1, \dots, \xi_n$ are independent stationary random fields with covariance functions $f_1, \dots, f_n$, respectively. Then ξ has the covariance function $\sum_{k=1}^n a_k f_k$.
3. Analogous to item 2, construct a random field $\xi(t) = \xi_1(t) \cdot \ldots \cdot \xi_n(t)$ where $\xi_1, \dots, \xi_n$ are independent stationary random fields with covariance functions $f_1, \dots, f_n$, respectively. Then ξ hat the covariance function $\prod_{k=1}^n f_k$.
4. This follows directly from Definition 9.26.
5. See [435]. □

Theorem 9.14. *Let $(E, \mathcal{B})$ be a measurable space endowed with a finite measure μ and $\{C_\nu, \nu \in E\}$ be a family of positive semi-definite functions $C_\nu : \mathbb{R}^d \times \mathbb{R}^d \to \mathbb{C}$. The function $C : \mathbb{R}^d \times \mathbb{R}^d \to \mathbb{C}$ is positive semi-definite if it is constructed by means of the following operations.*

1. Substitution: $C(s,t) = C_\nu(g(s), g(t))$ *for any mapping $g : \mathbb{R}^d \to \mathbb{R}^d$ and $\nu \in E$.*
2. Kernel approach: $C(s,t) = \langle g(s), g(t)\rangle_{\mathbb{L}}$ *where $\mathbb{L}$ is a Hilbert space over $\mathbb{C}$ with scalar product $\langle\cdot,\cdot\rangle_{\mathbb{L}}$ and $g : \mathbb{R}^d \to \mathbb{L}$ is an arbitrary mapping.*

3. Integration: $C(s,t) = \int_E C_\nu(s,t)\mu(d\nu)$ *where* $C_\nu(s,t)$ *is a* μ*-integrable function with respect to* $\nu \in E$. *In particular, this holds for* $C(s,t) = \sum_{\nu=0}^{\infty} a_\nu C_\nu(s,t)$ *whenever* $E = \mathbb{Z}_+$ *and this series absolutely converges.*
4. Scale mixtures: $C(s,t) = \int_0^\infty C_\nu(ys,yt)\gamma(dy)$ *for some* $\nu \in E$ *and finite measure* γ *on* $[0,\infty)$ *if this integral exists and is finite for* $s,t \in \mathbb{R}^d$.

Proof. 1. This statement follows immediately from Definition 9.26.
2. By Definition 9.26, for any $n \in \mathbb{N}$, $z_i \in \mathbb{C}$, $t_i \in \mathbb{R}^d$ and $i = 1,\dots,n$ we have

$$\sum_{i,j=1}^{n} C(t_i,t_j)z_i\bar{z_j} = \left\langle \sum_{i=1}^{n} g(t_i)z_i, \sum_{j=1}^{n} g(t_j)z_j \right\rangle_{\mathbb{L}} = \left\| \sum_{i=1}^{n} g(t_i)z_i \right\|_{\mathbb{L}}^2 \geq 0\,.$$

3. This follows from Theorem 9.13, 4, since $\int_E C_\nu(s,t)\mu(d\nu)$ is a limit of its integral sums. Each of these sums is a linear combination of positive semi-definite functions $C_{\nu_i}(s,t)$, $i = 1,\dots,n$, which is positive semi-definite by Theorem 9.13, 2. The same reasoning is true for series expansions.
4. For any $y \in [0,\infty)$, $C_\nu(ys,yt)$ is positive semi-definite by case 1. Then case 3 is used. □

Apart from Theorems 9.13 and 9.14, several other methods to construct positive semi-definite functions can be found in the literature. One of the easiest ways is, e.g., by means of specifying their spectral densities. For more sophisticated methods, see for example [14].

Remark 9.6. 1. Let $\mathbb{L}$ be a Hilbert space over $\mathbb{C}$ with scalar product $\langle\cdot,\cdot\rangle_{\mathbb{L}}$. Assume that $\mathbb{L}$ consists of functions $f : T \to \mathbb{C}$. A function $K : T \times T \to \mathbb{C}$ is a *reproducing kernel* of $\mathbb{L}$ if $K(t,\cdot) \in \mathbb{L}$ for all $t \in T$ and $\langle f, K(t,\cdot)\rangle_{\mathbb{L}} = f(t)$ for all $f \in \mathbb{L}$, $t \in T$. By Aronszajn's Theorem [16], K is a reproducing kernel of a Hilbert space $\mathbb{L}$ iff it is positive semi-definite, i.e. K is a covariance function. In particular, reproducing kernel Hilbert spaces are used to obtain new (series) representations of random functions, see for example [1, Chap. 3].
2. Covariance functions for *space-time random fields* $\xi = \{\xi(x,t),\ x \in \mathbb{R}^d,\ t \in \mathbb{R}\}$ can be constructed by formula

$$C((x,s),(y,t)) = C_1(x,y) + C_2(s,t), \quad x,y \in \mathbb{R}^d, \quad s,t \in \mathbb{R},$$

where C_1 is a space covariance and C_2 is a time covariance component. The same holds for

$$C((x,s),(y,t)) = C_1(x,y) \cdot C_2(s,t), \quad x,y \in \mathbb{R}^d, \quad s,t \in \mathbb{R}$$

(the so-called *separable space-time models*). For more complex construction methods we refer to [192, 193, 344].
3. By means of Theorem 9.14, (3), functions like $e^{C(s,t)}$, $\cosh C(s,t)$ etc. of a positive semi-definite function C are also positive semi-definite as they

can be represented by Taylor series with positive coefficients, e.g., $\mathrm{e}^{C(s,t)} = \sum_{n=0}^{\infty} \frac{(C(s,t))^n}{n!}$, $s, t \in \mathbb{R}^d$.

9.7 Infinitely Divisible Random Functions

In this section, we introduce infinitely divisible random functions as integrals of non-random kernels with respect to independently scattered infinitely divisible random measures. It will be clear that they form a generalization of the class of moving averages introduced on page 285. In Sect. 9.7.4, it will be shown that a large class of random functions can be represented in this way. This is the so-called *spectral representation* of infinitely divisible random functions, first proven in [413].

9.7.1 *Infinitely Divisible Distributions*

We recall some classical notions and results concerning the infinite divisibility of random variables and vectors.

Definition 9.28. 1. The probability measure μ on $(\mathbb{R}^m, \mathcal{B}(\mathbb{R}^m))$ is *infinitely divisible* if for any $n \in \mathbb{N}$ there exists a probability measure μ_n on the same space such that $\mu = \mu_n * \cdots * \mu_n$ (n-fold), where $*$ stands for convolution.
2. A random vector ξ with values in $\mathbb{R}^m$ is *infinitely divisible* if its distribution $\mathbf{P}_\xi$ is an infinitely divisible probability measure. If ξ is infinitely divisible then for all n there exist i.i.d. random vectors ξ_{ni}, $i = 1, \ldots, n$, such that

$$\xi \stackrel{d}{=} \xi_{n1} + \ldots + \xi_{nn}. \tag{9.47}$$

Hence, the characteristic function $\varphi_\xi(s)$ of ξ satisfies the relation

$$\varphi_\xi(s) = (\varphi_n(s))^n \tag{9.48}$$

where φ_n is the characteristic function of ξ_{n1}.

Formulae (9.47) and (9.48) provide equivalent definitions of an infinitely divisible random vector.

Infinitely divisible laws form an important class because they arise as limit distributions for sums of independent random elements, see for example [530, Chap. 5]. Many widely used distributions are infinitely divisible, e.g., (for $m = 1$) degenerate, Gaussian, Gamma, Poisson, Cauchy, α-stable, geometric, negative binomial, etc.

Exercise 9.32. Give an example of a random variable which is not infinitely divisible.

Recall the *Lévy representation* of the characteristic function φ_ξ of an infinitely divisible random variable ξ:

$$\varphi_\xi(s) = \exp\left\{ias - \frac{\sigma^2 s^2}{2} + \int_{\mathbb{R}} \left(\mathrm{e}^{isx} - 1 - is\tau(x)\right) \mathrm{H}(dx)\right\} \tag{9.49}$$

where $a \in \mathbb{R}$, $\sigma^2 \geq 0$,

$$\tau(x) = x\mathbf{1}(|x| \leq 1), \tag{9.50}$$

and H is a *Lévy measure*, i.e. a measure on $(\mathbb{R}, \mathcal{B}(\mathbb{R}))$ satisfying $\mathrm{H}(\{0\}) = 0$ and $\int_{\mathbb{R}} \min\{1, x^2\}\mathrm{H}(dx) < \infty$. Analogous definitions and representation can be given, e.g., for ξ with values in real separable Hilbert spaces (see [413]). Notice that alongside with function τ in (9.50), other truncation functions can be used, see for example [473, pp. 196–197].

The triplet $(a, \sigma^2, \mathrm{H})$ specifies the properties of ξ. For instance, $\mathbf{E}|\xi|^p < \infty$ for $p > 0$ iff

$$\int_{|x|>1} |x|^p \mathrm{H}(dx) < \infty.$$

Infinitely divisible distributions play a crucial role in the theory of *Lévy processes*, see their definition in Example 9.7, case 4. Namely, the distribution of their increments is infinitely divisible, see for example [436, p. 32]. Lévy processes form a wide class of random functions describing many phenomena in econometrics, insurance, finance and physics, see for example [53, Parts IV and V], [129] and [454, Chaps. 6, 7, 10].

Definition 9.29. A (real-valued) function $\xi = \{\xi(t), t \in T\}$ is *infinitely divisible* if all its finite-dimensional distributions are infinitely divisible.

In order to construct a class of infinitely divisible moving averages we need to introduce integration with respect to infinitely divisible random measures.

9.7.2 Infinitely Divisible Random Measures

Let $(E, \mathcal{E})$ be a measurable space. One can also use any δ-ring of subsets of a set E instead of $\mathcal{E}$ but we shall not go into these details.

Definition 9.30. μ is called a *Lévy basis* on $(E, \mathcal{E})$ if $\mu = \{\mu(A), A \in \mathcal{E}\}$ is a set-indexed infinitely divisible random function defined on $(\Omega, \mathcal{A}, \mathbf{P})$ with the following properties.

1. $\mu(\cdot)$ is a *random signed measure* on $(E, \mathcal{E})$, i.e. for all pairwise disjoint sets $A_n \in \mathcal{E}, n \in \mathbb{N}$, it holds $\mu(\bigcup_{i=1}^{\infty} A_i) = \sum_{i=1}^{\infty} \mu(A_i)$ $\mathbf{P}$-almost surely.

2. μ is *independently scattered*, i.e. $\mu(A_1), \mu(A_2), \ldots$ are independent random variables for pairwise disjoint $A_n \in \mathcal{E}$, $n \in \mathbb{N}$.

If μ is a measure (i.e. μ is σ-additive and $\mu(A) \geq 0$ for all $A \in \mathcal{E}$) then the Lévy basis μ is called *non-negative*.

The terminology "Lévy basis" was introduced by Barndorff–Nielsen, see [54]. Before it was called "infinitely divisible independently scattered random measure".

Since for any $A \in \mathcal{E}$ the random variable $\mu(A)$ is infinitely divisible, its characteristic function $\varphi_{\mu(A)}$ has the Lévy representation

$$\log \varphi_{\mu(A)}(s) = isa(A) - \frac{s^2}{2}\sigma^2(A) + \int_{\mathbb{R}} \left(e^{isx} - 1 - is\tau(x)\right) \mathrm{H}_A(dx), \quad s \in \mathbb{R}. \tag{9.51}$$

Set $\mathrm{H}(A, B) = \mathrm{H}_A(B)$, $A \in \mathcal{E}$, $B \in \mathcal{B}(\mathbb{R})$. It can be shown (see [413, Proposition 2.1 and Lemma 2.3]) that

1. $a(\cdot)$ is a signed measure on $(E, \mathcal{E})$.
2. $\sigma^2(\cdot)$ is a measure on $(E, \mathcal{E})$.
3. $\mathrm{H}(\cdot, \cdot)$ can be extended to have a measure on $\mathcal{E} \otimes \mathcal{B}(\mathbb{R})$.

Introduce the *control measure* μ_c of μ on $(E, \mathcal{E})$ by way of

$$\mu_c(A) = \|a\|_{TV}(A) + \sigma^2(A) + \int_{\mathbb{R}} \min\{1, x^2\}\mathrm{H}(A, dx) \tag{9.52}$$

where $\|a\|_{TV}$ is the total variation of a measure a. Then a, σ^2 and H are absolutely continuous with respect to μ_c and have the respective densities

1. $a(dy)/\mu_c(dy) = \tilde{a}(y)$.
2. $\sigma^2(dy)/\mu_c(dy) = \tilde{\sigma}^2(y)$.
3. $\mathrm{H}(dy, B)/\mu_c(dy) = h(y, B)$ where $h(y, \cdot)$ is the Lévy measure for fixed $y \in E$ and $B \in \mathcal{B}(\mathbb{R})$.

Denoting the cumulant function $\log \varphi_\xi(s)$ of a random variable ξ by $K_\xi(s)$ we get

$$K_{\mu(A)}(s) = \int_A K_{\tilde{\mu}(y)}(s)\mu_c(dy), \quad A \in \mathcal{E}, \quad s \in \mathbb{R},$$

where the *spot variable* $\tilde{\mu}(y)$ is an infinitely divisible random variable with cumulant function

$$K_{\tilde{\mu}(y)}(s) = is\tilde{a}(y) - \frac{s^2}{2}\tilde{\sigma}^2(y) + \int_{\mathbb{R}} \left(e^{isx} - 1 - is\tau(x)\right) h(y, dx), \quad y \in E.$$

One says that $(\tilde{a}, \tilde{\sigma}^2, h)$ is a *characteristic triplet* of μ with respect to control measure μ_c.

Exercise 9.33. Show that

1. If $\mathbf{E}(\tilde{\mu}(y))$ and $\mathbf{var}(\tilde{\mu}(y))$ exist then

$$\mathbf{E}(\tilde{\mu}(y)) = \tilde{a}(y) + \int_{\mathbb{R}\setminus[-1,1]} x\, h(y, dx), \quad \mathbf{var}(\tilde{\mu}(y)) = \tilde{\sigma}^2(y) + \int_{\mathbb{R}} x^2\, h(y, dx).$$

2. If $(E, \mathcal{E}) = (\mathbb{R}^d, \mathcal{B}(\mathbb{R}^d))$, $\tilde{a}(y) \equiv \tilde{a}, \tilde{\sigma}^2 \equiv \tilde{\sigma}^2, h(y, dx) = h(dx)$ do not depend on $y \in \mathbb{R}^d$ and $\mu_c(dy) = \nu_d(dy)$, then μ is strictly stationary, i.e. all its finite-dimensional distributions are translation-invariant. Due to this observation, the above choice of parameters is often considered in various applications.

Example 9.5. 1. *Poisson Lévy basis.* If $\mu(A) \sim \text{Pois}(\Lambda(A))$, $A \in \mathcal{E}$, for a finite measure Λ on $(E, \mathcal{E})$, then $\log \varphi_{\mu(A)}(s) = \Lambda(A)(\mathrm{e}^{is} - 1)$, $\tilde{\mu}(y) \sim \text{Pois}(1)$, $s \in \mathbb{R}$, $y \in E$, and thus it is a Lévy basis with a characteristic triplet $(\tilde{a}, \tilde{\sigma}^2, h) = (1, 0, \delta_1(dx))$ and control measure $\mu_c = \Lambda$, where δ_x is the Dirac δ-measure concentrated at $x \in \mathbb{R}$.

2. *Gaussian Lévy basis.* If $\mu(A) \sim N(a(A), \sigma^2(A))$, $A \in \mathcal{E}$, then

$$\log \varphi_{\mu(A)}(s) = isa(A) - \frac{s^2}{2}\sigma^2(A)$$

and $\tilde{\mu}(y) \sim N(\tilde{a}(y), \tilde{\sigma}^2(y))$, $y \in E$. It is a Lévy basis with characteristic triplet $(\tilde{a}, \tilde{\sigma}^2, 0)$ with respect to some control measure μ_c.

3. *Gamma Lévy basis.* Choose $\mu_c = M\nu_d$, $M > 0$, and consider the characteristic triplet $(\tilde{a}, 0, h)$ with $h(y, dx) = \mathbf{1}(x \in (0, \infty))\frac{1}{x}\mathrm{e}^{-\theta x}\, dx$, $\tilde{a}(y) = \frac{1}{\theta}\left(1 - \mathrm{e}^{-\theta}\right)$ for $\theta \in (0, \infty)$. Using formulae from Exercise 9.33 we get $\mathbf{E}(\tilde{\mu}(y)) = \theta^{-1}$ and $\mathbf{var}(\tilde{\mu}(y)) = \theta^{-2}$. By (9.52), the factor M is equal to

$$M = \frac{1 + \theta - 2\theta\mathrm{e}^{-\theta} - \mathrm{e}^{-\theta}}{\theta^2} + \int_1^\infty \frac{1}{x}\mathrm{e}^{-\theta x} dx$$

where $dx = \nu_1(dx)$. Then we have $\mu(A) \sim \Gamma(\nu_d(A), \theta)$ where $A \in \mathcal{B}_0(\mathbb{R}^d)$ and $\Gamma(\nu_d(A), \theta)$ is the Gamma distribution with probability density function

$$f(x) = \frac{\theta^{\nu_d(A)}}{\Gamma(\nu_d(A))} x^{\nu_d(A)-1}\mathrm{e}^{-\theta x}\mathbf{1}_{x \in [0,\infty)},$$

see [240, p. 608]. In the last formula $\Gamma(\cdot)$ denotes the Gamma function. In this case μ is called *Gamma Lévy basis*.

4. *Symmetric stable case.* Let $\mu(A)$, $A \in \mathcal{E}$, be a *symmetric α-stable random variable (SαS)*, $\alpha \in (0, 2)$, i.e. its characteristic function is

$$\varphi_{\mu(A)}(s) = \mathrm{e}^{-c_A^\alpha |s|^\alpha}, \quad s \in \mathbb{R}.$$

Here $c_A > 0$. It is a Lévy basis with characteristic triplet $(0, 0, h)$ where h has a density

$$dh/dx = \begin{cases} c_1(A)x^{-1-\alpha}, & x > 0, \\ c_2(A)|x|^{-1-\alpha}, & x < 0, \end{cases}$$

for specified factors $c_1(A), c_2(A) \geq 0$, $c_1(A) + c_2(A) > 0$ with respect to some control measure μ_c, see [436, p. 80].

9.7.3 Infinitely Divisible Stochastic Integrals

Our goal is to introduce a stochastic integral $\int_E g(x)\mu(dx)$ of a deterministic function $g : E \to \mathbb{R}$ with respect to a Lévy basis μ defined on $(E, \mathcal{E})$ where E is a non-empty subset of a real separable Hilbert space equipped with a σ-algebra $\mathcal{E}$.

Definition 9.31. An $\mathcal{E}$-measurable function $f : E \to \mathbb{R}$ is *integrable with respect to the Lévy basis* μ with control measure μ_c if

1. There exists a sequence $(f_n)_{n\in\mathbb{N}}$ of simple functions $f_n(x) = \sum_{j=1}^{k_n} x_{nj}\mathbf{1}(x \in A_{nj})$, $n \in \mathbb{N}$, $A_{ni} \cap A_{nj} = \emptyset$, $i \neq j$, $i, j \in \{1, \dots, k_n\}$ converging to f μ_c-almost everywhere as $n \to \infty$.
2. For any $A \in \mathcal{E}$, the sequence $(\int_A f_n(x)\mu(dx))_{n\in\mathbb{N}}$ converges in probability. Here, $\int_A f_n(x)\mu(dx) := \sum_{j=1}^{k_n} x_{nj}\,\mu(A \cap A_{nj})$.

Then the limit (in probability) of the above sequence for $A = E$ is denoted by $\int_E f(x)\mu(dx)$. Namely,

$$\int_E f(x)\mu(dx) = \lim_{n\to\infty} \int_E f_n(x)\mu(dx).$$

It can be shown that this integral is well-defined in the sense that it does not depend on the choice of the sequence $(f_n)_{n\in\mathbb{N}}$ of simple functions approximating f, see [318, Sects. 7.3 and 8.3]. The following sufficient conditions of integrability of f are given in [240, Lemma 1]. For both necessary and sufficient conditions see [413, Theorem 2.7].

Theorem 9.15. *An $\mathcal{E}$-measurable function $f : E \to \mathbb{R}$ is integrable with respect to the Lévy basis μ with a triplet $(a, \sigma^2, \mathrm{H})$ if one has*

1. $\int_E |f(y)|\, \|a\|_{TV}(dy) < \infty$,
2. $\int_E f^2(y)\,\sigma^2(dy) < \infty$,
3. $\int_E \int_{\mathbb{R}} |f(y)x|\, H(dy, dx) < \infty$.

Then the cumulant function of $\xi = \int_E f d\mu$ equals

$$K_\xi(s) = \int_E K_{\tilde{\mu}(y)}(sf(y))\mu_c(dy), \quad s \in \mathbb{R}, \tag{9.53}$$

where $\tilde{\mu}$ is the spot variable corresponding to μ and μ_c is the control measure.

Exercise 9.34. Prove that if $\mathbf{E}\xi$ and $\mathbf{var}\,\xi$ exist, then one has

$$\mathbf{E}\xi = \int_E f(y)\mathbf{E}(\tilde{\mu}(y))\mu_c(dy),$$

$$\mathbf{var}\,\xi = \int_E f^2(y)\,\mathbf{var}(\tilde{\mu}(y))\mu_c(dy).$$

Example 9.6. If μ is a Gaussian Lévy basis, i.e. $\mu(A) \sim N(a(A), \sigma^2(A))$, $A \in \mathcal{E}$, then $\int_E f\,d\mu \sim N(\int_E f(y)a(dy), \int_E f^2(y)\sigma^2(dy))$.

9.7.4 Spectral Representation of Infinitely Divisible Random Functions

Let T be any index set and $(E, \mathcal{E})$ an arbitrary measurable space. Introduce the random function

$$\xi(t) = \int_E f_t(y)\mu(dy), \quad t \in T, \tag{9.54}$$

where $\{f_t, t \in T\}$ is a family of $\mathcal{E}$-measurable real-valued functions that are integrable with respect to the Lévy basis μ. It can be easily shown that ξ is an infinitely divisible random function. Indeed, it suffices to prove that $(\varphi_{t_1,\dots,t_n})^\gamma$ is a characteristic function for any $\gamma > 0$, $n \in \mathbb{N}$ and $t_1, \dots, t_n \in T$ where $\varphi_{t_1,\dots,t_n}$ is a characteristic function of the vector $\xi_{t_1,\dots,t_n} = (\xi(t_1), \dots, \xi(t_n))^\top$. Since $\varphi_{t_1,\dots,t_n}(s)$, $s = (s_1, \dots, s_n)^\top$, can be regarded as a characteristic function of the random variable $\eta = \sum_{j=1}^n s_j \xi(t_j)$ and η is clearly infinitely divisible by additivity of the integral (9.54) with respect to the kernel f_t, one can use (9.53) for the cumulant function of η to see that $(\varphi_\eta)^\gamma$ has representation (9.53) with control measure $\gamma\mu_c$. Hence $(\varphi_\eta)^\gamma$ is a characteristic function.

Random functions (9.54) are used, e.g., to model turbulence phenomena in liquid flows (see [54] and references therein) and spatial distributed claims in storm insurance [287], see Fig. 9.16. They are also instrumental in the construction of new classes of Cox point processes [240].

Example 9.7. 1. If μ is a Gaussian Lévy basis, then ξ is a Gaussian random function.
2. If μ is a Poisson Lévy basis, then ξ is a shot-noise random function.
3. Let μ be an *α-stable Lévy noise* ($0 < \alpha < 2$), i.e. an independently scattered random measure with control measure μ_c where $\mu(A)$ is an α-stable random variable with zero shift parameter ($b = 0$), skewness $\beta(A) = \beta$ and scale parameter $\sigma = (\mu_c(A))^{1/\alpha}$ (compare the notation in Sect. 9.2.7) for all sets $A \in \mathcal{E}$. Assume that $f_t \in L^\alpha_{\mu_c}(E)$ for all $t \in T$ if $\alpha \neq 1$ and

$$f_t \in \{f \in L^1_{\mu_c}(E) : \int_E |f(x)\log|f(x)||\,\mu_c(dx) < \infty\}$$

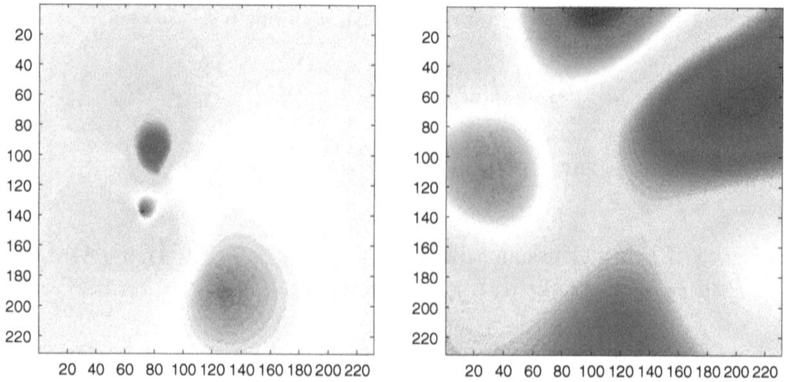

Fig. 9.16 Extrapolated yearly fluctuations of insurance claims around the mean in storm insurance (2005) in Vienna (*left*) and a simulated realization of a 1.3562-stable random field (9.54) with Gaussian kernel $f_t(x) = 0.982443 \exp\{-\|x-t\|_2^2/8.6944\}$ and skewness parameter $\beta = 0.2796$ modelling these fluctuations (*right*). The colour scale stretches from *red* for high (positive) values to *blue* for low (negative) values

for all $t \in T$ if $\alpha = 1$. Then $f_t, t \in T$, is integrable with respect to μ, and ξ from (9.54) is an *α-stable random function*.

4. Let μ_P be a Poisson random measure on $(0,\infty)\times\mathbb{R}\setminus\{0\}$ with intensity measure $\nu_1 \times \mathrm{H}$, see Sect. 4.1.1, Definition 4.2, where H is the Lévy measure on $\mathbb{R}\setminus\{0\}$. Let μ_G be a Gaussian (2-stable) random measure with Lebesgue control measure and skewness intensity $\beta \equiv 0$. Then the *Lévy process* with Lévy measure H, Gaussian part μ_G and drift γ is given by

$$\xi(t) = \int_0^t \int_{\mathbb{R}} x\,\mu_P(dy,dx) + t\left(\gamma - \int_{|x|<1} x\,\mathrm{H}(dx)\right) + \int_0^t \mu_G(dy)$$
$$= \int_{\mathbb{R}} \mathbf{1}(y \in [0,t])\,\mu(dy)$$

for $t \geq 0$. Here

$$\mu(dy) = \int_{\mathbb{R}} x\,\mu_P(dy,dx) + \left(\gamma - \int_{|x|<1} x\,\mathrm{H}(dx)\,dy\right) + \mu_G(dy).$$

Exercise 9.35. Show that

$$\gamma = \mathbf{E}\xi(1) - \int_{|x|\geq 1} x\,\mathrm{H}(dx)$$

whenever $\mathbf{E}\,\xi(1)$ exists and is finite.

Can any infinitely divisible random function ξ be represented as a stochastic integral (9.54)? The answer to this question is negative. However, the spectral representation (9.54) holds true for a large class of infinitely divisible random functions that satisfy some additional conditions, as the following result (see [413, Theorem 4.11]) shows.

Theorem 9.16. *Let $\xi = \{\xi(t),\ t \in T\}$ be an infinitely divisible L^p-separable real-valued random function ($p \geq 1$) satisfying one of the following three conditions.*

1. *ξ is symmetric.*
2. *$\mathbf{E}(\xi(t)) = 0$ for $t \in T$.*
3. *ξ is centered α-stable, $\alpha \in (0, 2)$, $p < \alpha$.*

Then there exists an uncountable Borel subset E of a Polish space equipped with a σ-algebra $\mathcal{E}$ and a Lévy basis μ on $(E, \mathcal{E})$ together with a family of μ-integrable functions $f_t : E \to \mathbb{R}$, $t \in T$, such that

$$\xi \stackrel{d}{=} \left\{ \int_E f_t(x)\mu(dx) : t \in T \right\}.$$

If ξ is α-stable, then μ is α-stable as well.

Remark 9.7. 1. In order to get a spectral representation of ξ not only in distribution, but almost surely, an enlargement of the probability space by "randomization" should be performed, see [413, Theorem 5.2].
2. For a $S\alpha S$ random field $\xi = \{\xi(t),\ t \in \mathbb{Z}^d\}$, a unique decomposition (in law) into a sum of two independent random fields ξ_C and ξ_D generated by conservative and dissipative Z^d-actions, respectively, is given in [428].

9.8 Elementary Statistical Inference for Random Fields

Let $\xi = \{\xi(t),\ t \in \mathbb{R}^d\}$, $\xi(t) = (\xi_1(t), \dots, \xi_m(t))^\top$, $m \geq 1$, be a stationary (vector-valued) random field with $\mathbf{E}\,\xi_i^2(o) < \infty$, mean $\mu = (\mu_1, \dots, \mu_m)^\top$ where $\mu_i = \mathbf{E}\,\xi_i(o)$ and cross-covariance function $C_{ij}(h) = \mathbf{cov}(\xi_i(o), \xi_j(h))$, $h \in \mathbb{R}^d$, $i, j = 1, \dots, m$.

In this section, we consider some non-parametric statistical assessment procedures for the estimation of μ, C_{ij} and related characteristics such as variogram, spectral density, asymptotic (cross) covariance matrix from a single realization of ξ. In our exposition we mainly follow [268, 387, 388].

9.8.1 Estimation of the Mean

To estimate $\mu = (\mu_1, \dots, \mu_m)^\top$ consider a sequence $(W_n)_{n \in \mathbb{N}}$ of bounded Borel sets $W_n \subset \mathbb{R}^d$ growing in the *Van Hove sense*. It means that

$$\lim_{n\to\infty} \nu_d(W_n) = \infty \text{ and } \lim_{n\to\infty} \frac{\nu_d(\partial W_n \oplus B_r(o))}{\nu_d(W_n)} = 0 \tag{9.55}$$

for any $r > 0$. We use $\oplus$ and $\ominus$ for Minkowski's addition and difference, respectively. Furthermore, suppose that the random fields $\xi_i = \{\xi_i(t), t \in \mathbb{R}^d\}$ are observable on subwindows $W_{ni} \subset W_n$, $\nu_d(W_{ni}) < \infty$ for all i and n. An unbiased estimator of the mean μ is given by $\hat{\mu}_n = (\hat{\mu}_{n1}, \ldots, \hat{\mu}_{nm})^\top$, $n \geq 1$, with

$$\hat{\mu}_{ni} = \int_{W_n} \xi_i(t)\, G_i(W_n, t)\, dt$$

for weight functionals $G_i : \mathcal{B}_0(\mathbb{R}^d) \otimes \mathbb{R}^d \to [0, \infty), i = 1, \ldots, m$, which satisfy

$$G_i(W, t) = 0, \quad t \in \mathbb{R}^d \setminus W_{ni}, \qquad \text{and} \qquad \int_{\mathbb{R}^d} G_i(W, t)\, dt = 1 \tag{9.56}$$

for any $W \in \mathcal{B}_0(\mathbb{R}^d)$. The simplest weight functional is the uniform weight

$$G_i(W, t) = \mathbf{1}(t \in W_{ni}) \,/\, \nu_d(W_{ni}), \quad i = 1, \ldots, m, \quad t \in \mathbb{R}^d. \tag{9.57}$$

Exercise 9.36. Use the stationarity of ξ and the Fubini theorem to show that $\hat{\mu}_n$ is an unbiased estimator of μ for any n.

Put

$$\Gamma_{nij}(t) = \int_{\mathbb{R}^d} G_i(W_n, y) G_j(W_n, y + t)\, dy \;\text{ for } i, j = 1, \ldots, m.$$

Note that $\Gamma_{nij}(t) = 0$ if $t \notin W_{ni} \oplus \check{W}_{nj}$. Here $\check{K}$ means the set $-K$.

Exercise 9.37. Prove, that for any $n \in \mathbb{N}$, it holds that

$$\mathbf{cov}(\hat{\mu}_{ni}, \hat{\mu}_{nj}) = \int_{\mathbb{R}^d} C_{ij}(t) \Gamma_{nij}(t)\, dt, \qquad i, j = 1, \ldots, m. \tag{9.58}$$

Hint: Use the Fubini theorem.

To study the asymptotic behavior of $\hat{\mu}_n$, we assume that there exist constants $c_1, \theta_{ij} \in (0, \infty)$ for all $i, j = 1, \ldots, m$ such that for any $t \in \mathbb{R}^d$

$$\sup_{t \in \mathbb{R}^d} G_i(W, t) \leq \frac{c_1}{\nu_d(W)}, \qquad \lim_{n\to\infty} \nu_d(W_n) \Gamma_{nij}(t) = \theta_{ij}\,. \tag{9.59}$$

Both conditions are evidently met, for example in the case (9.57).

Lemma 9.8. *Let conditions* (9.55), (9.56) *and* (9.59) *be satisfied. If C_{ij} is absolutely integrable on $\mathbb{R}^d$ then*

$$\lim_{n\to\infty} \nu_d(W_n)\,\mathbf{cov}(\hat{\mu}_{ni}, \hat{\mu}_{nj}) = \theta_{ij} \int_{\mathbb{R}^d} C_{ij}(t)\,dt, \qquad i, j = 1, \ldots, m. \tag{9.60}$$

Hence the estimator $\hat{\mu}_n$ is mean square consistent for μ, i.e.,

$$\lim_{n\to\infty} \mathbf{E}\,|\hat{\mu}_n - \mu|^2 = 0\,.$$

Proof. The first assertion follows from the inequality in (9.59), Exercise 9.37 and dominated convergence theorem. The second assertion is a consequence of (9.60) and the unbiasedness of $\hat{\mu}$ as $\lim_{n\to\infty} \nu_d(W_n) = \infty$ and thus $\lim_{n\to\infty} \mathbf{var}(\hat{\mu}_{ni}) = 0$, $i = 1, \ldots, m$. □

Under appropriate mixing and integrability conditions imposed on $\xi_1, \ldots, \xi_m$ (see for example [268, Sect. 1.7]) one can show that $\hat{\mu}_n = \hat{\mu}_n = (\hat{\mu}_{n1}, \ldots, \hat{\mu}_{nm})^\top$ is asymptotically normal, i.e.,

$$\sqrt{\nu_d(W_n)}(\hat{\mu}_{n1} - \mu_1, \ldots, \hat{\mu}_{nm} - \mu_m)^\top \xrightarrow{d} N(o, \Sigma), \quad n \to \infty, \tag{9.61}$$

$N(o, \Sigma)$ is an m-dimensional Gaussian random vector with mean zero and covariance matrix $\Sigma = (\sigma_{ij})$ where

$$\sigma_{ij} = \theta_{ij} \int_{\mathbb{R}^d} C_{ij}(t)\,dt, \qquad i, j = 1, \ldots, m.$$

9.8.2 Estimation of the Covariance Function and Related Characteristics

There exist many parametric estimation procedures for various second order characteristics of random functions. Most of them are based on the use of the method of least squares or the method of moments. Here we focus on non-parametric approaches.

Assume that each component of ξ is observed within the observation window W_n and (9.55) holds.

A classical estimator of the cross-covariance function C_{ij} within a compact set $K \subset \mathbb{R}^d$ is

$$\hat{C}_{nij}(h) = \frac{1}{\nu_d(W_n \cap (W_n - h))} \int_{W_n \cap (W_n - h)} \xi_i(t)\xi_j(t+h)\,dt - \hat{\mu}_{ni}\hat{\mu}_{nj} \tag{9.62}$$

where $i, j = 1, \ldots, m$, $n \in \mathbb{N}$ and $h \in K$. Alternatively, one can integrate over W_n instead of $W_n \cap (W_n - h)$ in (9.62) and normalize the integral by the volume of W_n. However, in this case we need the observations $\xi(t)$ for $t \in W_n \cup (W_n \oplus K)$.

Lemma 9.9. *If C_{ij} is absolutely integrable on $\mathbb{R}^d$ then the estimator* (9.62) *is asymptotically unbiased as $n \to \infty$.*

Proof. Add $\pm\mu_i\mu_j$ to $\hat{C}_{nij}(h)$. By stationarity of ξ and the Fubini theorem, we get

$$\mathbf{E}\,\hat{C}_{nij}(h) = C_{ij}(h) - \mathbf{cov}(\hat{\mu}_{ni}, \hat{\mu}_{nj}).$$

By Lemma 9.8, $\mathbf{cov}(\hat{\mu}_{ni}, \hat{\mu}_{nj}) \to 0$ as $n \to \infty$ which completes the proof. □

For simplicity, consider the case $m = 1$ in more detail. We shall write C for C_{11} and assume that ξ is a mean-zero field with variogram $\gamma : \mathbb{R}^d \to \mathbb{R}$ and spectral density $f : \mathbb{R}^d \to \mathbb{R}$. Introduce

$$\hat{C}_n(h) = \frac{1}{\nu_d(W_n)} \int_{W_n} \xi(t)\xi(t+h)\,dt \tag{9.63}$$

and

$$\hat{C}_n^*(h) = \frac{1}{\nu_d(W_n \cap (W_n - h))} \int_{W_n \cap (W_n - h)} \xi(t)\xi(t+h)\,dt, \qquad h \in K. \tag{9.64}$$

Stationarity of ξ and a straightforward application of the Fubini theorem yield that both estimators (9.63) and (9.64) are unbiased. Under certain additional assumptions on ξ, these estimators are a.s. consistent (also uniformly in $h \in K$) and asymptotically normally distributed, see [268, Chap. 4].

Since in practice one works with finite sums rather than with integrals, we give a discrete version of the estimator (9.64) used in geostatistical literature, see [122, 136, 504]. Let the random field ξ be observed at a finite number of spatial locations $t_1, \dots, t_k \in W_n$. Introduce

$$\hat{C}_n^{**}(h) = \frac{1}{N_h} \sum_{i,j:\,t_i - t_j = h} \xi(t_i)\xi(t_j), \qquad h \in K, \tag{9.65}$$

where N_h is the number of pairs of points $(t_i, t_j) \in W_n^2$ such that $t_i - t_j = h$. If the points t_i do not lie on a regular grid one can expect that the number N_h is either zero or very small for most values of h. Hence one considers all pairs of points (t_i, t_j) in (9.65) such that $t_i - t_j \approx h$, i.e. the vector $t_i - t_j$ lies in a small neighborhood of h. Unfortunately, $\hat{C}_n^{**}$ is not positive semi-definite. We refer to [136, p. 71] and [216, Sect. 4.1] for the asymptotic properties of $\hat{C}_n^{**}$.

The use of the variogram γ instead of C is very popular in geostatistics, see for example [300, 490]. To assess γ, the estimator

$$\hat{\gamma}_n(h) = \frac{1}{2\nu_d(W_n \cap (W_n - h))} \int_{W_n \cap (W_n - h)} (\xi(t) - \xi(t+h))^2\,dt, \quad h \in K, \tag{9.66}$$

is employed, see [504, p. 50]. Again, it is easily seen that $\hat{\gamma}_n$ is unbiased. A discretized version of (9.66) is

$$\hat{\gamma}_n^*(h) = \frac{1}{2N_h} \sum_{i,j:t_i-t_j=h} \left(\xi(t_i) - \xi(t_j)\right)^2, \qquad h \in K, \tag{9.67}$$

see for example [504, p. 47]. Asymptotic properties of $\hat{\gamma}_n^*$ are close to those of $\hat{C}_n^{**}$, see [136, p. 71]. Robust versions of (9.67) based on the trimmed mean and the sample median are given in papers [52, 181, 182] and in [136, pp. 74–83], [122, pp. 39–44]. Alternatively, the relation $\gamma(h) = C(0) - C(h)$, $h \in \mathbb{R}^d$ (cf. (9.43)) can be used to get the estimator $\hat{\gamma}_n^{**}(h) = \tilde{C}(0) - \tilde{C}(h)$, $h \in \mathbb{R}^d$, where $\tilde{C}$ is any estimator of C.

The spectral density f of ξ can be estimated, e.g., by means of *periodogram*

$$\hat{f}_n(h) = \frac{1}{(2\pi)^d \nu_d(W_n)} \left| \int_{W_n} \exp\{i\langle t, h\rangle\}\xi(t)\, dt \right|^2, \qquad h \in K. \tag{9.68}$$

If f is continuous this estimator is asymptotically unbiased as $n \to \infty$, cf. [425, p. 132] for the case $T = \mathbb{Z}$ and [180, Sect. 5.3] for the two-dimensional case. However, its variance does not vanish with increasing n which makes it of limited use in applications. To improve the asymptotic behavior of its variance, smoothed versions of the periodogram can be used:

$$\hat{f}_n^*(h) = \int_{\mathbb{R}^d} G_n(h-t)\hat{f}_n(t)\, dt, \qquad h \in K, \tag{9.69}$$

where $G_n : \mathbb{R}^d \to \mathbb{R}_+$ is a square integrable smoothing kernel such that G_n approximates the Dirac delta function as $n \to \infty$ and $\int_{\mathbb{R}^d} G_n(t)\, dt = 1$ for all $n \in \mathbb{N}$. Examples of the smoothing kernel G_n (Bartlett's, Parzen's, Zhurbenko's kernels) for $T = \mathbb{Z}$ can be found, e.g., in [472, pp. 444–445]. Under certain (regularity) assumptions on ξ and G_n, $n \in \mathbb{N}$, smoothed periodograms are asymptotically unbiased and consistent as $n \to \infty$, cf. [425, pp. 134–135] for the case $T = \mathbb{Z}$. Under further assumptions on ξ, $\hat{f}_n^*$ is asymptotically normally distributed, cf. [244] and [425, pp. 155–157]. For an overview of results on the estimation of spectral density (block estimators, tapering data, etc.) see [425, Chap. 5], [216, Chap. 4]. The estimation of cumulant densities of ξ of any order is considered in [425 and 523, Chap. 6].

9.8.3 Estimation of the Asymptotic Covariance Matrix

Since explicit formulae for σ_{ij}, $i, j = 1, \ldots, m$, are in general unknown, we are interested in the estimation of Σ in (9.61). To this end, we assume that

$$\int_{\mathbb{R}^d} |C_{ij}(t)|\,dt < \infty, \qquad i,j = 1,\ldots,m. \tag{9.70}$$

Consider $W_n = [-n,n)^d$, $n \in \mathbb{N}$. Assume that $(b_n)_{n\in\mathbb{N}}$ is any sequence of non-random numbers such that

$$b_n \to \infty \quad \text{and} \quad b_n = o(n^\gamma), \quad n \to \infty, \tag{9.71}$$

for some $\gamma > 0$.

9.8.3.1 The Estimator Involving Local Averaging

For $z \in W_n \cap \mathbb{Z}^d$, $n, b_n \in \mathbb{N}$ and $i,j = 1,\ldots,m$ set

$$K_z(b_n) = \{t \in \mathbb{Z}^d : \|z - t\|_\infty \le b_n\}, \quad D_z = D_z(W_n, b_n) = W_n \cap K_z(b_n),$$

where $\|t\|_\infty = \max_{i=1,\ldots,d} |t_i|$ for $t \in \mathbb{R}^d$. Following [101] we introduce the estimator $\hat{\Sigma}_n = (\hat{\sigma}_{nij})_{i,j=1}^m$, $n \in \mathbb{N}$, with the elements

$$\hat{\sigma}_{nij} = \frac{1}{\nu_d(W_n)} \sum_{z \in W_n \cap \mathbb{Z}^d} \nu_d(D_z) \left(\frac{S_i(D_z)}{\nu_d(D_z)} - \frac{S_i(W_n)}{\nu_d(W_n)}\right)\left(\frac{S_j(D_z)}{\nu_d(D_z)} - \frac{S_j(W_n)}{\nu_d(W_n)}\right),$$

here $S_i(D_z) = \int_{\tilde{D}_z} \xi_i(t)\,dt$, $\tilde{D}_z = \cup_{y\in D_z}[y, y+1]^d$ for $i,j = 1,\ldots,m$ and $z \in \mathbb{Z}^d$. The notation $[y, y+1]^d$ is used for a unit cube with lower vertex $y \in \mathbb{Z}^d$. Note that this estimator differs from the traditional one which is used in the case of independent observations. Here we deal with dependent summands and use the averaged variables $S_i(D_z)/\nu_d(D_z)$. Under specified conditions on the field ξ and the sequence $(b_n)_{n\in\mathbb{N}}$ one can prove (see [101, Theorem 2]) the L^1-consistency of the estimator $\hat{\Sigma}_n$, i.e.,

$$\lim_{n\to\infty} \mathbf{E}\,|\hat{\sigma}_{nij} - \sigma_{ij}| = 0$$

for all $i,j = 1,\ldots,m$.

9.8.3.2 A Covariance-Based Estimator

The random matrix $\hat{\Sigma}_n^* = (\hat{\sigma}_{nij}^*)_{i,j=1}^m$, $n \in \mathbb{N}$, is determined by means of the estimator (9.62) of the cross-covariance function C_{ij}, $i,j = 1,\ldots,m$, as a matrix with elements

$$\hat{\sigma}_{nij}^* = \frac{1}{\nu_d(W_n)} \int_{[-b_n,b_n]^d} \hat{C}_{nij}(t)\,\nu_d(W_n \cap (W_n - t))\,dt.$$

For estimators with more general weights G_i see [388].

Proposition 9.1 ([388]). *Let $\xi = \{\xi(t),\ t \in \mathbb{R}^d\}$ be a stationary square integrable vector-valued random field satisfying* (9.70)*. Then, for any sequence of non-random numbers $(b_n)_{n\in\mathbb{N}}$ such that* (9.71) *holds with $\gamma = 1/2$, the estimator $\hat{\Sigma}_n^*$ is asymptotically unbiased. If additionally $\mathbf{E}\,\xi_i^4(o) < \infty$ for $i = 1,\dots,m$ and*

$$\frac{1}{b_n^{2d}} \int_{[-b_n,b_n]^d} \int_{[-b_n,b_n]^d} \int_{\mathbb{R}^d} \left|\mathbf{cov}\left(\xi_i(o)\xi_j(t_1), \xi_i(t)\xi_j(t_2+t)\right)\right| \times dt\, dt_1 dt_2 < \infty, \tag{9.72}$$

$$\sup_{t_1,t_2\in\mathbb{R}^d} \int_{\mathbb{R}^d} \left|\mathbf{E}\big((\xi_i(o)-\mu_i)(\xi_i(t)-\mu_i)\xi_j(t_1)\xi_j(t_2)\big)\right| dt < \infty \tag{9.73}$$

for all $i, j = 1,\dots,m$ then $\hat{\Sigma}_n^*$ is mean-square consistent as $n \to \infty$.

Conditions (9.72) and (9.73) are evidently satisfied if the random field ξ has finite correlation range, i.e. $\mathbf{cov}(\xi(s), \xi(t)) = 0$ whenever $\|s-t\| \geq r_0$, for some $r_0 > 0$. In this case, all covariances (including C_{ij} and those in relations (9.72), (9.73)) have compact support.

9.8.3.3 The Subwindow Estimator

The calculation of estimators $\hat{\Sigma}_n$ and $\hat{\Sigma}_n^*$ is very time-consuming. The *subwindow estimator* described below is more efficient for applications.

Let $V_n = [-b_n, b_n)^d \subset W_n = [-n,n)^d$, $n \geq 1$, and $b_n \to \infty$ as $n \to \infty$. Consider subwindows $V_{n,k} = V_n \oplus \{h_{n,k}\}$ where $h_{n,k} \in \mathbb{R}^d$, $k = 1,\dots,N(n)$ and $(N(n))_{n\in\mathbb{N}}$ is an increasing sequence of positive integers. Assume that $\cup_{k=1}^{N(n)} V_{n,k} \subseteq W_n$ for each $n \in \mathbb{N}$ and there exists some $r > 0$ such that

$$V_{n,k} \cap V_{n,l} \subset \partial V_{n,k} \oplus B_r(0) \quad \text{for } k,l \in \{1,\dots,N(n)\} \text{ with } k \neq l.$$

Denote by

$$\hat{\mu}_{ni}^{(k)} = \frac{1}{\nu_d(V_n)} \int_{V_{n,k}} \xi_i(t)\,dt, \quad k = 1,\dots,N(n),$$

the estimator of μ_i based on observations within $V_{n,k}$, and by

$$\bar{\mu}_{ni} = \frac{1}{N(n)} \sum_{k=1}^{N(n)} \hat{\mu}_{ni}^{(k)}, \quad n \in \mathbb{N},\ i = 1,\dots,m,$$

the average of these estimators. Define the estimator $\hat{\Sigma}_n^{**} = (\hat{\sigma}_{nij}^{**})_{i,j=1}^m$ for the covariance matrix Σ by setting

$$\hat{\sigma}_{nij}^{**} = \frac{\nu_d(V_n)}{N(n)-1}\sum_{k=1}^{N(n)}(\hat{\mu}_{ni}^{(k)} - \bar{\mu}_{ni})(\hat{\mu}_{nj}^{(k)} - \bar{\mu}_{nj}). \tag{9.74}$$

The following result describes the asymptotic behaviour of $\hat{\Sigma}_n^{**}$ as $n \to \infty$.

Theorem 9.17 ([388]). *Let $\xi = \{\xi(t),\ t \in \mathbb{R}^d\}$ be a stationary square integrable vector-valued random field satisfying* (9.70). *Then the estimator $\hat{\Sigma}_n^{**}$ is asymptotically unbiased. If additionally*

$$\int_{\mathbb{R}^{3d}} |c_{ij}^{(2,2)}(x,y,z)|\, dx\, dy\, dz < \infty, \quad i,j = 1,\dots,m, \tag{9.75}$$

where the fourth-order cumulant function

$$\begin{aligned} c_{ij}^{(2,2)}(x,y,z) &= \mathbf{E}([\xi_i(0)-\mu_i][\xi_j(x)-\mu_j][\xi_i(y)-\mu_i][\xi_j(z)-\mu_j]) \\ &\quad - C_{ij}(x)C_{ij}(z-y) - C_{ii}(y)C_{jj}(x-z) - C_{ij}(z)C_{ji}(x-y), \end{aligned}$$

then $\hat{\Sigma}^{**}$ introduced in (9.74) *is mean-square consistent.*

Relation (9.75) holds for a random field ξ with finite dependence range, see Sect. 10.1.2. For strictly stationary vector-valued Gaussian time series, the corresponding sufficient condition for (9.75) to hold is the absolute summability of its cross-covariance functions, see [85, Condition 2.6.1]. In other cases this question is non-trivial.

9.9 Simulation of Random Fields

In many applications one has to (efficiently) simulate random fields with certain parameters which are either known or have been estimated from the (image) data. These simulation techniques can be rather involved. In this section, we give a brief survey of the simulation methods for widely used subclasses of infinitely divisible random fields such as Gaussian, shot-noise, stable ones. Special attention is paid to a general approach permitting to simulate all infinitely divisible random fields having the integral representation (9.54), see Sect. 9.9.3.

9.9.1 Gaussian Random Fields

There exist many ways to simulate Gaussian random fields, see [437] for an overview of the subject. They can be divided into two major classes: exact methods and approximative ones.

Exact methods yield a realization of the required Gaussian random field with specified mean and covariance function. These are, for instance, the *direct simulation* and the *circulant embedding*. The direct method simulates the Gaussian random vector with zero mean and a given covariance matrix using its Cholesky decomposition, see [280, 338]. Due to computational restrictions of the Cholesky decomposition, it is employed on modern PCs (with 8 GB RAM) mainly for Gaussian processes ($d = 1$) for up to $10,000$ simulation points or in two dimensions ($d = 2$) for an image of maximal size 100×100. The circulant embedding method (see [115,150,195,521]) relieves us from these restrictions allowing to simulate two-dimensional realizations of size $1,000 \times 1,000$. For larger images, an approximative circulant embedding can be used (see [521]).

We shall describe the circulant embedding in more detail. Let $\xi = \{\xi(t), t \in \mathbb{R}^d\}$, $d \geq 1$, be a centered stationary Gaussian random field. A field ξ has to be simulated within the window $W = [0,1]^d$ on a square regular grid G with n^d points which is equivalent to simulating a Gaussian random vector $\eta \sim N(o, A)$ with n^d components constructed by ordering the rows of the values of ξ at G in a certain way and putting it row by row into the vector η.

Definition 9.32. The matrix $A = (a_{kj})_{k,j=0}^{m-1}$ is *Toeplitz* if $a_{kj} = a_{k-j}$ for all $k, j = 0, \ldots, m-1$. It is *block Toeplitz* if this relation holds only within square blocks on the main diagonal of A, whereas all other elements of A are zero.

Definition 9.33. The matrix $A = (a_{kj})_{k,j=0}^{m-1}$ is *circulant* if its columns are consecutive permutations of a vector $a = (a_0, \ldots, a_{m-1})^\top$, i.e.,

$$A = \begin{pmatrix} a_0 & a_{m-1} & \ldots & a_2 & a_1 \\ a_1 & a_0 & \ldots & a_3 & a_2 \\ \ldots & \ldots & \ldots\ldots & \ldots \\ a_{m-2} & a_{m-3} & \ldots & a_0 & a_{m-1} \\ a_{m-1} & a_{m-2} & \ldots & a_1 & a_0 \end{pmatrix}.$$

It is *block circulant* if A has the above form within square blocks on the main diagonal of A whereas all other elements of A are zero.

Exercise 9.38. 1. Show that if A and B are $(m \times m)$ circulant matrices then $A + B$, AB are circulant and $AB = BA$.
2. Let $\hat{F}_m = (e^{-2\pi ijk/m})_{k,j=0}^{m-1}$ be the discrete Fourier transform matrix ($i^2 = -1$). Show that a symmetric circulant $m \times m$ -matrix A generated by a vector a has a decomposition $A = Q\Lambda Q^*$, where $Q = m^{-1/2}\hat{F}_m$ is unitary, Q^* is the conjugate transpose of Q, and $\Lambda = \operatorname{diag}(\hat{F}_m a)$. Demonstrate that the eigenvalues of A are given by $\hat{F}_m a$.

The covariance matrix A of η consisting of the values of C at certain points can be made Toeplitz in case $d = 1$ and block Toeplitz for $d = 2$ (for $d \geq 3$, nested block Toeplitz matrices are employed, see [521]). It can be shown that an

$m \times m$ (block) Toeplitz matrix A can be embedded into a $p \times p$ symmetric (block) circulant matrix B with $p = 2^q \geq 2(m-1)$. The embedding means that A can be viewed as a submatrix of B. In many cases, p can be chosen such that B is positive semi-definite, see [521, Propositon 2] for sufficient conditions. If one fails to find a positive semi-definite circulant embedding, approximative circulant embedding described in [521] can be used.

Suppose that matrix B is positive semi-definite. The idea is to simulate a larger p-dimensional random vector $\zeta \sim N(o, B)$ in an efficient way and then to pick up the coordinates from ζ which correspond to η. For simplicity, consider the one-dimensional case $d = 1$. Simulate

$$\zeta = Q\Lambda^{1/2}Q^*\rho$$

where $\rho \sim N(o, I)$ is a p-dimensional standard Gaussian vector. The matrix Λ and subsequent multiplications by Q and Q^* can be efficiently computed by the fast Fourier transform, see [408, Chap. 12]. Since Q is unitary, it can be easily verified that $\zeta \sim N(o, B)$ as stated above.

Approximative methods can be classified into three groups. Methods of the first group yield Gaussian realizations with an approximated covariance function. For example, we refer to methods in [471] (see also [151]) based on the spectral representation of stationary Gaussian processes [1, Theorem 5.4.2], on the Karhunen–Loève expansion [1, pp. 71–73], *sequential method* [276]). Methods of the second group simulate random fields with an exact covariance function and an approximated (non-Gaussian) joint distribution. These are, e.g., all methods based on CLT such as the *tessellation method* [320, p. 191] and the *spectral method* described in more detail below. The third group contains methods (such as, e.g., *turning bands*, see [320, p. 192]) that yield an approximation of the target random field regarding both the covariance function and the joint distribution.

The so-called *spectral method* for the simulation of a centered stationary Gaussian random field $\xi = \{\xi(t),\ t \in \mathbb{R}^d\}$ with covariance function C makes use of cosine waves, see Sect. 9.2.4. To obtain a realization of ξ, one simulates independent cosine waves ξ_k, $k = 1, \ldots, n$, where the distribution of the auxiliary random vector ζ is chosen to be the *spectral measure* μ of C, see Definition 9.27 and Theorem 9.6. One can show that each cosine wave ξ_k has the covariance function C. Then the finite-dimensional distributions of $S_n = \left\{S_n(t) = \frac{1}{\sqrt{n}}\sum_{k=1}^n \xi_k(t),\ t \in \mathbb{R}^d\right\}$ converge weakly to the finite-dimensional distributions of ξ as $n \to \infty$, see Exercise 9.8. As a measure of the accuracy of approximation, the Kolmogorov distance $\sup_{x\in\mathbb{R}} |F_{S_n}(x) - F_\xi(x)|$ between the cumulative distribution functions F_{S_n} of $S_n(0)$ and F_ξ of $\xi(0)$ can be chosen. The number n of cosine waves sufficient to perform simulations with a desired accuracy $\varepsilon > 0$ can be found by the Berry–Esseen inequality, see for example [472, p. 374] and cf. [495]. For other approaches to measure the quality of approximation see [320, pp. 197–199].

9.9.2 Shot-Noise Random Fields

Another simple example of infinitely divisible random fields is Poisson shot-noise ξ introduced in Sect. 9.2.5, see (9.11). Its simulation in an rectangular observation window $W \subset \mathbb{R}^d$ is straightforward. Namely, suppose that the response function g has a compact support such that $\operatorname{supp} g \subset B_r(o)$ for some $r > 0$. First, the homogeneous Poisson process Π_λ of germs is simulated in an enlarged window $W \oplus B_r(o)$ to account for edge effects, cf. Sect. 3.2.2. Then shifted response functions $g(\cdot - x_i)$ are placed at the points x_i of Π_λ within $W \oplus B_r(o)$. Their values are summed up at any location $t \in W$ to get the value of $\xi(t)$.

9.9.3 Infinitely Divisible Random Fields

There exists vast literature on the simulation methods for many particular classes of infinitely divisible random functions, see [288] for a survey. Here we describe a general method to simulate any infinitely divisible random field

$$\xi(t) = \int_{\mathbb{R}^k} f_t(x)\,\mu(dx), \quad t \in W = [-t_0, t_0]^d,$$

where $t_0 > 0$ and μ is a Lévy basis. Denote by $\operatorname{supp}(f_t)$ the support of f_t for each $t \in W$ and assume that it is compact and

$$\bigcup_{t \in W} \operatorname{supp}(f_t) \subset [-A, A]^k$$

for some $A > 0$. Then

$$\xi(t) = \int_{[-A,A]^k} f_t(x)\,\mu(dx), \quad t \in W. \tag{9.76}$$

Approximate sample paths of ξ using the approximation of f_t by

$$\tilde{f}_t^{(n)} = \sum_{i=1}^{m(n)} a_i g_{t,i}, \quad t \in W, \quad n \in \mathbb{N},$$

where $m(n) \in \mathbb{N}$, $a_i \in \mathbb{R}$ and $g_{t,i} : \mathbb{R}^k \to \mathbb{R}$ is μ-integrable, $i = 1, \ldots, m(n)$.
Due to the linearity of stochastic integral we get

$$\tilde{\xi}^{(n)}(t) = \int_{[-A,A]^k} \tilde{f}_t^{(n)}(x)\,\mu(dx) = \sum_{i=1}^{m(n)} a_i \int_{[-A,A]^k} g_{t,i}(x)\,\mu(dx), \quad t \in W.$$

One can consider $\tilde{\xi}^{(n)}$ as an approximation of ξ.

Let $g_{t,i}$ be simple functions such that

$$\int_{[-A,A]^k} g_{t,i}(x)\,\mu(dx) = \sum_{j=1}^{l} g_{t,i}(x_j)\mu(\Delta_j), \quad i = 1,\dots,m(n), \quad t \in W,$$

for some $x_j \in [-A,A]^k$, $l \in \mathbb{N}$ and a partition $\{\Delta_j\}_{j=1}^{l}$ of $[-A,A]^k$. Then

$$\tilde{\xi}^{(n)}(t) = \sum_{i=1}^{m(n)} \sum_{j=1}^{l} a_i g_{t,i}(x_j)\mu(\Delta_j)$$

which can be simulated if $\mu(\Delta_j)$, $j = 1,\dots,l$, can be simulated.

Example 9.8. Let μ be an α-stable Lévy basis with Lebesgue control measure and constant skewness intensity β. Then

$$\mu(\Delta_j) \sim S_\alpha(\nu_k(\Delta_j)^{1/\alpha}, \beta, 0), \quad j = 1,\dots,l,$$

see [432, p. 119]. Furthermore, $\mu(\Delta_j)$, $j = 1,\dots,l$, are independent since μ is independently scattered. A method to simulate α-stable random variables can be found in [114].

The approximation of the random field ξ by $\tilde{\xi}^{(n)}$ as described above implies that $\tilde{\xi}^{(n)}$ is close to ξ whenever $\tilde{f}_t^{(n)}$ is close to f_t in a sense. Assume that f_t, $\tilde{f}_t^{(n)}$ belong to $L^s_{\nu_k}([-A,A]^k)$ for all $t \in W$ and some $s > 0$. We use

$$\mathrm{Err}_s(\xi(t), \tilde{\xi}^{(n)}(t)) := \left\| f_t(x) - \tilde{f}_t^{(n)}(x) \right\|_{L^s} := \left(\int_{[-A,A]^k} |f_t(x) - \tilde{f}_t^{(n)}(x)|^s \, \nu_k(dx) \right)^{1/s}$$

to measure the approximation quality of $\tilde{\xi}^{(n)}$. Let us now motivate the choice of the error measure by several examples.

Example 9.9. 1. *α-stable random fields.* For any $t \in W$, $\tilde{\xi}^{(n)}(t)$ converges in probability to the α-stable random field $\xi(t)$ from (9.76) iff

$$\mathrm{Err}_\alpha(\xi(t), \tilde{\xi}^{(n)}(t)) = \int_{\mathbb{R}^k} |f_t(x) - \tilde{f}_t^{(n)}(x)|^\alpha \mu_c(dx) \to 0, \quad n \to \infty,$$

where μ_c is a control measure, see [432, p. 126]. It can also be shown that the above condition is sufficient for the weak convergence of all finite-dimensional distributions of $\tilde{\xi}^{(n)}(t)$ to those of $\xi(t)$. Since $\xi(t)$ and $\tilde{\xi}^{(n)}(t)$ are jointly α-stable random variables for all $t \in W$, $\xi(t) - \tilde{\xi}^{(n)}(t)$ is also an α-stable random variable with scale parameter $\sigma_{\xi(t)-\tilde{\xi}^{(n)}(t)} = \mathrm{Err}_\alpha(\xi(t), \tilde{\xi}^{(n)}(t))$, see [432, p. 122]. Furthermore, for $\alpha \neq 1$ one can show that the error $\mathbf{E}|\xi(t) - \tilde{\xi}^{(n)}(t)|^p$, $0 < p < \alpha$, is proportional to $\mathrm{Err}_\alpha^p(\xi(t), \tilde{\xi}^{(n)}(t))$. If $\alpha = 1$ and the skewness $\beta \neq 0$ for at least one $t \in W$ then $\mathrm{Err}_{3/2}(\xi(t), \tilde{\xi}^{(n)}(t))$ can be used to measure the approximation error, see [288].

2. *Square integrable infinitely divisible random fields*. Let ξ be an infinitely divisible random field with a finite second moment and spot variable $\tilde{\mu}(y)$. If

$$\mathbf{var}(\tilde{\mu}(y)) := c_1 < \infty, \quad y \in \mathbb{R}^k, \quad \text{and} \quad \int_{[-A,A]^k} (\mathbf{E}(\tilde{\mu}(y)))^2 \, \nu_k(dy) := c_2 < \infty$$

then we get

$$\left(\mathbf{E}(\xi(t) - \tilde{\xi}^{(n)}(t))^2\right)^{1/2} \leq (c_1 + c_2)^{1/2} \, \mathrm{Err}_2(\xi(t), \tilde{\xi}^{(n)}(t)).$$

Exercise 9.39. Find the constants c_1 and c_2 from the last example for the Gamma Lévy random field.

We see that the problem of approximating the random field ξ reduces to an approximation problem of the corresponding kernel functions. The goal is then to find a set of functions $(\tilde{f}_t^{(n)})_{t \in \mathbb{R}^k}$ such that $\mathrm{Err}_s(\xi(t), \tilde{\xi}^{(n)}(t))$ can be made arbitrarily small. These sets of functions can be, e.g., step functions, Haar wavelets, etc., see [288].

Let us consider the case of step function approximation more closely. For any $n \in \mathbb{N}$ and $j = (j_1, \ldots, j_k)^\top \in \mathbb{Z}^k$ with $-n \leq j_1, \ldots, j_k < n$, let

$$\xi_j = \left(\xi\left(j_1 \frac{A}{n}, \cdots, j_k \frac{A}{n}\right)\right),$$

$$\Delta_j = \left[j_1 \frac{A}{n}, (j_1 + 1)\frac{A}{n}\right) \times \cdots \times \left[j_k \frac{A}{n}, (j_k + 1)\frac{A}{n}\right).$$

Introduce the step function

$$\tilde{f}_t^{(n)}(x) := \sum_{|j| \leq n} f_t(\xi_j) \mathbf{1}(x \in \Delta_j)$$

where $|j| \leq n$ means $-n \leq j_i < n$ for $i = 1, \ldots, k$. Then we have

$$\tilde{\xi}^{(n)}(t) = \int_{[-A,A]^k} \tilde{f}_t^{(n)}(x) \mu(dx) = \sum_{|j| \leq n} f_t(\xi_j) \mu(\Delta_j). \tag{9.77}$$

The following result provides error bounds $\mathrm{Err}_s(\xi(t), \tilde{\xi}^{(n)}(t))$ for functions f_t which are Hölder-continuous.

Theorem 9.18 ([288]). *Assume that $0 < s \leq 2$, the control measure is the Lebesgue measure ν_d and the functions f_t are Hölder-continuous for all $t \in W$, i. e.*

$$|f_t(x) - f_t(y)| \leq C_t \cdot ||x - y||_2^{\gamma_t}, \quad x, y \in [-A, A]^k, \quad t \in W,$$

for some $0 < \gamma_t \leq 1$ and $C_t > 0$. Then for any $t \in W$ and all $n \in \mathbb{N}$ one has

$$\mathrm{Err}_s(\xi(t), \tilde{\xi}^{(n)}(t)) \leq \left(\frac{2^k C_t k}{1 + \gamma_t s}\right)^{1/s} A^{\gamma_t + k/s} n^{-\gamma_t}. \tag{9.78}$$

Suppose that the conditions of Theorem 9.18 are statisfied. If the support of f_t is not compact, we first approximate

$$\xi(t) = \int_{\mathbb{R}^k} f_t(x)\mu(dx)$$

by

$$\xi_K(t) = \int_{[-K,K]^k} f_t(x)\mu(dx).$$

For $K > 0$ large enough, the approximation error is small since

$$\mathrm{Err}_s(\xi(t), \xi_K(t)) = \left(\int_{\mathbb{R}^k \setminus [-K,K]^k} |f_t(x)|^s dx\right)^{1/s} \to 0, \quad K \to \infty,$$

and $f_t, \tilde{f}_t^{(n)} \in L^s([-A, A]^k, \nu_d)$.

Remark 9.8. Theorem 9.18 provides a pointwise estimate of the approximation error for each $t \in W$. One obtains a uniform error bound as follows.

Assume that $\gamma := \inf_{t \in W} \gamma_t > 0$. Then for each $t \in W$, f_t is Hölder-continuous with parameters γ and some constant $C_t^* > 0$. Set $C := \sup_{t \in [-t_0, t_0]^k} C_t^*$. Then the approximation error $\mathrm{Err}_s(\xi(t), \tilde{\xi}^{(n)}(t))$ can be estimated by (9.78) with C_t and γ_t replaced by C and γ.

Remark 9.9. Assume that $0 < s \leq 2$ and the functions f_t are differentiable with $||\nabla f_t(x)||_2 \leq C_t$ for all $x \in [-A, A]^k$ and $t \in W$. Then for any $t \in W$, (9.78) holds for all $n \geq 1$ with $\gamma_t = 1$.

The above simulation method is illustrated by two numerical examples (see Fig. 9.17). Fix the simulation window $[-1, 1]^2$ and the resolution 200×200. Let the accuracy of approximation not exceed $\varepsilon > 0$. Take the bisquare kernel

$$f_t(x) = \begin{cases} b \cdot (a^2 - \|x - t\|_2^2)^2, & \|x - t\|_2 \leq a, \\ 0, & \text{otherwise} \end{cases}$$

in (9.76) with $a = 0.2$, $b = 5$, $\varepsilon = 0.05$ in the case of Gamma Lévy basis ($\theta = 0.01$) and $a = 0.2$, $b = 1000$, $\varepsilon = 0.1$ in the case of symmetric 1.5-stable

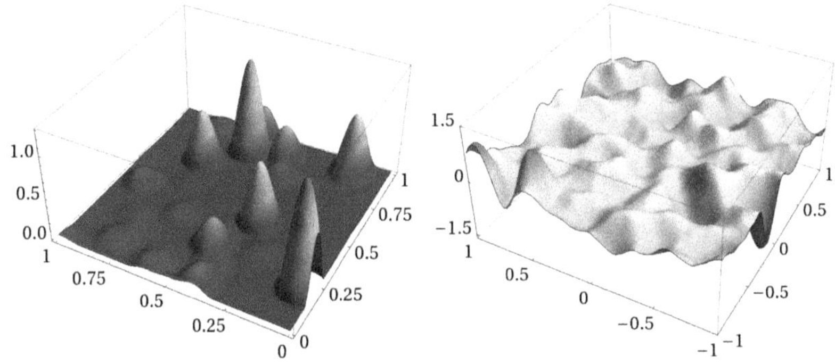

Fig. 9.17 Simulated realizations of a Gamma Lévy field (*left*) and a symmetric 1.5-stable random field (*right*)

basis ($\beta = 0$), see Example 9.5. Simulation of realizations shown in Fig. 9.17 requires 982 ms (26, 138 ms, respectively) of computational time on a Pentium Dual-Core CPU E5400, 2.70 GHz. We refer the reader to [288] for further performance issues.

Note that most non-Gaussian fractional fields are obtained by integration of deterministic kernels with respect to a random infinitely divisible measure. In [128] generalized shot noise series are used to obtain approximations of such fields, including linear and harmonizable fractional stable fields. Almost sure and L^p -norm rates of convergence, relying on asymptotic developments of the deterministic kernels, are presented as a consequence of an approximation result concerning series of symmetric random variables. The general framework is illustrated by simulations of classical fractional fields.

Chapter 10
Central Limit Theorems for Weakly Dependent Random Fields

Alexander Bulinski and Evgeny Spodarev

Abstract This chapter is a primer on the limit theorems for dependent random fields. First, dependence concepts such as mixing, association and their generalizations are introduced. Then, moment inequalities for sums of dependent random variables are stated which yield e.g. the asymptotic behaviour of the variance of these sums which is essential for the proof of limit theorems. Finally, central limit theorems for dependent random fields are given. Applications to excursion sets of random fields and Newman's conjecture in the absence of finite susceptibility are discussed as well.

10.1 Dependence Concepts for Random Fields

This section reviews several important dependence concepts of random variables and random fields such as mixing and m-dependence (already touched upon in Sect. 4.3 for point processes), association (both positive and negative), quasi-association, etc. Special attention is paid to association of random elements with values in partially ordered spaces and Fortuin–Kastelleyn–Ginibre inequalities.

A. Bulinski (✉)
Moscow State University, Moscow, Russia
e-mail: bulinski@mech.math.msu.su

E. Spodarev
Ulm University, Ulm, Germany
e-mail: evgeny.spodarev@uni-ulm.de

E. Spodarev (ed.), *Stochastic Geometry, Spatial Statistics and Random Fields*, Lecture Notes in Mathematics 2068, DOI 10.1007/978-3-642-33305-7_10,

10.1.1 Families of Independent and Dependent Random Variables

We consider a real-valued random function $\xi = \{\xi(t), t \in T\}$ defined on a probability space $(\Omega, \mathcal{A}, \mathbf{P})$ and a set T, i.e. $\xi(t) : \Omega \to \mathbb{R}$ is a *random variable* for any $t \in T$. Recall the following basic concept.

Definition 10.1. A family $\xi = \{\xi(t), t \in T\}$ consists of *independent random variables* if for each finite set $J \subset T$ and any collection of sets $B_t \in \mathcal{B}(\mathbb{R})$, $t \in J$, one has

$$\mathbf{P}\left(\bigcap_{t \in J} \{\xi(t) \in B_t\}\right) = \prod_{t \in J} \mathbf{P}(\xi(t) \in B_t). \tag{10.1}$$

If (10.1) does not hold then ξ is named a family of *dependent random variables.*

The independence of events $\{A_t, t \in T\}$ can be defined as independence of random variables $\{\mathbf{1}(A_t), t \in T\}$.

Exercise 10.1. Prove that validity of (10.1) is equivalent to the following statement. For all finite disjoint sets $I = \{s_1, \dots, s_k\} \subset T$, $J = \{t_1, \dots, t_m\} \subset T$ (with all possible values $k, m \in \mathbb{N}$) and any bounded Borel functions $f : \mathbb{R}^k \to \mathbb{R}$, $g : \mathbb{R}^m \to \mathbb{R}$

$$\mathbf{cov}(f(\xi(s_1), \dots, \xi(s_k)), g(\xi(t_1), \dots, \xi(t_m))) = 0. \tag{10.2}$$

Definition 10.1 can be easily extended to comprise random elements $\xi(t) : \Omega \to S_t$ where $(S_t, \mathcal{B}_t)$ are any measurable spaces and $\xi(t) \in \mathcal{A} \mid \mathcal{B}_t$ for each $t \in T$. Note that (10.1) is the particular case of the independence notion for arbitrary family of σ-algebras (for every $t \in T$ we use σ-algebras $\sigma\{\xi(t)\} = \{(\xi(t))^{-1}(B) : B \in \mathcal{B}_t\}$). Due to the Theorem 9.1 one can construct a collection $\{\xi(t), t \in T\}$ of independent random variables on some probability space $(\Omega, \mathcal{A}, \mathbf{P})$ (defined on an arbitrary set T and taking values in any measurable spaces $(S_t, \mathcal{B}_t)$) having given laws $\mu_t = \mathrm{Law}(\xi(t))$, $t \in T$. Many interesting stochastic models can be described by families of dependent random variables which are constructed by means of independent ones.

10.1.2 Mixing Coefficients and m-Dependence

There are many ways to describe the dependence structure of the (existing) family of random variables. Further on we concentrate on the study of real random functions $\xi = \{\xi(t), t \in T\}$ with $T = \mathbb{Z}^d$ or $T = \mathbb{R}^d$ $(d \geq 1)$. The investigation of stochastic processes (i.e. $d = 1$) has the following advantage. There is a total order on the real line $\mathbb{R}$ and we can operate with the "past" and the "future" of a process ξ whereas for $d > 1$, that is for *random fields*, one can introduce only partial orders on $\mathbb{R}^d$ (and $\mathbb{Z}^d$). In the latter case, well-known concepts of dependence involving the events related to the "past" and the "future" lose their meaning. Nevertheless, it is natural to assume that the dependence between collections of random variables $\{\xi(t),\ t \in I\}$ and $\{\xi(t),\ t \in J\}$ is "rather small" for $I, J \subset T$ such that the distance between I and J is "large enough". For example, the notion of *m-dependence* ($m > 0$) means that σ-algebras $\sigma_\xi(I)$ and $\sigma_\xi(J)$ are independent if

$$\operatorname{dist}(I, J) := \inf\{\rho(s,t) : s \in I, t \in J\} \geq m \tag{10.3}$$

where ρ is a metric in $\mathbb{R}^d$. $\sigma_\xi(I)$ is the σ-algebra in $\mathcal{A}$ generated by $\{\xi(t),\ t \in I\}$. One says also that ξ has a *finite dependence range*.

A stationary centered random field $\xi = \{\xi(t),\ t \in \mathbb{R}^d\}$ with covariance function C is said to be *long range dependent* (LRD) if

$$\int_{\mathbb{R}^d} |C(t)|\, dt = \infty.$$

Otherwise it is *short range dependent* (SRD).

Exercise 10.2. Show that the Gaussian random field with generalized Cauchy covariance introduced in Sect. 9.2.2 is a LRD random field iff $0 < \alpha\beta \leq d$.

It turns useful to consider a concept of *$m(U)$-dependent* random field $\xi = \{\xi(t),\ t \in \mathbb{R}^d\}$ assuming that $\sigma_\xi(V)$ and $\sigma_\xi(W)$ are independent for $V, W \subset U$ whenever $\operatorname{dist}(V, W) \geq m(U)$. Here $m(U)$ denotes a positive-valued function on subsets of $\mathbb{R}^d$, $d \geq 1$ (Fig. 10.1).

However, the generalization of the mixing coefficients known for stochastic processes (see for example [264]) is not straightforward. In contrast with the case $d = 1$, one has to take into account not only the distance between subsets I, $J \subset T = \mathbb{Z}^d$ (I and J for $d = 1$ belong to the "past" and "future", respectively) but also some other characteristics, for example their cardinalities $|I|$ and $|J|$ (see for example [83, 99, 159]). For instance, one can define for $\xi = \{\xi(t),\ t \in \mathbb{Z}^d\}$ the mixing coefficients

$$\begin{aligned} \alpha_{k,m}(r) := \ & \sup\{\alpha(\sigma_\xi(I), \sigma_\xi(J)) : I, J \subset \mathbb{Z}^d, \\ & \operatorname{dist}(I, J) \geq r,\ |I| \leq k,\ |J| \leq m\} \end{aligned} \tag{10.4}$$

where $k, m \in \mathbb{N} \cup \{\infty\}$, $r \in \mathbb{R}_+$ and

$$\alpha(\mathcal{B}, \mathcal{D}) = \sup\{|\mathbf{P}(AB) - \mathbf{P}(A)\mathbf{P}(B)| : A \in \mathcal{B},\ B \in \mathcal{D}\}$$

for σ-algebras $\mathcal{B}, \mathcal{D} \subset \mathcal{A}$. In a similar way we can introduce an analogue of the coefficient $\alpha_{k,m}$ for random fields $\xi = \{\xi(t),\ t \in \mathbb{R}^d\}$ by taking $I, J \subset \mathbb{R}^d$ in (10.4) and employing the diameters of I and J ($\operatorname{diam}(I) := \sup\{\rho(x, y) : x, y \in I\}$) instead of their cardinalities.

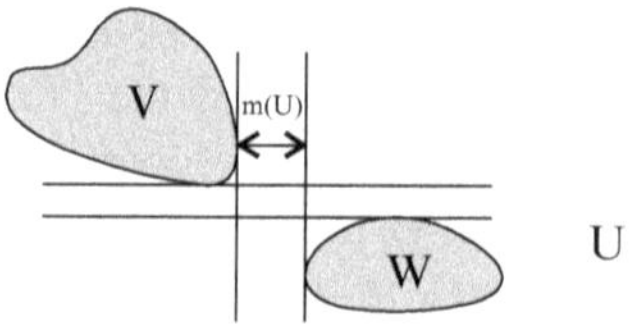

Fig. 10.1 $m(U)$ corresponds here to the sup-norm in $\mathbb{R}^d$ $(d = 2)$

Exercise 10.3. Prove that $\alpha(\mathcal{B}, \mathcal{D}) \leq 1/4$ for any σ-algebras $\mathcal{B}, \mathcal{D} \subset \mathcal{A}$. Give an example of $\mathcal{B}$ and $\mathcal{D}$ such that $\alpha(\mathcal{B}, \mathcal{D}) = 1/4$.

Note that the calculation or estimation of various mixing coefficients for random fields is by no means a trivial problem. Moreover, the class of well-studied models involving mixing random fields is rather meagre as compared with that of Gaussian fields or $m(U)$-dependent ones.

10.1.3 Association, Positive Association and Negative Association

Now we concentrate on stochastic models whose descriptions are based on the covariance function properties. Following the seminal paper by Newman [379] we are interested in generalizations of the independence notion in which approximate uncorrelatedness implies approximate independence in a sufficiently quantitative sense leading to useful limit theorems for sums of dependent random variables.

We start with the definition of associated random variables. It is underlined in [379] that there are two almost independent bodies of literature on this subject. One has originated from the works of Lehmann [327], Esary, Proschan and Walkup [167] and Sarkar [434] and is oriented towards mathematical statistics and reliability theory. Another has developed from the works by Harris [223] and Fortuin et al. [175] and is oriented towards percolation and statistical physics.

Introduce some notation. For $n \in \mathbb{N}$ let $\mathcal{M}(n)$ denote the class of bounded coordinate-wise nondecreasing Borel functions $f : \mathbb{R}^n \to \mathbb{R}$. Consider a real-valued random function $\xi = \{\xi(t), t \in T\}$ defined on a probability space $(\Omega, \mathcal{A}, \mathbf{P})$. Set $\xi_I = \{\xi(t), t \in I\}$ for $I \subset T$.

Definition 10.2. A family ξ is *associated* (we write $\xi \in \mathsf{A}$) if, for each finite set $I \subset T$ and any functions $f, g \in \mathcal{M}(|I|)$, one has

$$\mathbf{cov}(f(\xi_I), g(\xi_I)) \geq 0. \tag{10.5}$$

The notation $f(\xi_I)$ in (10.5) means that one considers any vector $\boldsymbol{\xi}_I$ in $\mathbb{R}^{|I|}$ constructed by ordering a collection $\{\xi(t), t \in I\}$. It is convenient to assume that $f(\xi_\emptyset) := 0$ for $I = \emptyset$. We say that a random vector $\eta = (\eta_1, \ldots, \eta_n)^\top$ is *associated* if a collection $\{\eta_1, \dots, \eta_n\} \in \mathsf{A}$. Thus a random vector $\eta \in \mathsf{A}$ iff any permutation of its components is associated. One says also that random variables $\eta_1, \dots, \eta_n$ are associated.

A probability measure Q on $(\mathbb{R}^n, \mathcal{B}(\mathbb{R}^n))$ is called *associated* if $Q = \mathbf{P}_\eta$ where a random vector $\eta \in \mathsf{A}$. In other words, Q is associated in view of (10.5) iff for any $f, g \in \mathcal{M}(n)$ one has

$$\int_{\mathbb{R}^n} f(x)g(x)Q(dx) \geq \int_{\mathbb{R}^n} f(x)Q(dx) \int_{\mathbb{R}^n} g(x)Q(dx). \tag{10.6}$$

Theorem 10.1 ([167]). *Any collection $\xi = \{\xi(t), t \in T\}$ consisting of independent random variables is associated.*

Proof. We start with T such that $|T| = 1$. Then inequality (10.5) is known as one of *Chebyshev*'s inequalities. To study this particular case of a single random variable η, let us take its independent copy ξ. For $f, g \in \mathcal{M}(1)$ (for bounded nondecreasing functions $f, g : \mathbb{R} \to \mathbb{R}$ which are automatically Borel functions) one has

$$\begin{aligned}\mathbf{cov}(f(\eta), g(\eta)) &= \mathbf{E}f(\eta)g(\eta) - \mathbf{E}f(\eta)\mathbf{E}g(\xi)\\ &= \frac{1}{2}\mathbf{E}(f(\eta) - f(\xi))(g(\eta) - g(\xi)) \geq 0 \end{aligned}\tag{10.7}$$

since the expression under the expectation is nonnegative for each $\omega \in \Omega$. Next we need the following simple

Lemma 10.1. *A union of mutually independent collections of associated random variables is associated.*

Proof. Assume that random vectors $\eta^k \in \mathsf{A}$ for each $k = 1, \dots, m$, and $\eta^1, \dots, \eta^m$ are independent. If $m = 1$ then the assertion is true. By induction, thus suppose that it holds for $\eta^1, \dots, \eta^{m-1}$. Let r be the dimension of $\nu := (\eta^1, \dots, \eta^{m-1})$ and n be the dimension of η^m. Set $Q = P_{\eta^m}$. Then for any $f, g \in \mathcal{M}(r + n)$ the Fubini theorem implies

$$\begin{aligned}\mathbf{cov}(f(\nu, \eta^m), g(\nu, \eta^m)) = &\int_{\mathbb{R}^n} \big(\mathbf{E}f(\nu, x)g(\nu, x) - \mathbf{E}f(\nu, x)\mathbf{E}g(\nu, x)\big) Q(dx)\\ &+ \int_{\mathbb{R}^n} \mathbf{E}f(\nu, x)\mathbf{E}g(\nu, x)Q(dx)\\ &- \int_{\mathbb{R}^n} \mathbf{E}f(\nu, x)Q(dx) \int_{\mathbb{R}^n} \mathbf{E}g(\nu, x)Q(dx).\end{aligned}$$

The integrand in the first term of the right-hand side is nonnegative by induction hypothesis as $\nu \in \mathsf{A}$ and for each $x \in \mathbb{R}^n$ the functions $f(\cdot, x), g(\cdot, x) \in \mathcal{M}(r)$. Note that $\mathbf{E}f(\nu, \cdot)$, $\mathbf{E}g(\nu, \cdot) \in \mathcal{M}(n)$ and $\eta^m \in \mathsf{A}$ (i.e., Q is associated). Hence the difference between the second and third terms is nonnegative in view of (10.6). □

Lemma 10.1 and the established Chebyshev's inequality prove Theorem 10.1. □

Remark 10.1 (see for example [96, Theorem 1.5]). The dominated convergence theorem allows to replace the boundedness assumptions concerning f and g in Definition 10.2 by the requirement that all expectations forming the covariance exist. A random vector $\eta = (\eta_1, \dots, \eta_n)^\top \in \mathsf{A}$ if condition (10.6) holds whenever both f and g taken from $\mathcal{M}(n)$ belong to either of the following classes of functions: (a) binary (being the indicators of some measurable sets), (b) continuous, (c) having bounded (partial) derivatives of any order. The functional class $\mathcal{M} = \cup_{n=1}^{\infty} \mathcal{M}(n)$ is closed under compositions. Therefore a family of functions belonging to $\mathcal{M}$ and having arguments taken from finite subsets of associated random variables is associated.

Exercise 10.4. Let $\eta_1, \dots, \eta_n$ be i.i.d. random variables and $\eta_{(1)}, \dots, \eta_{(n)}$ be the corresponding order statistics ($\eta_{(1)} = \min_{k=1,\dots,n} \eta_k, \dots, \eta_{(n)} = \max_{k=1,\dots,n} \eta_k$). Show that the random vector $(\eta_{(1)}, \dots, \eta_{(n)})^\top \in \mathsf{A}$.

Theorem 10.2 ([96, Theorem 3.8]). *Let $\xi = \{\xi(t), t \in \mathbb{R}^d\}$ be a shot-noise random field defined in Sect. 9.2.5 with g being a nonnegative integrable function. Then $\xi \in \mathsf{A}$.*

Theorem 10.3 ([407]). *A Gaussian process $\xi = \{\xi(t), t \in T\} \in \mathsf{A}$ iff*

$$\mathbf{cov}(\xi(s), \xi(t)) \geq 0 \quad \textit{for any} \quad s, t \in T.$$

Now we recall another important concept of dependence extending Definition 10.2.

Definition 10.3. A family $\xi = \{\xi(t), t \in T\}$ is *weakly* (or *positively*) *associated* ($\xi \in \mathsf{PA}$) if

$$\mathbf{cov}(f(\xi_I), g(\xi_J)) \geq 0 \tag{10.8}$$

for any finite disjoint sets $I, J \subset T$ and all functions $f \in \mathcal{M}(|I|), g \in \mathcal{M}(|J|)$.

It is useful to consider also the following counterpart of (10.8).

Definition 10.4. A family $\xi = \{\xi(t), t \in T\}$ is called *negatively associated* ($\xi \in \mathsf{NA}$) if

$$\mathbf{cov}(f(\xi_I), g(\xi_J)) \leq 0 \tag{10.9}$$

for any finite disjoint sets $I, J \subset T$ and all functions $f \in \mathcal{M}(|I|), g \in \mathcal{M}(|J|)$.

For NA systems the following result is a direct analogue of Theorem 10.3.

Theorem 10.4 ([273]). *A Gaussian process $\xi = \{\xi(t), t \in T\} \in \mathsf{NA}$ iff*

$$\mathbf{cov}(\xi(s), \xi(t)) \leq 0 \quad \textit{for any} \quad s \neq t, \quad s, t \in T.$$

Remark 10.2. The class PA is strictly larger than the class A (see [105]). However, in view of Theorem 10.3 any Gaussian process $\xi \in \mathsf{A}$ iff $\xi \in \mathsf{PA}$.

A family $\xi = \{\xi(t), t \in T\}$ consists of independent random variables iff simultaneously $\xi \in \mathsf{PA}$ and $\xi \in \mathsf{NA}$. To verify the last statement it is sufficient to consider the independence condition for $\xi(t)$, $t \in T$, in terms of their distribution functions and to take into account that if $(\xi(t_1), \ldots, \xi(t_n))^\top \in \mathsf{PA}$ and $g_k : \mathbb{R} \to \mathbb{R}_+$ are bounded nondecreasing functions for $k = 1, \ldots, n$ then

$$\mathbf{E} \prod_{k=1}^{n} g_k(\xi(t_k)) \geq \prod_{k=1}^{n} \mathbf{E} g_k(\xi(t_k)), \tag{10.10}$$

whereas if $\xi \in \mathsf{NA}$ then the analogue of (10.10) is true with the opposite sign of inequality. Relation (10.10) is obtained by induction with application of Remark 10.1 showing that $(g_1(\xi(t_1)), \ldots, g_k(\xi(t_n)))^\top \in \mathsf{A}$.

Remark 10.3. If there is a family of functions from $\mathcal{M}$ and their arguments belong to finite pairwise disjoint subsets of PA (or NA) random variables then these functions are PA (or NA).

Theorem 10.5 ([326]). *Let $\eta = (\eta_1, \ldots, \eta_n)^\top$ be an α-stable random vector. Then $\eta \in \mathsf{A}$ iff $\Gamma(S_-) = 0$ where*

$$S_- = \{(s_1, \ldots, s_n) \in S^{n-1} : s_i s_j < 0 \text{ for some } i, j \in \{1, \ldots, n\}\}$$

and $\eta \in \mathsf{NA}$ iff $\Gamma(S_+) = 0$ where

$$S_+ = \{(s_1, \ldots, s_n) \in S^{n-1} : s_i s_j > 0 \text{ for some } i, j \in \{1, \ldots, n\}, i \neq j\}.$$

Let $\mathcal{B}_0(S)$ consist of all bounded Borel subsets of a metric space S.

Theorem 10.6 ([104]). *A family $\{\mu(B), B \in \mathcal{B}_0(S)\} \in \mathsf{A}$ whenever μ is an independently scattered random measure.*

Proof. Let $B_1, \ldots, B_n \in \mathcal{B}_0(S)$. Then there is a finite number of pairwise disjoint sets $C_1, \ldots, C_r \in \mathcal{B}_0(S)$ such that every B_i is a union of some of C_j. Random variables $\mu(C_1), \ldots, \mu(C_r)$ are independent, hence Remark 10.1 (concerning increasing functions in associated random variables) implies that $(\mu(B_1), \ldots, \mu(B_n))^\top \in \mathsf{A}$. □

Let $\mathcal{R}$ be a ring of subsets of $\mathbb{R}^d$ consisting of the finite unions of the blocks having the form

$$C = (a_1, b_1] \times \ldots \times (a_d, b_d], \quad a_k \leq b_k, \quad k = 1, \ldots, d.$$

We give the following generalization of the result by Burton and Waymire [104].

Theorem 10.7 ([96, Theorem 3.19]). *Assume that μ is a random measure on the space $(\mathbb{R}^d, \mathcal{B}(\mathbb{R}^d))$ with a finite intensity measure $\mathbf{E}\mu$. Then $X = \{X(B), B \in \mathcal{R}\}$ introduced in* (9.15) *is associated.*

Further results concerning the association of point random fields can be found in the book [96].

10.1.4 Association on Partially Ordered Spaces

Definitions 10.2–10.4 admit natural extensions to partially ordered spaces as the only notion needed essentially was that of nondecreasing functions. As a motivation for such generalizations we mention the detailed proofs of interesting results by Barbato [51] concerning the association properties of strong solutions of stochastic differential equations (cf. [96, Chap. 1]). The connections between supermodular ordering and positive or negative association properties are considered in [126].

Let $(S, \mathcal{B})$ be a measurable space and $\leq_S$ a partial order on S (we write $\leq$ for the usual order on $\mathbb{R}$).

Definition 10.5. A function $f : S \to \mathbb{R}$ is *$\leq_S$-increasing* if $x, y \in S$ and $x \leq_S y$ imply $f(x) \leq f(y)$.

Definition 10.6. Let a space S be endowed with partial order $\leq_S$. A probability measure Q on the measurable space $(S, \mathcal{B})$ is called *positively correlated*, or *associated* (one writes $Q \in \mathsf{A}$) if

$$\int_S fg\, dQ \geq \int_S f\, dQ \int_S g\, dQ \tag{10.11}$$

for any bounded $\leq_S$-increasing $\mathcal{B}|\mathcal{B}(\mathbb{R})$-measurable functions $f, g : S \to \mathbb{R}$. A random element η defined on a probability space $(\Omega, \mathcal{A}, \mathbf{P})$ and taking values in S (i.e., η is $\mathcal{A}|\mathcal{B}$-measurable) is called *associated* if $\mathbf{P}_\eta \in \mathsf{A}$.

It is worthwhile to write $(S, \mathcal{B}, Q, \leq_S) \in \mathsf{A}$ to emphasize the role of partial order in the last definition. Clearly (10.6) is a particular case of (10.11) when $S = \mathbb{R}^n$ is endowed with usual partial order. For any probability measure Q on $(S, \mathcal{B})$ one can find the random element $\eta : \Omega \to S$ with $\mathbf{P}_\eta = Q$. Therefore relation (10.11) can be written in equivalent manner: $\mathbf{cov}(f(\eta), g(\eta)) \geq 0$ for any bounded $\leq_S$-increasing functions $f : S \to \mathbb{R}$ and $g : S \to \mathbb{R}$. We have the following analogue of statement (a) of Remark 10.1.

Theorem 10.8 ([334]). *One has* $(S, \mathcal{B}, Q, \leq_S) \in \mathsf{A}$ *iff*

$$Q(A \cap B) \geq Q(A)Q(B)$$

for any $\leq_S$*-increasing sets* $A, B \in \mathcal{B}$ *(i.e., indicators* $\mathbf{1}(A)$ *and* $\mathbf{1}(B)$ *are increasing functions on* $(S, \leq_S)$*).*

Proof. Necessity is clear. To establish sufficiency w.l.g. we can consider $\leq_S$-increasing $\mathcal{B}|\mathcal{B}(\mathbb{R})$-measurable $f : S \to \mathbb{R}$, $0 \leq f < 1$, and $g : S \to \mathbb{R}$ with the same properties (because for any random variables η, ζ and constants $a, b, c, d \in \mathbb{R}$ one has $\mathbf{cov}(a(\eta + b), c(\zeta + d)) = ac\ \mathbf{cov}(\eta, \zeta)$). For $n \in \mathbb{N}$ set

$$f_n(x) := \sum_{k=1}^{n} \frac{k}{n} \mathbf{1}\left(\frac{k-1}{n} \le f(x) < \frac{k}{n}\right) = \frac{1}{n} \sum_{k=1}^{n} \mathbf{1}\left(f(x) \ge \frac{k-1}{n}\right), \quad x \in S.$$

We define $g_n(x)$ in the same way. Obviously $f_n(x) \to f(x)$ and $g_n(x) \to g(x)$ for all $x \in S$ as $n \to \infty$. The function $\mathbf{1}(x : f(x) \ge v)$ is $\le_S$-increasing for every $v \in [0, 1)$. Therefore,

$$\int_S f_n g_n \, dQ \ge \int_S f_n \, dQ \int_S g_n \, dQ.$$

Due to Lebesgue's convergence theorem we come to (10.11). □

Now let $(S, \mathcal{B}, \le_S)$ be a partially ordered measurable space. For a subset $B \subset S$ introduce

$$C_B := \{ y \in S : y \ge_S x \ \text{ for some } \ x \in B \}. \tag{10.12}$$

This set (its indicator) is increasing. Let the partial order be *measurable*, that is for any $x \in S$, the sets $\{ y : y \ge_S x \}$ and $\{ y : y \le_S x \}$ belong to $\mathcal{B}$. Then for every $x \in S$

$$\{ y : y \le_S x \} \cap \{ y : y \ge_S x \} = \{ x \} \in \mathcal{B} \ \text{ and } \ C_{\{x\}} \in \mathcal{B}. \tag{10.13}$$

The following result can be regarded as "generalized Chebyshev's inequality", cf. (10.7).

Theorem 10.9 ([334]). *Let the measurable space $(S, \mathcal{B})$ be endowed with a measurable partial order $\le_S$. Then $(S, \mathcal{B}, \le_S)$ is totally ordered iff any probability measure on $\mathcal{B}$ is associated.*

Proof. First we verify the necessity. Assume that S is not totally ordered but every probability measure on $\mathcal{B}$ is associated. Then there exist two non-comparable elements $x, y \in S$. Take probability measure $\mathbf{P}$ such that $\mathbf{P}(\{ x \}) = \mathbf{P}(\{ y \}) = 1/2$. Introduce increasing sets $C_{\{x\}}$ and $C_{\{y\}}$ according to (10.12). Since $\mathbf{P} \in \mathsf{A}$ we get $\mathbf{P}(C_{\{x\}}C_{\{y\}}) \ge \mathbf{P}(C_{\{x\}})\mathbf{P}(C_{\{y\}}) \ge 1/4$ by Theorem 10.8. However, since $x \notin C_{\{y\}}$ and $y \notin C_{\{x\}}$, one has $\mathbf{P}(C_{\{x\}} \setminus \{ x \}) \le \mathbf{P}(S \setminus (\{ x \} \cup \{ y \})) = 0$. Similarly $\mathbf{P}(C_{\{y\}} \setminus \{ y \}) = 0$. Consequently

$$\mathbf{P}(C_{\{x\}}C_{\{y\}}) = \mathbf{P}((C_{\{x\}} \setminus \{ x \}) \cap (C_{\{y\}} \setminus \{ y \})) = 0.$$

We come to a contradiction.

To establish sufficiency let now S be totally ordered. We write $B^c := S \setminus B$ for $B \subset S$. Take a probability measure $\mathbf{P}$ on $\mathcal{B}$ and any increasing sets $C_1, C_2 \in \mathcal{B}$. We show that either $C_1 C_2^c = \emptyset$ or $C_1^c C_2 = \emptyset$. If this were not true then there would exist points $x \in C_1C_2^c$ and $y \in C_1^cC_2$. These points are different as $(C_1C_2^c) \cap (C_1^cC_2) = \emptyset$. But then neither $y \geq_S x$ nor $x \geq_S y$ which is impossible as S is totally ordered (indeed, if $y \geq_S x$ then $y \in C_1$ as C_1 is increasing, however $y \in C_1^cC_2 \subset C_1^c$ which is impossible, the case $x \geq_S y$ is analogous). Suppose at first that $C_1C_2^c = \emptyset$. Then

$$\mathbf{P}(C_1C_2) = \mathbf{P}(C_1) - \mathbf{P}(C_1C_2^c) = \mathbf{P}(C_1) \geq \mathbf{P}(C_1)\mathbf{P}(C_2).$$

The same situation occurs for $C_1^cC_2 = \emptyset$. Now we can complete the proof using Theorem 10.8. □

Example 10.1. Let $\leq$ and $\leq_{lex}$ be the usual partial and lexicographical orders in $\mathbb{R}^2$, respectively. Consider a random vector $\xi = (\xi_1, \xi_2)^\top$ with bounded components such that $\mathbf{cov}(\xi_1, \xi_2) < 0$. Then $(\mathbb{R}^2, \mathcal{B}(\mathbb{R}^2), \leq, \mathbf{P}_\xi) \notin \mathsf{A}$ in view of Remark 10.1. However, $(\mathbb{R}^2, \mathcal{B}(\mathbb{R}^2), \leq_{lex}, \mathbf{P}_\xi) \in \mathsf{A}$ due to Theorem 10.9. Thus we obtain an example of a probability space endowed with two different (partial) orders such that the first one does not provide an association whereas the second one does.

10.1.5 FKG-Inequalities and Their Generalizations

For a finite space L we assume that all its subsets (i.e., elements of 2^L) are measurable.

Definition 10.7. A partially ordered set L is called a *lattice*, if any two elements x and y in L have a *join* $x \vee y$ and a *meet* $x \wedge y$ (i.e., $x \leq_L x \vee y$, $y \leq_L x \vee y$, and $x \leq_L z$, $y \leq_L z$ imply $x \vee y \leq_L z$, while $x \wedge y$ is defined analogously). A lattice is called *distributive* if these operations satisfy the following conditions

$$x \wedge (y \vee z) = (x \wedge y) \vee (x \wedge z), \quad x \vee (y \wedge z) = (x \vee y) \wedge (x \vee z)$$

for all $x, y \in L$.

A typical example of a finite distributive lattice is a collection W of subsets of $\{1, \ldots, n\}$ such that $A, B \in W \implies A \cap B,\ A \cup B \in W$, with partial order $A \leq_W B \iff A \subset B$. By Birkhoff's theorem [73, p. 59], every finite distributive lattice L is *isomorphic* to some lattice W of subsets of $\{1, \ldots, n\}$ with partial order of inclusion. That is, there exists a bijection F from L to W such that $x \leq_L y$ iff $F(x) \subseteq F(y)$ for any $x, y \in L$.

To simplify the notation we write $Q(t)$ instead of $Q(\{t\})$ for a probability measure Q and $t \in L$.

Theorem 10.10 ([175]). *Let L be a finite distributive lattice. Let Q be a probability measure on $(L, 2^L)$ such that for any $x, y \in L$*

$$Q(x \vee y)Q(x \wedge y) \geq Q(x)Q(y). \tag{10.14}$$

Then $Q \in \mathsf{A}$, *i.e.* (10.11) *holds with* S *replaced by* L.

One usually refers to condition (10.14) as to *FKG lattice inequalities* (FKG stands for Fortuin–Kastelleyn–Ginibre). The proof of the above result based on the deep theorem by Holley [247] can be found in [96, Theorem 4.10].

Sufficient conditions for the Ising model (e.g., for two-body interaction) to satisfy the FKG inequalities are given in [175].

The results discussed above for a finite lattice can be extended to infinite sets if the potentials are defined in such a way that all the set-indexed series are absolutely convergent. In particular, it is true for the Ising model on $\mathbb{Z}^d$ when i and j are neighbours only if $|i - j| = 1$. FKG-inequalities and their generalizations remain a subject of constant interest in reliability theory, statistical physics, quantum physics, discrete mathematics. Interesting modifications appear when index sets are more general than $\mathbb{Z}^d$ (then the corresponding algebraic lattice may be non-distributive). For instance, the "right" FKG-inequalities differ from the usual ones on triangular nets (see [117]). Even if we consider a finite graph consisting of one triangle, the simplest example might fail to satisfy the classical FKG-inequalities. The modified FKG-inequalities were also applied to phase transition problems for the corresponding Potts model and to analyzing the existence of infinite clusters in percolation models. We mention also that Liggett [332] introduced the concept of conditional association leading to the so-called *downward FKG-property* for specified models, which lies strictly between regular FKG-inequalities and association.

The FKG-inequalities were generalized by Holley [247] and Preston [409]. We formulate the result by Ahlswede and Daykine which in its turn generalizes the former two. For sets $A, B \subset L$ define

$$A \vee B = \{x \vee y : x \in A, y \in B\}, \quad A \wedge B := \{x \wedge y : x \in A, y \in B\}.$$

For a function $f : L \to \mathbb{R}$ and $A \subset L$ introduce

$$f(A) = \sum_{x \in A} f(x).$$

Theorem 10.11 ([7]). *Let* f, g, h, k *be nonnegative functions defined on a finite distributive lattice* L *such that*

$$f(x)g(y) \le h(x \vee y)k(x \wedge y) \ \text{for any} \ x, y \in L. \tag{10.15}$$

Then

$$f(A)g(B) \le h(A \vee B)k(A \wedge B) \ \text{for any} \ A, B \subset L. \tag{10.16}$$

Notice the attractive similarity between the hypothesis (10.15) and the conclusion (10.16). Various applications of this theorem are considered in [6, 9]. For other aspects of the correlation inequalities see for example [63].

10.1.6 Quasi-association and Further Extensions

Now we revisit independence relation (10.2). Clearly a family $\xi = \{\xi(t), t \in T\}$ consists of independent random variables if (10.2) holds for sufficiently large class of functions f, g (and then for all Borel bounded functions). Moreover, the value $\mathbf{cov}(f(\xi(s_1), \ldots, \xi(s_k)), g(\xi(t_1), \ldots, \xi(t_m)))$ for a pair of test functions f, g indicates "how strong" is the dependence between the vectors $(\xi(s_1), \ldots, \xi(s_k))^\top$ and $(\xi(t_1), \ldots, \xi(t_m))^\top$. To clarify this idea, we consider the class of Lipschitz test functions. Recall that $f : \mathbb{R}^k \to \mathbb{R}$ is a *Lipschitz function* if

$$\mathrm{Lip}(f) := \sup_{x,y \in \mathbb{R}^k, x \neq y} \frac{|f(x) - f(y)|}{\|x - y\|_1} < \infty,$$

here $\|x - y\|_1 = \sum_{i=1}^k |x_i - y_i|$, $x = (x_1, \ldots, x_k)^\top$ and $y = (y_1, \ldots, y_k)^\top$.

Definition 10.8. A random field $\xi = \{\xi(j), j \in \mathbb{Z}^d\}$ is called *quasi-associated* ($\xi \in \mathsf{QA}$) if $\mathbf{E}\xi^2(j) < \infty$ for all $j \in \mathbb{Z}^d$, and, for any finite disjoint sets $I, J \subset \mathbb{Z}^d$ and for arbitrary bounded Lipschitz functions $f : \mathbb{R}^{|I|} \to \mathbb{R}$, $g : \mathbb{R}^{|J|} \to \mathbb{R}$, one has

$$|\mathbf{cov}(f(\xi_I), g(\xi_J)| \le \mathrm{Lip}(f)\,\mathrm{Lip}(g) \sum_{i \in I, j \in J} |\mathbf{cov}(\xi(i), \xi(j))|. \tag{10.17}$$

The extension of the above definition to random fields indexed by $t \in \mathbb{R}^d$ is proposed in [92].

Now we formulate the following elementary but important result.

Theorem 10.12 ([96, Theorem 5.3]). *Let $\xi = \{\xi(j), j \in \mathbb{Z}^d\}$ be a random field such that $\mathbf{E}\xi^2(j) < \infty$ for all $j \in \mathbb{Z}^d$. Then $\xi \in \mathsf{PA}$ or $\xi \in \mathsf{NA}$ implies that $\xi \in \mathsf{QA}$.*

One can obtain also the analogue of (10.17) using the "partial Lipschitz constants", see [96, p. 89].

Corollary 10.1. *Let a random vector $\xi = (\xi_1, \ldots, \xi_n)^\top \in \mathsf{A}$ (PA, NA) such that $\mathbf{E}\|\xi\|_2^2 < \infty$. Then for any $t_1, \ldots, t_n \in \mathbb{R}$ it holds*

$$\left| \mathbf{E}\, e^{i(t_1\xi_1 + \ldots + t_n\xi_n)} - \prod_{k=1}^n \mathbf{E}\, e^{it\xi_k} \right| \le 4 \sum_{1 \le j < k < n} |t_j t_k|\, |\mathbf{cov}(\xi_j, \xi_k)|. \tag{10.18}$$

It is interesting to compare the next statement with Theorems 10.3 and 10.4.

Theorem 10.13 ([466]). *Any Gaussian random field (having the covariance function of arbitrary signs) is quasi-associated.*

The *Cox–Grimmett coefficient* (see [135]) for a field $\xi = \{\xi(t), t \in \mathbb{Z}^d\}$ is introduced by formula

$$u(r) := \sup_{i \in \mathbb{Z}^d} \sum_{j:\|i-j\|_\infty \geq r} |\mathbf{cov}(\xi(i), \xi(j))|, \quad r \in \mathbb{Z}_+, \tag{10.19}$$

where $\|i\|_\infty := \max_{s=1,\dots,d} |i_s|$. If ξ is a wide-sense stationary random field satisfying the *finite susceptibility* condition by Newman, i.e.

$$\sum_{j \in \mathbb{Z}^d} |\mathbf{cov}(\xi(0), \xi(j))| < \infty, \tag{10.20}$$

then

$$u(r) = \sum_{j:\|j\|_\infty \geq r} |\mathbf{cov}(\xi(0), \xi(j))| < \infty \text{ for all } r \in \mathbb{Z}_+. \tag{10.21}$$

Following [98, 160] (for random fields and stochastic processes, respectively) we provide

Definition 10.9. A random field ξ is called (BL,θ)*-dependent* ($\xi \in (BL,\theta)$) if there exists a nonincreasing sequence $\theta = \{\theta_r\}_{r \in \mathbb{Z}_+}$ of nonnegative numbers $\theta_r \to 0$ as $r \to \infty$, such that for any finite disjoint sets $I, J \subset \mathbb{Z}^d$ and bounded Lipschitz functions $f : \mathbb{R}^{|I|} \to \mathbb{R}$, $g : \mathbb{R}^{|J|} \to \mathbb{R}$ one has

$$|\mathbf{cov}(f(\xi_I), g(\xi_J)| \leq \mathrm{Lip}(f)\,\mathrm{Lip}(g)(|I| \wedge |J|)\theta_r \tag{10.22}$$

where $r = \mathrm{dist}(I,J)$ and $a \wedge b = \min\{a,b\}$ for $a,b \in \mathbb{R}$.

Thus we obtain that if a random field $\xi \in \mathsf{QA}$ then $\xi \in (BL,\theta)$ with $\theta_r = u(r)$, $r \in \mathbb{Z}_+$. It is proved in [467] that the class of (BL,θ)-dependent random fields is strictly larger than QA.

We can generalize these definitions for a field $\xi = \{\xi(t), t \in \mathbb{R}^d\}$. For $\Delta > 0$ introduce a lattice $T(\Delta) = \{(j_1/\Delta, \dots, j_d/\Delta)^\top : (j_1, \dots, j_d)^\top \in \mathbb{Z}^d\}$.

Definition 10.10. A field $\xi = \{\xi(t), t \in \mathbb{R}^d\}$ is called (BL,θ)-dependent (we write $\xi \in (BL,\theta)$) if there exists a nonincreasing function $\theta : \mathbb{R}_+ \to \mathbb{R}_+$, $\theta(r) \to 0$ as $r \to \infty$ such that for all sufficiently large Δ, any finite disjoint $I, J \subset T(\Delta)$ and arbitrary bounded Lipschitz functions $f : \mathbb{R}^{|I|} \to \mathbb{R}$, $g : \mathbb{R}^{|J|} \to \mathbb{R}$

$$|\mathbf{cov}(f(\xi_I), g(\xi_J))| \leq \mathrm{Lip}(f)\,\mathrm{Lip}(g)\Delta^d(|I| \wedge |J|)\theta(r) \tag{10.23}$$

where $r = \mathrm{dist}(I,J)$.

Note that if $\xi = \{\xi(t), t \in \mathbb{R}^d\}$ is a wide-sense stationary random field with continuous covariance function C which is absolutely directly integrable in the Riemann sense then (see [92]) one can use in (10.23) the analogue of the Cox–Grimmett coefficient

$$\theta(r) = 2 \int_{\|t\|_2 \geq r} |C(t)|\, dt.$$

In the next section we illustrate the use of the dependence conditions introduced above in the limit theorems for random fields.

10.2 Moment and Maximal Inequalities for Partial Sums

Consider a partial sum of a stationary dependent random field over an unboundedly growing finite subset of $\mathbb{Z}^d$. This section defines the proper growth notion of such subsets in order to study the asymptotics of the variance and higher moments of these partial sums. Moreover, upper bounds for higher moments of these sums as well as of their maxima are given.

10.2.1 Regularly Growing Sets

Let $\xi = \{\xi(j),\ j \in \mathbb{Z}^d\}$ be a random field. For a finite set $U \subset \mathbb{Z}^d$ consider the *partial sum*

$$S(U) := \sum_{j \in U} \xi(j). \tag{10.24}$$

To study the limit behaviour of (normalized) partial sums $S(U_n)$ we have to specify the growth conditions for finite sets $U_n \subset \mathbb{Z}^d$ as $n \in \mathbb{N}$ increases.

First of all we recall the concept of "regular growth" for a sequence of sets in $\mathbb{R}^d$. Let $a = (a_1, \ldots, a_d)^\top$ be a vector with positive components. Introduce the parallelepiped

$$\Delta_0(a) = \{x \in \mathbb{R}^d : 0 < x_i \leq a_i,\ i = 1, \ldots, d\}.$$

For $j \in \mathbb{Z}^d$ define the shifted blocks

$$\Delta_j(a) = \Delta_0(a) \oplus \{ja\} = \{x \in \mathbb{R}^d : j_i a_i < x_i \leq (j_i + 1)a_i,\ i = 1, \ldots, d\}$$

where $ja = (j_1 a_1, \ldots, j_d a_d)^\top$ and $\oplus$ is the Minkowski addition. Clearly $\{\Delta_j(a),\ j \in \mathbb{Z}^d\}$ form a partition of $\mathbb{R}^d$. For a set $V \subset \mathbb{R}^d$ put

$$J_-(V, a) = \{j \in \mathbb{Z}^d : \Delta_j(a) \subset V\}, \quad J_+(V, a) = \{j \in \mathbb{Z}^d : \Delta_j(a) \cap V \neq \emptyset\},$$

$$V^-(a) = \bigcup_{j \in J_-(V,a)} \Delta_j(a), \quad V^+(a) = \bigcup_{j \in J_+(V,a)} \Delta_j(a).$$

Recall that for a measurable set $B \subset \mathbb{R}^d$ its Lebesgue measure is denoted by $\nu_d(B)$.

Definition 10.11. A sequence of sets $V_n \subset \mathbb{R}^d$ tends to infinity in the *Van Hove sense* (or is VH-growing) if, for any $a \in \mathbb{R}^d$ having positive components,

$$\nu_d(V^-(a)) \to \infty, \quad \frac{\nu_d(V^-(a))}{\nu_d(V^+(a))} \to 1 \text{ as } n \to \infty. \tag{10.25}$$

Exercise 10.5. Show that

$$V_n = (a^{(n)}, b^{(n)}] = \{ x \in \mathbb{R}^d : a_i^{(n)} < x_i \le b_i^{(n)}, i = 1, \dots, d \} \to \infty$$

in the Van Hove sense iff $\min_{1 \le i \le d}(b_i^{(n)} - a_i^{(n)}) \to \infty$ as $n \to \infty$.

For $\varepsilon > 0$ and $V \subset \mathbb{R}^d$ we define the ε-neighbourhood of V as follows

$$V^\varepsilon = \{ x \in \mathbb{R}^d : \rho(x, V) := \inf_{y \in V} \rho(x, y) \le \varepsilon \}$$

where ρ is the Euclidean metric.

Recall that the *boundary* of a set $V \subset \mathbb{R}^d$ is a set ∂V consisting of such points $z \in \mathbb{R}^d$ that in every neighbourhood of $\{z\}$ there exist a point $x \in V$ and a point $y \notin V$.

Lemma 10.2 ([431]). *Let $(V_n)_{n \in \mathbb{N}}$ be a sequence of bounded measurable sets in $\mathbb{R}^d$. Then this sequence is VH-growing iff for any $\varepsilon > 0$ one has*

$$\frac{\nu_d((\partial V_n)^\varepsilon)}{\nu_d(V_n)} \to 0 \textit{ as } n \to \infty. \tag{10.26}$$

The proof can be found in [95, p. 173].

For finite sets $U_n \subset \mathbb{Z}^d$ one can use an analogue of condition (10.26). Given $U \subset \mathbb{Z}^d$ and $p \in \mathbb{N}$, write

$$U^p = \{ j \in \mathbb{Z}^d \setminus U : \operatorname{dist}(j, U) := \inf_{i \in U} \operatorname{dist}(i, j) \le p \}$$

where dist is the metric on $\mathbb{Z}^d$ corresponding to the sup-norm in $\mathbb{R}^d$. Set $\partial U = U^1$. Therefore

$$\partial U = \{ j \in \mathbb{Z}^d \setminus U : \operatorname{dist}(j, U) = 1 \},$$

cf. Sect. 9.3. Recall that $|U|$ stands for the number of elements of a finite set $U \subset \mathbb{Z}^d$.

Definition 10.12. A sequence $(U_n)_{n \in \mathbb{N}}$ of finite subsets of $\mathbb{Z}^d$ is called *regularly growing* (or is *growing in a regular way*) if

$$|U_n| \to \infty \text{ and } |\partial U_n|/|U_n| \to 0 \text{ as } n \to \infty. \tag{10.27}$$

Now we formulate the result (see [95, Lemmas 3.1.5 and 3.1.6]) clarifying the relationship between VH-growing sets in $\mathbb{R}^d$ and regularly growing sets in $\mathbb{Z}^d$.

Lemma 10.3. *Let $(V_n)_{n\in\mathbb{N}}$ be a sequence of bounded sets in $\mathbb{R}^d$ such that $V_n \to \infty$ in the Van Hove sense. Then $U_n := V_n \cap \mathbb{Z}^d$ $(n \in \mathbb{N})$ form a regularly growing sequence in $\mathbb{Z}^d$. Conversely if $(U_n)_{n\in\mathbb{N}}$ is a sequence of regularly growing subsets of $\mathbb{Z}^d$ then $V_n := \cup_{j\in U_n} \Delta_j$ $(n \in \mathbb{N})$ form a VH-growing sequence where the cubes*

$$\Delta_j := \{ x \in \mathbb{R}^d : j_i < x_i \le j_i + 1, i = 1,\dots,d \}, \quad j = (j_1,\dots,j_d)^\top \in \mathbb{Z}^d.$$

10.2.2 Variances of Partial Sums

Theorem 10.14 ([74]). *Let $\xi = \{ \xi(j), \ j \in \mathbb{Z}^d \}$ be a wide-sense stationary random field such that*

$$\sum_{j\in\mathbb{Z}^d} \mathbf{cov}(\xi(0), \xi(j)) = \sigma^2$$

where this series is absolutely convergent. Then for any sequence $(U_n)_{n\in\mathbb{N}}$ of regularly growing subsets of $\mathbb{Z}^d$ and partial sums appearing in (10.24) *the following relation holds*

$$\frac{\mathbf{var}\, S(U_n)}{|U_n|} \to \sigma^2 \ \text{ as } \ n \to \infty. \tag{10.28}$$

Proof. Take arbitrary $p \in \mathbb{N}$ and introduce $G_n = U_n \cap (\partial U_n)^p$, $W_n = U_n \setminus G_n$. Then

$$\begin{aligned}\sigma^2|U_n| - \mathbf{var}\, S(U_n) &= \sum_{j\in U_n} \sum_{k\notin U_n} \mathbf{cov}(\xi(j), \xi(k)) \\ &= \sum_{j\in G_n} \sum_{k\notin U_n} \mathbf{cov}(\xi(j), \xi(k)) + \sum_{j\in W_n} \sum_{k\notin U_n} \mathbf{cov}(\xi(j), \xi(k)) \\ &=: T_{1,n} + T_{2,n}.\end{aligned}$$

Note that $|G_n| \le |(\partial U_n)^p| \le (2p+1)^d |\partial U_n|$ and by (10.27)

$$\frac{|T_{1,n}|}{|U_n|} \le \frac{|G_n|}{|U_n|} \sum_{j\in\mathbb{Z}^d} |\,\mathbf{cov}(\xi(0), \xi(j))| \le c_0 (2p+1)^d |\partial U_n|/|U_n| \to 0, \quad n \to \infty,$$

where $c_0 = \sum_{j\in\mathbb{Z}^d} |\,\mathbf{cov}(\xi(0), \xi(j))|$.

Taking into account that $\mathrm{dist}(W_n, \mathbb{Z}^d \setminus U_n) \ge p$ and $|W_n| \le |U_n|$ we come to the inequality

$$\limsup_{n\to\infty} \frac{|T_{2,n}|}{|U_n|} \le \sum_{j\in\mathbb{Z}^d : \|j\|_\infty \ge p} |\,\mathbf{cov}(\xi(0), \xi(j))| \tag{10.29}$$

where $\|j\|_\infty = \max_{1\le i\le d} |j_i|$.

The *finite susceptibility* condition

$$\sum_{j\in\mathbb{Z}^d} |\mathbf{cov}(\xi(0), \xi(j))| < \infty \tag{10.30}$$

implies that the right-hand side of (10.29) can be made arbitrary small when p is large enough. □

Exercise 10.6. Prove that if a random field $\xi = \{\xi(j), j \in \mathbb{Z}^d\} \in (BL, \theta)$ then (10.30) holds.

An extension of Theorem 10.14 for (BL,θ)-dependent stationary random field $\xi = \{\xi(t), t \in \mathbb{R}^d\}$ is provided in [92]. Now we consider random fields without assuming the finite susceptibility condition. We start with

Definition 10.13. A function $L : \mathbb{R}^d_+ \to \mathbb{R}\setminus\{0\}$ is called *slowly varying* (at infinity) if for any vector $a = (a_1,\dots,a_d)^\top$ with positive components

$$\frac{L(a_1x_1,\dots,a_dx_d)}{L(x_1,\dots,x_d)} \to 1 \text{ as } x \to \infty, \tag{10.31}$$

that is when $x_1 \to \infty,\dots,x_d \to \infty$. For such function we write $L \in \mathcal{L}(\mathbb{R}^d)$.

A function $L : \mathbb{N}^d \to \mathbb{R}\setminus\{0\}$ is slowly varying (at infinity) if for any $a = (a_1,\dots,a_d)^\top \in \mathbb{N}^d$ relation (10.31) holds with supplementary condition $x \in \mathbb{N}^d$. We write $L \in \mathcal{L}(\mathbb{N}^d)$.

For $d = 1$ this is the classical definition introduced by Karamata (see for example [462]). Clearly, $f(x) = \prod_{i=1}^d \log(x_i \vee 1)$, where $x \in \mathbb{R}^d_+$, belongs to $\mathcal{L}(\mathbb{R}^d)$. For $x \in \mathbb{R}^d$ put $[x] = ([x_1],\dots,[x_d])^\top$, where $[\cdot]$ stands for the integer part of a number.

Exercise 10.7. Let $L \in \mathcal{L}(\mathbb{N}^d)$ and L be nondecreasing in each argument. Set $H(x) = L([\tilde{x}])$ for all $x \in \mathbb{R}^d_+$ where $\tilde{x}_i = x_i \vee 1$, $i = 1,\dots,d$. Prove that $H \in \mathcal{L}(\mathbb{R}^d)$.

Let $\xi = \{\xi(j), j \in \mathbb{Z}^d\}$ be a family of square-integrable random variables. Set

$$\mathsf{K}_\xi(n) = \sum_{j\in\mathbb{Z}^d:-n\le j\le n} \mathbf{cov}(\xi(o), \xi(j)), \quad n \in \mathbb{N}^d. \tag{10.32}$$

The sum is taken over $j = (j_1,\dots,j_d)^\top \in \mathbb{Z}^d$ such that $-n_1 \le j_1 \le n_1,\dots, -n_d \le j_d \le n_d$. Write $\mathsf{1} = (1,\dots,1)^\top$ for a vector in $\mathbb{R}^d$ with components equal to 1.

Theorem 10.15. *Let $\xi = \{\xi(j), j \in \mathbb{Z}^d\}$ be a wide-sense stationary random field with nonnegative covariance function R and $U_n = \{j \in \mathbb{Z}^d : \mathsf{1} \le j \le n\}$, $n \in \mathbb{N}^d$. If $\mathsf{K}_\xi(\cdot) \in \mathcal{L}(\mathbb{N}^d)$ then*

$$\mathbf{var}\, S(U_n) \sim \mathsf{K}_\xi(n)\, |U_n| \text{ as } n \to \infty.$$

Conversely, if $\mathbf{var}\, S(U_n) \sim L(n)|U_n|$ *as* $n \to \infty$ *with* $L \in \mathcal{L}(\mathbb{N}^d)$*, then* $L(n) \sim \mathsf{K}_\xi(n)$*,* $n \to \infty$*.*

Proof. Assume that $\mathsf{K}_\xi(\cdot) \in \mathcal{L}(\mathbb{N}^d)$. Due to stationarity of the field ξ under consideration we have $\mathbf{cov}(\xi(i), \xi(j)) = \mathsf{R}(i-j)$ for $i, j \in \mathbb{Z}^d$. Thus

$$\begin{aligned}\mathbf{var}\, S(U_n) &= \sum_{i,j \in U_n} \mathbf{cov}(\xi(i), \xi(j)) = \sum_{i,j \in U_n} \mathsf{R}(i-j) \\ &= \sum_{m \in \mathbb{Z}^d : -(n-1) \le m \le n-1} (n_1 - |m_1|) \dots (n_d - |m_d|) R(m) \\ &\le |U_n| \sum_{m \in \mathbb{Z}^d : -(n-1) \le m \le n-1} \mathsf{R}(m) \le \mathsf{K}_\xi(n)\, |U_n| \end{aligned} \tag{10.33}$$

as R is a nonnegative function.

Take any $c \in (0,1)$ and $n \ge \frac{1}{1-c}\mathsf{1}$ (i.e. $cn \le n - \mathsf{1}$, $n \in \mathbb{N}^d$). One has

$$\begin{aligned}\mathbf{var}\, S(U_n) &= \sum_{m \in \mathbb{Z}^d ; -(n-1) \le m \le n-1} (n_1 - |m_1|) \dots (n_d - |m_d|) R(m) \\ &\ge (1-c)^d |U_n| \sum_{m \in \mathbb{Z}^d : -cn \le m \le cn} \mathsf{R}(m) = (1-c)^d \mathsf{K}_\xi([cn])\, |U_n|.\end{aligned}$$

In view of Exercise 10.7 and (10.31) we write

$$(1-c)^d \mathsf{K}_\xi([cn])\, |U_n| \sim (1-c)^d \mathsf{K}_\xi(n)\, |U_n|, \quad n \to \infty, \quad n \in \mathbb{N}^d.$$

Therefore $\mathbf{var}\, S(U_n) \sim \mathsf{K}_\xi(n)\, |U_n|$ as $n \to \infty$, because c can be taken arbitrary close to zero.

Now suppose that $\mathbf{var}\, S(U_n) \sim L(n)\, |U_n|$ as $n \to \infty$ holds with $L \in \mathcal{L}(\mathbb{N}^d)$. Then for any $\varepsilon > 0$ and all n large enough, i.e. when all components of n are large enough, (10.33) yields

$$\mathsf{K}_\xi(n) \ge \frac{\mathbf{var}\, S(U_n)}{|U_n|} \ge (1-\varepsilon) L(n). \tag{10.34}$$

For given $q \in \mathbb{N}$, $q > 1$, $n_r \in \mathbb{N}$ and $m_r \in \mathbb{Z}$ such that $|m_r| \le n_r$, $r = 1, \dots, d$, one has

$$\frac{q}{q-1}\left(1 - \frac{|m_r|}{n_r q}\right) \ge \frac{q}{q-1}\left(1 - \frac{n_r}{n_r q}\right) = 1.$$

Consequently

$$\mathsf{K}_\xi(n) \le \left(\frac{q}{q-1}\right)^d \sum_{m\in\mathbb{Z}^d:-n\le m\le n} \mathsf{R}(m) \prod_{r=1}^d \frac{(n_r q - |m_r|)}{n_r q}$$

$$\le \left(\frac{q}{q-1}\right)^d \left(\prod_{r=1}^d n_r q\right)^{-1} \sum_{m\in\mathbb{Z}^d:-nq\le m\le nq} \mathsf{R}(m) \prod_{r=1}^d (n_r q - |m_r|)$$

$$= \left(\frac{q}{q-1}\right)^d \frac{\mathbf{var}\, S(U_{nq})}{|U_{nq}|} \sim \left(\frac{q}{q-1}\right)^d L(nq), \quad n\to\infty. \tag{10.35}$$

As q can be taken arbitrary large and $L\in\mathcal{L}(\mathbb{N}^d)$, combining (10.34) and (10.35) we complete the proof. □

10.2.3 Moment Inequalities for Partial Sums

There are a number of beautiful results concerning the moments of sums of independent (or dependent in a sense) random variables. One can refer to the classical theorems by Khinchin, Doob, Burkholder–Davis–Gandy, see, for example [125, 405, 472]. However, the structure of the index set was not so important there as for partial sums generated by a random field $\xi = \{\xi(j),\ j\in\mathbb{Z}^d\}$ consisting of dependent random variables where additional difficulties arise due to the spatial configuration of the index set of summands.

Recall that if $(\eta_n)_{n\in\mathbb{N}}$ is a sequence of i.i.d. random variables with mean zero and $\mathbf{E}|\eta_1|^s < \infty$ for some $s>2$, then

$$\mathbf{E}\,|\eta_1 + \ldots + \eta_n|^s = O(n^{s/2}) \ \text{ as } n\to\infty \tag{10.36}$$

which follows, e.g., from the Rosenthal inequality, cf. [405, Theorem 2.10].

Exercise 10.8. Show that this estimate is sharp, that is one cannot obtain in general for such sequence $(\eta_n)_{n\in\mathbb{N}}$ and $r < s/2$ the estimate

$$\mathbf{E}\,|\eta_1 + \ldots + \eta_n|^s = O(n^r) \ \text{ as } n\to\infty.$$

However for dependent summands there are new effects which we are going to discuss. We start with PA or NA random field $\xi = \{\xi(j),\ j\in\mathbb{Z}^d\}$. Let $\mathcal{U}$ be the class of blocks in $\mathbb{Z}^d$, that is of the sets

$$U = (a,b]\cap\mathbb{Z}^d = ((a_1,b_1]\times\ldots\times(a_d,b_d])\cap\mathbb{Z}^d,\ \ a_i,b_i\in\mathbb{Z},\ a_i<b_i,\ i=1,\ldots,d.$$

For $\tau>0$ and $n\in\mathbb{N}$ let us introduce the analogue of the Cox–Grimmett [135] coefficient

$$\Lambda_\tau(n) = \sup\Big\{\sum_{i\in U}\Big(\sum_{j\notin U} |\,\mathbf{cov}(\xi(i),\xi(j))|\Big)^\tau : U \in \mathcal{U},\ |U| = n\Big\} \le \infty.$$

Given a block $U \in \mathcal{U}$, $r > 1$ and $n \in \mathbb{N}$, set

$$a_r(U) = \mathbf{E}|S(U)|^r, \quad A_r(n) = \sup\{\,a_r(U) : U \in \mathcal{U},\ |U| = n\,\}.$$

Theorem 10.16 ([91]). *Let a centered random field $\xi = \{\,\xi(j),\ j \in \mathbb{Z}^d\,\} \in \mathsf{PA}$ be such that for some $r > 2$, $\delta > 0$ and $\mu \ge 0$ one has*

$$A_{r+\delta}(1) = \sup_{j\in\mathbb{Z}^d} \mathbf{E}|\xi(j)|^{r+\delta} < \infty, \tag{10.37}$$

$$\Lambda_{\delta/\varkappa}(n) = O(n^\mu) \ \ as \ \ n \to \infty, \tag{10.38}$$

here $\varkappa = \delta + (r+\delta)(r-2)$. Then $A_r(n) = O(n^\tau)$ where

$$\tau = \tau(r,\delta,\mu) = \begin{cases} r/2, & 0 \le \mu < (1+\delta/\varkappa)/2, \\ \varkappa(\mu \wedge 1)/(r+\delta-2), & otherwise. \end{cases} \tag{10.39}$$

Remark 10.4. Therefore the upper bound for $A_r(n)$ has the same form $O(n^{r/2})$ as in the case of i.i.d random summands when the dependence is small enough, namely, $0 \le \mu < (1+\delta/\varkappa)/2$. Moreover, the result of Theorem 10.16 is sharp as the following statement shows.

Theorem 10.17 ([91]). *For any $r > 2$, $\delta > 0$, $\mu \ge 0$ and $d \in \mathbb{N}$ there exists a random field $\xi = \{\,\xi(j),\ j \in \mathbb{Z}^d\,\}$ which satisfies all the conditions of Theorem* 10.16 *and such that*

$$a_r(U_n) \ge c\,(\nu_d(U_n))^\tau$$

where $U_n = (0,n]^d$ for $n \in \mathbb{N}$, $\tau = \tau(r,\delta,\mu)$ was defined in (10.39) *and $c > 0$ does not depend on n.*

Denote the length of edges of a parallelepiped $V = (a,b] \subset \mathbb{R}^d$ by $l_1(V), \dots, l_d(V)$ $(a_i < b_i, i = 1,\dots,d)$. We write also $l_i(U)$ when $U = V \cap \mathbb{Z}^d \in \mathcal{U}$. Set

$$\mathcal{U}' = \{\,U \in \mathcal{U} : l_k(U) = 2^{q_k},\ q_k \in \mathbb{Z}_+,\ k = 1,\dots,d\,\}.$$

Introduce the congruent blocks v, v' obtained from V by drawing a hyperplane orthogonal to the edge having length $l_0(V) = \max_{1\le k\le d} l_k(V)$. If there are several edges $l_i(V)$ having the length $l_0(V)$ we take that with the minimal i.

Remark 10.5. Analyzing the proof of Theorem 10.16 (see [95, Theorem 2.1.4]) one sees that the condition $\Lambda_{\delta/\varkappa}(n) = O(n^\mu)$ as $n \to \infty$ can be replaced by a weaker assumption that $\Lambda_1(1) < \infty$ and

$$\sup\Big\{\sum_{i\in v}\Big(\sum_{j\in v'}\mathbf{cov}(\xi(i),\xi(j))\Big)^{\delta/\varkappa} : U = V\cap\mathbb{Z}^d\in\mathcal{U}',\ |v| = n\Big\} = O(n^{\mu}),\quad n\to\infty,$$

where the blocks v, v' for V are described above.

Corollary 10.2 ([95, p. 116]). *Let a centered field* $\xi = \{\xi(j),\ j\in\mathbb{Z}^d\}\in\mathsf{PA}$, $A_p(1)<\infty$ *for some* $p\in(2,3]$ *and* $u(n) = O(n^{-\nu})$ *for some* $\nu>0$, *where* $u(n)$ *is the Cox–Grimmett coefficient introduced in* (10.19). *Then there exists* $r\in(2,p]$ *such that* $A_p(n) = O(n^{r/2})$ *as* $n\to\infty$.

The following example constructed by Birkel shows that even for associated sequence ($d = 1$) of random variables in Theorem 10.16 one cannot assume $A_r(1)<\infty$ (as for i.i.d. summands) instead of the hypothesis $A_{r+\delta}(1)<\infty$ to obtain the nontrivial estimate for $A_r(n)$ as $n\to\infty$. Namely, we give

Theorem 10.18 ([72]). *Let* $(\gamma_n)_{n\in\mathbb{N}}$ *be a monotone sequence of positive numbers such that* $\gamma_n\to 0$ *as* $n\to\infty$. *Then for any* $r>2$ *there exists a centered sequence* $(\xi_n)_{n\in\mathbb{N}}\in\mathsf{A}$ *such that*

i) $A_r(1)<\infty$,
ii) $u(n) = O(\gamma_n)$ *as* $n\to\infty$,
iii) $a_r((0,n])\geq cn^r$ *where* $c>0$ *does not depend on* n.

Now we turn to (BL,θ)-dependent random fields. Here we shall only pay attention to the conditions guaranteeing the "independent-type" behaviour for partial sums moments.

Introduce a function

$$\psi(x) = \begin{cases}(x-1)(x-2)^{-1}, & 2<x\leq 4,\\ (3-\sqrt{x})(\sqrt{x}+1), & 4<x\leq t_0^2,\\ ((x-1)\sqrt{(x-2)^2-3}-x^2+6x-11)(3x-12)^{-1}, & x>t_0^2,\end{cases}\tag{10.40}$$

where $t_0\approx 2.1413$ is the maximal root of the equation $t^3+2t^2-7t-4=0$. Note that $\psi(x)\to 1$ as $x\to\infty$ (Fig. 10.2).

Theorem 10.19 ([96, p. 120]). *Let* $\xi = \{\xi_j,\ j\in\mathbb{Z}^d\}$ *be a centered* (BL,θ)-*dependent random field such that there are* $p>2$ *and* $c_0>1$ *ensuring that* $A_p(1)<\infty$ *and*

$$\theta_r\leq c_0r^{-\lambda},\quad r\in\mathbb{N},\tag{10.41}$$

for $\lambda>d\psi(p)$ *with* ψ *defined in* (10.40). *Then there exist* $\delta>0$ *and* $C>1$ *depending only on* $d, p, A_p(1), c_0$ *and* λ *such that for any block* $U\in\mathcal{U}$ *one has*

$$\mathbf{E}|S(U)|^{2+\delta}\leq C|U|^{1+\delta/2}.\tag{10.42}$$

Remark 10.6. In [315] this result was extended to cover finite $U\subset\mathbb{Z}^d$ having arbitrary configuration. Now we briefly describe the main idea of this generalization.

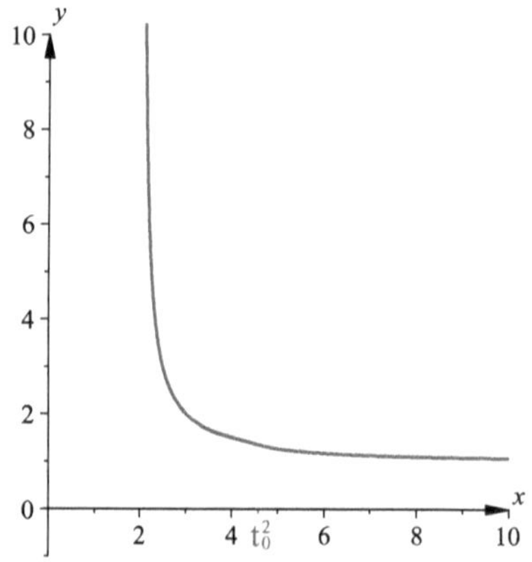

Fig. 10.2 Graph of the function $\psi(x)$

Let f be a probability density (with respect to the Lebesgue measure ν_d) on $\mathcal{B}(\mathbb{R}^d)$ such that

$$\sup_{x\in\mathbb{R}^d} f(x) \le \alpha \tag{10.43}$$

for some $\alpha > 0$. For unit vector $\nu \in \mathbb{R}^d$ and $\varepsilon > 0$ introduce the *concentration function*

$$Q(\nu,\varepsilon) = \sup_{b\in\mathbb{R}} \int_{x:b\le\langle\nu,x\rangle\le b+\varepsilon} f(x)\nu_d(dx)$$

where $\langle\nu,x\rangle = \sum_{k=1}^d \nu_k x_k$.

Theorem 10.20 ([331]). *Let* (10.43) *be fulfilled. Then for any* $d \ge 3$ *there exists* $C_d > 0$ *such that for any* $\varepsilon > 0$ *one can find the unit vector* $\nu = \nu(f,\varepsilon)$ *such that the relation* $Q(\nu,\varepsilon) \le C_d\alpha^{1/d}\varepsilon$ *is true.*

We shall say that sets $W_1, W_2 \subset \mathbb{R}^d$ are separated by a layer of width ε if there exist a unit vector $\nu \in \mathbb{R}^d$ and a number $b \in \mathbb{R}$ such that

$$W_1 \subset \{\, x \in \mathbb{R}^d : \langle x,\nu\rangle \le b \,\} \quad \text{and} \quad W_1 \subset \{\, x \in \mathbb{R}^d : \langle x,\nu\rangle \ge b+\varepsilon \,\}.$$

The following result can be viewed as useful analogue of Theorem 10.20 for discrete case.

Theorem 10.21 ([315]). *Let integer* $d \ne 2$ *and*

$$0 < a = \sum_{j\in\mathbb{Z}^d} a_j < \infty \text{ where } a_j \ge 0 \text{ for all } j \in \mathbb{Z}^d.$$

Then for any $\varepsilon \geq 1/2$ *there exists a partition* $\mathbb{Z}^d$ *into sets* U_1, U_2, U_3 *and* U_4 *with the following properties:*

i) U_1 *and* U_4 *are separated by a layer of width* 2ε*;*
ii) U_1 *and* $U_3 \cup U_4$ *as well as* $U_1 \cup U_2$ *and* U_4 *are separated by a layer of width* ε*;*
iii) for $i = 2, 3$ *and* $\mu_A(U) := \sum_{j \in U} a_j$ *where* $U \subset \mathbb{Z}^d$ *one has*

$$\mu_A(U_i) \leq K_d a_{\max}^{1/d} a^{1-1/d} \varepsilon,$$

here $a_{\max} = \max_{j \in \mathbb{Z}^d} a_j$ *and* K_d *depends on* d *only;*
iv) moreover,

$$|\mu_A(U_1 \cup U_2) - \mu_A(U_3 \cup U_4)| \leq \frac{1}{2} K_d a_{\max}^{1/d} a^{1-1/d} \varepsilon.$$

To construct the appropriate partition of $\mathbb{Z}^2$, the author of [315] employed the analogue of Theorem 10.20 established in [99] for $d = 2$.

The bisection method used to prove Theorem 10.19 and its generalization (see for example [95, p. 121]) allows (under specified conditions) to estimate the partial sums' moment of any order greater than two. The situation becomes simpler if one is interested in the moment of order $2r$ when $r > 1$ is an integer. In this case (see [36]) it is useful to consider the estimates of $\mathbf{cov}(F(\xi_I), G(\xi_J))$ for power-type "test functions" F and G and finite disjoint sets $I, J \subset \mathbb{Z}^d$. We provide here only such a result for the fourth moment of partial sums.

For $n \in \mathbb{N}$ and $\gamma \in \mathbb{R}$ set

$$B(n,\gamma) = \begin{cases} n^\gamma, & \gamma > 0, \\ \log(n \vee \mathrm{e}), & \gamma = 0, \\ 1, & \gamma < 0, \end{cases} \qquad h(\gamma) = \begin{cases} \gamma^{-1} \vee 1, & \gamma > 0, \\ 2, & \gamma = 0, \\ 2(|\gamma|^{-1} \vee 1), & \gamma < 0. \end{cases}$$

Theorem 10.22 ([468]). *Let* $\xi = \{\xi(j),\ j \in \mathbb{Z}^d\}$ *be a centered* (BL,θ)*-dependent random field such that* $A_p(1) = \sup_{j \in \mathbb{Z}^d} \mathbf{E}|\xi(j)|^p < \infty$ *for some* $p > 4$ *and* (10.41) *holds. Then for any* $U \in \mathcal{U}$ *one has the estimate*

$$\mathbf{E}S(U)^4 \leq \sum_{j \in U} \mathbf{E}\xi_j^4 + 12|U|^2(A_2(1) + \theta_1)^2 + |U| C_1(d,\lambda,p) c_0^\nu A_p(1)^{2/(p-2)} B(|U|,\gamma) \tag{10.44}$$

where

$$\nu = (p-4)/(p-2), \gamma = 3-(\lambda\nu/d), C_1(d,\lambda,p) = 192 d^2 3^{3d} (p-4)^{2/(p-2)} h(\gamma)\nu^{-1}.$$

If ξ is quasi-associated and for any $i \neq j$

$$|\,\mathbf{cov}(\xi(i), \xi(j))| \leq c_1 |i-j|^{-\varkappa} \tag{10.45}$$

with some $c_1, \varkappa > 0$, the same estimate (10.44) *holds upon replacement of c_0 with c_1 and λ with $\varkappa$.*

Corollary 10.3. *Assume that conditions of Theorem* 10.22 *are satisfied with any $p > 4$ (e.g. $|\xi_j| \leq b$ for all $j \in \mathbb{Z}^d$ and some constant $b > 0$) and $\lambda > 2d$. Then $A_4(n) = O(n^2)$ as $n \to \infty$. One can obtain the same estimate for $A_4(n)$ whenever $A_p(1) < \infty$ for some $p > 4$ and λ large enough.*

Indeed, $\gamma = 1$ means $\lambda v = 2d$. Now note that $v = v(p) \uparrow 1$ as $p \to \infty$. Setting $|U| = n$ in (10.44) we come to the desired statement. The second claim is verified in a similar way.

10.2.4 Bounds Based on Supermodular Results

Recall the following important

Definition 10.14. A function $f : \mathbb{R}^n \to \mathbb{R}$ is called *supermodular* if for any $x, y \in \mathbb{R}^n$ one has

$$f(x \vee y) + f(x \wedge y) \geq f(x) + f(y)$$

where $x \vee y = (x_1 \vee y_1, \ldots, x_n \vee y_n)^\top$ and $x \wedge y = (x_1 \wedge y_1, \ldots, x_n \wedge y_n)^\top$.

Exercise 10.9. Prove that the following functions are supermodular:

i) $\sum_{i=1}^n x_i$,
ii) $\max_{k=1}^n \sum_{i=1}^k x_i$,
iii) $g(h_1, \ldots, h_n)$ where $g : \mathbb{R}^n \to \mathbb{R}$ is supermodular and $h_i : \mathbb{R} \to \mathbb{R}$ are nondecreasing functions for $i = 1, \ldots, n$.
iv) $g \circ h$ where $h : \mathbb{R}^n \to \mathbb{R}$ is supermodular coordinatewise nondecreasing function and $g : \mathbb{R} \to \mathbb{R}$ is a nondecreasing convex function.

Definition 10.15. Let Q_1 and Q_2 be probability measures on $(\mathbb{R}^n, \mathcal{B}(\mathbb{R}^n))$. One says that Q_1 is less than Q_2 in the *supermodular order* (and writes $Q_1 \leq_{sm} Q_2$) if for any supermodular function $f : \mathbb{R}^n \to \mathbb{R}$

$$\int_{\mathbb{R}^n} f(x) Q_1(dx) \leq \int_{\mathbb{R}^n} f(x) Q_2(dx)$$

whenever both integrals exist. For random vectors $\xi = (\xi_1, \ldots, \xi_n)^\top$ and $\eta = (\eta_1, \ldots, \eta_n)^\top$ the notation $\xi \leq_{sm} \eta$ means that $\text{Law}(\xi) \leq_{sm} \text{Law}(\eta)$. If ξ and η are defined on the same probability space then $\xi \leq_{sm} \eta$ iff for any supermodular function $f : \mathbb{R}^n \to \mathbb{R}$

$$\mathbf{E}f(\xi) \le \mathbf{E}f(\eta) \tag{10.46}$$

whenever both expectations exist.

Definition 10.16. For a random vector $\xi = (\xi_1, \dots, \xi_n)^\top$, its *decoupled version* is a vector $\eta = (\eta_1, \dots, \eta_n)^\top$ such that $\mathrm{Law}(\xi_i) = \mathrm{Law}(\eta_i)$, $i = 1, \dots, n$, and the components of η are (mutually) independent. The decoupled version of a family $\{\xi(t),\ t \in T\}$ is a family[1] $\{\eta_t,\ t \in T\}$ consisting of independent random variables having $\mathrm{Law}(\xi(t)) = \mathrm{Law}(\eta_t)$ for each $t \in T$.

We shall assume without loss of generality that $\{\xi(t),\ t \in T\}$ and its decoupled version $\{\eta_t, t \in T\}$ are defined on the same probability space (the extension of the one where ξ was defined).

Now we can formulate the important result by Christofidis and Vaggelatou.

Theorem 10.23 ([126]). *Let $\eta = (\eta_1, \dots, \eta_n)^\top$ be the decoupled version of a random vector $\xi = (\xi_1, \dots, \xi_n)^\top$. If $\xi \in \mathsf{PA}$ then $\xi \ge_{sm} \eta$. If $\in \mathsf{NA}$ then $\xi \le_{sm} \eta$.*

Exercise 10.10. Is it possible that $\xi \ge_{sm} \eta$ where η is a decoupled version of a random vector ξ and $\xi \notin \mathsf{PA}$? Does there exist a random vector ξ such that $\xi \notin \mathsf{NA}$ and $\xi \le_{sm} \eta$ where η is its decoupled version?

Next we can easily obtain the following useful result by Shao.

Theorem 10.24 ([464]). *Let a sequence $(\xi_n)_{n\in\mathbb{N}} \in \mathsf{PA}$ and $(\eta_n)_{n\in\mathbb{N}}$ be its decoupled version. Set $S_n = \sum_{i=1}^n \xi_i$ and $T_n = \sum_{i=1}^n \eta_i$ for $n \in \mathbb{N}$. Then for any convex function $f : \mathbb{R} \to \mathbb{R}$ and each $n \in \mathbb{N}$*

$$\mathbf{E}f(S_n) \ge \mathbf{E}f(T_n) \tag{10.47}$$

whenever the expectations exist. If, moreover, f is nondecreasing, then for any $n \in \mathbb{N}$

$$\mathbf{E}f(M_n) \ge \mathbf{E}f(R_n) \tag{10.48}$$

whenever the expectations exist, here $M_n = \max_{1\le j\le n} S_j$ and $R_n = \max_{1\le j\le n} T_j$. If $(\xi_n)_{n\in\mathbb{N}} \in \mathsf{NA}$ then both assertions hold with reversed signs of inequalities.

Proof. We consider $(\xi_n)_{n\in\mathbb{N}} \in \mathsf{PA}$ because the NA case is analogous. If f is nondecreasing then (10.47) and (10.48) are derived from Theorem 10.23 and Exercise 10.9 (i), (ii) and (iv). Thus we have only to prove (10.47) without assuming that a convex f is nondecreasing. If f is convex and nonincreasing then we introduce $g(x) = f(-x)$, $x \in \mathbb{R}$. Obviously the function g is convex and nondecreasing. Note that $(-\eta_1, \dots, -\eta_n)^\top$ is a decoupled version of $(-\xi_1, \dots, -\xi_n)^\top \in \mathsf{PA}$. Consequently in view of Exercise 10.9 (i) and already discussed case we have

[1] Such family $\{\eta_t, t \in T\}$ exists due to Theorem 9.1.

$$\mathbf{E}f(S_n) = \mathbf{E}g\left(\sum_{k=1}^{n}(-\xi_k)\right) \geq \mathbf{E}g\left(\sum_{k=1}^{n}(-\eta_k)\right) = \mathbf{E}f(T_n)$$

if $\mathbf{E}g\left(\sum_{k=1}^{n}(-\xi_k)\right)$ and $\mathbf{E}g\left(\sum_{k=1}^{n}(-\eta_k)\right)$ exist or equivalently whenever $\mathbf{E}f(S_n)$ and $\mathbf{E}f(T_n)$ exist.

If convex f is neither nondecreasing nor nonincreasing (on $\mathbb{R}$) there exists at least one point $c \in \mathbb{R}$ such that $f\mathbf{1}(x < c)$ is nonincreasing and $f\mathbf{1}(x \geq c)$ is nondecreasing. Set

$$f^{(+)}(x) = f(c) + (f(x) - f(c))\mathbf{1}(x \geq c), \quad f^{(-)}(x) = f(x) - f^{(+)}(x), \quad x \in \mathbb{R},$$

If $\mathbf{E}\zeta$ exists for a random variable ζ, then for any event $A \in \mathcal{A}$ there exists $\mathbf{E}\zeta\mathbf{1}(A)$. The constant is integrable. Therefore if $\mathbf{E}f(S_n)$ and $\mathbf{E}f(T_n)$ exist we conclude that $\mathbf{E}f^{(+)}(S_n)$, $\mathbf{E}f^{(-)}(S_n)$, $\mathbf{E}f^{(+)}(T_n)$ and $\mathbf{E}f^{(-)}(T_n)$ exist as well. Thus by (10.47) already proved for convex nondecreasing f one has

$$\mathbf{E}f^{(+)}(S_n) \geq \mathbf{E}f^{(+)}(T_n), \quad n \in \mathbb{N}. \tag{10.49}$$

The sequence $(-\xi_n)_{n\in\mathbb{N}} \in \mathsf{PA}$ and its decoupled version is $(-\eta_n)_{n\in\mathbb{N}}$. Therefore, taking $g(x) = f^{(-)}(-x)$ for $x \in \mathbb{R}$, in view of (10.47) established for convex nondecreasing f we come to the inequality

$$\mathbf{E}f^{(-)}(S_n) = \mathbf{E}g(-S_n) \geq \mathbf{E}g(-T_n) = \mathbf{E}f^{(-)}(T_n), \quad n \in \mathbb{N}. \tag{10.50}$$

Clearly $\mathbf{E}f^{(-)}(S_n)$ and $\mathbf{E}f^{(-)}(T_n)$ exist when $\mathbf{E}f(S_n)$ and $\mathbf{E}f(T_n)$ exist. It remains to combine (10.49) and (10.50) as $f = f^{(+)} + f^{(-)}$. □

Exercise 10.11. Explain why the same reasoning that was used to prove (10.47) does not permit to obtain (10.48) for any convex f.

Using Theorem by Shao one can provide some analogues of the well-known Bernstein exponential inequality for negatively associated random variables. In this regard we refer to [95, Corollary 2.2.11].

Remark 10.7. Inequality (10.47) can be applied to random field $\xi = \{\xi(j), j \in \mathbb{Z}^d\} \in \mathsf{PA}$ or with opposite sign to $\xi \in \mathsf{NA}$. Namely, for any finite set $U \subset \mathbb{Z}^d$, any supermodular convex function $f : \mathbb{R}^U \to \mathbb{R}$ and decoupled version $\eta = \{\eta(j), j \in \mathbb{Z}^d\}$ of ξ one has

$$\mathbf{E}f(S(U)) \geq \mathbf{E}f(T(U)) \tag{10.51}$$

where $T(U) = \sum_{j\in U} \eta(j)$ whenever the expectations in (10.51) exist. Indeed we can enumerate the points of U and employ (10.47). However, it is impossible in general to prove inequality of the type (10.48) for random fields as the function

$$f(x_1, \ldots, x_n) = \sup_{I \in \Gamma} \sum_{i \in I} x_i, \quad x = (x_1, \ldots, x_n)^\top \in \mathbb{R}^n,$$

is not supermodular for arbitrary system Γ of subsets of $\{1, \ldots, n\}$. Let us provide a corresponding counterexample.

For a field $\xi = \{\xi(j), j \in \mathbb{Z}^d\}$ and $n = (n_1, \ldots, n_d)^\top \in \mathbb{N}^d$ introduce $M_n = \max_{j \in (0,n]} S((0,j])$ where the block $(0,n] = (0,n_1] \times \ldots \times (0,n_d]$. For the decoupled version $\eta = \{\eta(j), j \in \mathbb{Z}^d\}$ of ξ we define R_n in a similar way as M_n.

Theorem 10.25 ([98]). *Let $f : \mathbb{R} \to \mathbb{R}$ be any function with property $f(1) > f(0)$. Then for any integer $d > 1$ there exists a random field $\xi = \{\xi(j), j \in \mathbb{Z}^d\} \in \mathsf{NA}$ and a point $n \in \mathbb{N}^d$ such that*

$$\mathbf{E} f(M_n) > \mathbf{E} f(R_n).$$

Proof. It suffices to consider $d = 2$ and $n = (2,2)^\top$ as we always can take other random variables in ξ to be equal to zero a.s. Let ζ be a random variable such that $\mathbf{P}(\zeta = -1) = \mathbf{P}(\zeta = 1) = 1/2$. Set

$$\xi_{(1,1)} = 0, \quad \xi_{(1,2)} = \zeta, \quad \xi_{(2,1)} = -\zeta, \quad \xi_{(2,2)} = -3.$$

Then it is easily seen that $\{\xi_{(1,1)}, \xi_{(1,2)}, \xi_{(2,1)}, \xi_{(2,2)}\} \in \mathsf{NA}$. For $n = (2,2)^\top$ one has $M_n = 1$ a.s. since either $S_{(1,2)} = 1$ or $S_{(2,1)} = 1$ with other partial sums being at the same time nonpositive. Thus $\mathbf{E} f(M_n) = f(1)$.

The random variables $\eta_{(1,2)}$ and $\eta_{(2,1)}$ are independent and distributed as ζ. Clearly $\eta_{(1,1)} = 0$ and $\eta_{(1,1)} + \eta_{(1,2)} + \eta_{(2,1)} + \eta_{(2,2)} < 0$. Therefore $R_{(2,2)} = 0$ iff $\eta_{(1,2)} = \eta_{(2,1)} = -1$. Otherwise $R_{(2,2)} = 1$. Hence

$$\mathbf{E} f(R_n) = \frac{1}{4} f(0) + \frac{3}{4} f(1) < f(1) = \mathbf{E} f(M_n), \quad n = (2,2)^\top,$$

as $f(0) < f(1)$. This completes the proof. □

For random field $\xi = \{\xi(j), j \in \mathbb{Z}^d\} \in \mathsf{PA}$ the upper bounds for maximum of partial sums moments will be discussed further on.

10.2.5 The Móricz Theorem

There is a powerful method proposed by Móricz to obtain the inequalities for expectations of the maximum of partial sums using the corresponding estimates for moments of these sums.

Let $\mathcal{U}$ be a collection of blocks $U = (a,b] \cap \mathbb{Z}^d$, $a < b$, $a \in \mathbb{Z}^d$ as above.

Definition 10.17. A function $\varphi : \mathcal{U} \to \mathbb{R}$ is called *superadditive* if, for any blocks $U, U_1, U_2 \in \mathcal{U}$ such that $U = U_1 \cup U_2$ and $U_1 \cap U_2 = \emptyset$, one has $\varphi(U) \geq \varphi(U_1) + \varphi(U_2)$.

For example, the function $\varphi(U) = |U|^r$ is superadditive if $r \geq 1$. Let $\xi = \{\xi(j), \ j \in \mathbb{Z}^d\}$ be a real-valued random field. Introduce

$$S(U) = \sum_{j \in U} \xi(j), \quad M(U) = \max_{a<j\leq b} |S((a, j])|$$

for $U = (a, b] \cap \mathbb{Z}^d \in \mathcal{U}$ where $a < j \leq b$ means $a_i < j_i \leq b_i$ for $i = 1, \ldots, d$.

Theorem 10.26 ([371]). *Let $d \in \mathbb{N}$ and $\gamma \geq 1$. Suppose that there exist functions $\varphi : \mathcal{U} \to \mathbb{R}_+$ and $\psi : \mathbb{R}_+ \times \mathbb{Z}^d \to \mathbb{R}_+$ such that φ is superadditive, ψ is coordinatewise nondecreasing and, for any block $U \in \mathcal{U}$ with $m_i = b_i - a_i \geq 1$, $i = 1, \ldots, d$, one has*

$$\mathbf{E}|S(U)|^\gamma \leq \varphi(U)\psi^\gamma(\varphi(U), m_1, \ldots, m_d). \tag{10.52}$$

Then $\mathbf{E}|M(U)|^\gamma$, for any $U \in \mathcal{U}$, admits the following upper bound

$$\left(\frac{5}{2}\right)^d \varphi(U) \Bigg(\sum_{k_1=1}^{[\log_2 m_1]} \cdots \sum_{k_d=1}^{[\log_2 m_d]} \times \psi(2^{-k_1-\ldots-k_d}\varphi(U), [2^{-k_1}m_1], \ldots, [2^{-k_d}m_d]) \Bigg)^\gamma, \tag{10.53}$$

here $[\cdot]$ stands for the integer part of a number.

Corollary 10.4 ([371]). *Let $d \in \mathbb{N}$ and $\gamma \geq 1$, $\alpha \geq 1$. Assume that there exists a nonnegative and superadditive function $f : \mathcal{U} \to \mathbb{R}_+$ such that, for any block $U = (a, b] \cap \mathbb{Z}^d \in \mathcal{U}$, one has*

$$\mathbf{E}|S(U)|^\gamma \leq f^\alpha(U).$$

Then

$$\mathbf{E}M(U)^\gamma \leq \begin{cases} (5/2)^d (1 - 2^{(1-\alpha)/\gamma})^{-d\gamma} f^\alpha(U) & \text{if } \alpha > 1, \\ 5^d 2^{d(\gamma-1)} f(U)([\mathrm{Log}_2\, m_1] \ldots [\mathrm{Log}_2\, m_d])^\gamma & \text{if } \alpha = 1 \end{cases} \tag{10.54}$$

where $m_i = b_i - a_i$, $i = 1, \ldots, d$, and $\mathrm{Log}_2\, x := \log_2 (x \vee 2)$ for $x \in \mathbb{R}$.

Note that a very useful particular case of Corollary 10.4 is provided by the choice $f = c|U|$, $c = \text{const}$. Thus we can use here for instance Theorem 10.16 and Corollary 10.2.

Now we provide the simple proof, proposed by P.A. Yaskov, of the particular case of Theorem 10.26. The main idea here goes back to the dyadic representation of an integer number used in the proof of the famous Rademacher–Menshov inequality (see for example [336]).

Theorem 10.27. *Let $d \in \mathbb{N}$, $p \geq 1$ and $\alpha \geq 0$. Suppose that for any block $U = (a,b] \cap \mathbb{Z}^d \in \mathcal{U}$ and some superadditive function $f : \mathcal{U} \to \mathbb{R}_+$*

$$\mathbf{E}|S(U)|^p \leq f(U)|U|^\alpha. \tag{10.55}$$

Then, for every $U = (a,b] = (a_1,b_1] \times \ldots \times (a_d,b_d] \cap \mathbb{Z}^d \in \mathcal{U}$

$$\mathbf{E}M(U)^p \leq \begin{cases} ([\log_2 n_1]+1)^p \ldots ([\log_2 n_d]+1)^p f(U), & \alpha = 0, \\ \left(1 - 2^{-\alpha/p}\right)^{-dp} f(U)|U|^\alpha, & \alpha > 0, \end{cases}$$

where $n_i = b_i - a_i$, $1 \leq i \leq d$.

Proof. At first consider the case $d = 1$. Put for simplicity $(a,b] = (0,n]$. We shall write $l = [\log_2 n]$. From the dyadic expansion of $m \leq n$, i.e.

$$m = \sum_{k=0}^{l} b_k 2^{l-k}, \quad b_k \in \{0,1\}, \quad b_0 \neq 0,$$

we obtain $S((0,m]) = \sum_k S\left(\left(m^{(k-1)}, m^{(k)}\right]\right)$ where $m^{(k)} = b_0 2^l + \ldots + b_k 2^{l-k}$ with $m^{(-1)} = 0$. For example if $m = 7$ or 5, then, since $7 = 4+2+1$, $5 = 4+1$ and $S((0,0]) = 0$,

$$S((0,7]) = S((0,4]) + S((4,6]) + S((6,7]), \quad S((0,5]) = S((0,4]) + S((4,5]).$$

This yields

$$M((0,n]) = \max_{m \leq n} |S((0,m])| \leq \sum_{k=0}^{l} M_k,$$

where

$$M_k = \max_{i:(i+1)2^k \leq n} |S(I_{ik})|, \quad I_{ik} = \left(i2^k, (i+1)2^k\right].$$

Further, applying the triangular inequality for the norm $\|\xi\|_p = (\mathbf{E}|\xi|^p)^{1/p}$, we derive

$$\|M((0,n])\|_p \leq \sum_{k=0}^{l} \|M_k\|_p.$$

In addition,

$$\mathbf{E}M_k^p \leq \sum_{i:(i+1)2^k \leq n} \mathbf{E}|S(I_{ik})|^p \leq 2^{\alpha k} f((0,n]).$$

Finally, we obtain that $\mathbf{E}M((0,n])^p$ admits the upper bound

$$([\log_2 n] + 1)^p f((0, n]) \text{ as } \alpha = 0 \text{ and } \frac{n^\alpha f((0, n])}{\left(1 - 2^{-\alpha/p}\right)^p} \text{ if } \alpha > 0.$$

For $d > 1$ the proof is quite similar. The only difference is that we use the d-dimensional dyadic expansion of $S((0, m])$ with

$$\left(m^{(k-1)}, m^{(k)}\right] = \left(m_1^{(k_1-1)}, m_1^{(k_1)}\right] \times \ldots \times \left(m_d^{(k_d-1)}, m_d^{(k_d)}\right],$$

and consider the sum over $k : k_i \le [\log_2 n_i]$, $i = 1, \ldots, d$, defined by $m_1, \ldots, m_d$ in the same way as for $d = 1$. □

Remark 10.8. The proof above allows to derive a similar upper bound for $\mathbf{E}M(U)^p$ when there is the sum

$$f_1(U)|U|^{\alpha_1} + \ldots + f_N(U)|U|^{\alpha_N}$$

at the right-hand side of (10.55). The induction argument used by Móricz does not cover such case.

10.2.6 Rosenthal Type Inequalities

For $p \ge 1$ and a real-valued random variable η we write $\|\eta\|_p = (\mathbf{E}|\eta|^p)^{1/p}$. Let $\xi = \{\xi(t), t \in T\}$ be a family of (real-valued) random variables. For a finite set $U \subset T$ and numbers $a, b \ge 1$ introduce

$$S(U) = \sum_{t \in U} \xi(t), \quad Q(U, a, b) = \sum_{t \in U} \|\xi(t)\|_a^b.$$

The classical Rosenthal inequality (see for example [405, p. 83]) states that if $\xi(t)$, $t \in T$, are independent centered random variables such that $\mathbf{E}|\xi(t)|^p < \infty$ for all $t \in T$ and some $p > 1$, then for any finite set $U \subset T$ one has

$$\mathbf{E}|S(U)|^p \le 2^{p^2}(Q(U, p, p) \vee (Q(U, 1, 1))^p), \tag{10.56}$$

if, moreover, $\mathbf{E}|\xi(t)|^p < \infty$ for all $t \in T$ and some $p > 2$ then

$$\mathbf{E}|S(U)|^p \ge 2^{-p}(Q(U, p, p) \vee Q(U, 2, 2)^{p/2}). \tag{10.57}$$

Note that $x \mapsto |x|^p$ $(x \in \mathbb{R})$ is a convex function for $p \ge 1$. Therefore due to Theorem 10.24 we immediately come to the following statement.

Theorem 10.28 ([126]). *Let $\xi = \{\xi(j), j \in \mathbb{Z}^d\}$ be a centered random field and $\mathbf{E}|\xi(j)|^p < \infty$ for some $p > 1$, then for any finite set $U \subset \mathbb{Z}^d$ inequality* (10.56) *holds if* $\xi \in \mathsf{NA}$. *Let a centered field* $\xi \in \mathsf{PA}$ *and* $\mathbf{E}|\xi(j)|^p < \infty$ *for some* $p > 2$, *then* (10.57) *is valid.*

The next result is due to Vronski.

Theorem 10.29 ([501]). *Let a centered wide-sense stationary random field* $\xi = \{\xi(j),\ j \in \mathbb{Z}^d\} \in \mathsf{PA}$ *and*

i) $a < \mathbf{E}\xi(0)^2 < \infty$ *for some* $a > 0$,
ii) the finite susceptibility condition (10.20) *holds,*
iii) there exist an even integer k and some $\delta > 0$ *such that*

$$\mathbf{E}|\xi(j)|^{k+\delta} < \infty \text{ for any } j \in \mathbb{Z}^d,$$

$$\sum_{n=1}^{\infty} u(n)^{\delta/(k+\delta-2)} n^{d(k-1)-1} < \infty,$$

where $u(n)$ *is defined in* (10.21).

Then there exists $C = C(k, d, \delta, (a^{-2}U(n))_{n\in\mathbb{N}})$ *such that for any finite* $U \subset \mathbb{Z}^d$

$$\mathbf{E}|S(U)|^k \le C\Big(Q(U, k+\delta, k) \vee Q(U, 2+\delta, 2)^{k/2}\Big).$$

It would be desirable to find the optimal relationship between the conditions imposed on the moments of summands and the dependence structure of a field in the spirit of Theorems 10.16 and 10.17.

10.2.7 Estimates Obtained by Randomization Techniques

The idea of additional randomization to obtain the moment inequalities for maximum of absolute values of partial sums was developed by Peligrad [393], Shashkin [468] and Zhang and Wen [528].

Theorem 10.30 ([468]). *Let* $\xi = \{\xi(j),\ j \in \mathbb{Z}^d\}$ *be a centered random field such that* $A_2(1) < \infty$.

i) Assume that $\xi \in (BL, \theta)$ *and* (10.41) *holds. Then for any* $m \in \mathbb{N}$ *and* $U \in \mathcal{U}$ *one has*

$$\mathbf{E}M(U)^2 \le 3m^d\Big(\Big(\sum_{j\in U}\mathbf{E}|\xi(j)|\Big)^2 + 18\sum_{j\in U}\mathbf{E}\xi(j)^2 + 16c_0\,|U|\,m^{-\lambda}\Big). \quad (10.58)$$

ii) If $\xi \in \mathsf{QA}$ *and* (10.45) *holds, then* (10.58) *is true with* λ *and* c_0 *replaced by* $\varkappa$ *and* $c_1 = C(\varkappa)$, *respectively, where*

$$C(\varkappa) = \sum_{k \in \mathbb{Z}^d, k \neq 0} |k|^{-\varkappa}.$$

iii) If $\xi \in \mathsf{NA}$, then for any $U \in \mathcal{U}$

$$\mathbf{E}M(U)^2 \leq 2\Big(\sum_{j \in U} \mathbf{E}|\xi(j)|\Big)^2 + 100 \sum_{j \in U} \mathbf{E}\xi(j)^2.$$

Remark 10.9. In general this result provides only trivial estimate

$$M(U)^2 = O(|U|^2) \text{ as } |U| \to \infty.$$

However, it could be applied to "tails" of random variables arising after appropriate truncation. Thus it could lead to useful bounds.

We have not a possibility to discuss here the proof of Theorem 10.30 based on the employment of auxiliary random field $\{\varepsilon_j,\ j \in \mathbb{Z}^d\}$ consisting of i.i.d. random variables taking values 0 and 1 and such that this field and ξ are independent. We refer to [95, p. 153].

10.2.8 Estimates for the Distribution Functions of Partial Sums

Recall one of the Kolmogorov inequalities (see for example [405, p. 52]). Let $\eta_1, \ldots, \eta_n$ be independent random variables having finite variances and zero means. Then for any $x \in \mathbb{R}$ one has

$$\mathbf{P}(\max_{1 \leq k \leq n} T_k \geq x) \leq 2\mathbf{P}(T_n \geq x - \sqrt{2\ \mathbf{var}\ T_n}) \tag{10.59}$$

where $T_k = \sum_{i=1}^k \eta_i$ and $\mathbf{var}\ T_n = \sum_{i=1}^n \mathbf{E}\eta_i^2$. In particular, for any $\lambda > 0$

$$\mathbf{P}(\max_{1 \leq k \leq n} T_k \geq \lambda \sqrt{\mathbf{var}\ T_n}) \leq 2\mathbf{P}(T_n \geq (\lambda - \sqrt{2})\sqrt{\mathbf{var}\ T_n}).$$

The analogue of this inequality was obtained by Newman and Write for associated random variables.

Theorem 10.31 ([380]). *Let $(\xi_1, \ldots, \xi_n)^\top \in \mathsf{PA}$ and the components be centered and square integrable. Then $\mathbf{E}L_n^2 \leq \mathbf{E}S_n^2$ where $L_n = \max_{1 \leq k \leq n} S_k$ and $S_k = \xi_1 + \ldots + \xi_k$, $k = 1, \ldots, n$. If, moreover, $(\xi_1, \ldots, \xi_n)^\top \in \mathsf{A}$ then for $M_n = \max_{1 \leq k \leq n} |S_k|$ and any $\lambda > 0$*

$$\mathbf{P}(M_n \geq \lambda \sqrt{\mathbf{var}\ T_n}) \leq 2\mathbf{P}(|S_n| \geq (\lambda - \sqrt{2})\sqrt{\mathbf{var}\ S_n}).$$

This inequality was extended in [381] for partial sums generated by array $\{\xi(j), j \in \mathbb{N}^2\} \in \mathsf{PA}$. However, their method works only for dimension $d = 2$ of index set $\mathbb{N}^d$.

To formulate the maximal inequality for random field $\xi = \{\xi(j), j \in \mathbb{Z}^d\}$ with arbitrary $d \in \mathbb{N}$ we need some notation. For $U \in \mathcal{U}$ set

$$M_U = \max\{|S(W)| : W \in \mathcal{U}, W \subset U\}.$$

Theorem 10.32 ([100]). *Let a centered wide-sense stationary random field $\xi = \{\xi(j), j \in \mathbb{Z}^d\} \in \mathsf{A}$ be such that $A_p(1) < \infty$ for some $p > 2$ and the Cox–Grimmett coefficient $u(n) = O(n^{-\lambda})$ as $n \to \infty$ for some $\lambda > 0$. Then, for any $\tau \in (0,1)$, there exists $x_0 > 0$ such that for all $U \in \mathcal{U}$ and $x \geq x_0$ one has*

$$\mathbf{P}(M_U \geq x\sqrt{|U|}) \leq 2\mathbf{P}(|S(U)| \geq \tau x\sqrt{|U|}).$$

The proof can be found in [95, p. 104].

10.3 Limit Theorems for Partial Sums of Dependent Random Variables

Finally we are able to state and give ideas of the proof of a central limit theorem (CLT) for partial sums of (BL,θ)-dependent stationary random fields together with some corollaries and applications. Extensions of this CLT to random fields without finite susceptibility property are considered as well.

10.3.1 Generalization of the Newman Theorem

For partial sums of multiindexed random variables we use the notation introduced in (10.24).

Theorem 10.33 ([96, p. 178]). *Let $\xi = \{\xi(j), j \in \mathbb{Z}^d\}$ be a (BL,θ)-dependent strictly stationary centered square-integrable random field. Then for any sequence of regularly growing sets $U_n \subset \mathbb{Z}^d$ one has*

$$S(U_n)/\sqrt{|U_n|} \xrightarrow{d} N(0,\sigma^2), \quad as\ n \to \infty, \tag{10.60}$$

here σ^2 was defined in Theorem 10.14.

Proof. We divide the proof into several steps.

Step 1. First of all, we explain why, instead of proving (10.60), one can operate with normalized sums taken over finite unions of fixed blocks which form a partition of $\mathbb{R}^d$. For any bounded $V \subset \mathbb{R}^d$ set

$$S(V) = \sum_{j \in V \cap \mathbb{Z}^d} \xi(j).$$

Then obviously $S(U) = S(V)$ where $U = V \cap \mathbb{Z}^d$.

If $(U_n)_{n \in \mathbb{N}}$ is a sequence of regularly growing subsets of $\mathbb{Z}^d$ (see Definition 10.12) then in view of Lemma 10.3 we conclude that $V_n = \cup_{j \in U_n} \Delta_j \to \infty$ in the Van Hove sense as $n \to \infty$ where $\Delta_j = (j, j + \mathbf{1}] \subset \mathbb{R}^d$, $j \in \mathbb{Z}^d$, a vector $\mathbf{1}$ in $\mathbb{R}^d$ has all components equal to 1. Consequently we have to prove that for arbitrary fixed $t \in \mathbb{R}$

$$\mathbf{E} \exp\{\, itS(V_n)/\sqrt{\nu_d(V_n)}\,\} \to \exp\{\, -\sigma^2 t^2/2\,\} \text{ as } n \to \infty, \tag{10.61}$$

here $i^2 = -1$ and $|U_n| = \nu_d(V_n)$.

Take $a = (a_1, \dots, a_d)^\top$ with $a_k = mr$, $k = 1, \dots, d$, where $m, r \in \mathbb{N}$ will be specified later. Consider $V_n^-(a)$ introduced before Definition 10.12. Note that $|\exp\{\, ix\,\} - \exp\{\, iy\,\}| \le |x - y|$ for any $x, y \in \mathbb{R}$. Therefore

$$\begin{aligned}
&|\mathbf{E} \exp\{\, it\nu_d(V_n)^{-1/2} S(V_n)\,\} - \mathbf{E} \exp\{\, it\nu_d(V_n^-(a))^{-1/2} S(V_n^-(a))\,\}| \\
&\quad \le |t|\nu_d(V_n)^{-1/2} \mathbf{E}|S(V_n) - S(V_n^-(a))| \\
&\qquad + |t| \left|\nu_d(V_n)^{-1/2} - \nu_d(V_n^-(a))^{-1/2}\right| \mathbf{E}|S(V_n^-(a))|.
\end{aligned} \tag{10.62}$$

The Lyapunov inequality, and estimate (10.17) yield

$$\mathbf{E}|S(V_n) - S(V_n^-(a))| \le (\mathbf{E}S(V_n \setminus V_n^-(a))^2)^{1/2} \le \left((\mathbf{E}\xi_0^2 + \theta_1)\nu_d(V_n \setminus V_n^-(a))\right)^{1/2}$$

and since $V_n^-(a) \subset V_n \subset V_n^+(a)$ and $(V_n)_{n \in \mathbb{N}}$ is VH-growing sequence, it holds

$$\nu_d(V_n \setminus V_n^-(a)))/\nu_d(V_n) \to 0 \text{ as } n \to \infty. \tag{10.63}$$

It is easily seen that

$$\begin{aligned}
&\left|\nu_d(V_n)^{-1/2} - \nu_d(V_n^-(a))^{-1/2}\right| \\
&= \frac{\nu_d(V_n) - \nu_d(V_n^-(a))}{(\nu_d(V_n)\nu_d(V_n^-(a)))^{1/2}(\nu_d(V_n)^{1/2} + \nu_d(V_n^-(a))^{1/2})} \le \frac{\nu_d(V_n) - \nu_d(V_n^-(a))}{(\nu_d(V_n)\nu_d(V_n^-(a)))^{1/2}}.
\end{aligned}$$

Taking into account (10.62), (10.63) and the inequalities

$$\mathbf{E}|S(V_n^-(a))| \le (\mathbf{E}(S(V_n^-(a)))^2)^{1/2} \le ((\mathbf{E}\xi_0^2 + \theta_1)\nu_d(V_n^-(a)))^{1/2} \tag{10.64}$$

we conclude that to prove (10.60) it suffices to verify that

$$\mathbf{E}\exp\{\, it\nu_d(V_n^-(a))^{-1/2}S(V_n^-(a))\,\} \to \exp\{\,-\sigma^2t^2/2\,\} \text{ as } n\to\infty. \quad (10.65)$$

Step 2. This step is to construct for each $j \in \mathbb{Z}^d$ the auxiliary block $\tilde{\Delta}_j(a,m)$ inside $\Delta_j(a)$ in such a way that the law of $\nu_d(V_n^-(a))^{-1/2}S(V_n^-(a))$ is close enough to the one of $\nu_d(W_n(a,m))^{-1/2}S(W_n(a,m))$ for large n where

$$W_n(a,m) = \bigcup_{j\in J^-(V_n,a)} \tilde{\Delta}_j(a,m)$$

and at the same time the asymptotical behaviour of $\nu_d(W_n(a,m))^{-1/2}S(\tilde{\Delta}_j(a,m))$ is similar to that which the normalized sums of independent random variables demonstrate whenever parameters a and m (that is r and m) are chosen in appropriate way. For $r,m\in\mathbb{N}$, $r>2$ and $j=(j_1,\dots,j_d)^\top\in\mathbb{Z}^d$ introduce

$$\tilde{\Delta}_j(a,m) = \{\, x\in\mathbb{R}^d : mrj_k + m < x_k \le mr(j_k+1) - m,\ k=1,\dots,d\,\}.$$

Thus, for any fixed $t\in\mathbb{R}$, we show that $\mathbf{E}\exp\{\, it\nu_d(V_n^-(a))^{-1/2}S(V_n^-(a))\,\}$ can be approximated by $\mathbf{E}\exp\{\, it\nu_d(W_n(a,m))^{-1/2}S(W_n(a,m))\,\}$ when n is large enough. Clearly

$$\begin{aligned}\nu_d(V_n^-(a)\setminus W_n(a,m)) &= |J^-(V_n,a)|(\nu_d(\Delta_0(a)) - \nu_d(\tilde{\Delta}_0(a)))\\ &= \nu_d(V_n^-(a))\frac{\nu_d(\Delta_0(a)) - \nu_d(\tilde{\Delta}_0(a))}{\nu_d(\Delta_0(a))}\\ &\le \nu_d(V_n^-(a))\frac{(rm)^d - ((r-2)m)^d}{(rm)^d}\\ &\le \nu_d(V_n^-(a))\frac{2d}{r}.\end{aligned}$$

The same reasoning that was used to prove (10.62) and (10.64) leads for any $\varepsilon>0$, $t\in\mathbb{R}$ and $m,n\in\mathbb{N}$ to the estimate

$$\begin{aligned}&\left|\mathbf{E}\exp\{\, it\nu_d(V_n^-(a))^{-1/2}S(V_n^-(a))\,\} - \mathbf{E}\exp\{\,\nu_d(W_n(a,m))^{-1/2}S(W_n(a,m))\,\}\right|\\ &\le 2|t|\left(2d(\mathbf{E}\xi_0^2+\theta_1)/r\right)^{1/2} < \varepsilon\end{aligned}$$

if r is large enough.

Step 3. Now we reduce the problem to the study of auxiliary independent random variables. Write $N_n = |J^-(V_n,a)|$ and enumerate a family of random variables

$$\nu_d(\tilde{\Delta}_j(a,m))^{-1/2}S(\tilde{\Delta}_j(a,m)), \quad j \in J^-(V_n,a),$$

to obtain the random variables $\eta_{n,1},\dots,\eta_{n,N_n}$ (obviously $\eta_{n,k} = \eta_{n,k}(a,m)$). Then $\nu_d(W_n) = N_n\nu_d(\tilde{\Delta}_0(a,m))$ and

$$\mathbf{E}\exp\{it\nu_d(W_n)^{-1/2}S(W_n)\} = \mathbf{E}\exp\left\{itN_n^{-1/2}\sum_{k=1}^{N_n}\eta_{n,k}\right\}.$$

Note that $|\mathbf{E}\exp\{it\eta\}| \le 1$ for all $t \in \mathbb{R}$ and a (real-valued) random variable η. Recall that the covariance for complex-valued random variables η and ζ is defined by formula

$$\mathbf{cov}(\eta,\zeta) := \mathbf{E}\eta\overline{\zeta} - \mathbf{E}\eta\mathbf{E}\overline{\zeta} \tag{10.66}$$

where $\overline{\zeta}$ stands for the random variable conjugate to ζ. Thus for all n large enough (i.e. for $N_n > 1$)

$$\left|\mathbf{E}\exp\left\{itN_n^{-1/2}\sum_{k=1}^{N_n}\eta_{n,k}\right\} - \prod_{k=1}^{N_n}\mathbf{E}\exp\left\{itN_n^{-1/2}\eta_{n,k}\right\}\right|$$
$$\le \sum_{r=1}^{N_n-1}\left|\mathbf{cov}\left(\exp\left\{itN_n^{-1/2}\eta_{n,r}\right\}, \exp\left\{-itN_n^{-1/2}\sum_{k=r+1}^{N_n}\eta_{n,k}\right\}\right)\right|. \tag{10.67}$$

Obviously $\operatorname{dist}(\tilde{\Delta}_j(a,m),\tilde{\Delta}_k(a,m)) \ge 2m$ for all $j,k \in \mathbb{Z}^d$, $j \ne k$. Furthermore, $\operatorname{Lip}(\cos(\cdot)) = \operatorname{Lip}(\sin(\cdot)) = 1$. The Lipschitz constant of the composition of Lipschitz functions is estimated by the product of their Lipschitz constants. Taking into account (BL,θ)-dependence, for any $\varepsilon > 0$, $n,r \in \mathbb{N}$ and $r > 2$, the right-hand side of (10.67) admits the estimate

$$4t^2N_n\frac{\theta_{2m}\nu_d(\Delta_0(a,m))}{N_n\nu_d(\Delta_0(a,m))} = 4t^2\theta_{2m} < \varepsilon \tag{10.68}$$

when m is large enough. The factor 4 appeared as we used the Euler formula $\exp i\alpha = \cos\alpha + i\sin\alpha$ for $\alpha \in \mathbb{R}$, $i^2 = -1$, and separated four summands in (10.66). We used also the relation

$$\min\left\{\left|\Delta_r(a,m)\cap\mathbb{Z}^d\right|, \left|\bigcup_{k=1}^{N_n}\Delta_k(a,m)\cap\mathbb{Z}^d\right|\right\}$$
$$= \left|\Delta_0(a,m)\cap\mathbb{Z}^d\right| = \nu_d(\Delta_0(a,m)).$$

Introduce a decoupled version $(\zeta_{n,1},\dots,\zeta_{n,N_n})^\top$ of a vector $(\eta_{n,1},\dots\eta_{n,N_n})^\top$. Then

$$\prod_{k=1}^{N_n} \mathbf{E}\exp\{itN_n^{-1/2}\eta_{n,k}\} = \prod_{k=1}^{N_n} \mathbf{E}\exp\{itN_n^{-1/2}\zeta_{n,k}\}$$
$$= \mathbf{E}\exp\left\{itN_n^{-1/2}\sum_{k=1}^{N_n}\zeta_{n,k}\right\}. \tag{10.69}$$

Step 4. Now we explain why it suffices to prove the CLT for independent copies of random variables $\eta_{n,1},\dots,\eta_{n,N_n}$. Note that

$$N_n^{-1/2}\sum_{k=1}^{N_n}\zeta_{N,k} \xrightarrow{d} \zeta \sim \mathcal{N}(0,\sigma^2(a,m)) \text{ as } n\to\infty \tag{10.70}$$

where $\xrightarrow{d}$ means the convergence in law, $\mathcal{N}(0,\sigma^2(a,m))$ stands for the normal distribution with parameters 0 and

$$\sigma^2(a,m) = \mathbf{var}\,\zeta_{n,k} = \mathbf{var}\,\eta_{n,k} = \mathbf{var}\,S(\Delta_0(a,m))/\nu_d(\Delta_0(a,m)),\ n\in\mathbb{N},\ k=1,\dots,N_n.$$

Relation (10.70) is the simple variant of the CLT for an array of i.i.d square-integrable random variables. Theorem 10.14 yields that for each $m\in\mathbb{N}$ and $a = mr$

$$\sigma^2(a,m)\to\sigma^2 \text{ as } r\to\infty.$$

Thus to guarantee (for any fixed $t\in\mathbb{R}$) the validity of (10.65) we take arbitrary $\varepsilon>0$ and find $m\in\mathbb{N}$ large enough to ensure (10.68). Then we choose $r\in\mathbb{N}$ large enough to obtain

$$|\exp\{-\sigma^2(a,m)t^2/2\} - \exp\{-\sigma^2t^2/2\}| < \varepsilon.$$

The proof is complete in view of (10.67)–(10.70). □

Exercise 10.12. Let the conditions of Theorem 10.33 be satisfied and $\sigma^2>0$. Show that, for any sequence of regularly growing sets $U_n\subset\mathbb{Z}^d$, one has

$$S(U_n)/\sqrt{\mathbf{var}\ S(U_n)} \xrightarrow{d} N(0,1),\quad n\to\infty.$$

Remark 10.10. Let the conditions of Theorem 10.33 be satisfied. Then (10.60) holds. Assume that $\sigma^2\neq 0$ and $(\hat{\sigma}_n)_{n\in\mathbb{N}}$ is a sequence of consistent estimates for σ, i.e. $\hat{\sigma}_n \xrightarrow{d} \sigma$ as $n\to\infty$ (or $(\hat{\sigma}_n^2)_{n\in\mathbb{N}} \xrightarrow{d} \sigma^2$ as $n\to\infty$ because the function $f(x)=\sqrt{x}$, $x\in\mathbb{R}_+$, is continuous). Then in view of Slutsky's lemma we conclude that

$$\frac{S(U_n) - \mathbf{E}\xi(o)\,|U_n|}{\hat{\sigma}_n\sqrt{|U_n|}} \xrightarrow{d} \mathcal{N}(0,1),\quad n\to\infty. \tag{10.71}$$

10.3.2 Corollaries of the CLT

Cramér–Wald device permits to obtain the following result from Theorem 10.33.

Corollary 10.5 ([96, p. 180]). *Let $\xi = \{\xi(j),\ j \in \mathbb{Z}^d\}$ be a (BL,θ)-dependent, strictly stationary random field taking values in $\mathbb{R}^k$ and such that $\mathbf{E}\|\xi(o)\|^2 < \infty$. Then, for any sequence of regularly growing sets $U_n \subset \mathbb{Z}^d$, one has*

$$|U_n|^{-1/2} \sum_{j \in U_n} \xi(j) \xrightarrow{d} \mathcal{N}(o, C) \ \ as\ n \to \infty.$$

Here C is the $(k \times k)$-matrix having the elements

$$C_{l,m} = \sum_{j \in \mathbb{Z}^d} \mathbf{cov}\,(\xi_l(o), \xi_m(j)), \quad l, m = 1, \dots, k.$$

The classical Newman's CLT can be deduced from Theorem 10.33. To clarify it we introduce for $n \in \mathbb{N}$, $j = (j_1, \dots, j_d)^\top$ and $k = (k_1, \dots, k_d)^\top$

$$B_k^{(n)} := \{\, j \in \mathbb{Z}^d : nk_l < j_l \le n(k_l + 1), l = 1, \dots, d\,\}, \ \ \eta^{(n)}(k) := n^{-d/2} S\left(B_k^{(n)}\right).$$

Theorem 10.34 ([378]). *Let $\xi = \{\xi(j),\ j \in \mathbb{Z}^d\}$ be a centered, strictly stationary, associated random field such that $\mathbf{E}\xi(o)^2 < \infty$ and*

$$\sigma^2 = \sum_{j \in \mathbb{Z}^d} \mathbf{cov}\,(\xi(o), \xi(j)) < \infty.$$

Then the finite-dimensional distributions of the field $\{\eta^{(n)}(k),\ k \in \mathbb{Z}^d\}$ converge, as $n \to \infty$, to the corresponding ones of the field $\zeta = \{\zeta(k),\ k \in \mathbb{Z}^d\}$ consisting of independent $\mathcal{N}(0, \sigma^2)$ random variables.

Proof. Let $m \in \mathbb{N}$ and $k_1, \dots, k_m \in \mathbb{Z}^d$. For $t = (t_1, \dots, t_m)^\top \in \mathbb{R}^m$ with $\|t\|_2^2 = t_1^2 + \dots + t_m^2$ and $i^2 = -1$, we have

$$\left| \mathbf{E} e^{i \sum_{r=1}^m t_r \eta^{(m)}(k_r)} - e^{\frac{\sigma^2 \|t\|^2}{2}} \right|$$
$$\le \left| \mathbf{E} e^{i \sum_{r=1}^m t_r \eta^{(m)}(k_r)} - \prod_{r=1}^m \mathbf{E} e^{i t_r \eta^{(n)}(k_r)} \right| + \left| \prod_{r=1}^m \mathbf{E} e^{i t_r \eta^{(n)}(k_r)} - e^{\frac{\sigma^2 \|t\|^2}{2}} \right|.$$

The second term in the right-hand side of the last inequality converges to zero as $\eta^{(n)}(k) \xrightarrow{d} \zeta(k)$ for any $k \in \mathbb{Z}^d$ due to Theorem 10.33 (the field ξ is (BL,θ)-dependent with $\theta_m = u(m)$ where $u(m)$ is the Cox–Grimmett coefficient). For the first term by Corollary 10.1 one has

$$\begin{aligned}
&\left|\mathbf{E}\exp\left\{i\sum_{r=1}^{m} t_r \eta^{(m)}(k_r)\right\} - \prod_{r=1}^{m}\mathbf{E}\exp\left\{i t_r \eta^{(n)}(k_r)\right\}\right| \\
&\le 4 \sum_{1\le r,v\le m, r\neq v} |t_r t_v|\, \mathbf{cov}(\eta^{(n)}(k_r), \eta^{(n)}(k_v)) \\
&\le 4\|t\|_\infty^2 n^{-d}\left(\mathbf{var}\sum_{r=1}^{m} S(B_{k_r}^{(n)}) - \sum_{r=1}^{m}\mathbf{var}\, S(B_{k_r}^{(n)})\right)
\end{aligned}$$

where $\|t\|_\infty := \max_{1\le r\le m}|t_r|$. Note that $\cup_{r=1}^{m} B_{k_r}^{(n)}$ tends to infinity in the Van Hove sense as $n \to \infty$. Therefore, by virtue of Theorem 10.14

$$n^{-d}\left(\mathbf{var}\sum_{r=1}^{m} S(B_{k_r}^{(n)}) - \sum_{r=1}^{m}\mathbf{var}\, S(B_{k_r}^{(n)})\right) \to 0 \text{ as } n \to \infty.$$

The proof is complete. □

10.3.3 Application to the Excursion Sets

Definition 10.18. Let ξ be a measurable real-valued function on $\mathbb{R}^d$ and $T \subset \mathbb{R}^d$ be a measurable subset. Then for each $u \in \mathbb{R}$

$$A_u(\xi, T) = \{t \in T : \xi(t) \ge u\}$$

is called the *excursion set* of ξ in T over the level u.

For a real-valued random field $\xi = \{\xi(t), t \in \mathbb{R}^d\}$ we assume the measurability of $\xi(\cdot)$ as a function on $\mathbb{R}^d \times \Omega$ endowed with the σ-algebra $\mathcal{B}(\mathbb{R}^d) \otimes \mathcal{A}$. Thus

$$\nu_d(A_u(\xi, T)) = \int_T \mathbf{1}(\xi(t) \ge u)\, dt$$

is a random variable for each $u \in \mathbb{R}$ and any measurable set $T \subset \mathbb{R}^d$.

We assume for random field $\xi = \{\xi(t), t \in \mathbb{R}^d\}$ (see [97]) one of the following conditions.

(A) Let ξ be quasi-associated and strictly stationary such that $\xi(o)$ has a bounded density. Assume that a covariance function is continuous and there exists some $\alpha > 3d$ such that

$$|\mathbf{cov}(\xi(o), \xi(t))| = O\left(\|t\|_2^{-\alpha}\right) \text{ as } \|t\|_2 \to \infty. \tag{10.72}$$

(B) Let ξ be Gaussian and stationary. Suppose that its continuous covariance function satisfies (10.72) for some $\alpha > d$.

Note that continuity of a covariance function of ξ implies the existence of measurable modification of this field. We shall only consider such versions of ξ. Clearly one can write $\|\cdot\|_\infty$ in (10.72) instead of $\|\cdot\|_2$ as all the norms in $\mathbb{R}^d$ are equivalent. We also exclude the trivial case when $\xi(t) = \text{const}$ a.s. for all $t \in \mathbb{R}$. Introduce

$$S_{\mathbf{u}}(\xi, W_n) = (\nu_d(A_{u_1}(\xi, W_n)), \dots, \nu_d(A_{u_r}(\xi, W_n)))^\top, \quad n \in \mathbb{N}, \tag{10.73}$$

where $\mathbf{u} = (u_1, \dots, u_r)^\top \in \mathbb{R}^r$.

Theorem 10.35 ([97]). *Let $\xi = \{\xi(t),\ t \in \mathbb{R}^d\}$ be a random field satisfying condition* (A). *Then, for each $\mathbf{u} = (u_1, \dots, u_r)^\top \in \mathbb{R}^r$ and any VH-growing sequence $(W_n)_{n\in\mathbb{N}}$ of subsets of $\mathbb{R}^d$, one has*

$$\nu_d(W_n)^{-1/2}(S_{\mathbf{u}}(\xi, W_n) - \nu_d(W_n)\,\mathbf{P}(\mathbf{u})) \xrightarrow{d} V_{\mathbf{u}} \sim \mathcal{N}(o, \Sigma(\mathbf{u})) \text{ as } n \to \infty \tag{10.74}$$

where

$$\mathbf{P}(\mathbf{u}) = (\mathbf{P}(\xi(o) \geq u_1), \dots, \mathbf{P}(\xi(o) \geq u_r))^\top \tag{10.75}$$

and $\Sigma(\mathbf{u}) = (\sigma_{lm}(\mathbf{u}))_{l,m=1}^r$ is an $(r \times r)$-matrix having the elements

$$\sigma_{lm}(\mathbf{u}) = \int_{\mathbb{R}^d} \mathbf{cov}(\mathbf{1}(\xi(o) \geq u_l), \mathbf{1}(\xi(t) \geq u_m))\,dt. \tag{10.76}$$

The following theorem is a generalization of the result [268, p. 80].

Theorem 10.36 ([97]). *Let $\xi = \{\xi(t),\ t \in \mathbb{R}^d\}$ be a random field satisfying condition* (B) *and $\xi(o) \sim \mathcal{N}(a, \tau^2)$. Then, for each $\mathbf{u} = (u_1, \dots, u_r)^\top \in \mathbb{R}^r$ and any sequence $(W_n)_{n\in\mathbb{N}}$ of VH-growing subsets of $\mathbb{R}^d$, one has*

$$\nu_d(W_n)^{-1/2}(S_{\mathbf{u}}(\xi, W_n) - \nu_d(W_n)\,\Psi(\mathbf{u})) \xrightarrow{d} V_{\mathbf{u}} \sim \mathcal{N}(o, \Sigma(\mathbf{u})) \text{ as } n \to \infty. \tag{10.77}$$

Here $\Psi(\mathbf{u}) = (\Psi((u_1 - a)/\tau), \dots, \Psi((u_r - a)/\tau))^\top$ and $\Sigma(\mathbf{u}) = (\sigma_{lm}(\mathbf{u}))_{l,m=1}^r$ is an $(r \times r)$-matrix having the elements

$$\sigma_{lm}(\mathbf{u}) = \frac{1}{2\pi}\int_{\mathbb{R}^d}\int_0^{\rho(t)} g_{lm}(r)\,dr\,dt \tag{10.78}$$

where

$$g_{lm}(r) = \frac{1}{\sqrt{1-r^2}} \times \exp\left\{-\frac{(u_l - a)^2 - 2r(u_l - a)(u_m - a) + (u_m - a)^2}{2\tau^2(1-r^2)}\right\} \tag{10.79}$$

and $\rho(t) = \mathbf{corr}(\xi(o), \xi(t))$. *If* $\Sigma(\mathbf{u})$ *is nondegenerate, we obtain by virtue of* (10.77)

$$\nu_d(W_n)^{-1/2}\Sigma(\mathbf{u})^{-1/2}(S_{\mathbf{u}}(\xi, W_n) - \nu_d(W_n)\,\Psi(\mathbf{u})) \xrightarrow{d} \mathcal{N}(o, \mathrm{I}), \quad n \to \infty,$$

here I *is the* $(r \times r)$*-unit matrix.*

Theorems 10.35–10.36 are generalized to hold for a large subclass of (PA or NA) stochastically continuous stationary random fields (possibly without the finite second moment) in the paper [286]. Examples of fields belonging to this subclass are infinitely divisible random fields with an integral spectral representation (see Sect. 9.7.4) satisfying some conditions, max- and α-stable random fields with the corresponding tail behaviour of their kernel functions.

A functional limit theorem (the so-called invariance principle, where the level u is not fixed and interpreted as a new variable) for the volume of the excursion sets of stationary associated random fields with smooth realizations is proved in [353]. A limit theorem for the perimeter of the excursion sets of smooth stationary Gaussian random fields in two dimensions is established in [313]. The corresponding functional limit theorem has been recently given in [354]. The asymptotic behaviour of other related functionals of Gaussian random fields is considered in [314].

10.3.4 The Newman Conjecture and Some Extensions of CLT

In [379] Newman formulated the conjecture concerning the CLT for strictly stationary associated random field $\xi = \{\xi(j),\ j \in \mathbb{Z}^d\}$ with finite second moment. Namely, he wrote that instead of the finite susceptibility condition (10.20) it is sufficient (when one proves the CLT for partial sums over growing blocks) to use the hypothesis that the function $\mathsf{K}_\xi(\cdot)$ introduced in (10.32) is slowly varying.

Unfortunately, this elegant hypothesis is not true. The first counterexample was constructed by Herrndorf in the case $d = 1$.

Theorem 10.37 ([243]). *There exists a strictly stationary associated, centered sequence* $(\xi_j)_{j\in\mathbb{Z}}$ *such that* $\mathsf{K}_\xi(n) \sim \log n$ *as* $n \to \infty$ *and* $S_n/\sqrt{n\mathsf{K}_\xi(n)}$ *do not have any nondegenerate limit law as* $n \to \infty$ *(as usual,* $S_n = \xi_1 + \ldots + \xi_n$*).*

This theorem can be obtained as a corollary from the next result by Shashkin.

Theorem 10.38 ([469]). *Let* L *be a nondecreasing slowly varying function such that* $L(n) \to \infty$ *as* $n \to \infty$*. Then, for any positive unbounded sequence* $(b_n)_{n\in\mathbb{N}}$*, there exists a strictly stationary random sequence* $(\xi_j)_{j\in\mathbb{N}} \in \mathsf{A}$ *with properties*

i) $\mathbf{E}\xi_1 = 0$ *and* $\mathbf{E}\xi_1^2 = 1$*;*
ii) $\mathsf{K}_\xi(n) \sim L(n)$ *as* $n \to \infty$*;*
iii) $(S_n/\sqrt{nb_n})_{n\in\mathbb{N}}$ *do not have any nondegenerate limit in law as* $n \to \infty$*.*

Note that if $(\xi_j)_{j\in\mathbb{N}} \in \mathsf{A}$ then $\mathsf{K}_\xi(\cdot)$ is a nondecreasing function. Thus the assumption in Theorem 10.38 that L is nondecreasing looks natural.

However, it is possible to establish the CLT without the finite susceptibility assumption. To formulate it we recall the following

Definition 10.19. A family of real-valued random variables $\{\,\eta(t),\ t \in T\,\}$ is *uniformly integrable* if $\lim_{c\to\infty}\sup_{t\in T}\mathbf{E}|\eta(t)|\mathbf{1}(|\eta(t)|\geq c) = 0$.

Exercise 10.13. Let a family $\{\,\xi(n),\ n \in \mathbb{N}^d\,\}$ be uniformly integrable. Prove that, for any nonrandom family $\{\,c_n, n \in \mathbb{N}^d\,\}$ such that $c_n \to \infty$ as $n = (n_1,\dots,n_d)^\top \to \infty$ (i.e. $n_1 \to \infty,\dots,n_d \to \infty$), the following relation holds

$$\mathbf{E}|\xi(n)|\mathbf{1}(|\xi(n)|\geq c_n) \to 0,\ \ n\to\infty. \tag{10.80}$$

We also propose the extension of well-known result (see for example [69, Theorems 3.5–3.6]).

Lemma 10.4. *Let $\{\,\eta(n),\ n \in \mathbb{Z}^d\,\}$ be a family of uniformly integrable random variables. Suppose that $\eta(n) \xrightarrow{d} \eta$ as $n \to \infty$. Then η is integrable and*

$$\mathbf{E}\eta(n) \to \mathbf{E}\eta,\ \ n\to\infty. \tag{10.81}$$

Moreover, if $\{\,\eta(n),\ n \in \mathbb{Z}^d\,\}$ is a family of nonnegative random variables such that $\eta(n) \xrightarrow{d} \eta$ as $n \to \infty$ where η is integrable and (10.81) *holds then the family $\{\,\eta(n),\ n \in \mathbb{Z}^d\,\}$ is uniformly integrable.*

Set $\langle n\rangle = n_1\cdot\ldots\cdot n_d$, $U_n = [-n,n] = \prod_{k=1}^n[-n_k,n_k]$, $S_n := S(U_n)$ for $n\in\mathbb{N}^d$. The following result is a generalization of one by Lewis [329] established for sequences of random variables.

Theorem 10.39 ([93]). *Let $\xi = \{\,\xi(j),\ j \in \mathbb{Z}^d\,\}$ be a strictly stationary* PA *random field, $0 < \mathbf{E}\xi(o)^2 < \infty$ and $\mathsf{K}_\xi(\cdot) \in \mathcal{L}(\mathbb{N}^d)$. Then ξ satisfies CLT, that is*

$$\frac{S_n - \mathbf{E}S_n}{\sqrt{\mathbf{var}\ S_n}} \xrightarrow{d} \zeta \sim \mathcal{N}(0,1)\ \ as\ \ n\to\infty \tag{10.82}$$

iff the family $\left\{(S_n-\mathbf{E}S_n)^2/(\langle n\rangle\mathsf{K}_\xi(n)),\ n\in\mathbb{N}^d\right\}$ is uniformly integrable.

Proof. Necessity. Assume that (10.82) holds. Then

$$(S_n-\mathbf{E}S_n)^2/\,\mathbf{var}\,S_n \xrightarrow{d} \zeta^2\ \text{ as }\ n\to\infty$$

(if $\eta_n \xrightarrow{d} \eta$ and h is a continuous function then $h(\eta_n) \xrightarrow{d} h(\eta)$). Obviously $(S_n-\mathbf{E}S_n)^2/\,\mathbf{var}\,S_n \geq 0$, $\mathbf{E}(S_n-\mathbf{E}S_n)^2/\,\mathbf{var}\,S_n = 1 = \mathbf{E}\zeta^2$, $n\in\mathbb{N}^d$. Lemma 10.4 yields that the family $\{\,(S_n-\mathbf{E}S_n)^2/\,\mathbf{var}\,S_n\,\}$ is uniformly integrable. Due to Theorem 10.15, one has $\mathbf{var}\,S_n \sim \langle n\rangle\mathsf{K}_\xi(n)$ as $n\to\infty$. Therefore a family $\{\,(S_n-\mathbf{E}S_n)^2/(\langle n\rangle\mathsf{K}_\xi(n)),\ n\in\mathbb{N}^d\,\}$ is uniformly integrable as well.

Sufficiency. We shall assume that K_ξ is unbounded, as otherwise (10.82) follows from Theorem 10.33. With a slight abuse of notation, set $\mathsf{K}_\xi(t) := \mathsf{K}_\xi([t] \vee \mathbf{1})$ for $t \in \mathbb{R}_+^d$ and $\mathbf{1} = (1,\dots,1)^\top \in \mathbb{R}^d$ where $[t] \vee 1$ is understood componentwise. Then this extension of function K_ξ belongs to $\mathcal{L}(\mathbb{R}^d)$ as K_ξ is coordinate-wise nondecreasing on $\mathbb{N}^d$. □

Lemma 10.5. *There exists a family* $\{ q_n,\ n \in \mathbb{Z}^d \}$ *of nonrandom vectors* $q_n \in \mathbb{N}^d$, $q_n = (q_n^{(1)},\dots,q_n^{(d)})^\top$ *such that*

$$q_n^{(k)} \le n_k,\ \ q_n^{(k)}/n_k \to 0 \ \textit{for}\ k = 1,\dots,d,\ \ q_n \to \infty \ \textit{as}\ n \to \infty, \tag{10.83}$$

$$\mathsf{K}_\xi(n)/\mathsf{K}_\xi(q_n) \to 1,\ \ n \to \infty. \tag{10.84}$$

Proof. For any $L = (L^{(1)},\dots,L^{(d)})^\top \in \mathbb{N}^d$ we can choose $N_0(L)$ such that for any $n \ge N_0(L)$

$$\frac{\mathsf{K}_\xi(n_1,\dots,n_d)}{\mathsf{K}_\xi\left(\frac{n_1}{L^{(1)}},\dots,\frac{n_d}{L^{(d)}}\right)} - 1 \le \frac{1}{\langle L \rangle}.$$

Take a sequence $(L(r))_{r\in\mathbb{N}}$ such that $L(r) \in \mathbb{N}^d$ for all $r \in \mathbb{N}$ and $L(r) < L(r+1)$ for all $r \in \mathbb{N}$. Here $(a^{(1)},\dots,a^{(d)})^\top < (b^{(1)},\dots,b^{(d)})^\top$ means $a^{(k)} < b^{(k)}$ for $k = 1,\dots,d$. Set $M_0(1) = N_0(L(1))$ and $M_0(r+1) = (M_0(r) \vee N_0(L(r+1))) + \mathbf{1}$ for $r \in \mathbb{N}$ where

$$(a^{(1)},\dots,a^{(d)})^\top \vee (b^{(1)},\dots,b^{(d)})^\top = (a^{(1)} \vee b^{(1)},\dots,a^{(d)} \vee b^{(d)})^\top.$$

Then $M_0(r) < M_0(r+1)$ for all $r \in \mathbb{N}$. If $n = (n_1,\dots,n_d)^\top \ge M_0(r)$ then

$$\frac{\mathsf{K}_\xi(n_1,\dots,n_d)}{\mathsf{K}_\xi(\frac{n_1}{L^{(1)}(r)},\dots,\frac{n_d}{L^{(d)}(r)})} - 1 \le \frac{1}{\langle L(r) \rangle}.$$

Introduce the nonrandom family of numbers $\{ \varepsilon_n,\ n \in \mathbb{N}^d \}$ in the following way. Put $\varepsilon_{n_k} = 1/L^{(k)}(r)$ for $k = 1,\dots,d$ and $n = (n_1,\dots,n_d)^\top \in \mathbb{N}^d$ such that $M_0(r) \le n < M_0(r+1)$, $r \in \mathbb{N}$.

Take any $\varepsilon > 0$ and find $r_0 \in \mathbb{N}$ such that $1/\langle L(r_0) \rangle < \varepsilon$. Then for any n such that $M_0(r) \le n < M_0(r+1)$ where $r \ge r_0$ one has

$$1 \le \frac{\mathsf{K}_\xi(n_1,\dots,n_d)}{\mathsf{K}_\xi(n_1\varepsilon_{n_1},\dots,n_d\varepsilon_{n_d})} = \frac{\mathsf{K}_\xi(n_1,\dots,n_d)}{\mathsf{K}_\xi(n_1/L^{(1)}(r),\dots,n_d/L^{(d)}(r))}$$
$$\le 1 + \frac{1}{\langle L(r) \rangle} \le 1 + \frac{1}{\langle L(r_0) \rangle} \le 1 + \varepsilon.$$

Now we can take

$$q_n = (n_1\varepsilon_{n_1},\dots,n_d\varepsilon_{n_d})^\top \vee ([\log n_1],\dots,[\log n_d])^\top$$

to guarantee (10.83)–(10.84). The proof of the lemma is complete. □

It is not difficult to find a nonrandom family $\{ p_n,\ n \in \mathbb{N}^d \}$ with p_n taking values in $\mathbb{N}$ such that

$$q_n^{(k)} \le p_n^{(k)} \le n_k,\ \ q_n^{(k)}/p_n^{(k)} \to 0 \text{ and } p_n^{(k)}/n_k \to 0, \tag{10.85}$$

for $k = 1, \dots, d$, as $n \to \infty$. Now we use the well-known sectioning device by Bernstein. For $n, j \in \mathbb{N}^d$ and p_n, q_n introduced above, consider the blocks $U_n^{(j)} \subset \mathbb{N}^d$ with elements $r = (r_1, \dots, r_d)^\top$ such that

$$(j_k - 1)(q_n^{(k)} + p_n^{(k)}) < r_k \le (j_k - 1)q_n^{(k)} + j_k p_n^{(k)},\ k = 1, \dots, d.$$

Let $J_n = \{ j \in \mathbb{N}^d : U_n^{(j)} \subset U_n \}$. Set $W_n = \cup_{j \in J_n} U_n^{(j)}$, $G_n := U_n \setminus W_n$, $n \in \mathbb{N}^d$.

In other words, W_n consists of the union of "big rooms" (having the "size" $p_n^{(k)}$ along the k-th coordinate axis for $k = 1, \dots, d$) and separated by "corridors" belonging to G_n. See Fig. 10.3. We write $v_n = \sqrt{\langle n \rangle \mathsf{K}_\xi(n)}$, $n \in \mathbb{N}^d$. Then for each $t \in \mathbb{R}$ and $n \in \mathbb{N}^d$ one has

$$\begin{aligned}
\left| \mathbf{E} \exp\left\{ \frac{it}{v_n} S_n \right\} - e^{\frac{-t^2}{2}} \right| &\le \left| \mathbf{E} \exp\left\{ \frac{it}{v_n} S_n \right\} - \mathbf{E} \exp\left\{ \frac{it}{v_n} \sum_{j \in J_n} S(U_n^{(j)}) \right\} \right| \\
&+ \left| \mathbf{E} \exp\left\{ \frac{it}{v_n} \sum_{j \in J_n} S(U_n^{(j)}) \right\} - \prod_{j \in J_n} \mathbf{E} \exp\left\{ \frac{it}{v_n} S(U_n^{(j)}) \right\} \right| \\
&+ \left| \prod_{j \in J_n} \mathbf{E} \exp\left\{ \frac{it}{v_n} S(U_n^{(j)}) \right\} - e^{\frac{-t^2}{2}} \right| =: \sum_{m=1}^{3} Q_m
\end{aligned}$$

where $i^2 = -1$ and $Q_m = Q_m(n, t)$. As $|e^{ix} - e^{iy}| \le |x - y|$ for $x, y \in \mathbb{R}$, then using the Lyapunov inequality, we get

$$Q_1 \le \frac{|t|}{v_n} \mathbf{E} |S(G_n)| \le \frac{|t|}{v_n} \left(\mathbf{E} S(G_n)^2 \right)^{1/2}.$$

The covariance function of a field $\xi \in \mathsf{PA}$ is nonnegative. Thus using the wide sense stationarity of ξ one has

$$\begin{aligned}
\mathbf{E} S(G_n)^2 &\le \sum_{j \in G_n} \sum_{-n \le r \le n} \mathbf{cov}\,(\xi(j), \xi(r)) \le |G_n| \mathsf{K}_\xi(n) \\
&\le \mathsf{K}_\xi(n) \sum_{k=1}^{d} (m_n^{(k)} q_n^{(k)} + p_n^{(k)} + q_n^{(k)}) \prod_{1 \le l \le d,\, l \ne k} n_l
\end{aligned}$$

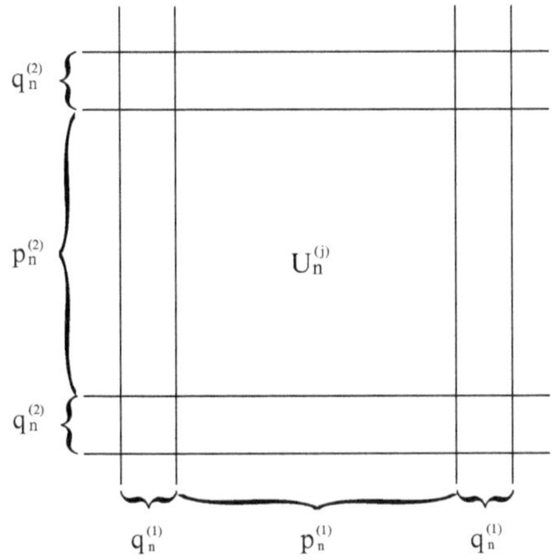

Fig. 10.3 Bernstein's sectioning technique for $d = 2$

where $m_n^{(k)} = [n_k/(p_n^{(k)} + q_n^{(k)})]$, $k = 1, \dots, d$. Due to (10.83)–(10.85) we come to the inequality

$$\frac{\mathbf{E}S(G_n)^2}{\langle n\rangle \mathsf{K}_\xi(n)} \le \sum_{k=1}^{d} \frac{m_n^{(k)} q_n^{(k)} + p_n^{(k)} + q_n^{(k)}}{n_k} \to 0, \quad n \to \infty.$$

Therefore, $Q_1(n,t) \to 0$ for each $t \in \mathbb{R}$ as $n \to \infty$.

For any $n \in \mathbb{N}^d$, the family $\{ S(U_n^{(j)}),\ j \in J_n \} \in \mathsf{PA}$ in view of Remark 10.3. Let us enumerate this family to obtain a collection $\{ \eta_{n,s},\ s = 1, \dots, M_n \}$ where $M_n = |J_n| = m_n^{(1)} \cdot \ldots \cdot m_n^{(d)}$. Now Theorem 10.12 and Corollary 10.1 yield

$$\begin{aligned} Q_2 &\le \sum_{s=1}^{M_n - 1} \left| \mathbf{cov}\left(\exp\left\{ \frac{it}{v_n} \eta_{n,s} \right\}, \exp\left\{ -\frac{it}{v_n} \sum_{l=s+1}^{M_n} \eta_{n,s} \right\} \right) \right| \\ &\le \frac{4t^2}{v_n^2} \sum_{1 \le s,u \le M_n,\, s \ne u} \mathbf{cov}(\eta_{n,s}, \eta_{n,u}) \le \frac{4t^2}{\langle n\rangle \mathsf{K}_\xi(n)} \sum_{j \in U_n} \sum_{r \in U_n:\, \|r-j\|_\infty > q_n} \mathbf{cov}(\xi(j), \xi(r)). \end{aligned}$$

Obviously, for any $j \in U_n$

$$\{ r \in U_n : \|r - j\|_\infty > q_n \} \subset \{ r : j - n \le r \le j + n \} \setminus \{ r : \|r - j\|_\infty \le q_n \}.$$

Consequently,

$$\sum_{j \in U_n} \sum_{r \in U_n:\, \|r-j\|_\infty > q_n} \mathbf{cov}(\xi(j), \xi(r)) \le \langle n\rangle (\mathsf{K}_\xi(n) - \mathsf{K}_\xi(q_n)),$$

and (10.84) implies that $Q_2(n,t) \to 0$ as $n \to \infty$ (for each $t \in \mathbb{R}$).

Now for each $n \in \mathbb{N}^d$ introduce $(\zeta_{n,1}, \ldots, \zeta_{n,M_n})^\top$ as a decoupled version of the vector $\frac{1}{v_n}(\eta_{n,1}, \ldots, \eta_{n,M_n})^\top$. By Theorem 10.15 for every $s = 1, \ldots, M_n$

$$\mathbf{var}\, \zeta_{n,s} = \mathbf{var}\, \zeta_{n,1} \sim \langle p_n \rangle \mathsf{K}_\xi(p_n)/\langle n \rangle \mathsf{K}_\xi(n), \quad n \to \infty.$$

Therefore,

$$\sum_{s=1}^{M_n} \mathbf{var}\, \zeta_{n,s} = M_n\, \mathbf{var}\, \zeta_{n,1} \to 1, \quad n \to \infty$$

because $M_n \langle p_n \rangle = \prod_{k=1}^d [n_k/(p_n^{(k)} + q_n^{(k)})] p_n^{(k)} \sim \langle n \rangle$ and $\mathsf{K}_\xi(p_n)/\mathsf{K}_\xi(n) \to 1$ as $n \to \infty$. Furthermore, for any $\varepsilon > 0$, using the strict stationarity of ξ, one has

$$\sum_{s=1}^{M_n} \mathbf{E}\zeta_{n,s}^2 \mathbf{1}(|\zeta_{n,s}| > \varepsilon) = \frac{M_n}{\langle n \rangle \mathsf{K}_\xi(n)} \mathbf{E}\eta_{n,1}^2 \{\, \eta_{n,1}^2 > \varepsilon^2 \langle n \rangle \mathsf{K}_\xi(n) \,\}$$

$$= \frac{M_n \langle p_n \rangle \mathsf{K}_\xi(p_n)}{\langle n \rangle \mathsf{K}_\xi(n)} \mathbf{E} \frac{S(U_n^{(1)})^2}{\langle p_n \rangle \mathsf{K}_\xi(p_n)} \mathbf{1}$$

$$\left(\frac{S(U_n^{(1)})^2}{\langle p_n \rangle \mathsf{K}_\xi(p_n)} > \varepsilon^2 \frac{\langle n \rangle \mathsf{K}_\xi(n)}{\langle p_n \rangle \mathsf{K}_\xi(p_n)} \right) \to 0 \text{ as } n \to \infty$$

in view of (10.80) since it holds $\langle n \rangle \mathsf{K}_\xi(n)/\langle p_n \rangle \mathsf{K}_\xi(p_n) \to \infty$ as $n \to \infty$ and the family $\{\, S(U_n^{(1)})^2/\langle p_n \rangle \mathsf{K}_\xi(p_n) \,\}$ is uniformly integrable. Indeed, $\{\, S_{p_n}^2/\langle p_n \rangle \mathsf{K}_\xi(p_n),\ n \in \mathbb{N}^d \,\}$ is a subfamily of $\{\, S_n^2/\langle n \rangle \mathsf{K}_\xi(n),\ n \in \mathbb{N}^d \,\}$. Thus, by the Lindeberg theorem (see for example [472, p. 329])

$$\sum_{s=1}^{M_n} \zeta_{n,s} \xrightarrow{d} \zeta \sim \mathcal{N}(0,1), \quad n \to \infty.$$

Consequently, for each $t \in \mathbb{R}$ one has $Q_3(n,t) \to 0$ as $n \to \infty$. □

10.3.5 Concluding Remarks

There are interesting generalizations of the CLT for random fields. Namely, the weak invariance principle (or functional CLT) and strong invariance principles are discussed in [96, Chap. 5]. In [465] one can find the proof of strong invariance principle for PA or NA (positively or negatively associated) random fields with power-type decreasing property of the Cox–Grimmett coefficient. The law of the iterated logarithm (LIL) and the functional LIL for PA or NA random fields are established in [96, Chap. 6]. In the same book, various statistical problems (e.g. the estimation of unknown density of a field) or the study of various functionals of random fields arising as solutions of partial differential equations (PDE) are given as well.

CLT for mixing random fields are given in the monograph [83] and papers [147, 148, 493]. Linear random fields are studied in [41, 390]. Rates of convergence in the CLT for vector-valued random fields can be found in [94]. Convergence of partial sums of weakly dependent random variables to infinitely divisible laws is established in [149] by the Lindeberg method. In this paper, convergence to Lévy processes for associated variables is studied as well. Convergence to stable laws is considered [269, 270]. CLTs for random elements with values in abstract spaces (e.g., generalized Hölder spaces) are treated in [412]. Weak convergence of random measures generated by Bernoulli associated random variables is proved in [33]. The application of limit theorems for random fields to the asymptotic theory of M-estimators is considered in [272].

Chapter 11
Strong Limit Theorems for Increments of Random Fields

Ulrich Stadtmüller

Abstract After reconsidering the oscillating behaviour of sums of i.i.d. random variables we study the oscillating behavior for sums over i.i.d. random fields under exact moment conditions. This summarizes several papers published jointly with A. Gut (Uppsala).

11.1 Introduction

We shall consider a classical scenario, namely i.i.d. random variables X, X_i, $i \in \mathbb{N}$ and we shall impose appropriate moment conditions later on. As usually we denote by

$$S_n = \sum_{j=1}^{n} X_j, \qquad n \in \mathbb{N}$$

the partial sums of these random variables and begin with an overview of almost sure limit theorems on S_n. In all kind of statistics and questions averages play an important role and averages are just of the form S_n/n and almost sure limit theorems deal with such averages. Typically in this situation there is an equivalence between such limit results and appropriate moment conditions. In order to demonstrate this we begin with the *strong law of large numbers (SLLN)*. Fore more details see, e.g., the book [210].

Theorem 11.1 (SLLN (Kolmogorov)).

$$\frac{S_n}{n} \xrightarrow{a.s.} 0 \iff \mathbf{E}|X| < \infty \text{ and } \mathbf{E}X = 0.$$

U. Stadtmüller (✉)
Ulm University, Ulm, Germany
e-mail: ulrich.stadtmueller@uni-ulm.de

E. Spodarev (ed.), *Stochastic Geometry, Spatial Statistics and Random Fields*, Lecture Notes in Mathematics 2068, DOI 10.1007/978-3-642-33305-7_11,

Remark 11.1. The $\Rightarrow$ direction is formulated somewhat sloppily, here and throughout it should be read as follows: if $\limsup_{n\to\infty} \frac{|S_n|}{n} < \infty$ then $\mathbf{E}|X| < \infty$, hence by the converse conclusion the limit exists and is then $\mathbf{E}X$.

This result was extended as follows.

Theorem 11.2 (Marcinkiewicz–Zygmund '37). *For* $0 < r < 2$

$$\frac{S_n}{n^{1/r}} \xrightarrow{a.s.} 0 \iff \mathbf{E}(|X|^r) < \infty \text{ and } (\mathbf{E}X = 0 \text{ provided } r \geq 1).$$

By the CLT it follows that the result fails to hold for $r = 2$. Next, we go on with the speed of convergence in the SLLN, the famous *law of iterated logarithm (LIL)*.

Theorem 11.3 (LIL (Hartmann–Wintner '44, Strassen '66)). *It holds that*

$$\limsup_{n\to\infty} \frac{S_n}{\sqrt{2n\log\log n}} \overset{a.s.}{=} 1 \iff \mathbf{E}|X|^2 = 1,\ \mathbf{E}X = 0.$$

Remark 11.2. Obviously under these moment conditions

$$\liminf_{n\to\infty} \frac{S_n}{\sqrt{2n\log\log n}} \overset{a.s.}{=} -1$$

and any point in $[-1, 1]$ is an a.s. limit point of the sequence $(S_n/\sqrt{2n\log\log n})$. A corresponding remark applies to related theorems below.

A somewhat different topic are limit laws for increments of sums of i.i.d. random variables. That is we shall study the almost sure oscillation behaviour of partial sums. This oscillation behaviour is interesting itself. We shall start with the following result where we denote by $\log^+ x = \max\{1, \log x\}, x > 0$.

Theorem 11.4 (Chow '73, Lai '74 [123, 124, 319]). *For* $0 < \alpha < 1$ *we have a SLLN*

$$\frac{S_{n+n^\alpha} - S_n}{n^\alpha} \xrightarrow{a.s.} 0 \iff \mathbf{E}(|X|^{1/\alpha}) < \infty,\ \mathbf{E}X = 0,$$

and a law of single logarithm (LSL)

$$\limsup_{n\to\infty} \frac{S_{n+n^\alpha} - S_n}{\sqrt{2n^\alpha \log n}} \overset{a.s.}{=} \sqrt{1-\alpha} \iff \mathbf{E}(|X|^{2/\alpha}(\log^+|X|)^{-1/\alpha}) < \infty$$

$$\text{and } \mathbf{E}X = 0,\ \mathbf{E}X^2 = 1.$$

Remark 11.3. 1. Increments of sums can be considered as special weighted sums of random variables.

$$S_{n+n^\alpha} - S_n = \sum_{k=1}^{\infty} w_{kn} X_k \text{ with } w_{kn} = \mathbf{1}(k \in (n, n+n^\alpha]).$$

In case $\alpha = 1/2$ these weight are related to the so-called *Valiron weights*

$$w_{kn} = \frac{1}{\sqrt{2\pi n}} \exp\left\{-\frac{1}{2n}(k-n)^2\right\}$$

being centered at n and having standard deviation $\sqrt{n}$. It was shown in the papers by Chow and Lai that analogous limit results with the same moment conditions occur for sums having such weights and also weights being asymptotically equivalent to those of the Valiron mean like, e.g. Euler- or Borel-means of i.i.d. random variables (observe that local CLT's apply to the corresponding weight sequences like $w_{kn} = e^{-n}n^k/k!$.).

2. The result above has been extended by Bingham and Goldie [70] to a larger class of span sizes e.g. span sizes $a(n)$ where $a(\cdot)$ is a regularly varying function of order $\alpha \in (0,1)$ in short $a(\cdot) \in RV(\alpha)$. For a short introduction to this notion, a measurable function $L : (0,\infty) \to (0,\infty)$ is called slowly varying if

$$\frac{L(\lambda t)}{L(\lambda)} \to 1 \quad as\ \lambda \to \infty \ \text{ for all } t > 0.$$

Typical examples are $L() \equiv c > 0$ or $L(t) = c(\log(1+t)^\alpha$ with constants $\alpha \in \mathbb{R}$ and $c > 0$. A measurable function $f : (0,\infty) \to (0,\infty)$ is regularly varying with index $\alpha \in \mathbb{R}$ if there exists a slowly varying function L such that $f(t) = t^\alpha L(t)$ which is equivalent to

$$\frac{f(\lambda t)}{f(\lambda)} \to t^\alpha \quad as\ \lambda \to \infty \ \text{ for all } t > 0.$$

See the book [71] for the notion of regular variation and many results and applications.

3. Interesting are also the limiting cases of the results from above such as $\alpha = 1$ which was dealt with in [211] and as $\alpha = 0$ containing e.g. the Erdös–Rényi laws (see e.g. [138]) where $a(n) = c \log n$ and the limit depends on the complete distribution function and not just on its moments. For related intermediate results see also [321].

11.2 Classical Laws for Random Fields

Now consider random fields $\{X_{\mathbf{n}}, \mathbf{n} \in \mathbb{N}^d\}$ containing i.i.d. random variables $X_{\mathbf{n}}$ with a multi-index $\mathbf{n} \in \mathbb{N}^d$ and having the same distribution as X. Again our goal is to derive strong limit theorems under exact moment conditions. In Chap. 10, limit theorems for random fields relaxing these assumptions in various directions to sufficient sets of conditions for limit theorems are discussed. As before we consider partial sums (where inequalities are understood componentwise)

$$S_{\mathbf{n}} = \sum_{\mathbf{j} \le \mathbf{n}} X_{\mathbf{j}} \quad \text{and} \quad T_{(\mathbf{n}),(\mathbf{n}+\mathbf{a}(\mathbf{n}))} = \sum_{\mathbf{n} \le \mathbf{j} \le \mathbf{n}+a(\mathbf{n})} X_{\mathbf{j}}$$

where $\mathbf{a}(\mathbf{n}) = (a^{(1)}(n_1), \dots, a^{(d)}(n_d))$ and $a^{(i)}(\cdot) \in RV(\alpha_i)$ with $\alpha = (\alpha_1, \dots, \alpha_d) \in [0,1]^d$, that is, in case $d = 2$ we consider sums over an area like

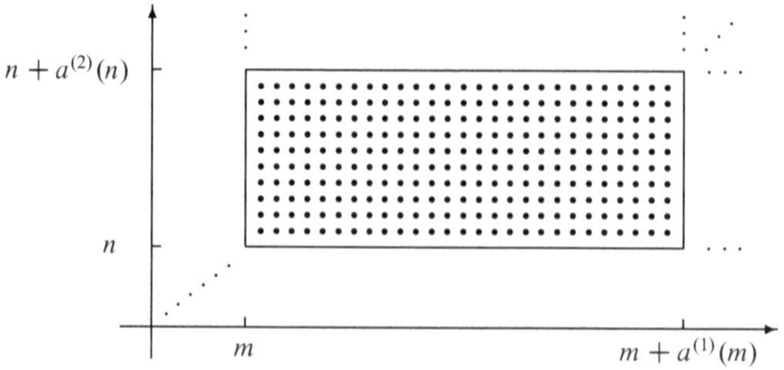

The result could be applied to deal with noise in pictures described by a smooth surface-function on $\mathbb{R}^2_+$.
Many classical limit laws for partial sums $S_{\mathbf{n}}$ have been carried over to the random field case. Most of the standard inequalities in probability can be transferred to the multiindex case, see e.g. [306,474]. Denote by $|\mathbf{n}| = n_1 \cdot \ldots \cdot n_d$, $|\mathbf{n}^{\alpha}| = n_1^{\alpha_1} \cdot \ldots \cdot n_d^{\alpha_d}$. Then we begin with the analogous results to the classical limit theorems described above.

Theorem 11.5 (SLLN (Smythe '74 [477])).

$$\frac{S_{\mathbf{n}}}{|\mathbf{n}|} \xrightarrow{a.s.} 0 \iff \mathbf{E}\big(|X|(\log^+ |X|)^{d-1}\big) < \infty \ \textit{and}\ \mathbf{E}X = 0.$$

Next, there is also a Marcinkiewicz–Zygmund-type result (MZ).

Theorem 11.6 (MZ (Klesov '02 [307], Gut-S. [214])). *For* $1/2 < \alpha_1 \le \alpha_2, \dots \le \alpha_d \le 1$ *and* $p = \operatorname{argmax}\{\alpha_j = \alpha_1\}$

$$\frac{S_n}{|\mathbf{n}^{\alpha}|} \xrightarrow{a.s.} 0 \iff \mathbf{E}(|X|^{1/\alpha_1}(\log^+ |X|)^{p-1}) < \infty \ \textit{and}\ \mathbf{E}X = 0.$$

And, also a law of iterated logarithm (LIL) holds.

Theorem 11.7 (LIL (Wichura '73 [515])).

$$\limsup_{\mathbf{n}\to\infty} \frac{S_{\mathbf{n}}}{\sqrt{2|\mathbf{n}| \log\log |\mathbf{n}|}} \overset{a.s.}{=} \sqrt{d} \iff \mathbf{E}\left(|X|^2 \frac{(\log^+ |X|)^{d-1}}{\log^+ \log^+ |X|}\right) < \infty$$

$$\textit{and}\ \mathbf{E}X = 0,\ \mathbf{E}X^2 = 1.$$

Why are there stronger moment assumptions than in the ordinary case, i.e. the case $d = 1$? Splitting the random variables in a part with large values and the rest, the part with large values should appear only finitely often, this determines the moment condition. That is, having the Borel–Cantelli Lemma in mind one considers sums of the following type with some function $\Phi : [1, \infty] \to (0, \infty)$, $\Phi(x) \to \infty$ as $x \to \infty$ such that

$$\sum_n \mathbf{P}(|X_n| > \Phi(n)) = \sum_n \mathbf{P}(\Phi^{-1}(|X|) > n) < \infty \iff \mathbf{E}\left(\Phi^{-1}(|X|)\right) < \infty,$$

where the latter equivalence comes from a simple argument comparing a sum and a integral. Now, in the multiindex case let $\Phi : \mathbb{N} \to \mathbb{N}$ and we obtain

$$\sum_{\mathbf{n}} \mathbf{P}(|X_{\mathbf{n}}| > \Phi(\mathbf{n})) = \sum_k \sum_{\Phi(\mathbf{n})=k} \mathbf{P}(|X| > k) = \sum_k d(k)\mathbf{P}(|X| > k) < \infty.$$

The function $d(\cdot)$ is typically rather irregular but its partial sums $D(m) = \sum_{k \le m} d_k$ has often a regular behaviour and satisfies $D(m)/m \to \infty$ as $m \to \infty$. E.g., if $\Phi(x) = x$ we have D_m is of order $m(\log m)^{d-1}$ $(d \ge 1)$. If d_k appears as a weight in a sum with "nice" additional weights we can nevertheless assume that d_k behaves like $(\log k)^{d-1}$. This means in particular that $d_k = 1$ in the scalar case $d = 1$. Hence we end with the natural extension of the scalar moment condition $\mathbf{E}|X| < \infty$ to the moment condition $\mathbf{E}(|X|(\log^+|X|)^{d-1}) < \infty$ which in the random field case with $d > 1$ is stronger than in the classical case $d = 1$.

11.3 Chow-Type Laws for Random Fields

We shall study here Chow-type laws for random fields with i.i.d. random variables where we obtain again limit results under some exact moment condition. Limit theorems under dependence assumptions and sufficient conditions are given in Chap. 10. We begin with SLLN for a rectangular window.

Theorem 11.8 (S., Thalmaier [479]). *Again, assume* $0 < \alpha_1 \le \alpha_2 \le \cdots \le \alpha_d < 1$, $p = \operatorname{argmax}\{\alpha_j = \alpha_1\}$ *and* $\alpha_i \le \lambda_i \le \min\{1, 2\alpha_i\}$, $1 \le i \le d$ *then*

$$\frac{1}{|\mathbf{n}^\alpha|} \max_{\mathbf{n} \le \mathbf{r} \le \mathbf{n}+\mathbf{n}^\lambda} T_{(\mathbf{n}),(\mathbf{n}+\mathbf{r})} \xrightarrow{a.s.} 0 \iff \mathbf{E}\left(|X|^{\frac{1}{\alpha_1}} \left(\log^+|X|\right)^{p-1}\right) < \infty,\ \mathbf{E}X = 0.$$

Remark 11.4. In particular we find under this moment assumption that

$$\frac{T_{(\mathbf{n}),(\mathbf{n}+\mathbf{n}^\alpha)}}{|\mathbf{n}^\alpha|} \xrightarrow{a.s.} 0$$

holds.

Concerning the oscillation behaviour for random fields a lot is known provided the moment generating function of X exists in $B_r(0)$ for some $r > 0$. Denote by J a half-open interval, by $|J|$ its volume and by

$$D(|\mathbf{n}|;k) = \max_{J \subset [0,\mathbf{n}],\ |J| \leq k} S_J .$$

Then the following result was shown.

Theorem 11.9 (Steinebach '83 [483]). *Let (k_n) be a sequence of integers such that $k_{|\mathbf{n}|}/\log|\mathbf{n}| \to \infty$ and $\mathbf{E}(\exp(tX)) < \infty$ for some $t \in B_r(0)$, $\mathbf{E}X = 0,\ \mathbf{E}X^2 = 1$ then*

$$\lim_{\mathbf{n}\to\infty} \frac{D(|\mathbf{n}|, k_{|\mathbf{n}|})}{\sqrt{2d\, k_{|\mathbf{n}|} \log|\mathbf{n}|}} = 1 \ \ a.s..$$

Remark 11.5. Some further situations are treated in [483], as e.g, an Erdős–Rényi law, this is the case where $k_{|\mathbf{n}|}/\log|\mathbf{n}| \to c$ with some $c > 0$.

Next, we shall discuss the following problem: If we do not assume the existence of a moment generating function, is it still possible to give this type of limit theorems under certain weaker moment conditions or can we even characterize the limit behaviour by a moment condition for the iid random variables involved?
So, the next result we discuss is the analogue of the LSL ?

Theorem 11.10 (A. Gut, U.S. [212, 213]). *Assume that $0 < \alpha_1 \leq \alpha_2 \leq \cdots \leq \alpha_d < 1,\ p = \text{argmax}\{\alpha_j = \alpha_1\}$ then*

$$\limsup_{\mathbf{n}\to\infty} \frac{T_{(\mathbf{n}),(\mathbf{n}+\mathbf{n}^\alpha)}}{\sqrt{2|\mathbf{n}^\alpha|\log|\mathbf{n}|}} \overset{a.s.}{=} \sqrt{1-\alpha_1} \iff \mathbf{E}\left(|X|^{\frac{2}{\alpha_1}}(\log^+ X|)^{p-1-\frac{1}{\alpha_1}}\right) < \infty$$
$$\text{and } \mathbf{E}X = 0,\ \mathbf{E}X^2 = 1.$$

This theorem complements the strong law given above in Theorem 11.8.

11.4 Proofs

We shall discuss the proof of Theorem 11.10 to some extend but not in all details. The typical pattern in proving results of the LIL type requires two truncations; the first one in order to match the Kolmogorov exponential bounds (see e.g. [210, Sect. 8.2], and the second one in order to match the moment requirements in Theorem 11.10. The proof follows that in [212].

Toward this end, let δ be small, let

$$b_{\mathbf{n}} = b_{|\mathbf{n}|} = \frac{\sigma\delta}{\varepsilon} \frac{\sqrt{|\mathbf{n}^\alpha|}}{\log|\mathbf{n}|} \tag{11.1}$$

and set

$$X'_{\mathbf{n}} = X_{\mathbf{n}}\mathbf{1}(|X_{\mathbf{n}}| \le b_{\mathbf{n}}), \quad X''_{\mathbf{n}} = X_{\mathbf{n}}\mathbf{1}(b_{\mathbf{n}} < |X_{\mathbf{n}}| \le \delta\sqrt{|\mathbf{n}^\alpha|\log|\mathbf{n}|}),$$
$$X'''_{\mathbf{n}} = X_{\mathbf{n}}\mathbf{1}(|X_{\mathbf{n}}| \ge \delta\sqrt{|\mathbf{n}^\alpha|\log|\mathbf{n}|}).$$

In the following all objects with primes or multiple primes refer to the respective truncated summands.

Since truncation destroys centering, we obtain, using standard procedures and noticing that $\mathbf{E}X = 0$ that

$$\begin{aligned}|\mathbf{E}X'_{\mathbf{k}}| &= |-\mathbf{E}(X_{\mathbf{k}}\mathbf{1}(|X_{\mathbf{k}}| > b_{\mathbf{k}}))| \le \mathbf{E}(|X|\mathbf{1}(|X_{\mathbf{k}}| > b_{\mathbf{k}}))\\ &\le \frac{\mathbf{E}(X^2(\log^+|X|)^{1-\alpha/2}\mathbf{1}(|X| > b_{\mathbf{k}}))}{b_{\mathbf{k}}(\log b_{\mathbf{k}})^{1-\alpha/2}},\end{aligned}$$

so that

$$\begin{aligned}|\mathbf{E}T'_{\mathbf{n},\mathbf{n}+\mathbf{n}^\alpha}| &\le \sum_{\mathbf{n}\le\mathbf{k}\le\mathbf{n}+\mathbf{n}^\alpha} \frac{\mathbf{E}(X^2(\log^+|X|)^{1-\alpha/2}\mathbf{1}(|X| > b_{\mathbf{k}}))}{b_{\mathbf{k}}(\log b_{\mathbf{k}})^{1-\alpha/2}}\\ &\le C\sqrt{|\mathbf{n}^\alpha|(\log|\mathbf{n}|)^\alpha}\cdot\mathbf{E}(X^2(\log^+|X|)^{1-\alpha/2}\mathbf{1}(|X| > b_{\mathbf{n}}))\\ &= o(\sqrt{|\mathbf{n}^\alpha|\log|\mathbf{n}|}) \text{ as } \mathbf{n}\to\infty. \end{aligned} \tag{11.2}$$

Moreover,

$$\mathbf{var}\, X_{\mathbf{n}} \le \mathbf{E}(X_{\mathbf{n}}^2) = \mathbf{E}X^2 = \sigma^2,$$

so that

$$\mathbf{var}(T'_{\mathbf{n},\mathbf{n}+\mathbf{n}^\alpha}) \le |\mathbf{n}^\alpha|\sigma^2. \tag{11.3}$$

Next we use Kolmogorov's upper exponential bounds (see e.g. [210, Lemma 8.2.1]) dealing with independent random variables (Y_k) with $\mathbf{E}Y_k = 0$ for all k, $\mathbf{var}\, Y_k = \sigma_k^2$ and $s_n^2 = \sum_{k=1}^n \sigma_k^2$. The goal is to have inequalities for the probability that the sum exceeds some threshold are close to that for Gaussian random variables. We begin with the upper bound.
If for $n \ge 1$ and some $c_n > 0$ it holds that $|Y_k| \le c_n s_n$ for $1 \le k \le n$ then we have for $0 < x < 1/c_n$

$$\mathbf{P}(\sum_{k=1}^n Y_k > x\,c_n) \le \exp\left\{-\frac{x^2}{2}\left(1 - \frac{x\,c_n}{2}\right)\right\}.$$

There is also a corresponding lower bound (see e.g. [210, Lemma 8.2.2]). Suppose in addition that $\gamma > 0$. Then there exist constants $x(\gamma)$ and $\kappa(\gamma)$ such that for $x(\gamma) \le x \le \kappa(\gamma)/c_n$ we have

$$\mathbf{P}(\sum_{k=1}^n Y_k > x\,c_n) \le \exp\left\{-\frac{x^2}{2}(1+\gamma)\right\}.$$

In the application it is important to choose x and c_n suitably.
First we use the upper bound. Here we choose $x = \varepsilon(1-\delta)\sqrt{2\log|\mathbf{n}|}$ and $c_{\mathbf{n}} = 2\delta/x$, (note $|X'_{\mathbf{k}}| = o\left(c_n\sqrt{\mathbf{var}(T_{\mathbf{n}+\mathbf{n}^\alpha})}\right)$ for $\mathbf{n} \le \mathbf{k} \le \mathbf{n}+\mathbf{n}^\alpha$). The inequality together with (11.2) and (11.3) now yield

$$\begin{aligned}
&\mathbf{P}(|T'_{\mathbf{n},\mathbf{n}+\mathbf{n}^\alpha}| > \varepsilon\sqrt{2|\mathbf{n}^\alpha|\log|\mathbf{n}|})\\
&\le \mathbf{P}\left(|T'_{\mathbf{n},\mathbf{n}+\mathbf{n}^\alpha} - \mathbf{E}T'_{\mathbf{n},\mathbf{n}+\mathbf{n}^\alpha}| > \varepsilon(1-\delta)\sqrt{2|\mathbf{n}^\alpha|\log|\mathbf{n}|}\right)\\
&\le \mathbf{P}\left(|T'_{\mathbf{n},\mathbf{n}+\mathbf{n}^\alpha} - \mathbf{E}\,T'_{\mathbf{n},\mathbf{n}+\mathbf{n}^\alpha}| > \frac{\varepsilon(1-\delta)}{\sigma}\sqrt{2\,\mathbf{var}(T'_{\mathbf{n},\mathbf{n}+\mathbf{n}^\alpha})\log|\mathbf{n}|}\right)\\
&\le \exp\left\{-\frac{2\varepsilon^2(1-\delta)^2}{2\sigma^2}\log|\mathbf{n}|(1-\delta)\right\}\\
&= |\mathbf{n}|^{-\frac{\varepsilon^2(1-\delta)^3}{\sigma^2}}. \qquad (11.4)
\end{aligned}$$

In order to apply the lower exponential bound (see e.g. [210, Lemma 8.2.2], we first need a lower bound for the truncated variances:

$$\begin{aligned}
\mathbf{var}\,X'_{\mathbf{n}} &= \mathbf{E}(X'_{\mathbf{n}})^2 - (\mathbf{E}(X'_{\mathbf{n}}))^2 = \mathbf{E}X^2 - \mathbf{E}(X^2\mathbf{1}(|X_{\mathbf{n}}| \ge b_{\mathbf{n}})) - (\mathbf{E}(X'_{\mathbf{n}}))^2\\
&\ge \sigma^2 - 2\mathbf{E}(X^2\mathbf{1}(|X_{\mathbf{n}}| \ge b_{\mathbf{n}})) \ge \sigma^2(1-\delta)
\end{aligned}$$

for $|\mathbf{n}|$ large, so that

$$\mathbf{var}(T'_{\mathbf{n},\mathbf{n}+\mathbf{n}^\alpha}) \ge |\mathbf{n}^\alpha|\sigma^2(1-\delta) \text{ for } |\mathbf{n}| \text{ large.} \qquad (11.5)$$

Next we conclude that for any $\gamma > 0$,

$$\begin{aligned}
&\mathbf{P}(|T'_{\mathbf{n},\mathbf{n}+\mathbf{n}^\alpha}| > \varepsilon\sqrt{2|\mathbf{n}^\alpha|\log|\mathbf{n}|})\\
&\ge \mathbf{P}\left(|T'_{\mathbf{n},\mathbf{n}+\mathbf{n}^\alpha} - \mathbf{E}T'_{\mathbf{n},\mathbf{n}+\mathbf{n}^\alpha}| > \varepsilon(1+\delta)\sqrt{2|\mathbf{n}^\alpha|\log|\mathbf{n}|}\right)\\
&\ge \mathbf{P}\left(|T'_{\mathbf{n},\mathbf{n}+\mathbf{n}^\alpha} - \mathbf{E}T'_{\mathbf{n},\mathbf{n}+\mathbf{n}^\alpha}| > \frac{\varepsilon(1+\delta)}{\sigma\sqrt{(1-\delta)}}\sqrt{2\,\mathbf{var}(T'_{\mathbf{n},\mathbf{n}+\mathbf{n}^\alpha})\log|\mathbf{n}|}\right)\\
&\ge \exp\left\{-\frac{2\varepsilon^2(1+\delta)^2}{2\sigma^2(1-\delta)}\log|\mathbf{n}|(1+\gamma)\right\}\\
&= |\mathbf{n}|^{-\frac{\varepsilon^2(1+\delta)^2(1+\gamma)}{\sigma^2(1-\delta)}} \text{ for } |\mathbf{n}| \text{ large.} \qquad (11.6)
\end{aligned}$$

Hence, roughly we obtain $\mathbf{P}(|T'_{\mathbf{n},\mathbf{n}+\mathbf{n}^\alpha}| > \varepsilon\sqrt{2|\mathbf{n}^\alpha|\log|\mathbf{n}|}) = |\mathbf{n}|^{-\frac{\varepsilon^2}{2\sigma^2}}$ for any $\varepsilon > 0$.

11.4.1 *Sufficiency: The Upper Bound*

We begin by taking care of the double- and triple primed contributions, after which we provide a convergent upper Borel–Cantelli sum for the single primed contribution over a suitably chosen subset of points in $\mathbb{Z}^d$. After this we apply the first Borel–Cantelli lemma for this subset, and then "fill the gaps" in order to include arbitrary windows.

11.4.1.1 The Term $T''_{\mathbf{n},\mathbf{n}+\mathbf{n}^\alpha}$

In this subsection we establish the fact that

$$\limsup_{\mathbf{n}\to\infty} \frac{|T''_{\mathbf{n},\mathbf{n}+\mathbf{n}^\alpha}|}{\sqrt{|\mathbf{n}^\alpha|\log|\mathbf{n}|}} \leq \frac{\delta}{1-\alpha} \quad \text{a.s.} \tag{11.7}$$

In order for $|T''_{\mathbf{n},\mathbf{n}+\mathbf{n}^\alpha}|$ to surpass the level $\eta\sqrt{|\mathbf{n}^\alpha|\log|\mathbf{n}|}$ it is necessary that at least $N \geq \eta/\delta$ of the X''s are nonzero, which, by stretching the truncation bounds to the extremes, implies that

$$\begin{aligned}
&\mathbf{P}(|T''_{\mathbf{n},\mathbf{n}+\mathbf{n}^\alpha}| > \eta\sqrt{|\mathbf{n}^\alpha|\log|\mathbf{n}|})\\
&\leq \binom{|\mathbf{n}^\alpha|}{N}\Big(\mathbf{P}\Big(b_{\mathbf{n}} < |X| \leq \delta\sqrt{(|\mathbf{n}|+|\mathbf{n}^\alpha|)\log(|\mathbf{n}|+|\mathbf{n}^\alpha|)}\Big)\Big)^N\\
&\leq |\mathbf{n}|^{\alpha N}\left(\mathbf{P}\left(|X| > C|\mathbf{n}|^{\alpha/2}/\log|\mathbf{n}|\right)\right)^N\\
&\leq C|\mathbf{n}|^{\alpha N}\left(\frac{\mathbf{E}|X|^{2/\alpha}(\log^+|X|)^{d-1-1/\alpha}}{(|\mathbf{n}|^{\alpha/2}/\log|\mathbf{n}|)^{2/\alpha}(\log|\mathbf{n}|)^{d-1-1/\alpha}}\right)^N\\
&= C\frac{(\log|\mathbf{n}|)^{N((3/\alpha)+1-d)}}{|\mathbf{n}|^{N(1-\alpha)}}.
\end{aligned}$$

Since the sum of the probabilities converges whenever $N(1-\alpha) > 1$, considering that, in addition, $N\delta \geq \eta$, we have shown that

$$\sum_{\mathbf{n}} \mathbf{P}\left(|T''_{\mathbf{n},\mathbf{n}+\mathbf{n}^\alpha}| > \eta\sqrt{|\mathbf{n}^\alpha|\log|\mathbf{n}|}\right) < \infty \quad \text{for all } \eta > \frac{\delta}{1-\alpha},$$

which establishes (11.7) via the first Borel–Cantelli lemma.

11.4.1.2 The Term $T'''_{\mathbf{n},\mathbf{n}+\mathbf{n}^\alpha}$

Next we show that

$$\lim_{\mathbf{n}\to\infty} \frac{|T'''_{\mathbf{n},\mathbf{n}+\mathbf{n}^\alpha}|}{\sqrt{|\mathbf{n}^\alpha|\log|\mathbf{n}|}} = 0 \quad \text{a.s.} \tag{11.8}$$

This one is easier, since in order for $|T'''_{\mathbf{n},\mathbf{n}+\mathbf{n}^\alpha}|$ to surpass the level $\eta\sqrt{|\mathbf{n}^\alpha|\log|\mathbf{n}|}$ infinitely often it is necessary that infinitely many of the X'''s are nonzero. However, via an appeal to the first Borel–Cantelli lemma, the latter event has zero probability since

$$\sum_{\mathbf{n}} \mathbf{P}\left(|X_\mathbf{n}| > \eta\sqrt{|\mathbf{n}^\alpha|\log|\mathbf{n}|}\right) = \sum_{\mathbf{n}} \mathbf{P}\left(|X| > \eta\sqrt{|\mathbf{n}^\alpha|\log|\mathbf{n}|}\right) < \infty$$

iff our moment condition holds by the following Lemma in [479].

Lemma 11.1. *Let* $0 < \alpha_1 \le \alpha_2 \le \cdots \le \alpha_d < 1$, $p = \operatorname{argmax}\{\alpha_j = \alpha_1\}$ *and suppose that* $\{X_\mathbf{k},\ \mathbf{k} \in \mathbb{Z}^d\}$ *are independent random variables with mean zero, then*

$$\sum_{\mathbf{n}} \mathbf{P}(|X_\mathbf{n}| > |\mathbf{n}|^\alpha(\log|\mathbf{n}|)) < \infty \quad \Longleftrightarrow \quad \mathbf{E}\left(|X|^{1/\alpha_1}(\log^+|X|)^{d-1-1/\alpha_1}\right) < \infty.$$

The proof of this Lemma contains the details of the arguments described at the end of Sect. 11.2.

11.4.1.3 The Term $T'_{\mathbf{n},\mathbf{n}+\mathbf{n}^\alpha}$

As for $T'_{\mathbf{n},\mathbf{n}+\mathbf{n}^\alpha}$ we have to resort to subsequences. Set $\lambda_1 = 1$, $\lambda_2 = 2$, and, further,

$$\lambda_i = \left(\frac{i}{\log i}\right)^{1/(1-\alpha)}, \ i = 3, 4, \ldots, \text{ and } \Lambda = \{\lambda_i, i \ge 1\}.$$

Our attention here is on the subset of points $\mathbf{n} = (n_1, n_2, \ldots, n_d) \in \mathbb{Z}^d$ such that $n_k \in \Lambda$, i.e $n_k = [\lambda_{i_k}]$, for all $k = 1, 2, \ldots, d$, in short $\mathbf{n} \in \boldsymbol{\lambda}$.

Suppose that $\mathbf{n} \in \boldsymbol{\lambda}$ and set $i = \prod_{k=1}^d i_k$. This implies, in particular, that $i_k \le i$ and that $\log i_k \le \log i$ for all k, so that

$$|\mathbf{n}| = \prod_{k=1}^d \lambda_k = \left(\frac{\prod_{k=1}^d i_k}{\prod_{k=1}^d \log i_k}\right)^{1/(1-\alpha)} \ge \frac{i^{1/(1-\alpha)}}{(\log i)^{d/(1-\alpha)}}.$$

With this in mind, the estimate (11.4) over the subset $\boldsymbol{\lambda}$ now yields

$$\begin{aligned}\sum_{\{\mathbf{n}\in\boldsymbol{\lambda}\}} \mathbf{P}\left(|T'_{\mathbf{n},\mathbf{n}+\mathbf{n}^\alpha}| > \varepsilon\sqrt{2|\mathbf{n}^\alpha|\log|\mathbf{n}|}\right) &\leq \sum_{\{\mathbf{n}\in\boldsymbol{\lambda}\}} |\mathbf{n}|^{-\frac{\varepsilon^2(1-\delta)^3}{\sigma^2}} \\ &\leq \sum_i \sum_{|\prod_{k=1}^d i_k|=i} |\mathbf{n}|^{-\frac{\varepsilon^2(1-\delta)^3}{\sigma^2}} \\ &\leq \sum_i d(i)\left(\frac{i^{1/(1-\alpha)}}{(\log i)^{d/(1-\alpha)}}\right)^{-\frac{\varepsilon^2(1-\delta)^3}{\sigma^2}} \\ &\leq C + \sum_{i\geq i_0} d(i) i^{-\frac{\varepsilon^2((1-\delta)^3-2\delta)}{\sigma^2(1-\alpha)}} < \infty \quad (11.9)\end{aligned}$$

for $\varepsilon > \sigma\sqrt{\frac{1-\alpha}{(1-\delta)^3-2\delta}}$, (where i_0 was chosen such that $(\log i)^{d(1-\delta)^3} \leq i^\delta$ and $d(i) \leq i^{\delta\frac{\varepsilon^2(1-\delta)^3}{\sigma^2(1-\alpha)}}$ for $i \geq i_0$).

11.4.1.4 Joining the Contributions

We first note that an application of the first Borel–Cantelli lemma to (11.9) provides an upper bound for lim sup $T'_{\mathbf{n},\mathbf{n}+\mathbf{n}^\alpha}$ as $\mathbf{n} \to \infty$ through the subset $\boldsymbol{\lambda}$. More precisely,

$$\limsup_{\substack{\mathbf{n}\to\infty \\ \{\mathbf{n}\in\boldsymbol{\lambda}\}}} \frac{|T'_{\mathbf{n},\mathbf{n}+\mathbf{n}^\alpha}|}{\sqrt{2|\mathbf{n}^\alpha|\log|\mathbf{n}|}} \leq \sigma\sqrt{\frac{1-\alpha}{(1-\delta)^3-2\delta}} \quad \text{a.s.} \qquad (11.10)$$

Joining this with (11.7) and (11.8) now yields

$$\limsup_{\substack{\mathbf{n}\to\infty \\ \{\mathbf{n}\in\boldsymbol{\lambda}\}}} \frac{|T_{\mathbf{n},\mathbf{n}+\mathbf{n}^\alpha}|}{\sqrt{2|\mathbf{n}^\alpha|\log|\mathbf{n}|}} \leq \sigma\sqrt{\frac{1-\alpha}{(1-\delta)^3-2\delta}} + \frac{\delta}{1-\alpha} \quad \text{a.s.,}$$

which, due to the arbitrariness of δ, tells us that

$$\limsup_{\substack{\mathbf{n}\to\infty \\ \{\mathbf{n}\in\boldsymbol{\lambda}\}}} \frac{|T_{\mathbf{n},\mathbf{n}+\mathbf{n}^\alpha}|}{\sqrt{2|\mathbf{n}^\alpha|\log|\mathbf{n}|}} \leq \sigma\sqrt{1-\alpha} \quad \text{a.s.} \qquad (11.11)$$

11.4.1.5 From the Subsequence to the Full Sequence

We shall omit the complete details here. This can be based on symmetrization and the Lévy-inequality and some technical details similar to techniques applied to the one-dimensional case. Desymmetrization follows the usual patterns.

11.4.2 Necessity

If even somewhat less, namely

$$\mathbf{P}\left(\frac{|T_{(\mathbf{n}),(\mathbf{n}+\mathbf{n}^\alpha)}|}{\sqrt{2|\mathbf{n}^\alpha|\log|\mathbf{n}|}} < \infty\right)$$

holds then, by the zero-one law, the probability that the lim sup is finite is 0 or 1, hence, being positive it equals 1. Consequently (cf. [319, p. 438] or [476, 477]),

$$\limsup_{n\to\infty} \frac{|X_{\mathbf{n}}|}{\sqrt{|\mathbf{n}^\alpha|\log|\mathbf{n}|}} < \infty \quad \text{a.s.,}$$

from which it follows via the second Borel–Cantelli lemma and the i.i.d. assumption that

$$\infty > \sum_{\mathbf{n}} \mathbf{P}\left(|X_{\mathbf{n}}| > \sqrt{|\mathbf{n}^\alpha|\log|\mathbf{n}|}\right) = \sum_{\mathbf{n}} \mathbf{P}\left(|X| > \sqrt{|\mathbf{n}^\alpha|\log|\mathbf{n}|}\right),$$

which verifies the moment condition by Lemma 11.1 above.

An application of the sufficiency part finally tells us that $\sigma^2 = \mathbf{var}\, X = 1$.

11.5 Boundary Cases

We shall add some comments on the limit cases $\alpha = 1$ and $\alpha = 0$ in the span sizes of the windows. Here we consider $d = 1$ and $a(n) = n/L(n)$ with a slowly varying function $L(\cdot)$ (under some mild additional assumptions).

Theorem 11.11 (Gut, Jonsson, S. [211]). *Let* $d(n) = \log L(n) + \log\log n$ *and* $f(n) = \min\{n, a(n)\,d(n)\}$ *then*

$$\limsup_{n\to\infty} \frac{S_{n+n/L(n)} - S_n}{\sqrt{2a(n)d(n)}} \overset{a.s.}{=} 1 \iff \mathbf{E}\left(f^{-1}(X^2)\right) < \infty,$$

$$\mathbf{E}X^2 = 1,\ \mathbf{E}X = 0.$$

Example 11.1. 1. If $L(n) = \log n$ then

$$\limsup_{n\to\infty} \frac{S_{n+n/\log n} - S_n}{\sqrt{4n \log\log n/\log n}} \stackrel{a.s.}{=} 1 \iff \mathbf{E}\left(X^2 \frac{\log^+|X|}{\log^+\log^+|X|}\right) < \infty,$$

$$\mathbf{E}X^2 = 1,\ \mathbf{E}X = 0.$$

2. $L(n) = \log\log n$ then

$$\limsup_{n\to\infty} \frac{S_{n+\frac{n}{\log\log n}} - S_n}{\sqrt{2n}} \stackrel{a.s.}{=} 1 \iff \mathbf{E}X^2 < \infty,$$

$$\mathbf{E}X^2 = 1,\ \mathbf{E}X = 0.$$

Random field extensions have been given in [215]. In the case $\alpha = 0$ and $L(n) = c\log n$ we are in the area of Erdős–Rényi theorems, where the limit depends in contrary to the results above on the complete distribution of the underlying random variables, see e.g. the book of Csörgő and Révész [138] in the case $d = 1$ and for the multi-index case see e.g. the paper of Steinebach, [483].

Acknowledgements This contribution is based on joint work with my colleague Allan Gut (Uppsala) with whom I enjoyed very much to work on this topic, I am very grateful for this partnership.

11.6 Exercises

1. Let $\Phi : (0,\infty) \to (0,\infty)$ be a nondecreasing, right continuous and unbounded function with generalized inverse $\Phi^{(-1)}(y) = \inf\{x : \Phi(x) \geq y\}$.

   (a) Show that

   $$\Phi(\Phi^{(-1)}(y)-) \leq y \leq \Phi(\Phi^{(-1)}(y)) \text{ and } \Phi^{(-1)}(\Phi(x)) \leq x \leq \Phi^{(-1)}(\Phi(x)+)$$

   where the plus or minus signs indicate one sided limits as e.g. $\Phi^{(-1)}(\Phi(x)+) = \lim_{y\searrow\Phi(x)} \Phi^{(-1)}(y)$, and that for a random variable X

   $$\mathbf{P}(\Phi(|X|) < t) = \mathbf{P}(|X| < \Phi^{(-1)}(t)).$$

   (b) Show the following equivalence

   $$\sum_{n=1}^{\infty} \mathbf{P}(|X| > \Phi(n)) < \infty \quad \Leftrightarrow \quad \mathbf{E}(\Phi^{(-1)}(|X|)) < \infty$$

2. Show that $\sum_{k,\ell\in\mathbf{N},k\cdot\ell\le n} 1 = (1+o(1))\cdot n\ \log n$ as $n\to\infty$.
3. Show that: If $f(x) = x^\alpha(\log x)^\beta a(x)$ is invertible for $x > x_0 > 1$ with $\alpha > 0$, $\beta\in\mathbb{R}$ and some function $a(x)\to 1$, $x\to\infty$, then there exists some function $b(.)$ on $(f(x_0),\infty)$ with $b(y)\to 1$, as $y\to\infty$, such that for $y > f(x_0)$

$$f^{(-1)}(y) = \frac{(\alpha^\beta y)^{1/\alpha}}{(\log y)^{\beta/\alpha}} b(y).$$

4. Show that for $w_{k,n} = e^{-n}\frac{n^k}{k!}$ it holds that (so-called local CLT)

$$w_{n+\nu\sqrt{n},n} = \frac{1}{\sqrt{2\pi n}}\exp\left(-\frac{\nu^2}{2}\right)\left(1+O\left(\frac{1}{\sqrt{n}}\right)\right) \quad \text{for fixed } \nu \text{ and } n\to\infty.$$

5. Verify the calculations leading to (11.4) and (11.6).
6. Verify the calculations in Sect. 11.4.1.3

Chapter 12
Geometry of Large Random Trees: SPDE Approximation

Yuri Bakhtin

Abstract In this chapter we present a point of view at large random trees. We study the geometry of large random rooted plane trees under Gibbs distributions with nearest neighbour interaction.

In the first section of this chapter, we study the limiting behaviour of the trees as their size grows to infinity. We give results showing that the branching type statistics is deterministic in the limit, and the deviations from this law of large numbers follow a large deviation principle. Under the same limit, the distribution on finite trees converges to a distribution on infinite ones. These trees can be interpreted as realizations of a critical branching process conditioned on non-extinction.

In the second section, we consider a natural embedding of the infinite tree into the two-dimensional Euclidean plane and obtain a scaling limit for this embedding. The geometry of the limiting object is of particular interest. It can be viewed as a stochastic foliation, a flow of monotone maps, or as a solution to a certain Stochastic PDE with respect to a Brownian sheet. We describe these points of view and discuss a natural connection with superprocesses.

12.1 Infinite Volume Limit for Random Plane Trees

In this section we discuss a biological motivation to study the geometry of large trees, introduce a relevant model (Gibbs distribution on trees) and study its behavior as the tree size grows to infinity, thus obtaining an infinite discrete random tree as a limiting object.

Y. Bakhtin (✉)
Georgia Institute of Technology, Atlanta, GA 30332, USA
e-mail: bakhtin@math.gatech.edu

E. Spodarev (ed.), *Stochastic Geometry, Spatial Statistics and Random Fields*, Lecture Notes in Mathematics 2068, DOI 10.1007/978-3-642-33305-7_12,

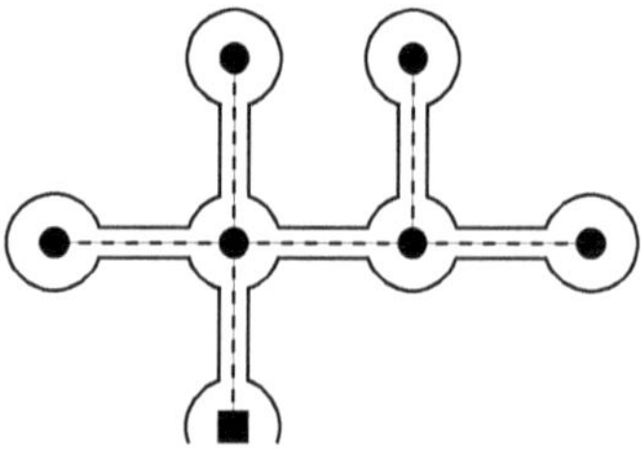

Fig. 12.1 An RNA secondary structure is shown in *solid line*. The *dashed lines* represent the edges of the encoding tree. The vertices of the tree are shown as *black circles* and the *black square* represents the root

12.1.1 Biological Motivation

The initial motivation for this study was the analysis of branching statistics for the secondary structures of large RNA molecules, and we begin with a description of the relevant mathematical models.

RNA molecules are much like DNA molecules, since the primary structure of an RNA molecule is a sequence of bases (nucleotides). One difference is that DNAs exist in the form of two complementary nucleotide strands coiled together into a double helix, whereas RNA is a single-stranded sequence of nucleotides, and there is a variety of three-dimensional configurations that RNA molecules can assume.

Nucleotides tend to pair up with complementary nucleotides, and this is exactly what makes the DNA double helix a stable structure. In a single stranded RNA, the nucleotides still have potential to get paired with other nucleotides, and in the absence of a complementary strand, they pair up to nucleotides of the same strand. The resulting shape or folding is called the secondary RNA structure. It consists of groups of paired nucleotides and groups of free unpaired nucleotides.

In Fig. 12.1 we see a schematic view of an RNA secondary structure. The groups of bases that got aligned and paired up are shown as pairs of parallel solid segments. Loops, i.e. groups of nucleotides that have no pair are shown as circular arcs.

If we ignore more complicated and rare situations where three groups of nucleotides aligned together can occur, then every RNA folding can be naturally encoded in a plane rooted tree. The procedure should be obvious from Fig. 12.1. The edges of the tree correspond to stacked base pairs and the vertices of the tree correspond to loops. The root of the tree is a special vertex that corresponds to the "external loop" formed by the ends of the sequence.

If one has a rooted tree on the plane then to reconstruct the corresponding RNA secondary structure one has simply to surround the tree by a contour beginning and terminating near the root.

Since the trees we consider are rooted trees, one can naturally interpret them as genealogical structures. For any vertex v of a tree, its *height* or generation number $h(v)$ is the distance from v to the root along the edges of the tree, i.e. the number of edges in the shortest path from v to the root. The shortest path is unique, and the first edge of this path connects v to its *parent* $p(v)$. The vertex v is called a *child* of $p(v)$. Each nonroot vertex has one parent and the *branching number* $\deg(v)$ of the vertex (also called its *out-degree*), i.e. the number of its children, is nonnegative.

In order to encode the RNA secondary structure, one has to fix the order of the child vertices for each vertex. Therefore, we shall consider two trees to be identical to each other if there is a one-to-one map between the vertex sets of the two trees that preserves the parent-child relation and the order of the child vertices. The classes of identical trees are called plane rooted trees or ordered rooted trees.

Now all the questions about RNA secondary structures can be restated using the tree terminology. The first problem we can study is to determine the typical behaviour of the branching type for a large tree. The branching type of a tree T is the vector $(\chi_0(T), \chi_1(T), \chi_2(T), \ldots)$, where $\chi_i(T)$ denotes the number of vertices with branching number $i \geq 0$ in T.

One approach is based on energy minimization. According to this approach, the typical trees are (close to) energy minimizers. To realize this approach one has to assign an energy value to every tree. We proceed to describe an energy model with nearest neighbour interaction used in [531]. We assume that the energy contribution of each vertex of the tree depends only on its branching number, i.e., on its nearest neighbourhood in the tree. We consider a sequence of numbers $(E_i)_{i=0,1,\ldots}$, and introduce the energy of the tree as

$$\mathsf{E}(T) = \sum_{v \in \mathbb{V}(T)} \mathsf{E}_{\deg(v)} = \sum_{i \geq 0} \chi_i(T) \mathsf{E}_i, \tag{12.1}$$

where $\mathbb{V}(T)$ is the set of vertices of the tree T. This is a very rough "low-resolution" model that ignores details of the nucleotide sequences. The concrete values of E_i can be found in [38] or [531].

We can now consider the following problem. For simplicity, let us fix a (large number) $D \in \mathbb{N}$ and for a given N, among trees on N vertices with branching numbers not exceeding D find trees that minimize $\mathsf{E}(\cdot)$. It follows from (12.1) that this problem is equivalent to:

$$\sum_{i=0}^{D} \chi_i E_i \to \min$$

$$\sum_{i=0}^{D} \chi_i = N, \tag{12.2}$$

$$\sum_{i=0}^{D} i \chi_i = N - 1, \tag{12.3}$$

$$\chi_i \geq 0, \quad i = 0, \ldots, D.$$

Restriction (12.2) means that the total number of vertices equals N. Restriction (12.3) means that the total number of child vertices equals $N - 1$ (since the root is the only vertex with no parent). Introducing $x_i = \chi_i/N$ and letting $N \to \infty$ we see that the limiting problem is

$$\mathsf{E}(x) \to \min$$

$$\sum_{i=0}^{D} x_i = 1, \tag{12.4}$$

$$\sum_{i=0}^{D} i x_i = 1, \tag{12.5}$$

$$x_i \geq 0, \quad i = 0, \ldots, D, \tag{12.6}$$

where

$$\mathsf{E}(x) = \sum_{i=0}^{D} x_i E_i. \tag{12.7}$$

Conditions (12.4)–(12.6) define a convex set Δ.

Exercise 12.1. Prove that Δ is, in fact, a $D-1$-dimensional simplex.

Since the function we have to minimize is linear, for most choices of $(E_i)_{i=0}^{D}$ the minimizers will have to be extremal points of Δ. In particular, it means that most coordinates of the minimizers will be 0. In fact, for the energy values suggested in [531], the minimizer is $(1/2, 0, 1/2, 0, 0, \ldots)$ which corresponds to the statistics of a binary tree (a tree that has only vertices with zero or two children).

This clearly explains why the fraction of vertices with high branching in RNA secondary structures is small. In fact, the energy minimization approach suggests that this fraction should be 0.

It turns out that most real RNA foldings contain branchings of higher orders, producing small but steady high branching frequencies. In [38] it is claimed that the failure to explain this by the energy minimization approach is due to the fact that the binary trees are too exotic, rare in the space of all possible trees, and one has to introduce a model that would take into account entropy considerations.

12.1.2 Gibbs Distribution on Trees. Law of Large Numbers and Large Deviation Principle for Branching Type

A natural candidate for such a model is the Boltzmann–Gibbs distribution on $\mathbb{T}_N = \mathbb{T}_N(D)$, the set of all trees on N vertices with branching not exceeding D. Let us fix an inverse temperature parameter $\beta \geq 0$ and define a probability measure P_N on $\mathbb{T}_N$ by

$$P_N(T) = \frac{\mathrm{e}^{-\beta \mathsf{E}(T)}}{Z_N},$$

where the normalizing factor (partition function) is defined by

$$Z_N = \sum_{T \in \mathbb{T}_N} \mathrm{e}^{-\beta \mathsf{E}(T)}.$$

For more information on Gibbs measures see Sect. 9.3.2.

In particular, if $\beta = 0$ or, equivalently, $E_i = 0$ for all i, then P_N is a uniform distribution on $\mathbb{T}_N$.

In this model, $\chi_0 = \chi_0(T), \chi_1 = \chi_1(T), \ldots$ become random variables (being functions of a random tree T), and the normalized branching type $\frac{1}{N}(\chi_0, \ldots, \chi_D)$ becomes a random vector. Let us describe the asymptotics of the normalized branching type as $N \to \infty$.

Let

$$J(x) = -\mathcal{H}(x) + \beta \mathsf{E}(x), \quad x \in \Delta, \tag{12.8}$$

where

$$\mathcal{H}(x) = -\sum_{i=0}^{D} x_i \log x_i$$

is the entropy of the probability vector $(x_0, \ldots, x_D) \in \Delta$, and $E(x)$ is the energy function defined in (12.7).

The following theorem plays the role of a law of large numbers for the branching type of a large random tree.

Theorem 12.1. *As $N \to \infty$,*

$$\left(\frac{\chi_0}{N}, \frac{\chi_1}{N}, \ldots, \frac{\chi_D}{N}\right) \to p$$

in probability, where p is a solution of the following optimization problem:

$$J(x) \to \min, \tag{12.9}$$

$$x \in \Delta. \tag{12.10}$$

Before we give a sketch of a proof of this theorem, let us make several comments. The function $J(x)$ is strictly convex on Δ. Therefore, the minimizer p is unique and it can be found using the method of Lagrange multipliers:

$$\log p_i + 1 + \beta E_i + \lambda_1 + i\lambda_2 = 0, \quad i = 0, 1, \ldots, D,$$

where λ_1 and λ_2 are the Lagrange multipliers. So we see that

$$p_i = C\rho^i \mathrm{e}^{-\beta E_i}, \quad i = 0, 1, \ldots, D, \tag{12.11}$$

where $C = \mathrm{e}^{-1-\lambda_1}$, and $\rho = \mathrm{e}^{-\lambda_2}$. Constants C and ρ can be uniquely defined from the two equations

$$\sum_{i=0}^{D} p_i = 1, \tag{12.12}$$

$$\sum_{i=0}^{D} i p_i = 1. \tag{12.13}$$

Since β is a finite number, we have

$$p_i > 0, \quad i = 0, 1, \ldots, D,$$

and the minimizer p belongs to the relative interior of Δ (with respect to the $D-1$ affine subspace it spans). Therefore, a typical large tree has a positive fraction of vertices with any given branching number. This explains why typical RNAs contain branchings of high degree. We refer the reader to [38] for more detailed analysis, and note here only that the phenomenon we are facing is typical for statistical mechanics models where there is always interplay and competition between the energy and entropy factors. If subsystems of a system are not independent, then the free energy of the system is not equal to the sum of the subsystem free energies, and an entropy correction is needed. This is precisely the content of definition (12.8), where J plays the role of free energy.

We shall now give a sketch of the proof of Theorem 12.1. The proof is based on the fact that trees with equal branching degree sequences have equal energy. Therefore,

$$P_N\left(\chi(T) = n\right) = \frac{\mathrm{e}^{-\beta \mathsf{E}(n)} C(N, n)}{Z_N}, \tag{12.14}$$

where $n = (n_0, \ldots, n_D)$ and $C(N, n)$ is the number of plane trees of order N with n_k nodes of branching degree k:

$$C(N, n) = \frac{1}{N}\binom{N}{n_0,\ n_1,\ n_2, \ldots,\ n_D} = \frac{(N-1)!}{n_0! n_1! n_2! \cdots n_D!} \tag{12.15}$$

if $n_1 + 2n_2 + \ldots + D N_D = N - 1$, and 0 otherwise (see e.g. [480, Theorem 5.3.10]). One can apply Stirling's formula to get

$$\begin{aligned} C(N, n) &= \exp\left\{ N \left(-\sum_{k=0}^{D} \frac{n_k}{N} \log \frac{n_k}{N} + \mathcal{O}\left(\frac{\log N}{N}\right)\right)\right\} \\ &= \exp\left\{ N \mathcal{H}\left(\frac{n}{N}\right) + \mathcal{O}(\log N)\right\} \end{aligned}$$

as $n \to \infty$, which holds true uniformly in n, see e.g. [166, Lemma I.4.4].

Plugging this into (12.14), we get

$$P_N\left(\frac{\chi(T)}{N}=\frac{n}{N}\right)=\frac{\exp\{\,-N\left[\beta E(\frac{n}{N})-\mathcal{H}(\frac{n}{N})\right]+\mathcal{O}(\log N)\,\}}{Z_N},$$
$$=\frac{\exp\{-NJ(\frac{n}{N})+\mathcal{O}(\log N)\}}{Z_N}.$$

Clearly, the expression in the numerator is maximal if n/N is close to p, and it is exponentially in N smaller for points n/N that are not close to p. Since the total number of points grows only polynomially in N, the desired asymptotics follows.

In fact, one can strengthen the law of large numbers of Theorem 12.1 and provide a large deviation principle (LDP). Roughly speaking, the LDP shows that probability for χ/N to deviate from p by at least ε decays exponentially in N, and gives the rate of decay in terms of a *large deviations rate function* the role of which is played by $J(x)-J(p)$, see [37, 38] for the details.

12.1.3 Infinite Volume Limit for Random Trees

We must note here that the above results concern only the branching type of a random tree which ignores a lot of details about the geometry of the tree. Completely different random trees can have the same branching types. So let us describe a result from [34] that takes into account the way the various generations are related to each other. We need more notation and terminology.

Each rooted plane tree can be uniquely encoded as a sequence of generations. Each generation is represented by its vertices and pointers to their parents. More precisely, by a generation we mean a monotone (nondecreasing) map $G:\{1,\ldots,k\}\to\mathbb{N}$, or, equivalently, the set of pairs $\{(i,G(i)) : i=1,\ldots,k\}$ such that if $i_1\le i_2$ then $G(i_1)\le G(i_2)$. We denote by $|G|=k$ the number of vertices in the generation, and for any $i=1,\ldots,k$, $G(i)$ denotes i's parent number in the previous generation.

For two generations G and G' we write $G\lhd G'$ and say that G' is a continuation of G if $G'(|G'|)\le|G|$. Each tree of height n can be viewed as a sequence of generations

$$1\lhd G_1\lhd G_2\lhd\ldots\lhd G_n\lhd 0,$$

where $1\lhd G_1$ means $G_1(|G_1|)=1$ (the 0-th generation consists of a unique vertex, the root) and $G_n\lhd 0$ means that the generation $n+1$ is empty. Infinite sequences $1\lhd G_1\lhd G_2\lhd\ldots$ naturally encode infinite trees.

For any plane tree T and any $n\in\mathbb{N}$, $\pi_n T$ denotes the neighbourhood of the root of radius n, i.e. the subtree of T spanned by all vertices with height not exceeding n.

For any n and sufficiently large N, the map π_n pushes the measure P_N on $\mathbb{T}_N$ forward to the measure $P_N\pi_n^{-1}$ on S_n, the set of all trees with height n. In other words, $\pi_n T$ is an S_n-valued random variable with distribution $P_N\pi_n^{-1}$.

Theorem 12.2. *1. There is a unique measure P on infinite rooted plane trees with branching bounded by D such that for any $n \in \mathbb{N}$, as $N \to \infty$,*

$$P_N \pi_n^{-1} \to P \pi_n^{-1}$$

in total variation.

2. Measure P defines a Markov chain on generations $(G_n)_{n \geq 0}$. The transition probabilities are given by

$$\mathbf{P}(G_{n+1} = g' \mid G_n = g) = \begin{cases} \frac{|g'|}{|g|} p_{i_1} \cdots p_{i_{|g|}}, & g \lhd g', \\ 0, & \textit{otherwise}, \end{cases} \tag{12.16}$$

where i_k, $k = 1, \ldots, |g|$ denotes the number of vertices in generation g' that are children of k-th vertex in generation g.

The transition probability formula can be understood as follows: given that the current generation has $k > 0$ vertices, the conditional probability that the first vertex produces i_1 children, the second vertex produces i_2 children, ..., the k-th vertex produces i_k children, equals

$$\frac{i_1 + \ldots + i_k}{k} p_{i_1} \ldots p_{i_k}. \tag{12.17}$$

To check that these are well-defined transition probabilities, we can write

$$\begin{aligned} \sum_{i_1,\ldots,i_k} \frac{i_1 + \ldots + i_k}{k} p_{i_1} \cdots p_{i_k} &= \frac{k}{k} \sum_{i_1,\ldots,i_k} i_1 p_{i_1} \cdots p_{i_k} \\ &= \left(\sum_{i_1} i_1 p_{i_1} \right) \left(\sum_{i_2} p_{i_2} \right) \cdots \left(\sum_{i_k} p_{i_k} \right), \end{aligned}$$

(where we used the symmetry with respect to index permutations) and notice that each of the factors on the right-hand side equals 1.

This process was first obtained in [294], under the name of the family tree of a critical branching process conditioned on nonextinction, a term resulting from the limiting procedure used in [294]. The probability vector p plays the role of the *branching distribution* for the underlying Galton–Watson process.

Let us give a sketch of the proof of Theorem 12.2 given in [34]. Take any $n \in \mathbb{N}$ and any two trees τ_1 and τ_2 of height n. The plan is to find

$$\lim_{N \to \infty} \frac{P_N \pi_n^{-1}\{\tau_1\}}{P_N \pi_n^{-1}\{\tau_2\}}.$$

Each tree contributing to $P_N \pi_n^{-1}\{\tau_j\}$, $j = 1, 2$ can be split into two parts. The first part consists of the vertices of first $n - 1$ generations of τ_j. The energy contribution of these vertices is entirely defined by τ, and we denote it by $\overline{E}(\tau_j)$. If we assume that this part has m_j vertices, and the n-th generation of τ_j has k_j vertices then the rest of the tree is a forest with k_j components and $N - m_j$ vertices. We can use a generalization of formula (12.15). The number of plane forests on N vertices with k components and $r_0, r_1, \ldots, r_D$ vertices with branching numbers, respectively, $0, 1, \ldots, D$ is (see e.g. Theorem 5.3.10 in [480])

$$\frac{k}{N}\binom{N}{r_0,\ r_1,\ \ldots,\ r_D}$$

if $r_0 + \ldots + r_D = N, r_1 + 2r_2 + \ldots + Dr_D = N - k$, and 0 otherwise.

Therefore,

$$\frac{P_N\pi_{n,N}^{-1}\{\tau_1\}}{P_N\pi_{n,N}^{-1}\{\tau_2\}} = \frac{\mathrm{e}^{-\beta\overline{E}(\tau_1)}\sum_{r\in\Delta(N,m_1,k_1)}\frac{k_1}{N-m_1}\binom{N-m_1}{r_0,\ r_1,\ \ldots,\ r_D}\mathrm{e}^{-\beta E(r)}}{\mathrm{e}^{-\beta\overline{E}(\tau_2)}\sum_{r\in\Delta(N,m_2,k_2)}\frac{k_2}{N-m_2}\binom{N-m_2}{r_0,\ r_1,\ \ldots,\ r_D}\mathrm{e}^{-\beta E(r)}}.$$

Here

$$\Delta(N,m,k) = \left\{r \in \mathbb{Z}_+^{D+1} : \sum_{i=1}^{D} r_i = N - m,\ \sum_{i=1}^{D} ir_i = N - m - k\right\},$$

and $\mathbb{Z}_+ = \mathbb{N} \cup \{0\}$.

As well as in the proof of Theorem 12.1 one can prove that the dominating contributions to the sums in the numerator and denominator come from r such that $\frac{r}{N-m_j}$ lies in a small neighbourhood of p (where p is the solution of (12.9), (12.10)) intersected with $\Delta(N, m_j, k_j)$. Moreover, there is a natural way to match points of these two small sets to each other and estimate the ratio of contributions from individual points. We omit the details of the computation and give only the result:

$$\lim_{N\to\infty}\frac{P_N\pi_n^{-1}\{\tau_1\}}{P_N\pi_n^{-1}\{\tau_2\}} = \frac{k_1\mathrm{e}^{-\beta\overline{E}(\tau_1)}\rho^{k_1}\mathrm{e}^{m_1 J(p)}}{k_2\mathrm{e}^{-\beta\overline{E}(\tau_2)}\rho^{k_2}\mathrm{e}^{m_2 J(p)}},$$

where ρ was introduced in (12.11).

It follows that for any $\tau \in S_n$ with k vertices of height n and m vertices of height less than n,

$$\lim_{N\to\infty} P_N\pi_n^{-1}\{\tau\} = Q_n k\mathrm{e}^{-\beta\overline{E}(\tau)}\rho^k\mathrm{e}^{mJ(p)},$$

where Q_n depends only on n. One can use the consistency of limiting distributions for all values of n to deduce that, in fact, $Q_n = C\mathrm{e}^{-J(p)}$ (where C was introduced in (12.11)) and does not depend on n. This results in

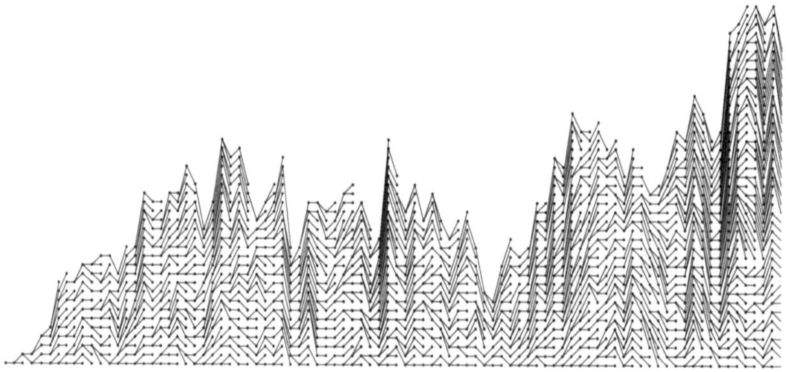

Fig. 12.2 Several first generations of a realization of an infinite random tree for $D = 2$, $E_0 = E_1 = E_2 = 0$

$$\lim_{N\to\infty} P_N \pi_n^{-1}\{\tau\} = Cke^{-\beta\overline{E}(\tau)}\rho^k e^{(m-1)J(p)},$$

which completely describes the limiting distribution on any finite number of generations. It now takes a straightforward calculation to see that the limiting process on generations is Markov and to derive the transition probabilities (12.16).

Notice that it is not a priori clear that the limiting process is Markov. The probability $P_N\pi_n^{-1}\{\tau\}$ depends on the "future", i.e., on the realization of the generations of the tree after the n-th one. However, this dependence disappears in the limit.

We can informally summarize this section as follows: large random trees under Gibbs distributions with nearest neighbour interaction asymptotically look and behave like realizations of Markov process on generations known as critical branching process conditioned on nonextinction. The limiting process has some interesting properties. Some of them like an "immortal particle" representation from [294] that represents the limiting tree via a unique infinite path with independent copies of critical Galton–Watson trees attached to it are well known. Some geometric properties of plane embeddings of these trees will be discussed in the next section.

Figure 12.2 shows several first generations of an infinite random tree realization for the case where $D = 2$, $E_0 = E_1 = E_2 = 0$, so that the branching distribution is $p_0 = p_1 = p_2 = 1/3$.

12.2 From Discrete to Continuum Random Trees

In this section, we study a scaling limit for the infinite random tree constructed in the previous section. The limits will be described in terms of diffusion processes and stochastic differential equations. We have to start with a brief and highly nonrigorous introduction to diffusion processes.

12.2.1 Diffusion Processes

First we have to recall a definition of a Markov process with continuous time. Suppose $\{\xi(t), t \geq 0\}$ is a stochastic process defined on a probability space $(\Omega, \mathcal{F}, \mathbf{P})$, with values in some metric space. For each $t > 0$ we denote

$$\mathcal{F}^{\xi}_{\leq t} = \sigma(\xi(s), 0 \leq s \leq t), \quad \mathcal{F}^{\xi}_{=t} = \sigma(\xi(t)), \quad \mathcal{F}^{\xi}_{\geq t} = \sigma(\xi(s), s \geq t).$$

These are sigma-algebras of events related to the past, present, and future with respect to t. A process is called Markov if for any $t > 0$, any bounded random variable $\zeta_{\geq t}$ measurable with respect to $\mathcal{F}^{\xi}_{\geq t}$,

$$\mathbf{E}(\zeta_{\geq t} \mid \mathcal{F}^{\xi}_{\leq t}) = \mathbf{E}(\zeta_{\geq t} \mid \mathcal{F}^{\xi}_{=t}).$$

Informally, a process is Markov if it exhibits instantaneous loss of memory and the future depends on the past only through the present.

Since the probabilistic properties of the evolution in the future depend only on the present, it is often convenient to work with transition probability kernels $P(s, x, t, A)$, describing the conditional probability for the process ξ to end up in set A at time $t \geq s$ given that $\xi(s) = x$. If the transition probabilities depend only on $t - s$, the process is called homogeneous and its transition kernel is denoted $P(x, t, A) = P(0, x, t, A)$.

It turns out that many $\mathbb{R}^d$-valued homogeneous Markov processes with continuous trajectories satisfy the following properties:

1. *Stochastic continuity*: for any x and any open set U containing x, $P(x, t, U^c) \to 0$ as $t \to 0$.
2. *Existence of local drift*: there is a vector $b(x) = (b_1(x), \ldots b_d(x))$ for any x, such that for any bounded open set U containing x, and any $i = 1, \ldots, d$

$$\int_U (y_i - x_i) P(x, t, dy) = b_i(x)t + o(t), \quad t \to 0. \tag{12.18}$$

3. *Existence of local covariance matrix*: there is a matrix $a(x) = (a_{ij}(x))^d_{i,j=1}$ for any x, such that for any bounded open set U containing x, and any $i, j = 1, \ldots, d$,

$$\int_U (y_i - x_i)(y_j - x_j) P(x, t, dy) = a_{ij}(x)t + o(t), \quad t \to 0. \tag{12.19}$$

We shall call these processes *diffusion processes*. Let us consider two simple examples. One example is a process with zero diffusion matrix a. In this case, the process is a solution of ODE $\dot{x} = b(x)$. If the drift is 0 and the diffusion matrix a is identical to the unit $d \times d$ matrix I_d, then the diffusion process is a standard d-dimensional Wiener process (we shall denote it by $W = (W_1, \ldots, W_d)$), i.e., for any s and $t > s$, the increment $W(t) - W(s)$ is a centered Gaussian random vector with covariance matrix $(t - s)I_d$, independent of $\mathcal{F}^W_{\le t}$.

Under mild restrictions on b and a, the distribution of the corresponding diffusion process coincides with that of the solution of the following stochastic differential equation:

$$d\xi(t) = b(\xi(t))\,dt + \sigma(\xi(t))\,dW(t),$$

where σ is a matrix function such that for each x, $\sigma(x)\sigma^T(x) = a(x)$. This equation has to be understood as an integral equation:

$$\xi(s) - \xi(0) = \int_0^s b(\xi(t))\,dt + \int_0^s \sigma(\xi(t))\,dW(t), \quad s > 0,$$

where the integral with respect to the Wiener process W is understood in the Itô sense.

Roughly speaking, a solution of a one-dimensional stochastic equation satisfies

$$\xi(t + \Delta t) - \xi(t) \approx b(\xi(t))\,\Delta t + \sigma(\xi(t))\,\Delta W(t), \tag{12.20}$$

where Δt is a small time interval and $\Delta W(t)$ is distributed as an increment of the Wiener process over this time interval. In other words, for small Δt the increment of ξ is approximately Gaussian with mean $b(\xi(t))\,\Delta t$ and variance $\sigma^2(\xi(t))\,\Delta t$.

The theory of Markov processes and stochastic equations can be found in many excellent books, see [164, 168, 266, 285]. One can roughly describe the diffusion process solving the equation as a stochastic perturbation (induced by the "noise" dW) of the deterministic dynamics defined by the vector field b. Typical realizations of diffusion processes are highly irregular. They are not smooth, and they are not even α-Hölder if $\alpha \ge 1/2$.

12.2.2 Convergence of the Process of Generation Sizes

We begin with a limit theorem for the process of the Markov infinite tree's generation sizes: $\xi_n = |G_n|$. We introduce moments of the distribution p:

$$B_n = \sum_{i=0}^{D} i^n p_i, \quad n \in \mathbb{N},$$

and its variance

$$\mu = B_2 - B_1^2 = B_2 - 1. \tag{12.21}$$

Let us now fix a positive time T and define

$$\eta_n(t) = \frac{1}{\mu n}(\xi_{[nt]} + \{nt\}(\xi_{[nt+1]} - \xi_{[nt]})), \quad t \in [0, T],$$

the rescaled process of linear interpolation between values of ξ_k given at integer times $k = 0, 1, \ldots$.

A proof of the following theorem showing that on average the generation sizes grow linearly can be found in [34], although the limiting process have appeared in the theory of random trees much earlier, see e.g. comments on Conjecture 7 in [8].

Theorem 12.3. *The distribution of η_n converges in the weak sense in uniform topology on $C([0,T])$ to the distribution of a diffusion process on $[0,\infty)$ with drift $b(x) = 1$ and diffusion $a(x) = x$ emitted from 0 at time 0.*

The limiting process can be viewed as a weak solution of the one-dimensional stochastic Itô equation

$$d\eta(t) = dt + \sqrt{\eta(t)}\, dW(t), \tag{12.22}$$

$$\eta(0) = 0, \tag{12.23}$$

i.e., this process can be realized on some probability space along with a Wiener process W so that equations (12.22), (12.23) hold true for all $t \geq 0$.

Proving weak convergence of distributions on a space like $C([0,T])$ is a nontrivial issue. It has been extensively studied, see [69, 168]. Most techniques involve verifying tightness of the sequence of measures and applying the Prokhorov theorem or one of its consequences. Checking the tightness condition usually requires significant efforts. Instead of going into details of a complete proof, let us see how the drift and diffusion coefficients can be guessed.

First let us compute the mean of the increment for the pre-limit process:

$$\begin{aligned}
\mathbf{E}(\xi_{j+1} \mid \xi_j = k) &= \frac{1}{k} \sum_{i_1,\ldots,i_k} (i_1 + \ldots + i_k)^2 p_{i_1} \cdots p_{i_k} \\
&= \frac{1}{k}\left[k \left(\sum_{i_1} i_1^2 p_{i_1}\right)\left(\sum_{i_2} p_{i_2}\right)^{k-1} + k(k-1)\left(\sum_{i_1} i_1 p_{i_1}\right)^2 \left(\sum_{i_2} p_{i_2}\right)^{k-2}\right] \\
&= \frac{1}{k}(kB_2 + k(k-1)) \\
&= B_2 + k - 1 = \mu + k,
\end{aligned} \tag{12.24}$$

so that

$$\mathbf{E}\left(\frac{\xi_{j+1}}{\mu n} - \frac{k}{\mu n} \,\middle|\, \frac{\xi_j}{\mu n} = \frac{k}{\mu n}\right) = \frac{1}{n}.$$

The role of a small time increment is played by $1/n$. The last formula shows that in time $1/n$ the process increases on average also by $1/n$. Comparing this to the definition of the drift (12.18), we conclude that the limiting drift $b(x)$ must be equal to 1 for all points x.

Similar computations show that

$$\begin{aligned}\mathbf{E}(\xi_{j+1}^2 \mid \xi_j = k) &= \frac{1}{k}\sum_{i_1,\ldots,i_k}(i_1+\ldots+i_k)^3 p_{i_1}\cdots p_{i_k}\\ &= \frac{1}{k}\big(kB_3 + 3k(k-1)B_2 + k(k-1)(k-2)\big)\\ &= B_3 + 3(k-1)B_2 + (k-1)(k-2). \qquad (12.25)\end{aligned}$$

Combining (12.24) and (12.25), we can compute that if we take a sequence of numbers k such that $k/(\mu n) \to x$ for some $x \geq 0$, then

$$\mathbf{E}\left(\left(\frac{\xi_{j+1}}{\mu n} - \frac{k}{\mu n}\right)^2 \,\middle|\, \frac{\xi_j}{\mu n} = \frac{k}{\mu n}\right) \to x \cdot \frac{1}{n}.$$

Comparing this with the definition of the diffusion matrix (12.19), we conclude that the limiting diffusion coefficient $a(x)$ is equal to x for all nonnegative x.

Of course, these computations do not prove the desired convergence of distributions of processes in $C([0,T])$, but they serve as a basis for a rigorous proof based on the martingale problem associated with a and b, see [34].

It follows from the Feller classification of boundary points for diffusions (see [266]) that the limiting process begins at 0, immediately drifts into the positive semiline and never touches 0 again staying positive all the time. The reason for this entrance and no exit character of point 0 is that the drift near 0 is sufficiently strong to dominate over the fluctuations generated by the diffusion coefficient.

Theorem 12.3 describes the behaviour of the process of population sizes and it ignores all the interesting details on how the generations are connected to each other.

12.2.3 Trees as Monotone Flows

This section is based on [35]. Our goal is to obtain a scaling limit for random trees that would take into account the genealogy geometry. To make sense of this limit, let us interprete trees as flows of monotone maps.

Let us describe the idea first. Suppose we have an infinite tree τ and consider generations n_1 and n_2 of the tree. Let the size of generation n_1 be k_1 and the size of generation n_2 be k_2. Then for $i \in \{1,\ldots,k_1\}$ we can find how many vertices in generation n_2 are descendants of vertices from $1,2,\ldots,i$ in generation n_1. Denoting this number by $m_{n_1,n_2}(i)$ we see that $m_{n_1,n_2}(i)$ is monotone nondecreasing in i, and maps $\{1,\ldots,k_1\}$ onto $\{1,\ldots,k_2\}$. Thus we have a family of consistent monotone maps m_{n_1,n_2} associated with the tree. It is natural to call this family a discrete monotone flow since n_1 and n_2 play the role of time. Obviously, two different trees will necessarily produce different flows, so that this procedure is a one-to-one encoding of trees by flows of monotone maps.

We are going to produce a continuum limit under an appropriate rescaling of the flows encoding random realizations of infinite random trees. To that end we must embed the discrete monotone flows defined above into a space of continuum monotone flows. Let us recall that Donsker's invariance principle for random walks involves constructing continuous trajectories out of realizations of discrete time random walks. This allows to embed the random walks into the space of continuous functions and prove convergence in that space, see e.g. [69] for details.

We have to develop an analogous procedure for random trees, and embed them into a space of monotone flows in continuous time.

We begin with considering a space of monotone maps. In our choice of the space and topology on it we have to take into account several things. In particular, the domains of monotone maps we consider do not have to coincide with each other.

Consider all points $z \geq 0$ and nonnegative nondecreasing functions f defined on $(-\infty, z]$ such that $f(x) = 0$ for all $x < 0$. Each of these functions has at most countably many discontinuities. We say that two such functions $f_1 : (-\infty, z_1] \to \mathbb{R}^+$, $f_2 : (-\infty, z_2] \to \mathbb{R}^+$ are equivalent if $z_1 = z_2$, $f_1(z_1) = f_2(z_2)$, and for each continuity point x of f_1, $f_1(x) = f_2(x)$. Although the roles of f_1 and f_2 seem to be different in this definition, it is easily seen to define a true equivalence relation. The set of all classes of equivalence will be denoted by $\mathbb{M}$. We would like to endow $\mathbb{M}$ with a metric structure, and to that end we develop a couple of points of view.

Sometimes, it is convenient to identify each element of $\mathbb{M}$ with its unique right-continuous representative. Sometimes, it is also convenient to work with graphs. The graph of a monotone function f defined on $(-\infty, z]$ is the set $G_f = \{(x, f(x)), x \leq z\}$. For each discontinuity point x of f one may consider the line segment $\bar{f}(x)$ connecting points $(x, f(x-))$ and $(x, f(x+))$. The continuous version of G_f is the union of G_f and all segments $\bar{f}(x)$. It is often convenient to identify an element of $\mathbb{M}$ with a continuous version of its graph restricted to $\mathbb{R}^2_+$, and we shall do so from now on calling the elements of $\mathbb{M}$ monotone graphs. Yet another way to look at monotone graphs is to think of them as monotone multivalued maps so that the image of each point is either a point $f(x)$ or a segment $\bar{f}(x)$.

The distance between $\Gamma_1 \in \mathbb{M}$ and $\Gamma_2 \in \mathbb{M}$ is defined via Hausdorff metric ρ_H of Γ_1 and Γ_2 as compact sets (graphs), see Sect. 1.2.3 for the definition of ρ_H.

Exercise 12.2. Prove that this metric turns $(\mathbb{M}, \rho_H)$ into a Polish (complete and separable) metric space.

We refer to [35] for several useful facts on the geometry of $(\mathbb{M}, \rho_H)$ such as criteria of convergence of sequences of monotone graphs. The main difference between the uniform distance on functions and ρ_H is that the former considers graphs of two functions to be close to each other if they differ just a little in the "vertical" direction, whereas the latter allows for proximity of graphs due to tweaking in the horizontal direction.

Let us introduce $z_j(\Gamma) = \sup\{x_j : (x_0, x_1) \in \Gamma\}$, $j = 0, 1$, and for two monotone graphs Γ_1 and Γ_2 with $z_1(\Gamma_1) = z_0(\Gamma_2)$, define their composition $\Gamma_2 \circ \Gamma_1$ as the set of all pairs (x_0, x_1) such that $(x_0, x_2) \in \Gamma_1$ and $(x_2, x_1) \in \Gamma_2$ for some x_2.

Let $T > 0$ and $\Delta_T = \{ (t_0, t_1) : 0 \le t_0 \le t_1 \le T \}$. We say that $(\Gamma^{t_0,t_1})_{(t_0,t_1)\in\Delta_T}$ is a (continuous) *monotone flow* on $[0, T]$ if the following properties are satisfied:

1. For each $(t_0, t_1) \in \Delta_T$, Γ^{t_0,t_1} is a monotone graph.
2. The monotone graph Γ^{t_0,t_1} depends on (t_0, t_1) continuously in ρ.
3. For each $t \in [0, T]$, $\Gamma^{t,t}$ is the identity map on $[0, \eta(t)]$ for some $\eta(t)$. The function η is called the *profile* of Γ.
4. For any $(t_0, t_1) \in \Delta_T$, $z_0(\Gamma^{t_0,t_1}) = \eta(t_0)$, $z_1(\Gamma^{t_0,t_1}) = \eta(t_1)$, where η is the profile of Γ.
5. If $(t_0, t_1) \in \Delta_T$ and $(t_1, t_2) \in \Delta_T$, then $\Gamma^{t_0,t_2} = \Gamma^{t_1,t_2} \circ \Gamma^{t_0,t_1}$.

It is easy to check that the space $\mathbb{M}[0, T]$ of all monotone flows on $[0, T]$ is a Polish space if equipped with uniform Hausdorff distance:

$$\rho_T(\Gamma_1, \Gamma_2) = \sup_{(t_0,t_1)\in\Delta_T} \rho_H(\Gamma_1^{t_0,t_1}, \Gamma_2^{t_0,t_1}). \tag{12.26}$$

Property 5 (consistency) implies that Property 2 (continuity) has to be checked only for $t_0 = t_1$.

Let us now embed infinite trees into $\mathbb{M}$. To any realization of an infinite tree τ we shall associate a continuous time monotone flow.

Recall that there are $\xi_n \ge 1$ vertices in the n-th generation of the tree. For $i \in \{ 1, \dots, \xi_n \}$, the i-th vertex of n-th generation is represented by the point $(n, i - 1)$ on the plane. The parent-child relation between two vertices of the tree is represented by a straight line segment connecting the representations of these vertices.

Besides these "regular" segments, we shall need some auxiliary segments that are not an intrinsic part of the tree but will be used in representing the discrete tree as a continuous flow. Suppose a vertex i in n-th generation has no children. Let j be the maximal vertex in generation $n + 1$ among those having their parents preceding i in generation n. Then an auxiliary segment of type I connects the points $(n, i - 1)$ and $(n + 1, j - 1)$. If vertex 1 in n-th generation has no children, points $(n, 0)$ and $(n + 1, 0)$ are connected by an auxiliary segment of type II. Auxiliary segments of type III connect points $(n, i - 1)$ and (n, i) for $1 \le i \le X_n - 1$.

Every bounded connected component of the complement to the union of the above segments on the plane is either a parallelogram with two vertical sides of length 1, or a triangle with one vertical side of length 1. One can treat both shapes as trapezoids (with one of the parallel sides having zero length in the case of triangle).

For each trapezoid, we shall establish a bijection with the unit square and define the monotone flow to act along the images of the "horizontal" segments of the square. A graphic illustration of the construction is given on Fig. 12.3, and we proceed to describe it precisely.

Each trapezoid L of this family has vertices $g_{0,0} = (n, i_{0,0})$, $g_{0,1} = (n, i_{0,1})$, $g_{1,0} = (n + 1, i_{1,0})$, $g_{1,1} = (n + 1, i_{1,1})$, where $i_{0,1} - i_{0,0} \in \{ 0, 1 \}$ and $i_{1,1} - i_{1,0} \in \{ 0, 1 \}$.

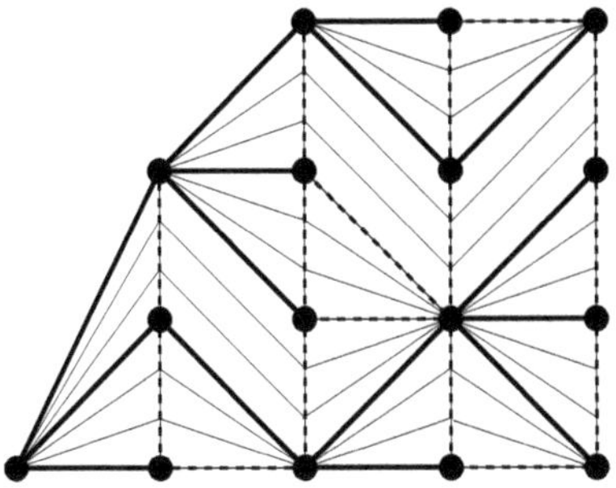

Fig. 12.3 Construction of the continuous monotone flow

Then, for every $\alpha \in (0,1)$ we define

$$g^L_m(\alpha) = g_{m,0} + \alpha(g_{m,1} - g_{m,0}), \quad m = 0, 1,$$

and

$$g^L(\alpha, s) = g^L_0(\alpha) + s(g^L_1(\alpha) - g^L_0(\alpha)), \quad s \in (0,1).$$

This definition introduces a coordinate system in L, i.e., a bijection between L and the unit square $(0,1) \times (0,1)$. We are going to use it to construct the monotone map associated with the tree for times t_0, t_1 assuming that there is $n \in \{0\} \cup \mathbb{N}$ such that $n < t_0 \le t_1 < n+1$. Let us take any x such that (x, t_0) belongs to one of the trapezoids L. Then there is a unique number $\alpha(x, t_0) \in (0,1)$ such that $g^L(\alpha(x,t_0), \{t_0\}) = x$, where $\{\cdot\}$ denotes the fractional part. We can define $g^{t_0,t_1}(x) = g^L(\alpha(x,t_0), \{t_1\})$. This strictly increasing function can be consistently and uniquely extended by continuity to points x such that (x, t_0) belongs either to a regular segment in the tree representation or an auxiliary segment of type I or II. This function g^{t_0,t_1} also uniquely defines a monotone graph $\tilde{\Gamma}^{t_0,t_1} = \tilde{\Gamma}^{t_0,t_1}(\tau)$ depending continuously on t_0, t_1. Next, if we allow t_0 and t_1 to take values n and $n+1$, then we can construct the associated monotone graph as the limit in $(\mathbb{M}, \rho_H)$ of the monotone graphs associated to the increasing functions defined above (as $t_0 \to n$ or $t_1 \to n+1$). Notice that the resulting monotone graphs may have intervals of constancy and shocks (i.e., contain horizontal and vertical segments). Now we can take any $(t_0, t_1) \in \Delta_\infty = \{(t_0, t_1) : 0 \le t_0 \le t_1\}$ and define

$$\tilde{\Gamma}^{t_0,t_1} = \tilde{\Gamma}^{[t_1],t_1} \circ \tilde{\Gamma}^{[t_1]-1,[t_1]} \circ \ldots \circ \tilde{\Gamma}^{[t_0]+1,[t_0]+2} \circ \tilde{\Gamma}^{t_0,[t_0]+1}$$

which results in a continuous monotone flow $(\tilde{\Gamma}^{t_0,t_1}(\tau))_{(t_0,t_1)\in\Delta_\infty}$.

To state our main result we need to introduce a rescaling of this family. For every $n \in \mathbb{N}$, we define

$$\Gamma^{t_0,t_1}_n(\tau) = \left\{ \left(\frac{x}{\mu n}, \frac{y}{\mu n} \right) : (x, y) \in \tilde{\Gamma}^{nt_0,nt_1}(\tau) \right\}, \quad (t_0, t_1) \in \Delta_\infty. \qquad (12.27)$$

Notice that this is exactly the same scaling as in Theorem 12.3. The time coordinate is rescaled by n, and the space coordinate is rescaled by μn.

For each $T > 0$ we can consider the uniform distance ρ_T on monotone flows in $\mathbb{M}[0,T]$ introduced in (12.26), and define the locally uniform (LU) metric on $\mathbb{M}[0,\infty)$ via

$$d(\Gamma_1,\Gamma_2) = \sum_{m=1}^{\infty} 2^{-m}(\rho_m(\Gamma_1,\Gamma_2)\wedge 1).$$

Theorem 12.4 ([35]). *Suppose p is a nonnegative vector satisfying (12.12) and (12.13). Let the infinite random tree τ be constructed according to transition probabilities (12.17). Let μ be defined by (12.21). The random monotone flow $\Gamma_n(\tau)$ defined in (12.27) converges as $n \to \infty$ in distribution in LU metric to a limiting flow Γ. The distribution of the limiting flow does not depend on p.*

This theorem is the first part of the main result of this chapter. It would not be complete without a description of the distribution of the limiting flow. This description can be obtained by tracing trajectories of individual points in the monotone flow.

Suppose $\Gamma \in \mathbb{M}[0,T]$, and η is the profile of Γ. Let $\zeta : \mathbb{R}^+ \times \Delta_T$ satisfy the following properties:

1. For any $(t_0,t_1) \in \Delta_T$, the function $\zeta(x,t_0,t_1)$ is monotone in $x \in [0,\eta(t_0)]$.
2. For any $(t_0,t_1) \in \Delta_T$, if $x \in [0,\eta(t_0)]$, then $(t_1,\zeta(x,t_0,t_1)) \in \Gamma^{t_0,t_1}$.
3. For all x, t_0, $\zeta(x,t_0,t_1)$ is continuous in t_1.

Then ζ and Γ are said to be compatible with each other, and U is said to be a *trajectory representation* of Γ. Clearly, the monotonicity implies that, given η, there is at most one monotone flow on $[0,T]$ compatible with ζ. Moreover, it is sufficient to know a trajectory representation $\zeta(x,t_0,t_1)$ for a dense set of points x, t_0, t_1 (e.g., rational points) to reconstruct the flow.

Although a trajectory representation for a monotone flow Γ with profile η is not unique (due to the presence of discontinuities in the monotone maps constituting the flow), there is a special representation $\zeta(x,t_0,t_1)$ that is right-continuous in $x \in [0,\eta(t_0)]$ for every $t_1 \geq t_0$:

$$\zeta(x,t_0,t_1) = \begin{cases} \sup\{\, y : (x,y) \in \Gamma^{t_0,t_1}\,\}, & x \in [0,\eta(t_0)] \\ x, & x > \eta(t_0). \end{cases}$$

The concrete way of defining $\zeta(x,t_0,t_1)$ for $x > \eta(t_0)$ is inessential for our purposes, and we often will simply ignore points (x,t_0,t_1) with $x > \eta(t_0)$.

It is often convenient to understand a monotone flow as a triple $\Gamma = (\Gamma,\eta,\zeta)$, where η is the profile of Γ, and ζ is one of the trajectory representations of Γ.

Theorem 12.5. *The distribution of the limiting monotone flow $\Gamma = (\Gamma,\eta,\zeta)$ of Theorem 12.4 is uniquely defined by the following properties:*

1. *The profile η is a weak solution of stochastic equation (12.22), (12.23).*
2. *For any $t_0 > 0$, any $m \in \mathbb{N}$, and any positive numbers $x_1 < \ldots < x_m$, on the event $\{\, x_m < \eta(t_0)\,\}$, the process*

$$(\zeta_1,\ldots,\zeta_m) = (\zeta(x_1,t_0,t),\ldots,\zeta(x_m,t_0,t)), \quad t \geq t_0.$$

is a weak solution of

$$d\zeta_k(t) = \frac{\zeta_k(t)}{\eta(t)}dt + \sum_{j=1}^{k} \sqrt{\zeta_k(t) - \zeta_{k-1}(t)}\, dW_j(t), \qquad (12.28)$$

$$\zeta_k(t_0) = x_k, \quad k = 1, \ldots, m,$$

where $(W_j)_{j=1}^m$ *are independent Wiener processes.*

Although the proofs of Theorems 12.4 and 12.5 are fairly technical, the derivation of the coefficients in the limiting stochastic equations is relatively easy and can be done in the spirit of our derivation of the coefficients for (12.22) describing the limiting behaviour of the profile of the random tree.

The trajectory description in terms of stochastic equations characterizes the distribution of the monotone flow uniquely, but its drawback is that it requires a separate stochastic system for each partition $x_1 < \ldots < x_m$ of the profile interval $[0, Z(t_0)]$. It turns out that one can write a single stochastic partial differential equation (SPDE) that describes the behaviour of trajectories for all partitions at once.

To write down this equation, we need a Browian sheet W on $\mathbb{R}_+^2$. It is a centered Gaussian random field indexed by bounded Borel subsets of $\mathbb{R}_+^2$, such that $\mathbf{cov}(W(A), W(B)) = |A \cap B|$, where $|\cdot|$ denotes the Lebesgue measure. It follows that $W(\cdot)$ is a finitely additive function on sets almost surely, the values of $W(\cdot)$ on disjoint sets are independent, the process $W(A \times [0, t])$ is a Gaussian martingale for any bounded Borel A. We refer to [505] for more background on the Brownian sheet and martingale measures in general.

The SPDE mentioned above is:

$$d\zeta(x, t_0, t) = \frac{\zeta(x, t_0, t)}{\eta(t)}\, dt + W([0, \zeta(x, t_0, t)] \times dt), \qquad (12.29)$$

$$\zeta(x, t_0, t_0) = x, \quad x \le \eta(t_0).$$

It must be understood as an integral equation

$$\zeta(x, t_0, t_1) = x + \int_{t_0}^{t_1} \frac{\zeta(x, t_0, t)}{\eta(t)}\, dt + \int_{t_0}^{t_1} \int_{\mathbb{R}} \mathbf{1}_{[0, \zeta(x, t_0, t)]}(y) W(dy \times dt),$$

where the right-hand side contains a stochastic integral with respect to the Brownian sheet, understood as an integral with respect to a martingale measure. Informally, the equation means that for small Δt, analogously to (12.20)

$$\zeta(x, t_0, t + \Delta t) - \zeta(x, t_0, t) \approx \frac{\zeta(x, t_0, t)}{\eta(t)}\, \Delta t + W([0, \zeta(x, t_0, t)] \times \Delta t),$$

where W is the Brownian sheet introduced above.

12.2.4 Structure of the Limiting Monotone Flow

The random monotone flow ζ constructed in the previous section being a distributional limit of trees can be called a continuum random tree. Instead of a discrete genealogy structure we have continual one and for any time t_0 and any interval $[a,b] \subset [0,\eta(t_0)]$ we can trace the progeny of this set for all future times. We can also do this for more general sets replacing $[a,b]$.

Other models of continual branching have appeared in the literature. Our model is most tighly connected to the Dawson–Watanabe superprocesses, and we comment on this in the end of this section.

Let us study the structure of the monotone flow ζ. The difference $\zeta(x_2,t_0,t_1) - \zeta(x_1,t_0,t_1)$ describes the mass of progeny generated by particles located between x_1 and x_2 at time t_0. It can happen that at some point this subpopulation becomes extinct, resulting in $\zeta(x_2,t_0,t_1) - \zeta(x_1,t_0,t_1) = 0$ for all times t_1 starting with the extinction time. Therefore, it is possible that monotone maps of the flow have intervals of constancy. In fact, this happens with probability one.

Another interesting effect which is well-known in the theory of superprocesses is that the monotone maps of the flow have discontinuities (shocks) with probability one. It means that infinitesimal mass at some time t_0 produces macroscopic progeny at $t_1 > t_0$. This can be seen using the following criterion of continuity:

Lemma 12.1. *If $f : [0,z] \to \mathbb{R}$ is a bounded variation function then its quadratic variation $\Psi(f)$ defined by*

$$\Psi(f) = \lim_{n\to\infty} \sum_{i=0}^{n} (f(z(i+1)/n) - f(zi/n))^2$$

satisfies:

$$\Psi(f) = \sum_{x\in\Delta(f)} (f(x+) - f(x-))^2,$$

where $\Delta(f)$ is the set of all discontinuity points of f. In particular, f is continuous on $[0,z]$ iff $\Psi(f) = 0$.

Let us compute the quadratic variation of $\zeta(\cdot,t_0,t_1)$ using tools of stochastic calculus. For $n \in \mathbb{N}$ we introduce (omitting dependence on n)

$$x_k = \frac{k}{n}\cdot\eta(t_0), \quad k = 0,\dots,n,$$

and

$$\gamma_k(t_1) = \zeta(x_k,t_0,t_1) - \zeta(x_{k-1},t_0,t_1), \quad k = 1,\dots,n.$$

Itô's formula (see, e.g., [168, Theorem 2.9, Chap. 5]) implies

$$\begin{aligned} d\gamma_k^2(t_1) &= 2\gamma_k(t_1)\,d\gamma_k(t_1) + \gamma_k(t_1)\,dt_1 \\ &= \frac{2\gamma_k^2(t_1)}{\eta(t_1)}\,dt_1 + 2\gamma_k(t_1) W([\zeta(x_{k-1},t_0,t_1),\zeta(x_k,t_0,t_1)] \times dt_1) + \gamma_k(t_1)\,dt_1. \end{aligned}$$

Let

$$\Psi_n(t_1) = \gamma_1^2(t_1) + \ldots + \gamma_n^2(t_1).$$

Then $\Psi_n(t_0) = (\eta(t_0))^2/n$, and

$$\begin{aligned} \Psi_n(t_1) &= \frac{(\eta(t_0))^2}{n} + 2\int_{t_0}^{t_1} \frac{\Psi_n(t)}{\eta(t)}\,dt \\ &\quad + 2\sum_{k=1}^{n}\int_{t_0}^{t_1} \gamma_k(t) W([\zeta(x_{k-1},t_0,t),\zeta(x_k,t_0,t)] \times dt) + \int_{t_0}^{t_1} \eta(t)\,dt. \end{aligned}$$

Let us define $\Psi(t) = \lim_{n\to\infty} \Psi_n(t)$ and $\nu = \inf\{\, t > t_0 : \Psi(t) > 0 \,\}$. If $\nu > t_0$, then taking the limit as $n \to \infty$ in both sides of the equation above at $t_1 = \nu$, we see that all terms converge to zero except for $\int_{t_0}^{\nu} \eta(t)\,dt$. To obtain the convergence to zero for the martingale stochastic integral term, it is sufficient to see that its quadratic variation in time converges to zero as $n \to \infty$. Recall that the quadratic variation of a stochastic process $(\xi_t)_{t\ge 0}$ is another process denoted by $([\xi]_t)_{t\ge 0}$ such that for each $t > 0$,

$$\sum_{i=1}^{m} \left(\xi\left(\frac{i}{m}t\right) - \xi\left(\frac{i-1}{m}t\right)\right)^2 \xrightarrow{\mathbf{P}} [\xi]_t, \quad m \to \infty.$$

For the martingale term above, the quadratic variation converges to 0 as $n \to \infty$ since

$$\left[\sum_{k=1}^{n}\int_{t_0}^{\cdot} \gamma_k(t) W([\zeta(x_{k-1},t_0,t),\zeta(x_k,t_0,t)] \times dt)\right]_{\nu} = \int_{t_0}^{\nu} \sum_{k=1}^{n} \gamma_k^3(t)\,dt.$$

Since $\int_{t_0}^{\nu} \eta(t)\,dt$ is a strictly positive random variable that does not depend on n, we obtain a contradiction which shows that $\nu = t_0$, so that $\Psi(t) > 0$ for any $t > t_0$, and the proof of a.s.-existence of shocks in monotone maps of the flow is finished.

These effects of extinction of subpopulations and creation of positive mass by infinitesimal elements lead to an interesting geometric picture. The area in space-time below the profile process η

$$\{\,(t,x) \in \mathbb{R}_+^2 : 0 \le x \le \eta(t)\,\}$$

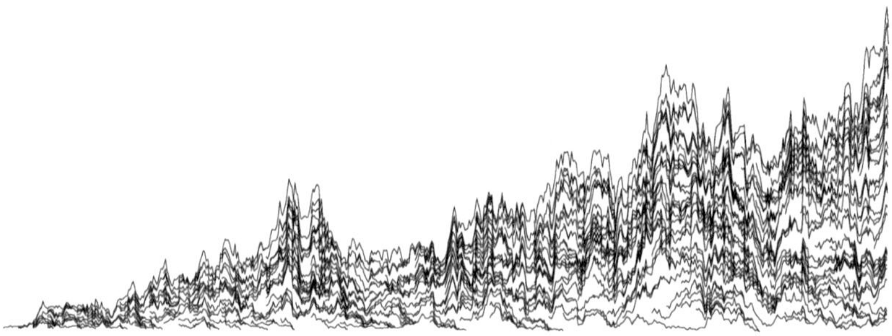

Fig. 12.4 Stochastic foliation constructed for 600 generations of a tree

is foliated by diffusion trajectories $\zeta(x, t_0, t_1)$. However, this *stochastic foliation* is very irregular. A most regular foliation would be represented by introducing space-time coordinates so that level sets for the space coordinate map coincide with the diffusion trajectories. This is exactly the situation with stochastic flows of diffeomorphisms generated by stochastic Itô or Stratonovich equations, see [316]. However, our monotone flow generated by an SPDE is very far from being a flow of diffeomorphisms. The creation of positive mass from infinitesimal elements is followed by extinction that results in blobs that make it impossible to introduce a reasonable global coordinate system in the stochastic foliation.

Figure 12.4 shows a realization of a pre-limit monotone flow for a large random tree. Every tenth generation is split into about ten subpopulations, their progenies are tracked and shown on the figure.

There is an important connection of our results to the theory of superprocesses. Superprocesses are measure-valued stochastic processes describing the evolution of populations of branching and migrating particles. The limiting SPDE that we have constructed is similar to the genealogy in the Dawson–Watanabe superprocess with no motion conditioned on nonextinction, see [169, 170]. Our approach is more geometric than the superprocess point of view. For the superprocess corresponding to our situation, the continual mass momentarily organizes itself into a finite random number of atoms of positive mass (corresponding to discontinuities of the monotone maps in our approach). The mass of these atoms evolves in time analogously to (12.28), but our approach helps to understand what happens inside the atoms by unfolding the details of the genealogy. We emphasize the ordering and the geometry of the stochastic foliation, the structures ignored in the superprocess approach. So, the dynamics we have constructed is richer than in the corresponding superprocess.

Combining all the results of this chapter we conclude that a typical embedding of a large ordered rooted tree in the plane if rescaled appropriately looks like a stochastic foliation described by SPDE (12.29). It would be interesting to obtain rigorously a direct convergence result that would not involve the intermediate infinite discrete tree. However, currently this kind of result is not available.

References

1. Adler, R.J., Taylor, J.E.: Random Fields and Geometry. Springer, New York (2007)
2. Affentranger, F.: The convex hull of random points with spherically symmetric distributions. Rend. Sem. Mat. Univ. Pol. Torino **49**, 359–383 (1991)
3. Affentranger, F.: Aproximación aleatoria de cuerpos convexos. Publ. Mat. **36**, 85–109 (1992)
4. Affentranger, F., Schneider, R.: Random projections of regular simplices. Discrete Comput. Geom. **7**, 219–226 (1992)
5. Agterberg, F.P.: Automatic contouring of geological maps to detect target areas for mineral exploration. J. Int. Ass. Math. Geol. **6**, 373–395 (1974)
6. Ahlswede, R., Blinovsky, V.: Lectures on Advances in Combinatorics. Springer, Berlin (2008)
7. Ahlswede, R., Daykin, D.E.: An inequality for the weights of two families of sets, their unions and intersections. Probab. Theor. Relat. Fields **43**, 183–185 (1978)
8. Aldous, D.: The continuum random tree. II. An overview. In: Barlow, M.T., Bingham, N.H. (eds.) Stochastic Analysis. London Mathematical Society Lecture Note Series, vol. 167. Cambridge University Press, Cambridge (1991)
9. Alon, N., Spencer, J.H.: The Probabilistic Method, 2nd edn. Wiley, New York (2000)
10. Alonso-Gutiérrez, D.: On the isotropy constant of random convex sets. Proc. Am. Math. Soc. **136**, 3293–3300 (2008)
11. Ambartzumian, R.V.: On an equation for stationary point processes. Dokl. Akad. Nauk Armjanskoi SSR **42**, 141–147 (1966)
12. Anandkumar, A., Yukich, J.E., Tong, L., Swami, A.: Energy scaling laws for distributed inference in random networks. IEEE J. Sel. Area. Comm., Issue on Stochastic Geometry and Random Graphs for Wireless Networks **27**, 1203–1217 (2009)
13. Anderson, D.N.: A multivariate Linnik distribution. Stat. Probab. Lett. **14**, 333–336 (1992)
14. Apanasovich, T.V., Genton, M.G.: Cross-covariance functions for multivariate random fields based on latent dimensions. Biometrika **97**, 15–30 (2010)
15. Araujo, A., Giné, E.: The Central Limit Theorem for Real and Banach Valued Random Variables. Wiley, New York (1980)
16. Aronszajn, N.: Theory of reproducing kernels. Trans. Am. Math. Soc. **68**, 337–404 (1950)
17. Artstein, Z., Vitale, R.A.: A strong law of large numbers for random compact sets. Ann. Probab. **3**, 879–882 (1975)
18. Askey, R.: Refinements of Abel summability for Jacobi series. In: More, C.C. (ed.) Harmonic Analysis on Homogeneous Spaces. Proceedings of Symposium in Pure Mathematics, vol. XXVI. AMS, Providence (1973)
19. Auneau, J., Jensen, E.B.V.: Expressing intrinsic volumes as rotational integrals. Adv. Appl. Math. **45**, 1–11 (2010)

E. Spodarev (ed.), *Stochastic Geometry, Spatial Statistics and Random Fields*, Lecture Notes in Mathematics 2068, DOI 10.1007/978-3-642-33305-7,

20. Avram, F., Bertsimas, D.: On central limit theorems in geometrical probability. Ann. Appl. Probab. **3**, 1033–1046 (1993)
21. Baccelli, F., Blaszczyszyn, B.: Stochastic Geometry and Wireless Networks. Now Publishers, Delft (2009)
22. Baddeley, A.: Fundamentals of point process statistics. In: Baddeley, A.J., Bárány, I., Schneider, R., Weil, W. (eds.) Stochastic Geometry - Lecture Notes in Mathematics, vol. 1892. Springer, Berlin (2007)
23. Baddeley, A.: Analysing spatial point patterns in R. Tech. rep., CSIRO Mathematics, Informatics and Statistics (2008). www.csiro.au/resources/pf16h.html
24. Baddeley, A., Bárány, I., Schneider, R., Weil, W.: Stochastic Geometry. Springer, Heidelberg (2004)
25. Baddeley, A., Berman, M., Fisher, N.I., Hardegen, A., Milne, R.K., Schuhmacher, D., Shah, R., Turner, R.: Spatial logistic regression and change-of-support for Poisson point processes. Electron. J. Stat. **4**, 1151–1201 (2010). doi: 10.1214/10-EJS581
26. Baddeley, A., Turner, R.: Practical maximum pseudolikelihood for spatial point patterns (with discussion). Aust. N. Z. J. Stat. **42**, 283–322 (2000)
27. Baddeley, A., Turner, R.: Spatstat: an R package for analyzing spatial point patterns. J. Stat. Software **12**, 1–142 (2005). www.jstatsoft.org
28. Baddeley, A.J.: Vertical sections. In: Weil, W., Ambartzumian, R.V. (eds.) Stochastic Geometry and Stereology (Oberwolfach 1983). Teubner, Berlin (1983)
29. Baddeley, A.J.: Time-invariance estimating equations. Bernoulli **6**, 783–808 (2000)
30. Baddeley, A.J., Gundersen, H.J.G., Cruz-Orive, L.M.: Estimation of surface area from vertical sections. J. Microsc. **142**, 259–276 (1986)
31. Baddeley, A.J., Jensen, E.B.V.: Stereology for Statisticians. Chapman & Hall, Boca Raton (2005)
32. Baddeley, A.J., van Lieshout, M.N.M.: Area-interaction point processes. Ann. Inst. Stat. Math. **47**, 601–619 (1995)
33. Bakhtin, Y.: Limit theorems for associated random fields. Theor. Probab. Appl. **54**, 678–681 (2010)
34. Bakhtin, Y.: Thermodynamic limit for large random trees. Random Struct. Algorithm **37**, 312–331 (2010)
35. Bakhtin, Y.: SPDE approximation for random trees. Markov Process. Relat. Fields **17**, 1–36 (2011)
36. Bakhtin, Y., Bulinski, A.: Moment inequalities for sums of dependent multiindexed random variables. Fundam. Prikl. Math. **3**, 1101–1108 (1997)
37. Bakhtin, Y., Heitsch, C.: Large deviations for random trees. J. Stat. Phys. **132**, 551–560 (2008)
38. Bakhtin, Y., Heitsch, C.: Large deviations for random trees and the branching of RNA secondary structures. Bull. Math. Biol. **71**, 84–106 (2009)
39. Baltz, A., Dubhashi, D., Srivastav, A., Tansini, L., Werth, S.: Probabilistic analysis for a vehicle routing problem. Random Struct. Algorithms **30**, 206–225 (2007)
40. Bandle, C.: Isoperimetric inequalities and applications. Monographs and Studies in Mathematics, vol. 7. Pitman (Advanced Publishing Program), Boston (1980)
41. Banys, P.: CLT for linear random fields with stationary martingale–difference innovation. Lithuanian Math. J. **17**, 1–7 (2011)
42. Bárány, I.: Random polytopes in smooth convex bodies. Mathematika **39**, 81–92 (1992). Corrigendum: Mathematika **51**, 31 (2005)
43. Bárány, I.: Random polytopes, convex bodies, and approximation. In: Baddeley, A.J., Bárány, I., Schneider, R., Weil, W. (eds.) Stochastic Geometry - Lecture Notes in Mathematics, vol. 1892. Springer, Berlin (2007)
44. Bárány, I.: Random points and lattice points in convex bodies. Bull. Am. Math. Soc. **45**, 339–365 (2008)
45. Bárány, I., Buchta, C.: Random polytopes in a convex polytope, independence of shape, and concentration of vertices. Math. Ann. **297**, 467–497 (1993)

46. Bárány, I., Fodor, F., Vigh, V.: Intrinsic volumes of inscribed random polytopes in smooth convex bodies. Adv. Appl. Probab. **42**, 605–619 (2009)
47. Bárány, I., Larman, D.G.: Convex bodies, economic cap coverings, random polytopes. Mathematika **35**, 274–291 (1988)
48. Bárány, I., Reitzner, M.: On the variance of random polytopes. Adv. Math. **225**, 1986–2001 (2010)
49. Bárány, I., Reitzner, M.: Poisson polytopes. Ann. Probab. **38**, 1507–1531 (2010)
50. Bárány, I., Vu, V.: Central limit theorems for Gaussian polytopes. Ann. Probab. **35**, 1593–1621 (2007)
51. Barbato, D.: FKG inequalities for Brownian motion and stochastic differential equations. Electron. Comm. Probab. **10**, 7–16 (2005)
52. Bardossy, A.: Introduction to Geostatistics. Tech. rep., University of Stuttgart, Stuttgart (1997)
53. Barndorff-Nielsen, O.E., Mikosch, T., Resnick, S.I. (eds.): Lévy processes. Theory and Applications. Birkhäuser, Boston (2001)
54. Barndorff-Nielsen, O.E., Schmiegel, J.: Spatio-temporal modeling based on Lévy processes, and its applications to turbulence. Russ. Math. Surv. **59**, 65–90 (2004)
55. Baryshnikov, Y., Eichelsbacher, P., Schreiber, T., Yukich, J.E.: Moderate deviations for some point measures in geometric probability. Annales de l'Institut Henri Poincaré - Probabilités et Statistiques **44**, 442–446 (2008)
56. Baryshnikov, Y., Penrose, M., Yukich, J.E.: Gaussian limits for generalized spacings. Ann. Appl. Probab. **19**, 158–185 (2009)
57. Baryshnikov, Y., Yukich, J.E.: Gaussian limits for random measures in geometric probability. Ann. Appl. Probab. **15**, 213–253 (2005)
58. Baryshnikov, Y.M., Vitale, R.A.: Regular simplices and Gaussian samples. Discrete Comput. Geom. **11**, 141–147 (1994)
59. Baumstark, V., Last, G.: Some distributional results for Poisson-Voronoi tessellations. Adv. Appl. Probab. **39**, 16–40 (2007)
60. Baumstark, V., Last, G.: Gamma distributions for stationary Poisson flat processes. Adv. Appl. Probab. **41**, 911–939 (2009)
61. Beardwood, J., Halton, J.H., Hammersley, J.M.: The shortest path through many points. Proc. Camb. Philos. Soc. **55**, 229–327 (1959)
62. Berg, C., Christensen, J.P.R., Ressel, P.: Harmonic Analysis on Semigroups. Springer, Berlin (1984)
63. van den Berg, J., Kahn, J.: A correlation inequality for connection events in percolation. Ann. Probab. **29**, 123–126 (2001)
64. Berman, M., Turner, T.R.: Approximating point process likelihoods with GLIM. Appl. Stat. **41**, 31–38 (1992)
65. Besag, J.: Statistical analysis of non-lattice data. The Statistician **24**, 179–195 (1975)
66. Besag, J.: Some methods of statistical analysis for spatial data. Bull. Int. Stat. Inst. **44**, 77–92 (1978)
67. Bickel, P., Yan, D.: Sparsity and the possibility of inference. Sankhyā **70**, 1–23 (2008)
68. Biermé, H., Meerschaert, M.M., Scheffler, H.P.: Operator scaling stable random fields. Stoch. Process. Appl. **117**, 312–332 (2007)
69. Billingsley, P.: Convergence of Probability Measures, 2nd edn. Wiley Series in Probability and Statistics: Probability and Statistics. Wiley, New York (1999)
70. Bingham, N.H., Goldie, C.M.: Riesz means and self-neglecting functions. Math. Z. **199**, 443–454 (1988)
71. Bingham, N.H., Goldie, C.M., Teugels, J.L.: Regular Variation. Cambridge University Press, Cambridge (1987)
72. Birkel, T.: Moment bounds for associated sequences. Ann. Probab. **16**, 1184–1193 (1988)
73. Birkhoff, G.: Lattice Theory, 3rd edn. AMS, Providence (1979)
74. Bolthausen, E.: On the central limit theorem for stationary mixing random fields. Ann. Probab. **10**, 1047–1050 (1982)

75. Böröczky, K.J., Fodor, F., Hug, D.: The mean width of random polytopes circumscribed around a convex body. J. Lond. Math. Soc. **81**, 499–523 (2010)
76. Böröczky, K.J., Fodor, F., Reitzner, M., Vígh, V.: Mean width of random polytopes in a reasonably smooth convex body. J. Multivariate Anal. **100** (2009)
77. Böröczky, K.J., Hoffmann, L.M., Hug, D.: Expectation of intrinsic volumes of random polytopes. Period. Math. Hungar. **57**, 143–164 (2008)
78. Böröczky, K.J., Hug, D.: Stability of the reverse Blaschke-Santaló inequality for zonoids and applications. Adv. Appl. Math. **44**, 309–328 (2010)
79. Böröczky, K.J., Schneider, R.: The mean width of circumscribed random polytopes. Can. Math. Bull. **53**, 614–628 (2010)
80. Bowman, F.: Introduction to Bessel functions. Dover, New York (1958)
81. Box, G.E.P., Jenkins, G.M., Reinsel, G.C.: Time Series Analysis, 4th edn. Wiley, Hoboken (2008)
82. Box, G.E.P., Muller, M.E.: A note on the generation of random normal deviates. Ann. Math. Stat. **29**, 610–611 (1958)
83. Bradley, R.C.: Introduction to Strong Mixing Conditions. Vol. 1, 2, 3. Kendrick Press, Heber City (2007)
84. Brémaud, P.: Markov Chains: Gibbs fields, Monte Carlo simulation, and queues. Texts in Applied Mathematics, vol. 31. Springer, New York (1999)
85. Brillinger, D.R.: Time Series. Data Analysis and Theory. Holt, Rinehart and Winston, Inc., New York (1975)
86. Brix, A., Kendall, W.S.: Simulation of cluster point processes without edge effects. Adv. Appl. Prob. **34**, 267–280 (2002)
87. Brockwell, P.J., Davis, R.A.: Time Series: Theory and Methods, 2nd edn. Springer, New York (1991)
88. Buchta, C.: Zufällige Polyeder - Eine Übersicht. In: Hlawka, E. (ed.) Zahlentheoretische Analysis - Lecture Notes in Mathematics, vol. 1114. Springer, Berlin (1985)
89. Buchta, C.: An identity relating moments of functionals of convex hulls. Discrete Comput. Geom. **33**, 125–142 (2005)
90. Buchta, C., Reitzner, M.: The convex hull of random points in a tetrahedron: Solution of Blaschke's problem and more general results. J. Reine Angew. Math. **536**, 1–29 (2001)
91. Bulinski, A.: Inequalities for moments of sums of associated multi-indexed variables. Theor. Probab. Appl. **38**, 342–349 (1993)
92. Bulinski, A.: Central limit theorem for random fields and applications. In: Skiadas, C.H. (ed.) Advances in Data Analysis. Birkhäuser, Basel (2010)
93. Bulinski, A.: Central limit theorem for positively associated stationary random fields. Vestnik St. Petersburg University: Mathematics **44**, 89–96 (2011)
94. Bulinski, A., Shashkin, A.: Rates in the central limit theorem for dependent multiindexed random vectors. J. Math. Sci. **122**, 3343–3358 (2004)
95. Bulinski, A., Shashkin, A.: Strong invariance principle for dependent random fields. In: Dynamics & Stochastics. IMS Lecture Notes Monograph Series, vol. 48, pp. 128–143. Institute of Mathematical Statistics, Beachwood (2006)
96. Bulinski, A., Shashkin, A.: Limit Theorems for Associated Random Fields and Related Systems. World Scientific, Singapore (2007)
97. Bulinski, A., Spodarev, E., Timmermann, F.: Central limit theorems for the excursion sets volumes of weakly dependent random fields. Bernoulli **18**, 100–118 (2012)
98. Bulinski, A., Suquet, C.: Normal approximation for quasi-associated random fields. Stat. Probab. Lett. **54**, 215–226 (2001)
99. Bulinski, A.V.: Limit Theorems under Weak Dependence Conditions. MSU. Moscow (1990) (in Russian)
100. Bulinski, A.V., Keane, M.S.: Invariance principle for associated random fields. J. Math. Sci. **81**, 2905–2911 (1996)
101. Bulinski, A.V., Kryzhanovskaya, N.: Convergence rate in CLT for vector-valued random fields with self-normalization. Probab. Math. Stat. **26**, 261–281 (2006)

102. Bulinski, A.V., Shiryaev, A.N.: Theory of Stochastic Processes. FIZMATLIT, Moscow (2005) (in Russian)
103. Bürgisser, P., Cucker, F., Lotz, M.: Coverage processes on spheres and condition numbers for linear programming. Ann. Probab. **38**, 570–604 (2010)
104. Burton, R., Waymire, E.: Scaling limits for associated random measures. Ann. Probab. **13**, 1267–1278 (1985)
105. Burton, R.M., Dabrowski, A.R., Dehling, H.: An invariance principle for weakly associated random vectors. Stoch. Process. Appl. **23**, 301–306 (1986)
106. Calka, P.: Mosaïques Poissoniennes de l'espace Euclidian. Une extension d'un résultat de R. E. Miles. C. R. Math. Acad. Sci. Paris **332**, 557–562 (2001)
107. Calka, P.: The distributions of the smallest disks containing the Poisson-Voronoi typical cell and the Crofton cell in the plane. Adv. Appl. Probab. **34**, 702–717 (2002)
108. Calka, P.: Tessellations. In: Kendall, W.S., Molchanov, I. (eds.) New Perspectives in Stochastic Geometry. Oxford Univerxity Press, London (2010)
109. Calka, P., Schreiber, T.: Limit theorems for the typical Poisson-Voronoi cell and the Crofton cell with a large inradius. Ann. Probab. **33**, 1625–1642 (2005)
110. Calka, P., Schreiber, T.: Large deviation probabilities for the number of vertices of random polytopes in the ball. Adv. Appl. Probab. **38**, 47–58 (2006)
111. Calka, P., Schreiber, T., Yukich, J.E.: Brownian limits, local limits and variance asymptotics for convex hulls in the ball. Ann. Probab. (2012) (to appear)
112. Carlsson, G.: Topology and data. Bull. Am. Math. Soc. (N.S.) **46**, 255–308 (2009)
113. Castaing, C., Valadier, M.: Convex Analysis and Measurable Multifunctions. Lecture Notes Mathematics, vol. 580. Springer, Berlin (1977)
114. Chambers, J.M., Mallows, C., Stuck, B.W.: A method for simulating stable random variables. J. Am. Stat. Assoc. **71**, 340–344 (1976)
115. Chan, G., Wood, A.T.A.: An algorithm for simulating stationary Gaussian random fields. J. R. Stat. Soc. Ser. C **46**, 171–181 (1997)
116. Chatterjee, S.: A new method of normal approximation. Ann. Probab. **36**, 1584–1610 (2008)
117. Chayes, L., Lei, H.K.: Random cluster models on the triangular lattice. J. Stat. Phys. **122**, 647–670 (2006)
118. Chazal, F., Guibas, L., Oudot, S., Skraba, P.: Analysis of scalar fields over point cloud data, Preprint (2007)
119. Chazal, F., Oudot, S.: Towards persistence-based reconstruction in euclidean spaces. ACM Symp. Comput. Geom. **232** (2008)
120. Chen, L., Shao, Q.M.: Normal approximation under local dependence. Ann. Probab. **32**, 1985–2028 (2004)
121. Chen, T., Huang, T.S.: Region based hidden Markov random field model for brain MR image segmentation. World Academy of Sciences. Eng. Technol. **4**, 233–236 (2005)
122. Chilès, J.P., Delfiner, P.: Geostatistics: Modeling Spatial Uncertainty. Wiley, New York (1999)
123. Chow, Y.S.: Delayed sums and Borel summability of independent, identically distributed random variables. Bull. Inst. Math. Acad. Sinica **1**, 207–220 (1973)
124. Chow, Y.S., Lai, T.L.: Some one-sided theorems on the tail distribution of sample sums with applications to the last time and largest excess of boundary crossings. Trans. Am. Math. Soc. **208**, 51–72 (1975)
125. Chow, Y.S., Teicher, H.: Probability Theory: Independence, Interchangeability, Martingales. Springer Texts in Statistics. Springer, New York (2003)
126. Christofidis, T.C., Vaggelatou, E.: A connection between supermodular ordering and positive/negative association. J. Multivariate Anal. **88**, 138–151 (2004)
127. Clyde, M., Strauss, D.: Logistic regression for spatial pair-potential models. In: Possolo, A. (ed.) Spatial Statistics and Imaging. IMS Lecture Notes Monograph Series, vol. 20, chap. II, pp. 14–30. Institute of Mathematical Statistics, Hayward, California (1991)
128. Cohen, S., Lacaux, C., Ledoux, M.: A general framework for simulation of fractional fields. Stoch. Proc. Appl. **118**, 1489–1517 (2008)
129. Cont, R., Tankov, P.: Financial Modeling with Jump Processes. Chapman & Hall, Boca Raton (2004)

130. Costa, J., Hero III, A.: Geodesic entropic graphs for dimension and entropy estimation in manifold learning. IEEE Trans. Signal Process. **58**, 2210–2221 (2004)
131. Costa, J., Hero III, A.: Determining intrinsic dimension and entropy of high-dimensional shape spaces. In: Krim, H., Yezzi, A. (eds.) Statistics and Analysis of Shapes. Birkhäuser, Basel (2006)
132. Cowan, R.: The use of the ergodic theorems in random geometry. Adv. in Appl. Prob. **10** (suppl.), 47–57 (1978)
133. Cowan, R.: Properties of ergodic random mosaic processes. Math. Nachr. **97**, 89–102 (1980)
134. Cox, D.R., Hinkley, D.V.: Theoretical Statistics. Chapman & Hall, London (1974)
135. Cox, J.T., Grimmett, G.: Central limit theorems for associated random variables and the percolation model. Ann. Probab. **12**, 514–528 (1984)
136. Cressie, N.A.C.: Statistics for Spatial Data, 2nd edn. Wiley, New York (1993)
137. Cruz-Orive, L.M.: A new stereological principle for test lines in threedimensional space. J. Microsc. **219**, 18–28 (2005)
138. Csörgő, M., Révész, P.: Strong Approximations in Probability and Statistics. Academic, New York (1981)
139. Dafnis, N., Giannopoulos, A., Guédon, O.: On the isotropic constant of random polytopes. Adv. Geom. **10**, 311–322 (2010)
140. Daley, D.J., Vere-Jones, D.: An introduction to the theory of point processes. Vol. I and II. Probab. Appl. (New York). Springer, New York (2003/2008)
141. Davy, P.J.: Stereology: A Statistical Viewpoint. Ph.D. thesis, Australian National University (1978)
142. Davy, P.J., Miles, R.E.: Sampling theory for opaque spatial specimens. J. R. Stat. Soc. Ser. B Stat. Meth. **39**, 56–65 (1977)
143. Davydov, Y., Molchanov, I., Zuyev, S.: Stable distributions and harmonic analysis on convex cones. C. R. Math. Acad. Sci. Paris **344**, 321–326 (2007)
144. Davydov, Y., Molchanov, I., Zuyev, S.: Strictly stable distributions on convex cones. Electron. J. Probab. **13**, 259–321 (2008)
145. Davydov, Y., Molchanov, I., Zuyev, S.: Stability for random measures, point processes and discrete semigroups. Bernoulli **17**, 1015–1043 (2011)
146. Davydov, Y.A.: Convergence of distributions generated by stationary stochastic processes. Theor. Probab. Appl. **13**, 691–696 (1968)
147. Dedecker, J.: A central limit theorem for stationary random fields. Probab. Theor. Relat. Fields **110**, 397–437 (1998)
148. Dedecker, J.: Exponential inequalities and functional central limit theorems for a random field. ESAIM Probab. Stat. **5**, 77–104 (2001)
149. Dedecker, J., Louhichi, S.: Convergence to infinitely divisible distributions with finite variance for some weakly dependent sequences. ESAIM: Prob. Stat. **9**, 38–73 (2005)
150. Dietrich, C.R., Newsam, G.N.: A fast and exact method for multidimensional Gaussian stochastic simulations. Water Resource Res. **29**, 2861–2869 (1993)
151. Dietrich, C.R., Newsam, G.N.: Fast and exact simulation of stationary Gaussian processes through circulant embedding of the covariance matrix. J. Sci. Comput. **18**, 1088–1107 (1997)
152. Diggle, P.J.: On parameter estimation and goodness-of-fit testing for spatial point patterns. Biometrika **35**, 87–101 (1979)
153. Diggle, P.J.: A point process modelling approach to raised incidence of a rare phenomenon in the vicinity of a prespecified point. J. Roy. Stat. Soc. Ser. A **153**, 349–362 (1990)
154. Diggle, P.J.: Statistical Analysis of Spatial Point Patterns, 2nd edn. Hodder Arnold, London (2003)
155. Dimitrakopoulos, R., Mustapha, H., Gloaguen, E.: High-order statistics of spatial random fields: Exploring spatial cumulants for modeling complex non-gaussian and non-linear phenomena. Math. Geosci. **42**, 65–99 (2010)
156. Donoho, D., Grimes, C.: Hessian eigenmaps: locally linear embedding techniques for high dimensional data. Proc. Natl. Acad. Sci. **100**, 5591–5596 (2003)

157. Donoho, D.L., Tanner, J.: Counting faces of randomly projected polytopes when the projection radically lowers dimension. J. Am. Math. Soc. **22**, 1–53 (2009)
158. Donoho, D.L., Tanner, J.: Exponential bounds implying construction of compressed sensing matrices, error-correcting codes, and neighborly polytopes by random sampling. IEEE Trans. Inform. Theor. **56**, 2002–2016 (2010)
159. Doukhan, P.: Mixing. Lecture Notes in Statistics, vol. 85. Springer, New York (1994)
160. Doukhan, P., Louhichi, S.: A new weak dependence condition and applications to moment inequalities. Stoch. Process. Appl. **84**, 313–342 (1999)
161. Drinkwater, M.J., Parker, Q.A., Proust, D., Slezak, E., Quintana, H.: The large scale distribution of galaxies in the Shapley supercluster. Publ. Astron. Soc. Aust. **21**, 89–96 (2004). doi:10.1071/AS03057
162. Dryden, I.L., Farnoosh, R., Taylor, C.C.: Image segmentation using Voronoi polygons and MCMC, with application to muscle fibre images. J. Appl. Stat. **33**, 609–622 (2006)
163. Dvoretzky, A., Robbins, H.: On the "parking" problem. MTA Mat. Kut. Int. Kől. (Publications of the Math. Res. Inst. of the Hungarian Academy of Sciences) **9**, 209–225 (1964)
164. Dynkin, E.B.: Markov processes. Vols. I, II. Die Grundlehren der Mathematischen Wissenschaften, Band 121/122. Academic, New York (1965)
165. Efron, B., Stein, C.: The jackknife estimate of variance. Ann. Stat. **9**, 586–596 (1981)
166. Ellis, R.S.: Entropy, large deviations, and statistical mechanics. Classics in Mathematics. Springer, Berlin (2006). Reprint of the 1985 original
167. Esary, J.D., Proschan, F., Walkup, D.W.: Association of random variables, with applications. Ann. Math. Stat. **38**, 1466–1474 (1967)
168. Ethier, S.N., Kurtz, T.G.: Markov processes. Characterization and convergence. Wiley Series in Probability and Mathematical Statistics: Probability and Mathematical Statistics. Wiley, New York (1986)
169. Evans, S.N.: Two representations of a conditioned superprocess. Proc. Roy. Soc. Edinb. Sect. A **123**, 959–971 (1993)
170. Evans, S.N., Perkins, E.: Measure-valued Markov branching processes conditioned on nonextinction. Israel J. Math. **71**, 329–337 (1990)
171. Federer, H.: Curvature measures. Trans. Am. Math. Soc. **93**, 418–491 (1959)
172. Feller, W.: An Introduction to Probability Theory and its Applications. Wiley, New York (1971)
173. Fleischer, F., Gloaguen, C., Schmidt, H., Schmidt, V., Schweiggert, F.: Simulation algorithm of typical modulated Poisson–Voronoi cells and application to telecommunication network modelling. Jpn. J. Indust. Appl. Math. **25**, 305–330 (2008)
174. Fleischer, F., Gloaguen, C., Schmidt, V., Voss, F.: Simulation of the typical Poisson-Voronoi-Cox-Voronoi cell. J. Stat. Comput. Simul. **79**, 939–957 (2009)
175. Fortuin, C., Kasteleyn, P., Ginibre, J.: Correlation inequalities on some partially ordered sets. Comm. Math. Phys. **22**, 89–103 (1971)
176. Foss, S., Zuyev, S.: On a Voronoi aggregative process related to a bivariate Poisson process. Adv. Appl. Probab. **28**, 965–981 (1996)
177. Gaetan, C., Guyon, X.: Spatial Statistics and Modeling. Springer, Berlin (2010)
178. Gatzouras, D., Giannopoulos, A.: A large deviations approach to the geometry of random polytopes. Mathematika **53**, 173–210 (2006)
179. Gatzouras, D., Giannopoulos, A., Markoulakis, N.: Lower bound for the maximal number of facets of a 0/1 polytope. Discrete Comput. Geom. **34**, 331–349 (2005)
180. Gelfand, A.E., Diggle, P.J., Fuentes, M., Guttorp, P.: Handbook of Spatial Statistics. CRC, Boca Raton (2010)
181. Genton, M.G.: Highly robust variogram estimation. Math. Geol. **30**, 213–221 (1998)
182. Genton, M.G.: Robustness problems in the analysis of spatial data. In: Moore, M. (ed.) Spatial Statistics: Methodological Aspects and Applications. Springer, New York (2001)
183. Georgii, H.O.: Canonical and grand canonical Gibbs states for continuum systems. Comm. Math. Phys. **48**, 31–51 (1976)
184. Georgii, H.O.: Gibbs Measures and Phase Transitions. Walter de Gruyter, Berlin (1988)

185. Gerstein, M., Tsai, J., Levitt, M.: The volume of atoms on the protein surface: calculated from simulation, using Voronoi polyhedra. J. Mol. Biol. **249**, 955–966 (1995)
186. Geyer, C.J.: Likelihood inference for spatial point processes. In: Barndorff-Nielsen, O.E., Kendall, W.S., van Lieshout, M.N.M. (eds.) Stochastic Geometry: Likelihood and Computation. Monographs on Statistics and Applied Probability, vol. 80, chap. 3, pp. 79–140. Chapman & Hall, Boca Raton (1999)
187. Gilbert, E.N.: Random subdivisions of space into crystals. Ann. Math. Stat. **33**, 958–972 (1962)
188. Gilbert, E.N.: The probability of covering a sphere with n circular caps. Biometrika **52**, 323–330 (1965)
189. Gloaguen, C., Fleischer, F., Schmidt, H., Schmidt, V.: Simulation of typical Cox-Voronoi cells with a special regard to implementation tests. Math. Meth. Oper. Res. **62**, 357–373 (2005)
190. Gloaguen, C., Voss, F., Schmidt, V.: Parametric distributions of connection lengths for the efficient analysis of fixed access networks. Ann. Telecommun. **66**, 103–118 (2011)
191. Gneiting, T.: Radial positive definite functions generated by Euclid's hat. J. Multivariate Anal. **69**, 88–119 (1999)
192. Gneiting, T.: Nonseparable, stationary covariance functions for space-time data. J. Am. Stat. Assoc. **97**, 590–600 (2002)
193. Gneiting, T., Genton, M.G., Guttorp, P.: Geostatistical space-time models, stationarity, separability and full symmetry. In: Finkenstaedt, B., Held, L., Isham, V. (eds.) Statistics of Spatio-Temporal Systems. Monographs in Statistics and Applied Probability, pp. 151–175. Chapman & Hall, Boca Raton (2007)
194. Gneiting, T., Sasvári, Z.: The characterization problem for isotropic covariance functions. Math. Geol. **31**, 105–111 (1999)
195. Gneiting, T., Ševčíková, H., Percival, D.B., Schlather, M., Jiang, Y.: Fast and exact simulation of large Gaussian lattice systems in R^2: Exploring the limits. J. Comput. Graph. Stat. **15**, 483–501 (2006)
196. Goldman, A.: Le spectre de certaines mosaïques Poissoniennes du plan et l'enveloppe convexe du pont Brownien. Probab. Theor. Relat. Fields **105**, 57–83 (1996)
197. Goldman, A.: Sur une conjecture de D. G. Kendall concernant la cellule de Crofton du plan et sur sa contrepartie brownienne. Ann. Probab. **26**, 1727–1750 (1998)
198. Goldman, A., Calka, P.: On the spectral function of the Poisson-Voronoi cells. Ann. Inst. H. Poincaré Probab. Stat. **39**, 1057–1082 (2003)
199. Götze, F., Hipp, C.: Asymptotic expansions for potential functions of i.i.d. random fields. Probab. Theor. Relat. Fields **82**, 349–370 (1989)
200. Goudsmit, S.: Random distribution of lines in a plane. Rev. Mod. Phys. **17**, 321–322 (1945)
201. Grimmett, G.: A theorem about random fields. Bull. Lond. Math. Soc. **5**, 81–84 (1973)
202. Grimmett, G.: Probability Theory and Phase Transition. Kluwer, Dordrecht (1994)
203. Grimmett, G.R., Stirzaker, D.R.: Probability and random processes. Oxford University Press, London (1982)
204. Groemer, H.: On the mean value of the volume of a random polytope in a convex set. Arch. Math. **25**, 86–90 (1974)
205. Groemer, H.: On the extension of additive functionals on classes of convex sets. Pac. J. Math. **75**, 397–410 (1978)
206. Gruber, P.: Convex and Discrete Geometry. Springer, Berlin (2007)
207. Gruber, P.M.: Comparisons of best and random approximations of convex bodies by polytopes. Rend. Circ. Mat. Palermo (2) Suppl. **50**, 189–216 (1997)
208. Gual-Arnau, X., Cruz-Orive, L.M.: A new expression for the density of totally geodesic submanifolds in space forms, with stereological applications. Differ. Geom. Appl. **27**, 124–128 (2009)
209. Guo, H., Lim, C.Y., Meerschaert, M.M.: Local Whittle estimator for anisotropic random fields. J. Multivariate Anal. **100**, 993–1028 (2009)
210. Gut, A.: Probability: A Graduate Course, Corr. 2nd printing. Springer, New York (2007)
211. Gut, A., Jonsson, F., Stadtmüller, U.: Between the LIL and LSL. Bernoulli **16**, 1–22 (2010)

212. Gut, A., Stadtmüller, U.: Laws of the single logarithm for delayed sums of random fields. Bernoulli **14**, 249–276 (2008)
213. Gut, A., Stadtmüller, U.: Laws of the single logarithm for delayed sums of random fields. II. J. Math. Anal. Appl. **346**, 403–414 (2008)
214. Gut, A., Stadtmüller, U.: An asymmetric Marcinkiewcz-Zygmund LLN for random fields. Stat. Probab. Lett. **79**, 1016–1020 (2009)
215. Gut, A., Stadtmüller, U.: On the LSL for random fields. J. Theoret. Probab. **24**, 422–449 (2011)
216. Guyon, X.: Random Fields on a Network: Modeling, Statistics, and Applications. Springer, New York (1995)
217. de Haan, L., Ferreira, A.: Extreme Value Theory: An Introduction. Springer, New York (2006)
218. de Haan, L., Pereira, T.: Spatial extremes: models for the stationary case. Ann. Stat. **34**, 146–168 (2006)
219. de Haan, L., Zhou, C.: On extreme value analysis of a spatial process. REVSTAT **6**, 71–81 (2008)
220. Hadwiger, H.: Vorlesungen über Inhalt, Oberfläche und Isoperimetrie. Springer, Berlin (1957)
221. Haenggi, M., Andrews, J.G., Baccelli, F., Dousse, O., Franceschetti, M.: Stochastic geometry and random graphs for the analysis and design of wireless networks. IEEE J. Sel. Area. Comm. **27**, 1029–1046 (2009)
222. Hall, P.: On the coverage of k-dimensional space by k-dimensional spheres. Ann. Probab. **13**, 991–1002 (1985)
223. Harris, T.E.: A lower bound for the critical probability in a certain percolation process. Proc. Camb. Philos. Soc. **56**, 13–20 (1960)
224. Hartzoulaki, M., Paouris, G.: Quermassintegrals of a random polytope in a convex body. Arch. Math. **80**, 430–438 (2003)
225. Heesch, D., Petrou, M.: Non-Gibbsian Markov random fields models for contextual labelling of structured scenes. In: Proceedings of British Machine Vision Conference, pp. 930–939 (2007)
226. Heinrich, L.: Asymptotic expansions in the central limit theorem for a special class of m-dependent random fields. I/II. Math. Nachr. **134**, 83–106 (1987); **145**, 309–327 (1990)
227. Heinrich, L.: Asymptotic Gaussianity of some estimators for reduced factorial moment measures and product densities of stationary Poisson cluster processes. Statistics **19**, 87–106 (1988)
228. Heinrich, L.: Goodness-of-fit tests for the second moment function of a stationary multidimensional Poisson process. Statistics **22**, 245–268 (1991)
229. Heinrich, L.: On existence and mixing properties of germ-grain models. Statistics **23**, 271–286 (1992)
230. Heinrich, L.: Normal approximation for some mean-value estimates of absolutely regular tessellations. Math. Meth. Stat. **3**, 1–24 (1994)
231. Heinrich, L.: Gaussian limits of multiparameter empirical K-functions of spatial Poisson processes (submitted) (2012)
232. Heinrich, L., Klein, S.: Central limit theorem for the integrated squared error of the empirical second-order product density and goodness-of-fit tests for stationary point processes. Stat. Risk Model. **28**, 359–387 (2011)
233. Heinrich, L., Klein, S., Moser, M.: Empirical mark covariance and product density function of stationary marked point processes - A survey on asymptotic results. Meth. Comput. Appl. Probab. (2012), online available via DOI 10.1007/s11009-012-9314-7
234. Heinrich, L., Liebscher, E.: Strong convergence of kernel estimators for product densities of absolutely regular point processes. J. Nonparametr. Stat. **8**, 65–96 (1997)
235. Heinrich, L., Lück, S., Schmidt, V.: Asymptotic goodness-of-fit tests for the Palm mark distribution of stationary point processes with correlated marks. Submitted (2012). An extended version is available at http://arxiv.org/abs/1205.5044
236. Heinrich, L., Molchanov, I.S.: Central limit theorem for a class of random measures associated with germ-grain models. Adv. Appl. Probab. **31**, 283–314 (1999)

237. Heinrich, L., Pawlas, Z.: Weak and strong convergence of empirical distribution functions from germ-grain processes. Statistics **42**, 49–65 (2008)
238. Heinrich, L., Prokešová, M.: On estimating the asymptotic variance of stationary point processes. Meth. Comput. Appl. Probab. **12**, 451–471 (2010)
239. Heinrich, L., Schmidt, H., Schmidt, V.: Central limit theorems for Poisson hyperplane tessellations. Ann. Appl. Probab. **16**, 919–950 (2006)
240. Hellmund, G., Prokešová, M., Jensen, E.V.: Lévy-based Cox point processes. Adv. Appl. Prob. **40**, 603–629 (2008)
241. Herglotz, G.: Über Potenzreihen mit positivem, reellen Teil im Einheitskreis. Ber. Verh. Sächs. Akad. Wiss. **63**, 501–511 (1911)
242. Hero, A.O., Ma, B., Michel, O., Gorman, J.: Applications of entropic spanning graphs. IEEE Signal Process. Mag. **19**, 85–95 (2002)
243. Herrndorf, N.: An example on the central limit theorem for associated sequences. Ann. Probab. **12**, 912–917 (1984)
244. Heyde, C.C., Gay, R.: Smoothed periodogram asymptotics and estimation for processes and fields with possible long–range dependence. Stoch. Process. Appl. **45**, 169–182 (1993)
245. Hille, E.: Functional analysis and semi-groups. American Mathematical Society Colloquium Publications, vol. 31. American Mathematical Society, Providence (1948)
246. Himmelberg, C.: Measurable relations. Fund. Math. **87**, 53–72 (1974)
247. Holley, R.: Remarks on the FKG inequalities. Comm. Math. Phys. **36**, 227–231 (1974)
248. Hsing, T.: On the asymptotic distribution of the area outside a random convex hull in a disk. Ann. Appl. Probab. **4**, 478–493 (1994)
249. Hueter, I.: Limit theorems for the convex hull of random points in higher dimensions. Trans. Am. Math. Soc. **351**, 4337–4363 (1999)
250. Hug, D.: Contributions to affine surface area. Manuscripta Math. **91**, 283–301 (1996)
251. Hug, D.: Curvature relations and affine surface area for a general convex body and its polar. Results Math. **29**, 233–248 (1996)
252. Hug, D.: Random mosaics. In: Baddeley, A.J., Bárány, I., Schneider, R., Weil, W. (eds.) Stochastic Geometry - Lecture Notes in Mathematics, vol. 1892. Springer, Berlin (2007)
253. Hug, D., Munsonius, O., Reitzner, M.: Asymptotic mean values of Gaussian polytopes. Contr. Algebra Geom. **45**, 531–548 (2004)
254. Hug, D., Reitzner, M.: Gaussian polytopes: variances and limit theorems. Adv. Appl. Probab. **37**, 297–320 (2005)
255. Hug, D., Reitzner, M., Schneider, R.: Large Poisson–Voronoi cells and Crofton cells. Adv. Appl. Probab. **36**, 667–690 (2004)
256. Hug, D., Reitzner, M., Schneider, R.: The limit shape of the zero cell in a stationary Poisson hyperplane tessellation. Ann. Probab. **32**, 1140–1167 (2004)
257. Hug, D., Schneider, R.: Large cells in Poisson-Delaunay tessellations. Discrete Comput. Geom. **31**, 503–514 (2004)
258. Hug, D., Schneider, R.: Large typical cells in Poisson-Delaunay mosaics. Rev. Roum. Math. Pures Appl. **50**, 657–670 (2005)
259. Hug, D., Schneider, R.: Asymptotic shapes of large cells in random tessellations. Geom. Funct. Anal. **17**, 156–191 (2007)
260. Hug, D., Schneider, R.: Typical cells in Poisson hyperplane tessellations. Discrete Comput. Geom. **38**, 305–319 (2007)
261. Hug, D., Schneider, R.: Large faces in Poisson hyperplane mosaics. Ann. Probab. **38**, 1320–1344 (2010)
262. Hug, D., Schneider, R.: Faces of Poisson-Voronoi mosaics. Probab. Theor. Relat. Fields **151**, 125–151 (2011)
263. Hug, D., Schneider, R.: Faces with given directions in anisotropic Poisson hyperplane tessellations. Adv. Appl. Probab. **43**, 308–321 (2011)
264. Ibragimov, I.A., Linnik, Y.V.: Independent and stationary sequences of random variables. Wolters-Noordhoff, Groningen (1971)

265. Illian, J., Penttinen, A., Stoyan, D., Stoyan, H.: Statistical Analysis and Modelling of Spatial Point Patterns. Wiley, Chichester (2008)
266. Itô, K., McKean, H.P. Jr.: Diffusion Processes and Their Sample Paths. Springer, Berlin (1974). 2nd printing, corrected, Die Grundlehren der mathematischen Wissenschaften, Band 125
267. Ivanoff, G.: Central limit theorems for point processes. Stoch. Process. Appl. **12**, 171–186 (1982)
268. Ivanov, A.V., Leonenko, N.N.: Statistical Analysis of Random Fields. Kluwer, Dordrecht (1989)
269. Jakubowski, A.: Minimal conditions in p–stable limit theorems. Stoch. Process. Appl. **44**, 291–327 (1993)
270. Jakubowski, A.: Minimal conditions in p–stable limit theorems II. Stoch. Process. Appl. **68**, 1–20 (1997)
271. Janzer, H.S., Raudies, F., Neumann, H., Steiner, F.: Image processing and feature extraction from a perspective of computer vision and physical cosmology. In: Arendt, W., Schleich, W. (eds.) Mathematical Analysis of Evolution, Information and Complexity. Wiley, New York (2009)
272. Jenish, N., Prucha, I.R.: Central limit theorems and uniform laws of large numbers for arrays of random fields. J. Econometrics **150**, 86–98 (2009)
273. Joag-Dev, K., Proschan, F.: Negative association of random variables, with applications. Ann. Stat. **11**, 286–295 (1983)
274. Jolivet, E.: Central limit theorem and convergence of empirical processes for stationary point processes. In: Bartfai, P., Tomko, J. (eds.) Point Processes and Queueing Problems. North-Holland, Amsterdam (1980)
275. Jolivet, E.: Upper bound of the speed of convergence of moment density estimators for stationary point processes. Metrika **31**, 349–360 (1984)
276. Journel, A.G., Huijbregts, C.J.: Mining Geostatistics. Academic, London (1978)
277. Judge, G.G., Griffiths, W.E., Hill, R.C., Lee, T.C.: The Theory and Practice of Econometrics. Wiley, New York (1980)
278. Kabluchko, Z., Schlather, M., de Haan, L.: Stationary max-stable fields associated to negative definite functions. Ann. Probab. **37**, 2042–2065 (2009)
279. Kac, M.: Can one hear the shape of a drum? Am. Math. Mon. **73**, 1–23 (1966)
280. Kailath, T.: A theorem of I. Schur and its impact on modern signal processing. In: Gohberg, I. (ed.) Schur Methods in Operator Theory and Signal Processing, Vol. I. Birkhäuser, Basel (1986)
281. Kallenberg, O.: An informal guide to the theory of conditioning in point processes. Int. Stat. Rev. **52**, 151–164 (1984)
282. Kallenberg, O.: Random Measures, 4th edn. Academic, New York (1986)
283. Kallenberg, O.: Foundations of Modern Probability, 2nd edn. Springer, New York (2002)
284. Kaltenbach, F.J.: Asymptotisches Verhalten zufälliger konvexer Polyeder. Doctoral thesis, Albert-Ludwigs-Universität, Freiburg i. Br. (1990)
285. Karatzas, I., Shreve, S.E.: Methods of mathematical finance. Applications of Mathematics (New York), vol. 39. Springer, New York (1998)
286. Karcher, W.: A central limit theorem for the volume of the excursion sets of associated stochastically continuous stationary random fields. Preprint (2012), submitted
287. Karcher, W., Scheffler, H.P., Spodarev, E.: Efficient simulation of stable random fields and its applications. In: Capasso, V., Aletti, G., Micheletti, A. (eds.) Stereology and Image Analysis. Ecs10: Proceedings of The 10th European Congress of ISS, vol. 4, pp. 63–72. ESCULAPIO Pub. Co., Bologna (2009)
288. Karcher, W., Scheffler, H.P., Spodarev, E.: Simulation of infinitely divisible random fields. Comm. Stat. Simulat. Comput. **42**, 215–246 (2013)
289. Karr, A.F.: Point Processes and their Statistical Inference. Dekker, New York (1986)
290. Kelley, C.T.: Solving Nonlinear Equations with Newton's Method. No. 1 in Fundamentals of Algorithms. SIAM, Philadelphia (2003)

291. Kelly, F.P.: Reversibility and Stochastic Networks. Wiley, Chichester (1979)
292. Kendall, D.G.: Foundations of a theory of random sets. In: Harding, E.F., Kendall, D.G. (eds.) Stochastic Geometry. Wiley, New York (1974)
293. Kent, J.T., Wood, A.T.A.: Estimating fractal dimension of a locally self-similar Gaussian process by using increments. J. R. Stat. Soc. B **59**, 679–699 (1997)
294. Kesten, H.: Subdiffusive behavior of random walk on a random cluster. Ann. Inst. H. Poincaré Probab. Stat. **22**, 425–487 (1986)
295. Kesten, H., Lee, S.: The central limit theorem for weighted minimal spanning trees on random points. Ann. Appl. Probab. **6**, 495–527 (1996)
296. Khoshnevisian, D.: Multiparameter Processes. An Introduction to Random Fields. Springer, New York (2002)
297. Kinderman, R., Snell, J.L.: Markov Random Fields and Their Applications. Contemporary Mathematics, vol. 1. AMS, Providence (1980)
298. Kingman, J.F.C.: Poisson Processes, Oxford Studies in Probability. Oxford University Press, London (1993)
299. Kirby, M.: Geometric Data Analysis: An Empirical Approach to Dimensionality Reduction and the Study of Patterns. Wiley, New York (2001)
300. Kitanidis, P.K.: Introduction to Geostatistics: Applications to Hydrogeology. Cambridge University Press, New York (1997)
301. Klain, D.A.: A short proof of hadwiger's characterization theorem. Mathematika **42**, 329–339 (1995)
302. Klain, D.A., Rota, G.C.: Introduction to Geometric Probability. Cambridge University Press, Cambridge (1997)
303. Klartag, B.: An isomorphic version of the slicing problem. J. Funct. Anal. **218**, 372–394 (2005)
304. Klartag, B.: On convex perturbations with a bounded isotropic constant. Geom. Funct. Anal. **16**, 1274–1290 (2006)
305. Klartag, B., Kozma, G.: On the hyperplane conjecture for random convex sets. Israel J. Math. **170**, 253–268 (2009)
306. Klesov, O.I.: The Hájek-Rényi inequality for random fields and the strong law of large numbers. Theor. Probab. Math. Stat. **22**, 63–71 (1981)
307. Klesov, O.I.: Strong law of large numbers for multiple sums of independent, identically distributed random variables. Math. Notes **38**, 1006–1014 (1985)
308. Kolmogorov, A.N.: Wienersche Spiralen und einige andere interessante Kurven in Hilbertschem Raum. (Dokl.) Acad. Sci. USSR **26**, 115–118 (1940)
309. Koo, Y., Lee, S.: Rates of convergence of means of Euclidean functionals. J. Theoret. Probab. **20**, 821–841 (2007)
310. Koralov, L.B., Sinai, Y.G.: Theory of Probability and Random Processes. Springer, New York (2009)
311. Kotz, S., Kozubowski, T.J., Podgorski, K.: The Laplace Distribution and Generalizations. Birkhäuser, Boston (2001)
312. Kovalenko, I.N.: A simplified proof of a conjecture of D. G. Kendall concerning shapes of random polygons. J. Appl. Math. Stoch. Anal. **12**, 301–310 (1999)
313. Kratz, M., Leon, J.: Central limit theorems for level functionals of stationary Gaussian processes and fields. J. Theor. Probab. **14**, 639–672 (2001)
314. Kratz, M., Leon, J.: Level curves crossings and applications for Gaussian models. Extremes **13**, 315–351 (2010)
315. Kryzhanovskaya, N.Y.: A moment inequality for sums of multi-indexed dependent random variables. Math. Notes **83**, 770–782 (2008)
316. Kunita, H.: Stochastic flows and stochastic differential equations. Cambridge Studies in Advanced Mathematics, vol. 24. Cambridge University Press, Cambridge (1997). Reprint of the 1990 original
317. Kutoyants, Y.A.: Statistical Inference for Spatial Poisson Processes. No. 134 in Lecture Notes in Statistics. Springer, New York (1998)

318. Kwapień, S., Woyczyński, W.A.: Random Series and Stochastic Integrals: Single and Multiple. Birkhäuser, Boston (1992)
319. Lai, T.L.: Limit theorems for delayed sums. Ann. Probab. **2**, 432–440 (1974)
320. Lantuejoul, C.: Geostatistical Simulation – Models and Algorithms. Springer, Berlin (2002)
321. Lanzinger, H., Stadtmüller, U.: Maxima of increments of partial sums for certain subexponential distributions. Stoch. Process. Appl. **86**, 307–322 (2000)
322. Lao, W., Mayer, M.: U-max-statistics. J. Multivariate Anal. **99**, 2039–2052 (2008)
323. Last, G.: Stationary partitions and Palm probabilities. Adv. Appl. Probab. **38**, 602–620 (2006)
324. Lautensack, C., Zuyev, S.: Random Laguerre tessellations. Adv. Appl. Probab. **40**, 630–650 (2008)
325. Lawson, A.B., Denison, D.G.T.: Spatial Cluster Modelling: An Overview. In: Spatial Cluster Modelling, pp. 1–19. Chapman & Hall, Boca Raton (2002)
326. Lee, M.L.T., Rachev, S.T., Samorodnitsky, G.: Association of stable random variables. Ann. Probab. **18**, 1759–1764 (1990)
327. Lehmann, E.L.: Some concepts of dependence. Ann. Math. Stat. **37**, 1137–1153 (1966)
328. Levina, E., Bickel, P.J.: Maximum likelihood estimation of intrinsic dimension. In: Saul, L.K., Weiss, Y., Bottou, L. (eds.) Advances in NIPS, vol. 17 (2005)
329. Lewis, T.: Limit theorems for partial sums of quasi-associated random variables. In: Szysskowicz, B. (ed.) Asymptotic Methods in Probability and Statistics. Elsevier, Amsterdam (1998)
330. Li, S.Z.: Markov Random Fields Modeling in Computer Vision. Springer, New York (2001)
331. Lifshits, M.A.: Partitioning of multidimensional sets. In: Rings and Modules. Limit Theorems of Probability Theory, No. 1, pp. 175–178. Leningrad University, Leningrad (1985) (in Russian)
332. Liggett, T.M.: Conditional association and spin systems. ALEA Lat. Am. J. Probab. Math. Stat. **1**, 1–19 (2006)
333. Lim, S.C., Teo, L.P.: Gaussian fields and Gaussian sheets with generalized Cauchy covariance structure. Stoch. Proc. Appl. **119**, 1325–1356 (2009)
334. Lindquist, B.H.: Association of probability measures on partially ordered spaces. J. Multivariate Anal. **26**, 11–132 (1988)
335. Litvak, A.E., Pajor, A., Rudelson, M., Tomczak-Jaegermann, N.: Smallest singular value of random matrices and geometry of random polytopes. Adv. Math. **195**, 491–523 (2005)
336. Loève, M.: Probability Theory. Van Nostrand Co. Inc., Princeton (1960)
337. Ludwig, M., Reitzner, M.: A classification of $sl(n)$ invariant values. Ann. Math. **172**, 1219–1267 (2010)
338. Ma, F., Wei, M.S., Mills, W.H.: Correlation structuring and the statistical analysis of steady–state groundwater flow. SIAM J. Sci. Stat. Comput. **8**, 848–867 (1987)
339. Maier, R., Mayer, J., Schmidt, V.: Distributional properties of the typical cell of stationary iterated tessellations. Math. Meth. Oper. Res. **59**, 287–302 (2004)
340. Maier, R., Schmidt, V.: Stationary iterated tessellations. Adv. Appl. Probab. **35**, 337–353 (2003)
341. Malyshev, V.A., Minlos, R.A.: Gibbs Random Fields. Kluwer, Dordrecht (1991)
342. Mandelbrot, B.B., Van Ness, J.: Fractional Brownian motion, fractional noises and applications. SIAM Rev. **10**, 422–437 (1968)
343. Matern, B.: Spatial Variation, 2nd edn. Springer, Berlin (1986)
344. Mateu, J., Porcu, E., Gregori, P.: Recent advances to model anisotropic space-time data. Stat. Meth. Appl. **17**, 209–223 (2008)
345. Matheron, G.: Random Sets and Integral Geometry. Wiley, New York (1975)
346. Matthes, K., Kerstan, J., Mecke, J.: Infinitely Divisible Point Processes. Wiley, Chichester (1978)
347. Mayer, M., Molchanov, I.: Limit theorems for the diameter of a random sample in the unit ball. Extremes **10**, 129–150 (2007)
348. Mecke, J.: Stationäre zufällige Maße auf lokalkompakten Abelschen Gruppen. Z. Wahrscheinlichkeitstheorie und verw. Gebiete **9**, 36–58 (1967)

349. Mecke, J.: Eine charakteristische Eigenschaft der doppelt Poissonschen Prozesse. Z. Wahrscheinlichkeitstheorie und verw. Gebiete **11**, 74–81 (1968)
350. Mecke, J.: Parametric representation of mean values for stationary random mosaics. Math. Operationsforsch. Stat. Ser. Stat. **15**, 437–442 (1984)
351. Mecke, J.: Random tessellations generated by hyperplanes. In: Ambartzumian, R.V., Weil, W. (eds.) Stochastic Geometry, Geometric Statistics, Stereology (Proc. Conf. Oberwolfach, 1983). Teubner, Leipzig (1984)
352. Mecke, J.: On the relationship between the 0-cell and the typical cell of a stationary random tessellation. Pattern Recogn. **32**, 1645–1648 (1999)
353. Meschenmoser, D., Shashkin, A.: Functional central limit theorem for the measure of level sets generated by a Gaussian random field. Stat. Probab. Lett. **81**, 642–646 (2011)
354. Meschenmoser, D., Shashkin, A.: Functional central limit theorem for the measure of level sets generated by a Gaussian random field. Theor. Probab. Appl. **57**(1), 168–178 (2012, to appear)
355. Metropolis, N., Rosenbluth, A.W., Rosenbluth, M.N., Teller, A.H., Teller, E.: Equation of state calculations by fast computing machines. J. Chem. Phys. **21**, 1087–1092 (1953)
356. Miles, R.E.: Random polygons determined by random lines in a plane. Proc. Natl. Acad. Sci. USA **52**, 901–907 (1964)
357. Miles, R.E.: Random polygons determined by random lines in a plane II. Proc. Natl. Acad. Sci. USA **52**, 1157–1160 (1964)
358. Miles, R.E.: On the homogeneous planar Poisson point process. Math. Biosci, **6**, 85–127 (1970)
359. Miles, R.E.: The various aggregates of random polygons determined by random lines in a plane. Adv. Math. **10**, 256–290 (1973)
360. Miles, R.E.: Some integral geometric formula, with stochastic applications. J. Appl. Probab. **16**, 592–606 (1979)
361. Miles, R.E., Davy, P.J.: Precise and general conditions for the validity of a comprehensive set of stereological fundamental formulae. J. Microsc. **107**, 211–226 (1976)
362. Milman, V.D., Pajor, A.: Isotropic position and inertia ellipsoids and zonoids of the unit ball of a normed n-dimensional space. In: Geometric Aspects of Functional Analysis. Lecture Notes in Mathematics, vol. 1376. Springer, Berlin (1989)
363. Molchanov, I.: Theory of Random Sets. Springer, London (2005)
364. Molchanov, I.: Convex geometry of max-stable distributions. Extremes **11**, 235–259 (2008)
365. Molchanov, I.: Convex and star-shaped sets associated with multivariate stable distributions. I. Moments and densities. J. Multivariate Anal. **100**, 2195–2213 (2009)
366. Molchanov, I.S.: Statistics of the Boolean Model for Practitioners and Mathematicians. Wiley, Chichester (1997)
367. Møller, J.: Random Johnson-Mehl tessellations. Adv. Appl. Probab. **24**, 814–844 (1992)
368. Møller, J.: Lectures on random Voronoi tessellations. In: Lecture Notes in Statistics, vol. 87. Springer, New York (1994)
369. Møller, J., Waagepetersen, R.P.: Statistical Inference and Simulation for Spatial Point Processes. Chapman & Hall, Boca Raton (2004)
370. Møller, J., Zuyev, S.: Gamma-type results and other related properties of Poisson processes. Adv. Appl. Probab. **28**, 662–673 (1996)
371. Móricz, F.: A general moment inequality for the maximum of the rectangular partial sums of multiple series. Acta Math. Hungar. **41**, 337–346 (1983)
372. Mourier, E.: L-random elements and L^*-random elements in Banach spaces. In: Proceedings of Third Berkeley Symposium on Mathematical Statistics and Probability, vol. 2, pp. 231–242. University of California Press, Berkeley (1955)
373. Moussa, A., Sbihi, A., Postaire, J.G.: A Markov random field model for mode detection in cluster analysis. Pattern Recogn. Lett. **29**, 1197–1207 (2008)
374. Müller, A., Stoyan, D.: Comparison Methods for Stochastic Models and Risks. Wiley, Chichester (2002)
375. Nagaev, A.V.: Some properties of convex hulls generated by homogeneous Poisson point processes in an unbounded convex domain. Ann. Inst. Stat. Math. **47**, 21–29 (1995)

376. Nagel, W., Weiss, V.: Crack STIT tessellations: characterization of stationary random tessellations stable with respect to iteration. Adv. Appl. Probab. **37**, 859–883 (2005)
377. Neveu, J.: Sur les mesures de Palm de deux processus ponctuels stationnaires. Z. Wahrscheinlichkeitstheorie und verw. Gebiete **34**, 199–203 (1976)
378. Newman, C.M.: Normal fluctuations and the FKG inequalities. Comm. Math. Phys. **74**, 119–128 (1980)
379. Newman, C.M.: Asymptotic independence and limit theorems for positively and negatively dependent random variables. In: Tong, Y.L. (ed.) Inequalities in Statistics and Probability. Institute of Mathematical Statistics, Hayward (1984)
380. Newman, C.M., Wright, A.L.: An invariance principle for certain dependent sequences. Ann. Probab. **9**, 671–675 (1981)
381. Newman, C.M., Wright, A.L.: Associated random variables and martingale inequalities. Z. Wahrscheinlichkeitstheorie und verw. Gebiete **59**, 361–371 (1982)
382. Nguyen, X.X., Zessin, H.: Punktprozesse mit Wechselwirkung. Z. Wahrscheinlichkeitstheorie und verw. Gebiete **37**, 91–126 (1976)
383. Nguyen, X.X., Zessin, H.: Integral and differential characterizations of the Gibbs process. Math. Nachr. **88**, 105–115 (1979)
384. Norberg, T.: Existence theorems for measures on continuous posets, with applications to random set theory. Math. Scand. **64**, 15–51 (1989)
385. Ohser, J., Stoyan, D.: On the second-order and orientation analysis of planar stationary point processes. Biom. J. **23**, 523–533 (1981)
386. Okabe, A., Boots, B., Sugihara, K., Chiu, S.N.: Spatial Tessellations: Concepts and Applications of Voronoi Diagrams. With a foreword by D. G. Kendall. Wiley, Chichester (2000)
387. Pantle, U., Schmidt, V., Spodarev, E.: Central limit theorems for functionals of stationary germ–grain models. Ann. Appl. Probab. **38**, 76–94 (2006)
388. Pantle, U., Schmidt, V., Spodarev, E.: On the estimation of integrated covariance functions of stationary random fields. Scand. J. Stat. **37**, 47–66 (2010)
389. Paroux, K.: Quelques théorèmes centraux limites pour les processus Poissoniens de droites dans le plan. Adv. Appl. Probab. **30**, 640–656 (1998)
390. Paulauskas, V.: On Beveridge–Nelson decomposition and limit theorems for linear random fields. J. Multivariate Anal. **101**, 621–639 (2010)
391. Paulauskas, V.: On Beveridge-Nelson decomposition and limit theorems for linear random fields. J. Multivariate Anal. **101**, 621–639 (2010)
392. Pawlas, Z.: Empirical distributions in marked point processes. Stoch. Process. Appl. **119**, 4194–4209 (2009)
393. Peligrad, M.: Maximum of partial sums and an invariance principle for a class of weak dependent random variables. Proc. Am. Math. Soc. **126**, 1181–1189 (1998)
394. Penrose, M.: Random Geometric Graphs. Oxford University Press, London (2003)
395. Penrose, M.D.: Gaussian limits for random geometric measures. Electron. J. Probab. **12**, 989–1035 (2007)
396. Penrose, M.D.: Laws of large numbers in stochastic geometry with statistical applications. Bernoulli **13**, 1124–1150 (2007)
397. Penrose, M.D., Wade, A.R.: Multivariate normal approximation in geometric probability. J. Stat. Theor. Pract. **2**, 293–326 (2008)
398. Penrose, M.D., Yukich, J.E.: Central limit theorems for some graphs in computational geometry. Ann. Appl. Probab. **11**, 1005–1041 (2001)
399. Penrose, M.D., Yukich, J.E.: Mathematics of random growing interfaces. J. Phys. A **34**, 6239–6247 (2001)
400. Penrose, M.D., Yukich, J.E.: Limit theory for random sequential packing and deposition. Ann. Appl. Probab. **12**, 272–301 (2002)
401. Penrose, M.D., Yukich, J.E.: Weak laws of large numbers in geometric probability. Ann. Appl. Probab. **13**, 277–303 (2003)
402. Penrose, M.D., Yukich, J.E.: Normal approximation in geometric probability. In: Barbour, A.D., Chen, L.H.Y. (eds.) Stein's Method and Applications. Lecture Note Series, vol. 5. Institute for Mathematical Sciences, National University of Singapore, 37–58 (2005)

403. Penrose, M.D., Yukich, J.E.: Limit theory for point processes on manifolds. Ann. Appl. Probab. (2013) (to appear). ArXiv:1104.0914
404. Petkantschin, B.: Integralgeometrie 6. Zusammenhänge zwischen den Dichten der linearen Unterräume im n-dimensionalen Raum. Abf. Math. Sem. Univ. Hamburg **11**, 249–310 (1936)
405. Petrov, V.V.: Limit Theorems of Probability Theory. Clarendon Press, Oxford (1995)
406. Pfanzagl, J.: On the measurability and consistency of minimum contrast estimates. Metrika **14**, 249–276 (1969)
407. Pitt, L.D.: Positively correlated normal variables are associated. Ann. Probab. **10**, 496–499 (1982)
408. Press, W.H., Teukolsky, S.A., Vetterling, W.T., Flannery, B.P.: Numerical Recipes in C. The Art of Scientific Computing, 2nd edn. Cambridge University Press, Cambridge (1997)
409. Preston, C.J.: Gibbs States on Countable Sets. Cambridge University Press, London (1974)
410. Quintanilla, J., Torquato, S.: Local volume fluctuations in random media. J. Chem. Phys. **106**, 2741–2751 (1997)
411. Rachev, S.T., Mittnik, S.: Stable Paretian Models in Finance. Wiley, New York (2000)
412. Račkauskas, A., Suquet, C., Zemlys, V.: Hölderian functional central limit theorem for multi–indexed summation process. Stoch. Process. Appl. **117**, 1137–1164 (2007)
413. Rajput, B.S., Rosinski, J.: Spectral representations of infinitely divisible processes. Probab. Theor. Relat. Fields **82**, 451–487 (1989)
414. Raynaud, H.: Sur l'enveloppe convexe des nuages de points aléatoires dans R^n. I. J. Appl. Probab. **7**, 35–48 (1970)
415. Redmond, C.: Boundary rooted graphs and Euclidean matching algorithms. Ph.D. thesis, Department of Mathematics, Lehigh University, Bethlehem, PA (1993)
416. Redmond, C., Yukich, J.E.: Limit theorems and rates of convergence for subadditive Euclidean functionals. Ann. Appl. Probab. **4**, 1057–1073 (1994)
417. Redmond, C., Yukich, J.E.: Limit theorems for Euclidean functionals with power-weighted edges. Stoch. Process. Appl. **61**, 289–304 (1996)
418. Reitzner, M.: Random polytopes and the Efron-Stein jackknife inequality. Ann. Probab. **31**, 2136–2166 (2003)
419. Reitzner, M.: Central limit theorems for random polytopes. Probab. Theor. Relat. Fields **133**, 483–507 (2005)
420. Reitzner, M.: Random polytopes. In: Kendall, W.S., Molchanov, I. (eds.) New Perspectives in Stochastic Geometry. Oxford University Press, London (2010)
421. Rényi, A., Sulanke, R.: Über die konvexe Hülle von n zufällig gewählten Punkten. Z. Wahrscheinlichkeitstheorie und verw. Gebiete **2**, 75–84 (1963)
422. Resnick, S.: Extreme Values, Regular Variation and Point Processes. Springer, New York (2008)
423. Rio, E.: Covariance inequality for strongly mixing processes. Ann. Inst. H. Poincaré Probab. Stat. **29**, 587–597 (1993)
424. Ripley, B.D.: The second-order analysis of stationary point processes. J. Appl. Probab. **13**, 255–266 (1976)
425. Rosenblatt, M.: Stationary Sequences and Random Fields. Birkhäuser, Boston (1985)
426. Rother, W., Zähle, M.: A short proof of the principal kinematic formula and extensions. Trans. Am. Math. Soc. **321**, 547–558 (1990)
427. Roweis, S., Saul, L.: Nonlinear dimensionality reduction by locally linear embedding. Science **290**, 2323–2326 (2000)
428. Roy, P.: Ergodic theory, Abelian groups and point processes induced by stable random fields. Ann. Probab. **38**, 770–793 (2010)
429. Rozanov, Y.A.: Markov Random Fields. Springer, New York (1982)
430. Rue, H., Held, L.: Gaussian Markov Random Fields: Theory and Applications. Chapman & Hall, Boca Raton (2005)
431. Ruelle, D.: Statistical Mechanics: Rigorous Results. World Scientific, Singapore (1969)
432. Samorodnitsky, G., Taqqu, M.: Stable non-Gaussian Random Processes. Stochastic Models with Infinite Variance. Chapman & Hall, Boca Raton (1994)

433. Santaló, L.A.: Integral Geometry and Geometric Probability. Addison-Wesley, Reading (1976)
434. Sarkar, T.K.: Some lower bounds of reliability. Tech. Rep. 124, Department of Operation Research and Statistics, Stanford University (1969)
435. Sasvári, A.: Positive Definite and Definitizable Functions. Akademie-Verlag, Berlin (1994)
436. Sato, K.I.: Lévy Processes and Infinitely Divisible Distributions. Cambridge University Press, Cambridge (1999)
437. Schlather, M.: Introduction to positive definite functions and to unconditional simulation of random fields. Tech. rep., Lancaster University (1999)
438. Schlather, M.: On the second-order characteristics of marked point processes. Bernoulli **7**, 99–117 (2001)
439. Schlather, M.: Models for stationary max-stable random fields. Extremes **5**, 33–44 (2002)
440. Schlather, M.: Construction of positive definite functions. Lecture at 15th Workshop on Stoch. Geometry, Stereology and Image Analysis, Blaubeuren (2009)
441. Schlather, M.: Some covariance models based on normal scale mixtures. Bernoulli **16**, 780–797 (2010)
442. Schneider, R.: Extremal properties of random mosaics. In: Bárány, I., Böröczky, K.J., Fejes Tóth, G., Pach, J. (eds.) Geometry - Intuitive, Discrete, and Convex. Springer, Berlin (to appear)
443. Schneider, R.: Approximation of convex bodies by random polytopes. Aequationes Math. **32**, 304–310 (1987)
444. Schneider, R.: Random approximation of convex sets. J. Microsc. **151**, 211–227 (1988)
445. Schneider, R.: Convex Bodies: The Brunn-Minkowski Theory. Cambridge University Press, Cambridge (1993)
446. Schneider, R.: Discrete aspects of stochastic geometry. In: Goodman, J.E., O'Rourke, J. (eds.) Handbook of Discrete and Computational Geometry. CRC Press, Boca Raton (1997)
447. Schneider, R.: Recent results on random polytopes. Boll. Unione Mat. Ital. **1**, 17–39 (2008)
448. Schneider, R.: Weighted faces of Poisson hyperplane tessellations. Adv. Appl. Probab. **41**, 682–694 (2009)
449. Schneider, R.: Vertex numbers of weighted faces in Poisson hyperplane mosaics. Discrete Comput. Geom. **44**, 599–607 (2010)
450. Schneider, R., Weil, W.: Translative and kinematic integral formulae for curvature measures. Math. Nachr. **129**, 67–80 (1986)
451. Schneider, R., Weil, W.: Stochastic and Integral Geometry. Springer, Berlin (2008)
452. Schneider, R., Wieacker, J.A.: Random polytopes in a convex body. Z. Wahrscheinlichkeitstheorie und verw. Gebiete **52**, 69–73 (1980)
453. Schoenberg, I.J.: Metric spaces and complete monotone functions. Ann. Math **39**, 811–841 (1938)
454. Schoutens, W.: Lévy Processes in Finance: Pricing Financial Derivatives. Wiley, Chichester (2003)
455. Schreiber, T.: Variance asymptotics and central limit theorems for volumes of unions of random closed sets. Adv. Appl. Probab. **34**, 520–539 (2002)
456. Schreiber, T.: Asymptotic geometry of high density smooth-grained Boolean models in bounded domains. Adv. Appl. Probab. **35**, 913–936 (2003)
457. Schreiber, T.: Limit theorems in stochastic geometry. In: Kendall, W.S., Molchanov. I. (eds.) New Perspectives in Stochastic Geometry. Oxford University Press, London (2010)
458. Schreiber, T., Penrose, M.D., Yukich, J.E.: Gaussian limits for multidimensional random sequential packing at saturation. Comm. Math. Phys. **272**, 167–183 (2007)
459. Schreiber, T., Yukich, J.E.: Variance asymptotics and central limit theorems for generalized growth processes with applications to convex hulls and maximal points. Ann. Probab. **36**, 363–396 (2008)
460. Schreiber, T., Yukich, J.E.: Limit theorems for geometric functionals of Gibbs point processes. Ann. Inst. H. Poincaré Probab. Stat. (2012, to appear)
461. Schütt, C.: Random polytopes and affine surface area. Math. Nachr. **170**, 227–249 (1994)

462. Seneta, E.: Regularly Varying Functions. Springer, Berlin (1976)
463. Seppäläinen, T., Yukich, J.E.: Large deviation principles for Euclidean functionals and other nearly additive processes. Probab. Theor. Relat. Fields **120**, 309–345 (2001)
464. Shao, Q.M.: A comparison theorem on moment inequalities between negatively associated and independent random variables. J. Theoret. Probab. **13**, 343–356 (2000)
465. Shashkin, A.: A strong invariance principle for positively or negatively associated random fields. Stat. Probab. Lett. **78**, 2121–2129 (2008)
466. Shashkin, A.P.: Quasi-associatedness of a Gaussian system of random vectors. Russ. Math. Surv. **57**, 1243–1244 (2002)
467. Shashkin, A.P.: A dependence property of a spin system. In: Transactions of XXIV Int. Sem. on Stability Problems for Stoch. Models, pp. 30–35. Yurmala, Latvia (2004)
468. Shashkin, A.P.: Maximal inequality for a weakly dependent random field. Math. Notes **75**, 717–725 (2004)
469. Shashkin, A.P.: On the Newman central limit theorem. Theor. Probab. Appl. **50**, 330–337 (2006)
470. Shepp, L.: Covering the circle with random arcs. Israel J. Math. **11**, 328–345 (1972)
471. Shinozuka, M., Jan, C.M.: Digital simulation of random processes and its applications. J. Sound Vib. **25**, 111–128 (1972)
472. Shiryaev, A.N.: Probability. Graduate Texts in Mathematics, vol. 95, 2nd edn. Springer, New York (1996)
473. Shiryaev, A.N.: Essentials of Stochastic Finance. Facts, Models, Theory. World Scientific, River Edge (1999)
474. Shorack, G.R., Smythe, R.T.: Inequalities for max $| S_{\mathbf{k}} | / b_{\mathbf{k}}$ where $\mathbf{k} \in N^r$. Proc. Am. Math. Soc. **54**, 331–336 (1976)
475. Siegel, A.F., Holst, L.: Covering the circle with random arcs of random sizes. J. Appl. Probab. **19**, 373–381 (1982)
476. Smythe, R.T.: Strong laws of large numbers for r-dimensional arrays of random variables. Ann. Probab. **1**, 164–170 (1973)
477. Smythe, R.T.: Sums of independent random variables on partially ordered sets. Ann. Probab. **2**, 906–917 (1974)
478. Speed, T.P., Kiiveri, H.T.: Gaussian Markov distributions over finite graphs. Ann. Stat. **14**, 138–150 (1986)
479. Stadtmüller, U., Thalmaier, M.: Strong laws for delayed sums of random fields. Acta Sci. Math. (Szeged) **75**, 723–737 (2009)
480. Stanley, R.P.: Enumerative combinatorics. Vol. 2. Cambridge Studies in Advanced Mathematics, vol. 62. Cambridge University Press, Cambridge (1999)
481. Steele, J.M.: Subadditive Euclidean functionals and nonlinear growth in geometric probability. Probab. Theor. Relat. Fields **9**, 365–376 (1981)
482. Steele, J.M.: Probability Theory and Combinatorial Optimization. SIAM, Philadelphia (1997)
483. Steinebach, J.: On the increments of partial sum processes with multidimensional indices. Z. Wahrscheinlichkeitstheorie und verw. Gebiete **63**, 59–70 (1983)
484. Sterio, D.C.: The unbiased estimation of number and sizes of aritrary particles using the disector. J. Microsc. **134**, 127–136 (1984)
485. Stevens, W.L.: Solution to a geometrical problem in probability. Ann. Eugenics **9**, 315–320 (1939)
486. Stoev, S.A.: On the ergodicity and mixing of max-stable processes. Stoch. Process. Appl. **118**, 1679–1705 (2008)
487. Stoev, S.A., Taqqu, M.S.: Extremal stochastic integrals: a parallel between max-stable processes and α-stable processes. Extremes **8**, 237–266 (2006)
488. Stoyan, D.: Fundamentals of point process statistics. In: Baddeley, A., Gregori, P., Mateu, J., Stoica, R., Stoyan, D. (eds.) Case Studies in Spatial Point Process Modeling. Lecture Notes in Statistics, vol. 185. Springer, New York (2006)
489. Stoyan, D., Kendall, W., Mecke, J.: Stochastic Geometry and Its Applications, 2nd edn. Wiley, New York (1995)

490. Stoyan, D., Stoyan, H., Jansen, U.: Umweltstatistik. Teubner, Stuttgart (1997)
491. Tchoumatchenko, K., Zuyev, S.: Aggregate and fractal tessellations. Probab. Theor. Relat. Fields **121**, 198–218 (2001)
492. Tenenbaum, J.B., de Silva, V., Langford, J.C.: A global geometric framework for nonlinear dimensionality reduction. Science **290**, 2319B–2323 (2000)
493. Tone, C.: A central limit theorem for multivariate strongly mixing random fields. Probab. Math. Stat. **30**, 215–222 (2010)
494. Torquato, S.: Random Heterogeneous Materials. Springer, New York (2002)
495. Tyurin, I.S.: Sharpening the upper bounds for constants in Lyapunov's theorem. Russ. Math. Surv. **65**, 586–588 (2010)
496. Uchaikin, V.V., Zolotarev, V.M.: Chance and Stability. Stable Distributions and Their Applications. VSP, Utrecht (1999)
497. Véhel, J.L.: Fractals in Engineering: From Theory to Industrial Applications. Springer, New York (1997)
498. Voss, F., Gloaguen, C., Fleischer, F., Schmidt, V.: Distributional properties of Euclidean distances in wireless networks involving road systems. IEEE J. Sel. Area. Comm. **27**, 1047–1055 (2009)
499. Voss, F., Gloaguen, C., Fleischer, F., Schmidt, V.: Densities of shortest path lengths in spatial stochastic networks. Stoch. Models **27**, 141–167 (2011)
500. Voss, F., Gloaguen, C., Schmidt, V.: Scaling limits for shortest path lengths along the edges of stationary tessellations. Adv. Appl. Prob. **42**, 936–952 (2010)
501. Vronski, M.A.: Rate of convergence in the SLLN for associated sequences and fields. Theor. Probab. Appl. **43**, 449–462 (1999)
502. Vu, V.: Sharp concentration of random polytopes. Geom. Funct. Anal. **15**, 1284–1318 (2005)
503. Vu, V.: Central limit theorems for random polytopes in a smooth convex set. Adv. Math. **207**, 221–243 (2006)
504. Wackernagel, H.: Multivariate Geostatistics. An Introduction with Applications, 3rd edn. Springer, Berlin (2003)
505. Walsh, J.B.: An introduction to stochastic partial differential equations. In: École d'été de probabilités de Saint-Flour, XIV—1984. Lecture Notes in Mathematics, vol. 1180, pp. 265–439. Springer, Berlin (1986)
506. Wang, Y., Stoev, S.A., Roy, P.: Ergodic properties of sum- and max- stable stationary random fields via null and positive group actions. Ann. Probab. (2012, to appear). ArXiv:0911.0610
507. Watkins, K.P., Hickman, A.H.: Geological evolution and mineralization of the Murchison Province, Western Australia. Bulletin 137, Geological Survey of Western Australia (1990). Published by Department of Mines, Western Australia. Available online from Department of Industry and Resources, State Government of Western Australia. www.doir.wa.gov.au
508. Weil, W.: An application of the central limit theorem for Banach-space-valued random variables to the theory of random sets. Z. Wahrscheinlichkeitstheorie und verw. Gebiete **60**, 203–208 (1982)
509. Weil, W.: Densities of quermassintegrals for stationary random sets. In: Ambartzumian, R.V., Weil, W. (eds.) Stochastic Geometry, Geometric Statistics, Stereology (Proc. Conf. Oberwolfach, 1983). Teubner, Leipzig (1984)
510. Weil, W.: Densities of mixed volumes for Boolean models. Adv. Appl. Prob. **33**, 39–60 (2001)
511. Weil, W.: Mixed measures and functionals of translative integral geometry. Math. Nachr. **223**, 161–184 (2001)
512. Weil, W., Wieacker, J.A.: Densities for stationary random sets and point processes. Adv. Appl. Prob. **16**, 324–346 (1984)
513. Weil, W., Wieacker, J.A.: Stochastic geometry. In: Gruber, P.M., Wills, J.M. (eds.) Handbook of Convex Geometry, Vol. B. North-Holland, Amsterdam (1993)
514. van de Weygaert, R.: Fragmenting the universe III. The construction and statistics of 3-D Voronoi tessellations. Astron. Astrophys. **283**, 361–406 (1994)
515. Wichura, M.J.: Some Strassen-type laws of the iterated logarithm for multiparameter stochastic processes with independent increments. Ann. Probab. **1**, 272–296 (1973)

516. Widom, B., Rowlinson, J.S.: New model for the study of liquid-vapor phase transitions. J. Chem. Phys. **52**, 1670–1684 (1970)
517. Wieacker, J.A.: Einige Probleme der polyedrischen Approximation. Diplomarbeit, Albert-Ludwigs-Universität, Freiburg i. Br. (1978)
518. Willinger, W., Paxson, V., Taqqu, M.S.: Self-similarity and heavy tails: Structural modeling of network traffic. A Practical Guide to Heavy Tails, pp. 27–53. Birkhäuser, Boston (1998)
519. Wills, J.M.: Zum Verhältnis von Volumen und Oberfläche bei konvexen Körpern. Archiv der Mathematik **21**, 557–560 (1970)
520. Winkler, G.: Image Analysis, Random Fields and Markov Chain Monte Carlo Methods, 2nd edn. Springer, Berlin (2003)
521. Wood, A.T.A., Chan, G.: Simulation of stationary Gaussian processes in $[0,1]^d$. J. Comput. Graph. Stat. **3**, 409–432 (1994)
522. Yoshihara, K.: Limiting behavior of U-statistics for stationary, absolutely regular processes. Z. Wahrscheinlichkeitstheorie und verw. Gebiete **35**, 237–252 (1976)
523. Yuan, J., Subba Rao, T.: Higher order spectral estimation for random fields. Multidimens. Syst. Signal Process. **4**, 7–22 (1993)
524. Yukich, J.E.: Probability Theory of Classical Euclidean Optimization Problems. Lecture Notes in Mathematics, vol. 1675. Springer, Berlin (1998)
525. Yukich, J.E.: Limit theorems for multi-dimensional random quantizers. Electron. Comm. Probab. **13**, 507–517 (2008)
526. Yukich, J.E.: Point process stabilization methods and dimension estimation. Proceedings of Fifth Colloquium of Mathematics and Computer Science. Discrete Math. Theor. Comput. Sci. 59–70 (2008) http://www.dmtcs.org/dmtcs-ojs/index.php/proceedings/issue/view/97/showToc
527. Zähle, M.: Curvature measures and random sets I. Math. Nachr. **119**, 327–339 (1984)
528. Zhang, L.X., Wen, J.: A weak convergence for functions of negatively associated random variables. J. Multivariate Anal. **78**, 272–298 (2001)
529. Ziegler, G.: Lectures on 0/1 polytopes. In: Kalai, G., Ziegler, G.M. (eds.) Polytopes – Combinatorics and Computation, DMV Seminars. Birkhäuser, Basel (2000)
530. Zolotarev, V.M.: Modern Theory of Summation of Random Variables. VSP, Utrecht (1997)
531. Zuker, M., Mathews, D.H., Turner, D.H.: Algorithms and thermodynamics for RNA secondary structure prediction: A practical guide. In: Barciszewski, J., Clark, B.F.C. (eds.) RNA Biochemistry and Biotechnology. NATO ASI Series. Kluwer, Dordrecht (1999)
532. Zuyev, S.: Strong Markov Property of Poisson Processes and Slivnyak Formula. Lecture Notes in Statistics, vol. 185. Springer, Berlin (2006)
533. Zuyev, S.: Stochastic geometry and telecommunication networks. In: Kendall, W.S., Molchanov, I. (eds.) New Perspectives in Stochastic Geometry. Oxford University Press, London (2010)

Index

E. Spodarev (ed.), *Stochastic Geometry, Spatial Statistics and Random Fields*, Lecture Notes in Mathematics 2068, DOI 10.1007/978-3-642-33305-7,

Edited by J.-M. Morel, B. Teissier; P.K. Maini

Editorial Policy (for Multi-Author Publications: Summer Schools / Intensive Courses)

1. Lecture Notes aim to report new developments in all areas of mathematics and their applications - quickly, informally and at a high level. Mathematical texts analysing new developments in modelling and numerical simulation are welcome. Manuscripts should be reasonably selfcontained and rounded off. Thus they may, and often will, present not only results of the author but also related work by other people. They should provide sufficient motivation, examples and applications. There should also be an introduction making the text comprehensible to a wider audience. This clearly distinguishes Lecture Notes from journal articles or technical reports which normally are very concise. Articles intended for a journal but too long to be accepted by most journals, usually do not have this "lecture notes" character.

2. In general SUMMER SCHOOLS and other similar INTENSIVE COURSES are held to present mathematical topics that are close to the frontiers of recent research to an audience at the beginning or intermediate graduate level, who may want to continue with this area of work, for a thesis or later. This makes demands on the didactic aspects of the presentation. Because the subjects of such schools are advanced, there often exists no textbook, and so ideally, the publication resulting from such a school could be a first approximation to such a textbook. Usually several authors are involved in the writing, so it is not always simple to obtain a unified approach to the presentation.
   For prospective publication in LNM, the resulting manuscript should not be just a collection of course notes, each of which has been developed by an individual author with little or no coordination with the others, and with little or no common concept. The subject matter should dictate the structure of the book, and the authorship of each part or chapter should take secondary importance. Of course the choice of authors is crucial to the quality of the material at the school and in the book, and the intention here is not to belittle their impact, but simply to say that the book should be planned to be written by these authors jointly, and not just assembled as a result of what these authors happen to submit.
   This represents considerable preparatory work (as it is imperative to ensure that the authors know these criteria before they invest work on a manuscript), and also considerable editing work afterwards, to get the book into final shape. Still it is the form that holds the most promise of a successful book that will be used by its intended audience, rather than yet another volume of proceedings for the library shelf.

3. Manuscripts should be submitted either online at www.editorialmanager.com/lnm/ to Springer's mathematics editorial, or to one of the series editors. Volume editors are expected to arrange for the refereeing, to the usual scientific standards, of the individual contributions. If the resulting reports can be forwarded to us (series editors or Springer) this is very helpful. If no reports are forwarded or if other questions remain unclear in respect of homogeneity etc, the series editors may wish to consult external referees for an overall evaluation of the volume. A final decision to publish can be made only on the basis of the complete manuscript; however a preliminary decision can be based on a pre-final or incomplete manuscript. The strict minimum amount of material that will be considered should include a detailed outline describing the planned contents of each chapter.
   Volume editors and authors should be aware that incomplete or insufficiently close to final manuscripts almost always result in longer evaluation times. They should also be aware that parallel submission of their manuscript to another publisher while under consideration for LNM will in general lead to immediate rejection.

4. Manuscripts should in general be submitted in English. Final manuscripts should contain at least 100 pages of mathematical text and should always include

   - a general table of contents;
   - an informative introduction, with adequate motivation and perhaps some historical remarks: it should be accessible to a reader not intimately familiar with the topic treated;
   - a global subject index: as a rule this is genuinely helpful for the reader.

   Lecture Notes volumes are, as a rule, printed digitally from the authors' files. We strongly recommend that all contributions in a volume be written in the same LaTeX version, preferably LaTeX2e. To ensure best results, authors are asked to use the LaTeX2e style files available from Springer's web-server at
   ftp://ftp.springer.de/pub/tex/latex/svmonot1/ (for monographs) and
   ftp://ftp.springer.de/pub/tex/latex/svmultt1/ (for summer schools/tutorials).
   Additional technical instructions, if necessary, are available on request from:
   lnm@springer.com.

5. Careful preparation of the manuscripts will help keep production time short besides ensuring satisfactory appearance of the finished book in print and online. After acceptance of the manuscript authors will be asked to prepare the final LaTeX source files and also the corresponding dvi-, pdf- or zipped ps-file. The LaTeX source files are essential for producing the full-text online version of the book. For the existing online volumes of LNM see:
   http://www.springerlink.com/openurl.asp?genre=journal&issn=0075-8434.
   The actual production of a Lecture Notes volume takes approximately 12 weeks.

6. Volume editors receive a total of 50 free copies of their volume to be shared with the authors, but no royalties. They and the authors are entitled to a discount of 33.3 % on the price of Springer books purchased for their personal use, if ordering directly from Springer.

7. Commitment to publish is made by letter of intent rather than by signing a formal contract. Springer-Verlag secures the copyright for each volume. Authors are free to reuse material contained in their LNM volumes in later publications: a brief written (or e-mail) request for formal permission is sufficient.

Addresses:

Professor J.-M. Morel, CMLA,
École Normale Supérieure de Cachan,
61 Avenue du Président Wilson, 94235 Cachan Cedex, France
E-mail: morel@cmla.ens-cachan.fr

Professor B. Teissier, Institut Mathématique de Jussieu,
UMR 7586 du CNRS, Équipe "Géométrie et Dynamique",
175 rue du Chevaleret,
75013 Paris, France
E-mail: teissier@math.jussieu.fr

For the "Mathematical Biosciences Subseries" of LNM:

Professor P. K. Maini, Center for Mathematical Biology,
Mathematical Institute, 24-29 St Giles,
Oxford OX1 3LP, UK
E-mail : maini@maths.ox.ac.uk

Springer, Mathematics Editorial I,
Tiergartenstr. 17,
69121 Heidelberg, Germany,
Tel.: +49 (6221) 4876-8259
Fax: +49 (6221) 4876-8259
E-mail: lnm@springer.com

Lightning Source UK Ltd.
Milton Keynes UK
UKHW020109061222
413411UK00003B/10

9 783642 333040